Preface

Integrated Foundations of Pharmacy

As a result of significant changes taking place within the profession of pharmacy, there is an increasing trend for universities to adopt an integrated approach to pharmacy education. There is now an overwhelming view that pharmaceutical science must be combined with the more practice-oriented aspects of pharmacy. This assists students to see the impact of, and relationships between, those subjects which make up the essential knowledge base for a practising pharmacist.

This series supports integrated pharmacy education, so that from day one of the course students can take a professionally relevant approach to their learning. This is achieved through the organization of content and the use of key learning features. Cross-references highlight related topics of importance, directing the reader to further information, and case studies explore the ways in which pharmaceutical science and practice impact upon patients' lives, allowing material to be addressed in a patient-centred context.

There are four books in the series, covering each main strand: *Pharmaceutical Chemistry, Pharmaceutics: The Science of Medicine Design, Pharmacy Practice*, and *Therapeutics and Human Physiology: How Drugs Work*. Each book is edited by a subject expert, with contributors from across pharmacy education. They have been carefully written to ensure an appropriate breadth and depth of knowledge for the first-year student.

Each book concludes with an overview of the subject and application to pharmacy, building on students' understanding of the concepts and bringing everything together. It applies the material to pharmacy practice in a variety of ways and places it in the context of all four pharmacy strands.

Pharmaceutical Chemistry

In this volume, we hope to persuade you that *Pharmaceutical Chemistry* is fundamental to the discipline of pharmacy. The chemistry we present here is illustrated using real drug molecules and the reactions that they undergo in the body, in the laboratory, and on storage in the pharmacy and in the home. Most of the reactions, mechanisms, and techniques described in this book may be found in organic chemistry textbooks, but here they are integrated with *Therapeutics and Human Physiology*, *Pharmaceutics*, and *Pharmacy Practice*. There is a chapter on pharmaceutical inorganic chemistry, an important area that is sometimes forgotten.

Do please *read* this book. It can be used as a reference book, especially in the run-up to examinations, but the authors would like to think that you will curl up on a winter's evening and just read it, and it has been written with that use in mind.

Acknowledgements

The editors wish to acknowledge the support of the contributing authors, all experts in the field of pharmacy or medicinal chemistry education, who have spent a considerable amount of time writing and reviewing their chapters. We would also like to extend our thanks to the staff at Oxford University Press, especially Jonathan Crowe, who has supported this project from the outset and displayed great patience during its halting progress.

Nicola Wood has reviewed much of the book from the student's perspective, and Alex White's perceptive reviews have also greatly influenced the final form of the book. We thank them and many others who have been involved in the review process.

Please return / renew by date shown.
You can renew at: **norlink.norfolk.gov.uk**
or by telephone: **0344 800 8006**
Please have your library card & PIN ready.

NORFOLK LIBRARY
AND INFORMATION SERVICE

NORFOLK ITEM

30129 076 890 889

PHARMACY PRACTICE

THERAPEUTICS AND
HUMAN PHYSIOLOGY

how drugs work

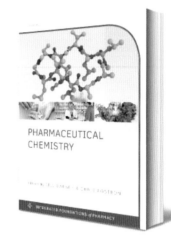

PHARMACEUTICAL
CHEMISTRY

PHARMACEUTICS

INTEGRATED FOUNDATIONS *of* PHARMACY

Pharmaceutical Chemistry

Edited by Jill Barber and Chris Rostron

OXFORD
UNIVERSITY PRESS

 INTEGRATED FOUNDATIONS *of* PHARMACY

UNIVERSITY PRESS

Great Clarendon Street, Oxford, OX2 6DP,
United Kingdom

Oxford University Press is a department of the University of Oxford.
It furthers the University's objective of excellence in research, scholarship,
and education by publishing worldwide. Oxford is a registered trade mark of
Oxford University Press in the UK and in certain other countries

British Library Cataloguing in Publication Data
Data available

ISBN 978–0–19–965530–4

Printed in Great Britain by
Bell & Bain Ltd., Glasgow

Contents

An introduction to the *Integrated Foundations of Pharmacy* series   x

About the editors   xii

Abbreviations   xiii

1 The importance of pharmaceutical chemistry JILL BARBER   1

1.1 Chemical structures and nomenclature   1

1.2 The human test tube   4

1.3 More test tubes: plants and microorganisms   9

1.4 Glass and plastic test tubes   14

1.5 From test tube to pharmaceutical   19

2 Organic structure and bonding ALASTAIR MANN   24

2.1 What is organic chemistry?   24

2.2 The shape of molecules   26

2.3 The electronic configuration of carbon   28

2.4 The shape of organic molecules   29

2.5 Intermolecular forces   38

2.6 Reaction types and the making and breaking of bonds   40

2.7 The principles of organic reaction mechanisms   42

3 Stereochemistry and drug action ROSALEEN J. ANDERSON AND ADAM TODD   48

3.1 Introduction   48

3.2 Constitutional isomerism   49

3.3 Conformational isomerism   51

3.4 Stereoisomerism   56

3.5 Protein folding diseases   80

4 Properties of aliphatic hydrocarbons ANDREW J. HALL   81

4.1 Nomenclature   82

4.2 Physical properties of aliphatic hydrocarbons   88

4.3 Alkanes – preparation and chemical properties   91

4.4 Alkenes – preparation and chemical properties   92

4.5 Alkynes – preparation and reactions   105

4.6 Hydrocarbons in pharmacy   112

Contents

5 Alcohols, phenols, ethers, organic halogen compounds, and amines CHRIS ROSTRON 117

5.1 The hydroxyl group 118

5.2 Polyhydric alcohols 124

5.3 Phenols 126

5.4 Ethers 130

5.5 Haloalkanes and other organic halogen compounds 133

5.6 Aromatic halogen compounds 135

5.7 Polyhalogen compounds 136

5.8 Amines 137

5.9 Quaternary ammonium compounds 139

6 The carbonyl group and its chemistry MATTHEW INGRAM 141

6.1 Carbonyl structure and nomenclature 141

6.2 The power of the carbonyl group 145

6.3 Reactions of carbonyl compounds – nucleophilic attack on carbon 149

6.4 α-substitution reactions 162

6.5 Carbonyls in the body 171

6.6 Carbonyls in drugs – opportunities and problems 172

7 Introduction to aromatic chemistry MIKE SOUTHERN 177

7.1 What is aromatic chemistry? 177

7.2 Why is aromatic chemistry important? 180

7.3 The chemistry of benzene 185

7.4 The synthesis of drugs 198

7.5 Aromatic chemistry in the body 203

8 Inorganic chemistry in pharmacy GEOFF HALL 210

8.1 Concepts in inorganic chemistry 211

8.2 Metals: introduction 213

8.3 Group 1 metals 214

8.4 Group 2 metals 217

8.5 Period 4 metals 218

8.6 Precious metals 222

8.7 Phosphorus 224

8.8 Sulfur 229

9 The chemistry of biologically important macromolecules ALEX WHITE 241

9.1 Small molecules versus large molecules 241

9.2 Nucleic acids and nucleosides 242

9.3 Proteins 249

9.4 Carbohydrates 256

9.5 Lipids 261

10 Origins of drug molecules TIM SNAPE 270

10.1 Drugs, dyes, and cleaning fluid: similarities and differences 270

10.2 Natural products as drugs and medicines 273

10.3 Semi-synthetic drugs 282

10.4 Synthetic drugs 285

10.5 Genetic engineering and fermentation (biotechnology) for the production of drugs 291

10.6 The principles of organic synthesis 292

10.7 Purification methods 293

11 Introduction to pharmaceutical analysis LARRY GIFFORD AND TONY CURTIS 296

11.1 Quality control of pharmaceuticals and formulated products 297

11.2 The electromagnetic spectrum 299

11.3 Ultraviolet-visible spectrophotometry 300

11.4 Infrared spectroscopy 303

11.5 Nuclear magnetic resonance (NMR) spectroscopy 307

11.6 Mass spectrometry 315

11.7 Chromatographic methods of analysis 319

11.8 Gas-liquid chromatography 323

12 The molecular characteristics of good drugs JILL BARBER 327

12.1 Rules in chemistry 327

12.2 The Rule of Five – an empirical rule 328

12.3 Structure-activity relationships 331

12.4 Stereochemistry 337

12.5 Purity 339

Glossary 341

Index 346

An introduction to the *Integrated Foundations of Pharmacy* series

The path to becoming a qualified pharmacist is incredibly rewarding, but it requires diligence. Not only will you need to assimilate knowledge from a range of disciplines, but you must also understand – and demonstrate – how to apply this knowledge in a practical, hands-on environment. This series is written to support you as you encounter the challenges of pharmacy education today.

There are a range of features used in the series, each carefully designed to help you master the material and to encourage to you to see the connections between the different strands of the discipline.

Mastering the material

Boxes

Additional material that adds interest or depth to concepts covered in the main text is provided in the Boxes.

> **BOX 2.1**
>
> **Bacteria versus archaea**
>
> Archaea are often found living in some of the most extreme environments found on Earth, in conditions where humans would not be able to survive. These

Key points

The important 'take home messages' that you must have a good grasp of are highlighted in the Key points. You may find these form a helpful basis for your revision.

> **KEY POINT**
>
> All our cells contain identical genetic information; however the cells differentiate to make up the various types of tissues, organs, and body systems that we are made of.

Self-check questions

Questions are provided throughout the chapters in order for you to test your understanding of the material. Take the time to complete these, as they will allow you to evaluate how you are getting on, and they will undoubtedly aid your learning. Answers are provided at the back of each volume.

> **SELF CHECK 1.2**
>
> Why may it be important for health care professionals to establish the family medical history?

Further reading

In this section, we direct you to additional resources that we encourage you to seek out, in your library or online. They will help you to gain a deeper understanding of the material presented in the text.

> **FURTHER READING**
>
> Boarder, M., Newby, D., and Navti, P. *Pharmacology for Pharmacy and the Health Sciences: A Patient-Centred Approach.* Oxford University Press, 2010.

Glossary

You will need to master a huge amount of new terminology as you study pharmacy. The glossaries in each volume should help you with this. Glossary terms are shown in pink.

> **ACE inhibitor**   Drug that inhibits angiotensin-converting enzyme.
> **Acetylcholine (Ach)**   The neurotransmitter at preganglionic autonomic neurons and postganglionic parasympathetic neurons and at the neuromuscular junction. It acts on nicotinic and muscarinic receptors. Unusually, acetylcholine is also released from sympathetic nerves that

Online resources

Visit the Online Resource Centre for related materials, including 10 multiple-choice questions for each chapter, with answers and feedback.

Go to: **www.oxfordtextbooks.co.uk/orc/ifp**

Seeing the connections

Integration boxes

Examples in these Integration boxes show how pharmacy practice, pharmaceutics, pharmaceutical chemistry, human physiology, and therapeutics are all closely interlinked.

> **INTEGRATION BOX 8.2**
>
> ## Sodium lauryl sulfate in toothpaste
>
> Not surprisingly, sodium lauryl sulfate (SLS) is an irritant, because it solubilizes proteins. Indeed, anecdotal evidence suggests that SLS in toothpastes exacerbates

Case studies

Case studies show how the science you learn at university will impact on how you might advise a patient. Reflection questions and sample answers encourage you to think critically about the points raised in the case study.

> **CASE STUDY 2.1**
>
> Angela has told Ravi she is pregnant and they are both thrilled at the prospect of being parents. However, Angela had an older brother who had cystic fibrosis (CF)

Cross references

Linking related sections across all four volumes in the series (as well as other sections within this volume), cross references will give you a good idea of just how integrated the subject is. Importantly, it will allow you to easily access material on the same subject, as viewed from the perspectives of the different strands of the discipline.

> The study of dosage forms is covered in the **Pharmaceutics** book within this series.

Lecturer support materials

For registered adopters of the volumes in this series, the Online Resource Centre also features figures in electronic format, available to download, for use in lecture presentations and other educational resources.

To register as an adopter, visit www.oxfordtextbooks. co.uk/orc/ifp, select the volume you are interested in, and follow the on-screen instructions.

Any comments?

We welcome comments and feedback about any aspect of the series. Just visit www.oxfordtextbooks. co.uk/orc/feedback and share your views.

About the editors

Editor, Dr Jill Barber, studied Natural Sciences at the University of Cambridge and completed a PhD in Bio-organic Chemistry at the same university. She then spent five years in some of the oldest universities in Europe, learning Biochemistry, German, and Renaissance Music. She settled in Manchester, with a permanent position in the School of Pharmacy and Pharmaceutical Sciences, where she teaches chemotherapy and its underlying chemistry and biochemistry. Her current research involves using mass spectrometry to quantify the proteins involved in the response to drugs, both in bacteria and in humans. She has also published several teaching-related research papers about the factors influencing student success. She is a Lady Grandmaster of the International Correspondence Chess Federation and enjoys singing and playing the trombone.

Series Editor, Dr Chris Rostron, graduated in Pharmacy from Manchester University and completed a PhD in Medicinal Chemistry at Aston University. He gained Chartered Chemist status in 1975. After a period of post-doctoral research he was appointed as a lecturer in Medicinal Chemistry at Liverpool Polytechnic. He is now an Honorary Research Fellow in the School of Pharmacy and Biomolecular Sciences at Liverpool John Moores University. Prior to this, he was an Academic Manager, and then a Reader in Medicinal Chemistry at the school. He was a member of the Academic Pharmacy Group Committee of the Royal Pharmaceutical Society of Great Britain and chairman for the past 5 years. He is currently chairman of the Academic Pharmacy Forum and deputy chair of the Education Expert Advisory Panel of the Royal Pharmaceutical Society. He is a past and present external examiner in Medicinal Chemistry at a number of Schools of Pharmacy both in the UK and abroad. In 2008, he was awarded honorary membership of the Royal Pharmaceutical Society of Great Britain for services to Pharmacy education.

Contributors

Professor Rosaleen J. Anderson, Faculty of Applied Science, University of Sunderland, UK

Dr Tony Curtis, School of Pharmacy, Keele University, UK

Dr Andrew J. Hall, Medway School of Pharmacy, UK

Dr Geoff Hall, Leicester School of Pharmacy, De Montfort University, UK

Dr Matthew Ingram, School of Pharmacy and Biomolecular Sciences, University of Brighton, UK

Professor Larry Gifford, School of Pharmacy, Keele University, UK

Dr Alastair Mann, Faculty of Science, Engineering and Computing, Kingston University, UK

Dr Chris Rostron, School of Pharmacy and Biomolecular Sciences, Liverpool John Moores University, UK

Dr Jill Barber, School of Pharmacy and Pharmaceutical Sciences, University of Manchester, UK

Dr Tim Snape, School of Pharmacy and Biomedical Sciences, University of Central Lancashire, UK

Dr Mike Southern, School of Chemistry, Trinity Bioscience Institute, Trinity College, Dublin, Ireland

Dr Adam Todd, School of Medicine, Pharmacy and Health, Durham University, UK

Dr Alex White, Cardiff School of Pharmacy and Pharmaceutical Sciences, Cardiff University, UK

Abbreviations

ABO	ABO blood groups		IPA	isopropyl alcohol
ACE	angiotensin converting enzyme		IR	infrared
ADP	adenosine diphosphate		IUD	intrauterine contraceptive devices
ADME	absorption, distribution, metabolism and excretion		IUPAC	International Union of Pure and Applied Chemistry
ADR	adverse drug reactions		MALDI	matrix assisted laser desorption ionization
AMP	adenosine monophosphate		MDMA	3,4-methylenedioxy-*N*-methylamphetamine (ecstasy)
APCI	atmospheric pressure chemical ionization			
API	active pharmaceutical ingredient		Me	methyl
ATP	adenosine triphosphate		MHRA	Medicines and Healthcare products Regulatory Agency
ATR	attenuated total reflectance			
AZT	azidothymidine		mRNA	messenger RNA
BNF	British Nation Formulary		MRSA	methicillin-resistant strains of *Staphylococcus aureus*
BP	British Pharmacopoeia or blood pressure			
CFC	chlorofluorocarbon		MS	mass spectrometry
CI	chemical ionization		NAS	nucleophilic aromatic substitution
CNS	central nervous system		NMR	nuclear magnetic resonance
CoA	coenzyme A		NSAID	non-steroidal anti-inflammatory drug
COMT	catechol-*O*-methyltransferase		ODS	octadecylsilane
CYP	cytochrome P450		OTC	over the counter
DMSA	dimercaptosuccinic acid		RMM	relative molecular mass
DNA	deoxyribonucleic acid		RNA	ribonucleic acid
DOSY	diffusion ordered spectroscopy		ROS	reactive oxygen species
EAS	electrophilic aromatic substitution		rRNA	ribosomal RNA
EDTA	ethylenediamine tetracetic acid		SAM	*S*-adenosylmethionine
EHC	emergency hormonal contraception		SAR	structure-activity relationship
EI	electron ionization		SFFC	spurious/falsely labelled/falsified/counterfeit
ESI	electrospray ionization			
Et	ethyl		SHU	Scoville heat units
FDA	Food and Drug Administration		SSRI	selective serotonin reuptake inhibitors
GABA	gamma-aminobutyric acid		SLS	sodium lauryl sulfate
GC	gas chromatography		THC	Δ^9-tetrahydrocanabinol
GLC	gas-liquid chromatography		TLC	thin layer chromatography
GORD	gastro-oesophageal reflux disease		TOF	time of flight
HPLC	high performance liquid chromatography		tRNA	transfer RNA
HTS	high throughput screening		USP	United States Pharmacopoeia
ICI	Imperial Chemical Industries		UV	ultraviolet
INR	international normalized ratio		WHO	World Health Organisation

The importance of pharmaceutical chemistry

JILL BARBER

Pharmacy is all about drugs: how drugs are made, how to get them into the body, how they work, their **metabolism**, their side-effects, their interactions with other drugs, and how we communicate with patients and other health care professionals about drugs. At the heart of the discipline of pharmacy is chemistry, because drugs are, of course, chemicals.

This chapter is an overview of the importance of pharmaceutical chemistry. Every living organism is like a test tube, carrying out huge numbers of chemical reactions; in this chapter we will explore some of these reactions. Every drug is made using chemical reactions, some of these are in a laboratory, but some are in nature; we will study both. Every drug needs to be made up into a **formulation**: perhaps a tablet, or a cream or an injectable solution, and sometimes the formulation process needs chemistry as well. Many drugs are metabolized, and drug metabolism is chemistry. We will briefly overview these processes.

Learning objectives

Having read this chapter you are expected to be able to:

➤ draw chemical structures the way organic chemists draw them

➤ give examples of chemical reactions that take place in the human body

➤ give examples of the importance of chemistry in the manufacture and formulation of drugs.

1.1 Chemical structures and nomenclature

Before we consider the importance of pharmaceutical chemistry in detail, it is important that you understand the chemical structures and some of the nomenclature used in this book. Most students who study organic chemistry at university (whether in a chemistry course or as part of a biological sciences course), get confused by two things.

• University chemists use 'old-fashioned' names for simple chemicals like ethanoic acid (they call it acetic acid).

- University chemists hardly ever label carbon atoms or count hydrogen atoms.

It is tempting to conclude either that school teachers are just wrong, or that organic chemists know very little about their own subject. Neither is true. When you started primary school you learnt to print your letters very carefully, using wide-lined paper. Some years later, you learnt how to do joined-up writing on unlined paper and eventually, you learnt to use a word processing package on a computer. For most students, writing is quite different at the age of 18 from the age of 5. It is similar with chemistry.

Trivial and IUPAC systematic nomenclature

The International Union of Pure and Applied Chemistry (IUPAC) has defined systematic names for organic compounds, and you may be familiar with many of these. Nevertheless, very common substances retain their trivial (non-systematic) names because lots of people, including cooks, gardeners and biologists, use these names. Some trivial names (such as valeric acid for 3-methylbutanoic acid) have already gone out of fashion but other very common trivial names (such as acetic acid) remain, at least for the time being. A few systematic names are so similar to other systematic names that they are inconvenient or even dangerous. Chemical laboratories are noisy places, so

trichloromethane can sound like dichloromethane, and ethanal can sound like ethanol. To prevent accidents we continue to use the trivial names for these chemicals: chloroform for trichloromethane and acetaldehyde for ethanal. Even IUPAC does not recommend systematic names when they might be dangerous! In this book we will use systematic names, except where trivial names are required either for safety or for communication with members of the public. This means that you will use more trivial names than when you were at school.

 In Chapter 4 'Properties of aliphatic hydrocarbons', the nomenclature of organic compounds is introduced.

Chemical structures

Consider a simple drug molecule, aspirin, as shown in Figure 1.1. All the structures (A–E) are correct but most chemists would normally use A or B.

Structures A and E (the professional structure and the college structure) are both right, but have three important differences.

- Structure A has no carbons represented by C. Carbons-7, -8 and -9 are represented by the ends or conjunctions of bonds.

- Structure A has only one hydrogen atom drawn in. Hydrogens are implied   you know that carbon

FIGURE 1.1 The chemical structure of aspirin, drawn in several different ways.

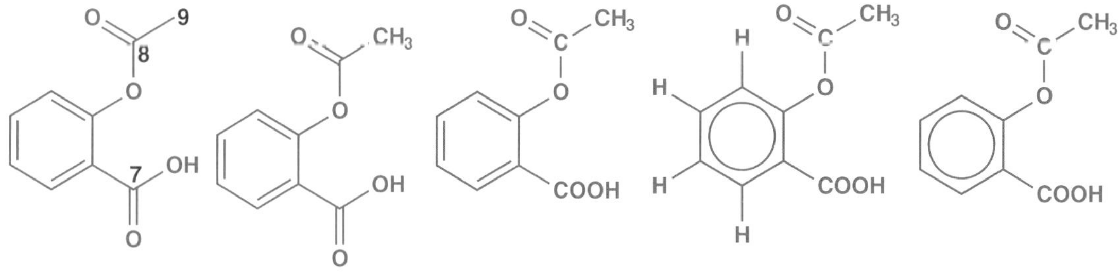

(A)   (B)   (C)   (D)   (E)

must have four bonds or charges; the hydrogen atoms on the benzene ring and at carbon-9 are not drawn in.

- Finally, in structure A, a Kekulé ring structure is drawn. You may have learnt that the two Kekulé structures are in very rapid equilibrium, so that all the bond lengths in a benzene ring are equal. This is absolutely correct, but it is much easier to draw mechanisms using a Kekulé structure, so most chemists use these most of the time.

So professional chemists abbreviate structures for speed and convenience, whereas school teachers draw in carbons and hydrogens to help less experienced students understand exactly which atoms are present in a given molecule. You will not usually be forced to abbreviate your structures, but you will need to recognize and understand structures drawn like Figure. 1.1A.

 Aspirin is used as a painkiller, as an anti-inflammatory and to reduce the risk of blood clots. Like many drugs, it is an aromatic molecule, containing a benzene ring. Chapter 7 'Introduction to aromatic chemistry', introduces compounds with benzene rings, and explains how their structures influence their physical and chemical properties. You can also learn lots more about aspirin in Chapter 3 'The biochemistry of cells', Chapter 7 'Communication systems in the body – autocoids and hormones', and Chapter 9 'Haematology' of the *Therapeutics and Human Physiology* book of this series.

SELF CHECK 1.1

Check that you understand how chemical structures can be pictured, by redrawing the following structures with all the hydrogens and carbons labelled: (a) the painkiller, paracetamol (Figure 1.2A), (b) metronidazole, an antibacterial agent used to treat some important infections of the stomach and intestines (Figure 1.2B), and (c) naproxen, another painkiller, available over the counter in the USA but only on prescription in Europe (Figure 1.2C).

Now have a look at Figure 1.3(A–C) and compare it with Figure 1.2(A–C). Do you see what has happened? A chemical structure can be drawn correctly from any angle and different people have their preferred angles. If you are drawing out a chemical reaction, it is usually easiest to put the reactive groups on the right if you are writing in English or another language that goes from left to right, but you do not have to.

SELF CHECK 1.2

The OCH_3 group on the naproxen molecule is sometimes written OMe instead. How could you represent $O-CH_2CH_3$ using the same convention?

 Chapter 3 'Stereochemistry and drug action', introduces the three-dimensional structures of molecules. Studying this chapter will help you to visualize molecules from different angles. It will also explain the importance of three-dimensional structure in drug action.

FIGURE 1.2   A, paracetamol; B, metronidazole; C, naproxen.

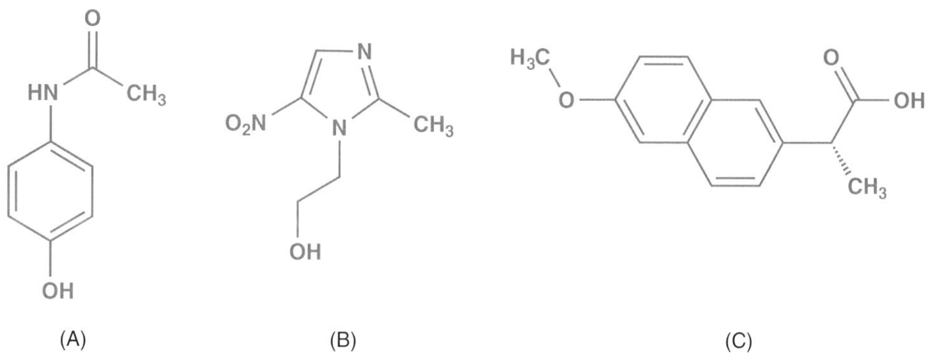

(A)

(B)

(C)

FIGURE 1.3   A, paracetamol; B, metronidazole; C, naproxen.

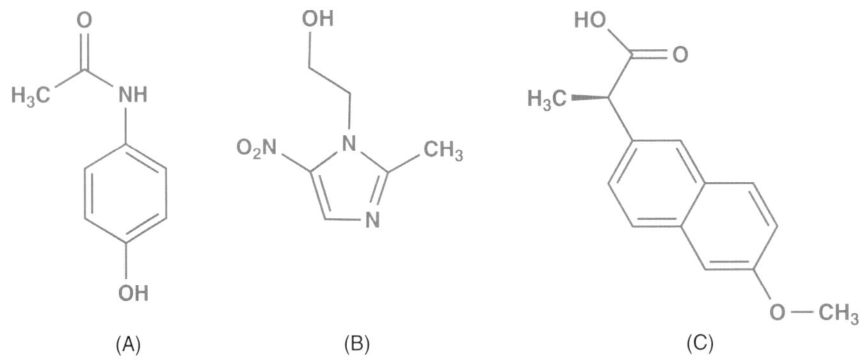

(A)   (B)   (C)

1.2 **The human test tube**

This section introduces some of the chemical reactions that take place in the body. The human body carries out simple chemistry on complex structures. The same chemical reactions can also take place outside the body, and they are discussed in greater detail in later chapters.

Vision

'I am fearfully and wonderfully made' wrote King David of Israel nearly 3000 years ago, despite the fact that he knew little or no chemistry! King David was in awe of the way the human body works, and one of the most exciting things a body can do is see. IIe would have been amazed to learn that vision is based on a simple chemical reaction, in which 11-*cis*-retinal is **isomerized** to *trans*-retinal (see Figure 1.4), catalysed by the action of light.

This reaction and the reverse reaction are fast. The human brain can detect changes in colour and light within a few milliseconds and distinguish movement at nearly 1000 frames per second. Films in the cinema and on television run at about 50–60 frames per second. Our brains are clever at filling gaps; were they not, our favourite movies would judder like old news footage.

Opsin is a protein involved in vision; retinal binds to it. Opsin comes in three forms, absorbing light in the red, green and yellow regions of the electromagnetic spectrum, and enabling us to see colours. The red and green opsins are very similar indeed and it is easy for

one to mutate so that it absorbs light in the wrong region. This leads to red-green colour blindness. Much of Chapter 11 is about spectroscopy, the use of instrumentation to detect different parts of the electromagnetic spectrum.

 The terms *cis*- and *trans*-retinal are used here. A *cis* double bond is a special case of a Z double bond, and a *trans* double bond is a special case of an E double bond. Chapter 3 'Stereochemistry and drug action', introduces stereochemistry, and these terms are discussed in more detail there.

Energy

Vision is remarkable, but energy metabolism is perhaps even more astonishing. Why do we get tired when we do not eat? Why is it possible to starve to death?

The answer, of course, is chemistry. Most of our food gets broken down to a small molecule called acetate, CH_3COO^-. In its protonated form, CH_3COOH, acetate is found in vinegar (acetic or ethanoic acid), but in the body it is normally deprotonated. Acetate takes part in numerous chemical reactions, which enable us to derive energy from food, as well as to make vital components of our bodies. Central to acetate metabolism is the citric acid cycle (Figure 1.5). You may have come across this cycle when learning biology.

Acetate is activated in the body to give the **thioester** acetyl coenzyme A, which reacts with oxaloacetate (a

FIGURE 1.4 The chemistry of vision. Light catalyses the conversion of 11-*cis*-retinal (bound to the protein opsin) to *trans*-retinal, starting a signalling cascade that leads to an image being perceived in the brain. The mechanism of this geometrical isomerization is still being investigated; however, we know that free rotation can only occur about single bonds, so the mechanism shown is possible.

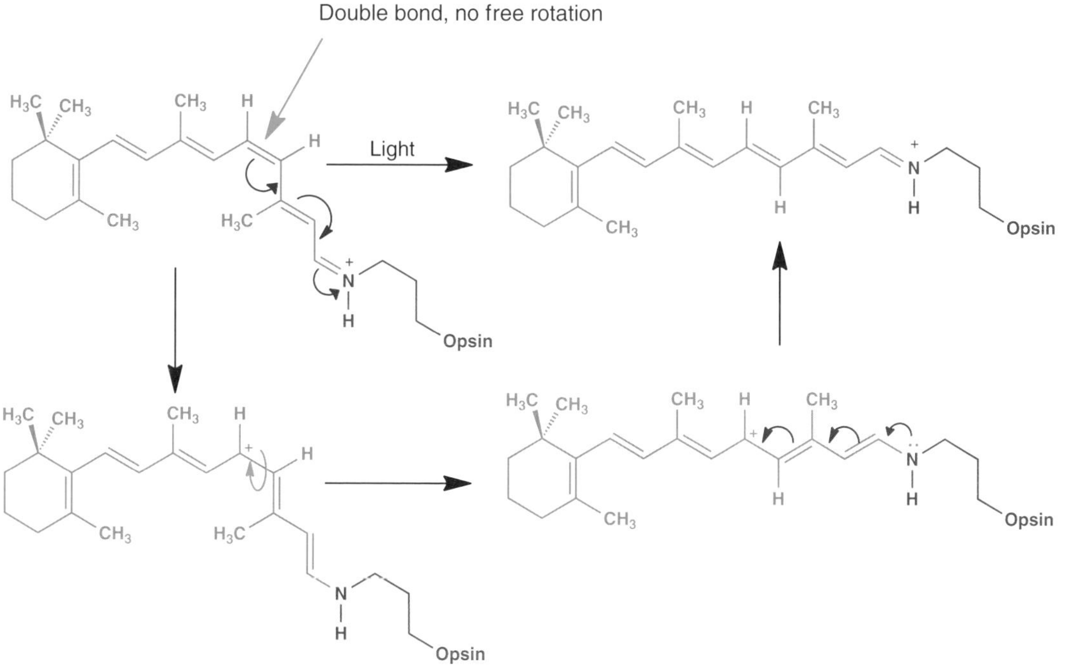

FIGURE 1.5 The formation of citrate from acetate in the citric acid cycle. The carbon-carbon bond forming reaction in which citrate is formed is discussed in Chapters 6 and 10. This reaction is catalysed by citrate synthase, an **enzyme** that speeds up the reaction. (If you find the structures a bit daunting at this stage, do not worry – all will become clear in the next few chapters.)

Lemon: This file is licensed under the Creative Commons Attribution-Share Alike 2.5 Generic license, copyright André Karwath. Vinegar: source Stockbyte.

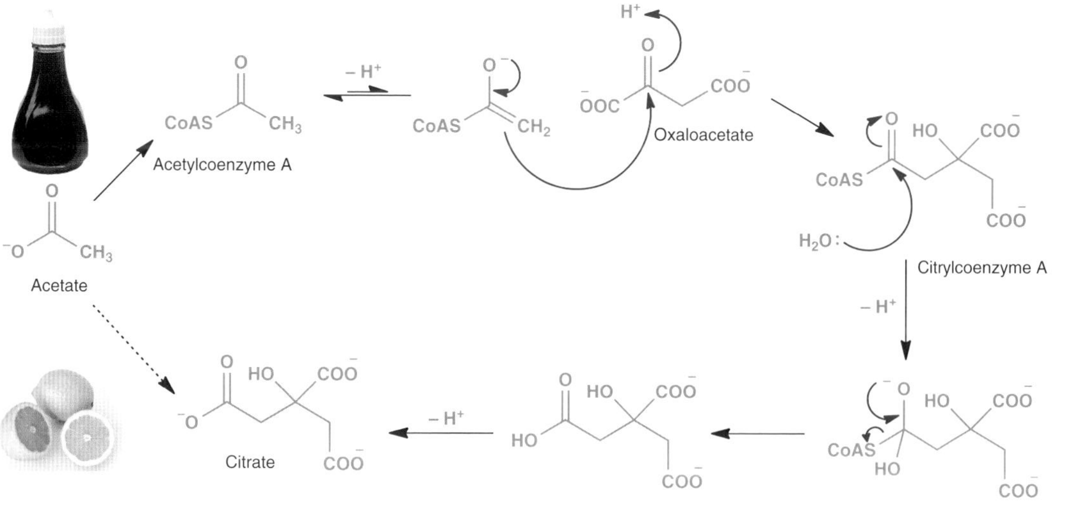

4-carbon molecule derived from two molecules of acetate). This reaction is a standard carbon-carbon bond forming reaction, yielding citryl coenzyme A, another thioester. Citryl coenzyme A is hydrolysed to give citrate. The concentration of citric acid (protonated citrate) can be as high as 0.3 M in lemons and limes, but is a lot lower in our bodies. Citrate then undergoes many more chemical reactions, eventually yielding proteins, nucleic acids, fats and all the other molecules we need to live.

 Citric acid is a tricarboxylic acid. It contains three carbonyl groups. The chemistry of the carbonyl group and its importance in drugs and in biological systems is introduced in Chapter 6 'The carbonyl group and its chemistry'. In Chapter 8 'Inorganic chemistry in pharmacy' of this book, the chemistry of sulfur, phosphorus and metals important in biology is introduced. Thioesters are very important in biological systems, but are seldom found in the chemical laboratory.

The citric acid cycle and coenzymes are covered in Chapter 3 'The biochemistry of cells' of the *Therapeutics and Human Physiology* book of this series.

Also the pK$_a$s of carboxylic acids are discussed in Chapter 6 'Acids and bases' of *Pharmaceutics*, another book of this series.

You will have noticed the curly arrows on the structures in Figure 1.5. These represent the movement of

a pair of electrons. If you are not familiar with curly arrows, do not worry; by the end of Chapter 2, you will be.

The citric acid cycle does not only make molecules though. It also converts food into energy using, of course, chemistry. The chemical reactions of the citric acid cycle indirectly generate **ATP** (adenosine triphosphate); ATP is the universal currency of energy. The body drives chemical reactions by hydrolysing ATP. This reaction is shown in Figure 1.6.

So if we do not eat, we get tired, because we cannot do the chemical reactions to make ATP, and if we do not eat for a long time, we starve, because we cannot replace the bits of us that are continuously made and replaced.

 Chapter 8 'Inorganic chemistry in pharmacy' introduces phosphorus chemistry, including the chemistry of phosphoesters, such as ATP. Also ΔG is discussed in Chapter 5 'Thermodynamics' of *Pharmaceutics*, while ATP hydrolysis is discussed in Chapter 3 'The biochemistry of cells' of *Therapeutics and Human Physiology*, both are books from this series.

FIGURE 1.6   The hydrolysis of ATP. This reaction has a large negative ΔG, and is able to drive less favourable reactions.

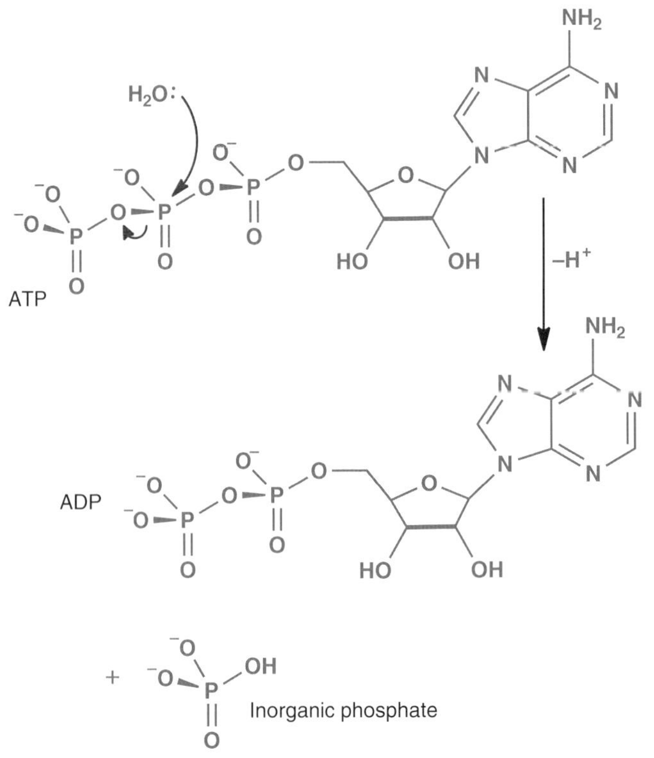

FIGURE 1.7 The conversion of citrate to isocitrate.

Citrate → Aconitase → Isocitrate

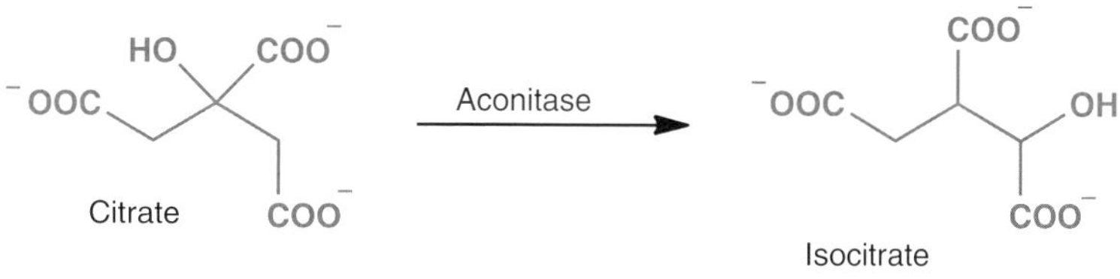

SELF CHECK 1.3

In the citric acid cycle, citrate is isomerized to isocitrate (see Figure 1.7). Redraw isocitrate with all its carbon and hydrogen atoms labelled.

SELF CHECK 1.4

The conversion of citrate to isocitrate is a two-stage process, catalysed by the enzyme aconitase (Figure 1.7). Draw a mechanism (use curly arrows) for the conversion of citrate to isocitrate. The intermediate is known as *cis-aconitate*. This is a much harder question. If you find it too difficult, try again after working through Chapter 5.

 Isocitrate is another intermediate in the citric acid cycle, which you will learn about in Chapter 3 of the *Therapeutics and Human Physiology* book of this series.

The liver

Even in ancient times, people thought that the liver had a role in well-being. We now know that this is because the liver carries out quite complex chemistry. The liver recognizes **xenobiotics** (substances that are not normally in the blood stream, see Box 1.1) and processes them. Oxidation is a common reaction in the liver, since oxidized products are often more easily excreted than the parent xenobiotics. For example, ethanol is oxidized in the liver (Figure 1.8).

BOX 1.1

Terms containing 'xeno' and 'philius'

Xenophilius Lovegood, a character in Harry Potter and the Deathly Hallows, has a name meaning love of the strange, or stranger, from the Greek 'xeno' meaning strange and 'philius' meaning love. He should help you remember xenobiotic and words such as hydrophilic (water loving).

Chapter 5 'Alcohols, phenols, ethers, organic halogen compounds and amines' in this book, introduces the chemistry of important functional groups such as hydroxyl groups, amines and halogen compounds. Oxidation and dehydration of alcohols such as ethanol are among the reactions described.

In the liver, ethanol is oxidized to acetate via acetaldehyde (ethanal), and acetate is used to produce energy. When it is metabolized properly, alcohol is just food, and is very high in calories; however, the real problems come when it is not metabolized properly. Oxidation of ethanol is slow, so most people can only process about one unit per hour. The ethanol that is not metabolized acts on the central nervous system, causing all the familiar effects of alcohol consumption (slow reactions, lack of inhibition, difficulty in controlling speech or movement). You can slow down the absorption of alcohol by eating, helping your liver to keep up, but that's pharmaceutics, not chemistry.

FIGURE 1.8 The oxidation of ethanol to acetate in the liver.

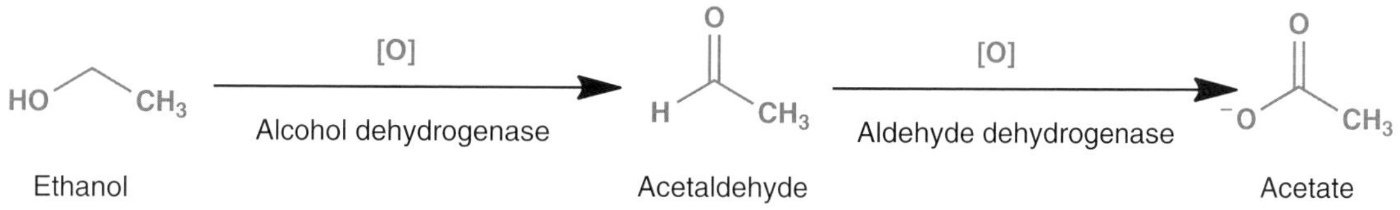

Ethanol → [O] Alcohol dehydrogenase → Acetaldehyde → [O] Aldehyde dehydrogenase → Acetate

The effects of alcohol are, however, less harmful than the effects of acetaldehyde. Acetaldehyde, like other aldehydes, causes the symptoms of hangover: nausea, vomiting, shortness of breath, and accelerated heart rate. The oxidation of alcohol is a two-stage process, as shown in Figure 1.8, and if you drink too much alcohol, acetaldehyde can accumulate, causing a hangover. The drug disulfuram (Antabuse), used in the treatment of chronic alcoholism, works by inhibiting the enzyme aldehyde dehydrogenase which converts acetaldehyde to acetate. Patients taking disulfuram (usually alcoholics) experience severe hangover symptoms within 30 minutes of consuming alcohol, which usually dissuades them from further alcohol consumption.

 More information on the metabolism of alcohols can be found in Chapter 5 'Alcohols, phenols, ethers, organic halogen compounds and amines' of this book.

SELF CHECK 1.5

Methanol is very toxic, because it is oxidized by the liver to become an aldehyde, which cannot be further oxidized. Draw the structure of this aldehyde. (It is known as formaldehyde, but its IUPAC name is methanal.)

SELF CHECK 1.6

How would you treat methanol poisoning? (This is a harder question.)

Protein synthesis

King David worried about his hair going grey but he did not know how hair is made. Hair, nails, muscles, tendons, ligaments are all made largely from proteins. Enzymes are also usually proteins and proteins are made by chemical reactions. Proteins are polymers of amino acids, and amino acids are just carboxylic acids with amino groups (see Figure 1.9). To make proteins, these amino acids are joined together (polymerized) using amide bonds. In proteins these bonds are known as peptide bonds. The carboxylic acid group of one amino acid and the amino group of another amino acid react together and straightforward carbonyl chemistry leads to the formation of an amide bond.

The amide bond is very strong, so it is absolutely perfect for hair and nails. Your hair does not fall out in the rain because this bond is so strong. To break it requires boiling in acid, or catalysis by enzymes.

 The chemistry of amides is covered in more detail in Chapter 6 'The carbonyl group and its chemistry'.

FIGURE 1.9 (A) General structure of an amino acid. (B) Formation of a peptide bond, shown in green. R can be any of 20 different groups. At its simplest, R is H, which gives the amino acid glycine; when R is CH_3, the amino acid is alanine.

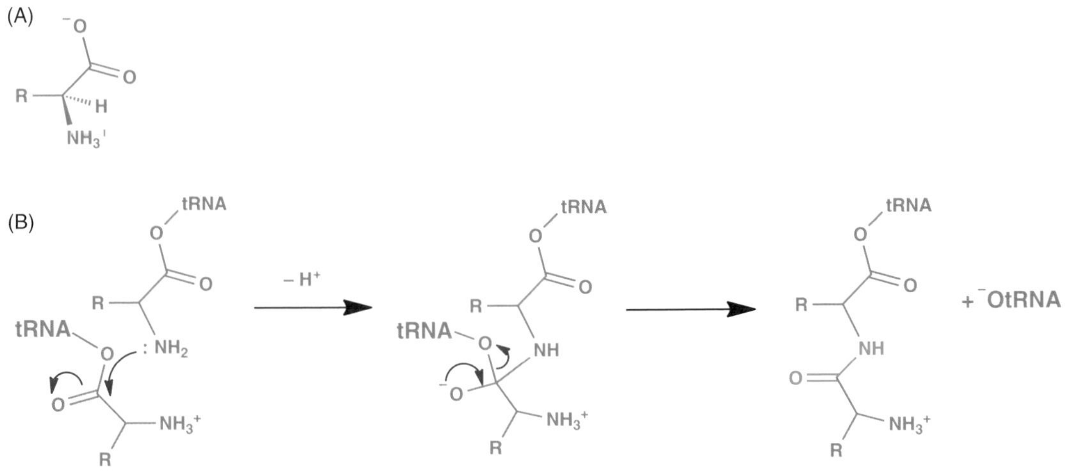

The formation of peptide bonds is very simple chemistry, yet the cell uses up to 40% of its energy making proteins, and the machinery that makes proteins can constitute up to 30% of the cell's dry weight (the weight of everything except the water). Look at Figure 1.9 and see if you can see why.

The point is that R can be any of 20 different groups, and to make a particular protein of perhaps 300 amino acids, the correct amino acid needs to be selected each time. The vast machinery of the ribosome, and other associated enzymes, is required to ensure that proteins are produced accurately.

 More information on the mechanism of operation of ribosomes is given in Chapter 2 'Molecular cell biology' of the *Therapeutics and Human Physiology* book in this series.

SELF CHECK 1.7

Draw the structure of the tripeptide glycine-alanine-glycine (usually abbreviated gly-ala-gly or GAG).

 You may have noticed that alanine is a **chiral** molecule. The four groups surrounding the carbon-2 are all different and this means that the mirror image of the structure shown cannot be superimposed on that structure. Chirality is discussed in detail in Chapter 3 'Stereochemistry and drug action'.

KEY POINT

Human cells carry out a vast array of chemical reactions that are vital to the body's normal processes, such as energy production, protein synthesis and protection against toxic xenobiotics.

1.3 **More test tubes: plants and microorganisms**

Like humans and other animals, plants and microorganisms depend on chemistry for their normal function. Sometimes this chemistry is adapted in surprising ways to produce compounds that we can use as drugs.

Antimalarial drugs

Every 45 seconds a child dies from malaria, a protozoal illness common in Africa, the Indian subcontinent and parts of South America. Like many tropical illnesses, research into its treatment has been badly underfunded and there are few effective drugs for treating the disease. The first effective drug against malaria was quinine (see Figure 1.10), isolated from the bark of the *Cinchona* tree, where it can accumulate at up to 13% dry weight. Quinine is a remarkable structure, but you will not be surprised to learn that it is made using (enzyme-catalysed) chemistry. The **biosynthesis** of

FIGURE 1.10 Quinine, artemisinin, and the Anopheles mosquito. When an infected mosquito bites a human, it passes the malaria parasite into the human blood stream, causing fever and sometimes death. This is why insect nets and insect repellent are very important in the prevention of malaria. Drugs such as quinine and artemisinin can be used to treat malarial infections.

Quinine   Artemisinin

FIGURE 1.11 Farnesyl pyrophosphate is converted into a cyclic structure (amorpha-4,11-diene) then oxidized to artemisinic acid in genetically engineered yeast. The remainder of the artemisinin synthesis is carried out in the laboratory. This work was funded by the Bill and Melinda Gates Foundation.

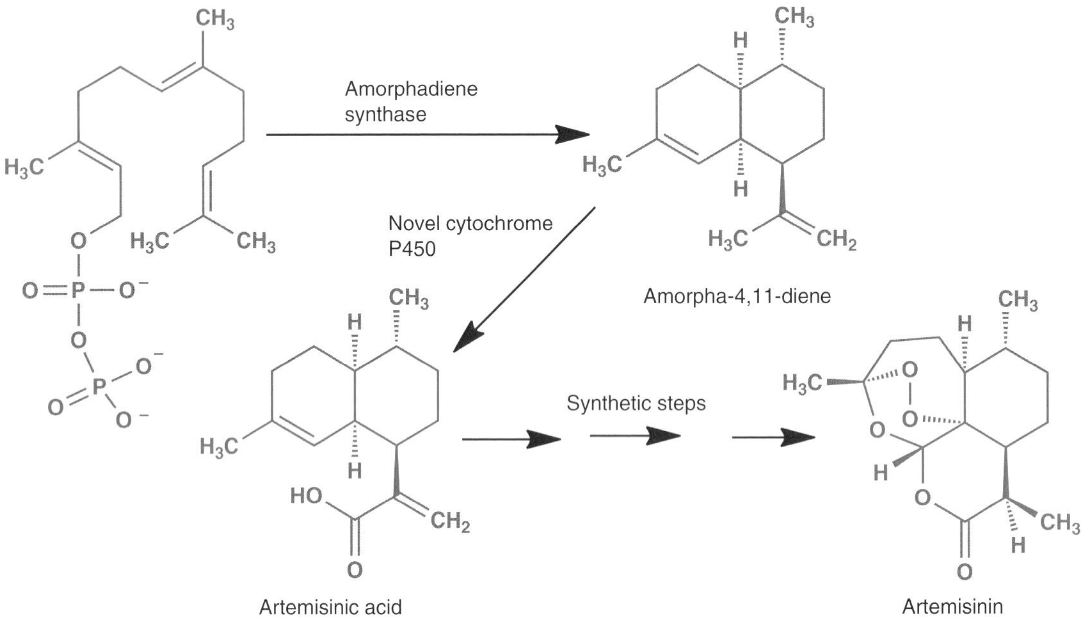

Amorphadiene synthase

Amorpha-4,11-diene

Novel cytochrome P450

Synthetic steps

Artemisinic acid

Artemisinin

quinine is far from straightforward, but you should pay some attention to the structure. To study chemistry effectively, you need to be able to draw structures, even complex structures, quickly and neatly. Quinine is a good structure to practise drawing (you can copy it – there is no need to memorize it).

Quinine changed the course of history, allowing Europeans to colonize much of Africa in Victorian times. Thousands of troops died during the Second World War because they were cut off from the supply of quinine, most of which came from Indonesia. The best organic chemists in the world tried to synthesize the drug, to develop an alternative supply, but this is the sort of molecule that plants make much more efficiently than chemists, and synthetic quinine is not commercially successful.

The malaria parasite readily develops resistance to drugs and quinine is not widely used today. The most important antimalarial drug now is artemisinin, which, traditionally, is extracted from the Chinese herb *Artemisia annua*. Modern production methods, however, use genetically modified yeast. The enzymes that catalyse the synthesis of artemisinic acid have been copied into ordinary baker's yeast; artemisinic acid is then chemically converted into artemisinin

(see Figure 1.11). This is very modern chemistry, at the interface with biology.

 In Chapter 10 'Origins of drug molecules', we ask how drugs are made. Some drugs are made by nature (plants, fungi or bacteria typically), some are made in the chemical laboratory, and an increasing number, like artemisinin, are made by a combination of chemistry, biology and genetic engineering.

Nystatin

Fungal infections can be very dangerous, but more usually they are just uncomfortable. Thrush and Athlete's Foot are among the most common. Soil bacteria are a rich source of drugs to treat such infections, a fact recognized by Elizabeth Lee Hazen, a microbiologist working in New York in the middle of the twentieth century. Hazen was clearly a generous person; when she discovered a promising strain of soil bacterium in a friend's garden, she did not name it after herself, but after the friend. The bacterium is called *Streptomyces noursei,* after the Nourse family. Hazen and her colleague Rachel Fuller Brown, a biochemist, isolated the first clinically useful antifungal agent from *S. noursei* cultures. They named this drug nystatin, in honour of their employer, the New York State Department of Public Health. After its launch in 1954, the drug was a huge success, and Hazen and Brown collected $13 m in royalties, which they donated to a trust fund for advancing women in science.

The large ring of nystatin (see Figure 1.12) is derived from acetate and propionate units, linked head-to-tail. These are modified so that one side of the molecule is covered in hydroxyl residues and attracts water (is hydrophilic) and the other side of the molecule contains hydrophobic alkene groups. This structure enables the molecule to interact with hydrophobic molecules in the centre of the membrane and with hydrophilic groups on the outside, thus disrupting the membrane. The disrupted membrane leaks and the fungal cell dies.

Linking together acetate and propionate molecules is a recurring theme in microbial chemistry. Figure 1.13 shows the reaction in which two acetate molecules are joined. It is very similar to the formation

 In Chapter 4 'Properties of aliphatic hydrocarbons', we consider the properties of hydrocarbons, including alkenes, like nystatin. The structure of the cell membrane can be seen in Chapter 2 'Molecular biology of cells' of the **Therapeutics and Human Physiology** book of this series.

of citrate as shown in Figure 1.5. This reaction can be repeated many times and is used to form fatty acids, drugs and other molecules. The microbial cell expends a lot of energy in these processes, which require many different enzymes; nevertheless, each enzyme catalyses a quite ordinary chemical reaction.

In 1966, the River Arno flooded, and the city of Florence was devastated. Numerous collections of artwork in churches, libraries and private collections were affected, and the science of restoration advanced under the threat of the loss of priceless art treasures. There was a fear that fungi would colonize the priceless painted wooden panels, so they were sprayed with nystatin!

FIGURE 1.12 (A) Nystatin. Acetate units are represented in blue and propionate units in purple. (B) Diagram of a cell membrane. The nystatin molecule is able to interact with the hydrophobic inside of the membrane and the hydrophilic outside.
Adapted from *Human Physiology: The Basis of Medicine*, 2nd edn, by Gillian Pocock and Christopher D. Richards (2006) by permission of Oxford University Press.

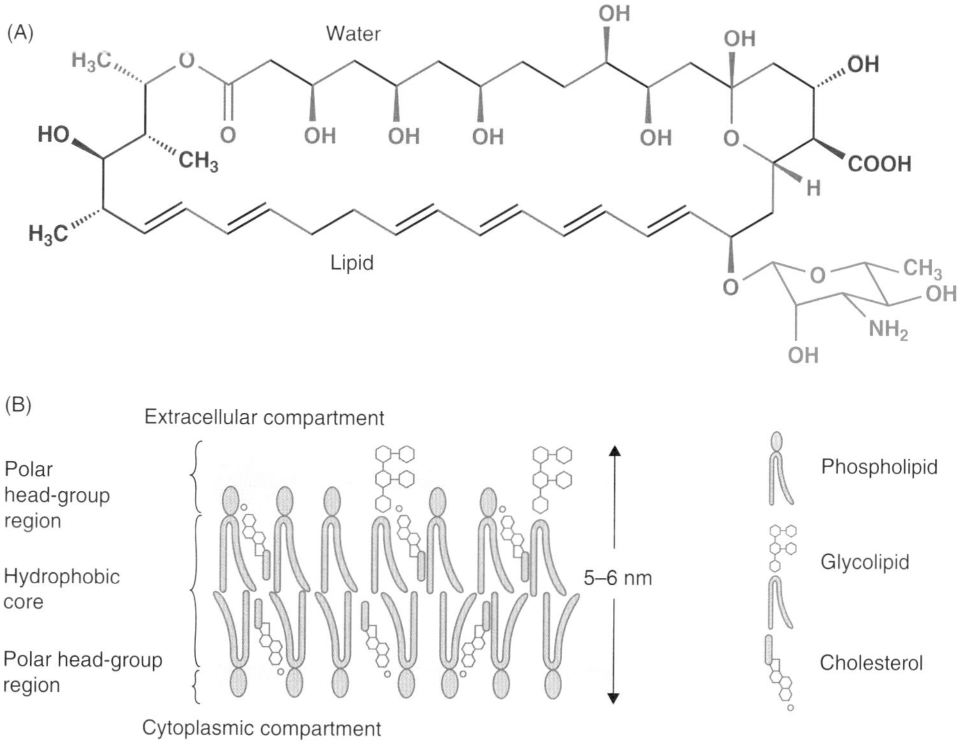

FIGURE 1.13 Head-to-tail polymerization of acetate units. Acetyl coenzyme A is activated, by conversion to malonyl coenzyme A. Both acetyl coenzyme A and malonyl coenzyme A become enzyme-bound and are able to react together, giving a four-carbon unit that can interact with another molecule of malonyl coenzyme A, and so on.

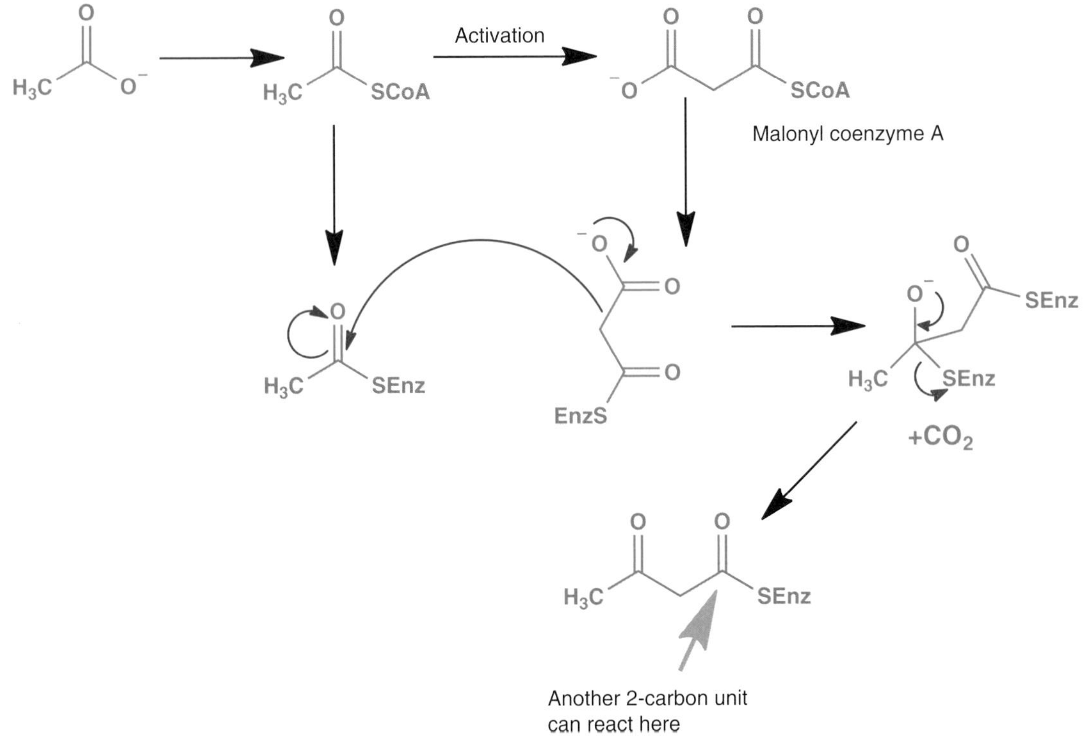

SELF CHECK 1.8

Can you draw the mechanism of addition of another acetate unit to the four-carbon unit shown in Figure 1.13?

SELF CHECK 1.9

Since penicillin is made from amino acids it contains amide bonds. Can you find them?

Penicillin

Penicillin is a hugely successful antibiotic made by fungi from three amino acids, as shown in Figure 1.14.

The discovery of penicillin is perhaps the most famous drug discovery story ever told. The Scottish microbiologist, Alexander Fleming, went on holiday leaving some bacteria growing on nutrient plates. When he returned, zones of bacteria had been killed, close to where a fungus had fallen on the plates. There is no doubt that the discovery of penicillin was a lucky accident, but Fleming was prepared to work very hard to isolate penicillin from the fungal culture he had discovered by accident.

Penicillin itself (now called penicillin G to distinguish it from other penicillins) has saved countless lives, but the fungi that produce penicillin can do something even cleverer than that. They can produce 6-aminopenicillanic acid, a molecule that can be modified by chemists to give numerous different penicillins, including amoxicillin, the yellow medicine usually given to children with respiratory tract infections.

In 1957, the American chemist John Sheehan synthesized penicillin V in the laboratory, but no laboratory synthesis has ever been as efficient as the fungal synthesis, so they are not used to produce penicillin. We can see why by looking at the molecule – as explained in Box 1.2.

FIGURE 1.14 Biosynthesis of penicillin G and 6-aminopenicillanic acid and semi-synthesis of amoxicillin.

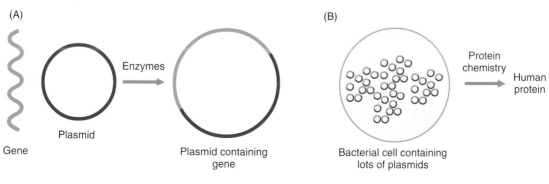

L-Cysteine

Penicillin G

L-Phenylalanine

D-Valine

6-Aminopenicillanic acid

Laboratory Chemistry

Amoxicillin

BOX 1.2

Nature is good at chirality

When nature makes a chiral molecule, it makes just one isomer, in the case of alanine, the one drawn in blue. Chemists will nearly always make a 50:50 mixture of the two possible isomers, the blue one and the purple one.

When you have four chiral centres, as in penicillin, instead of two possible isomers, there are 2^4, that is 16, only one of which works as a drug, but the fungus makes the right one every time. This is the main reason why so many drugs are still isolated from fungi, bacteria and plants.

 Chirality is discussed in Chapter 3, and penicillin is discussed in more detail in Chapter 10.

Insulin

Insulin is a small protein that most human beings and other mammals produce quite successfully in the pancreas. It enables us to take up glucose from blood into tissues. If we do not do this, the high levels of glucose become toxic. People with Type 1 diabetes (which normally begins in childhood) are unable to synthesize adequate amounts of insulin and need to take it as a medicine (usually as a subcutaneous injection). Initially, individuals were treated with pig insulin, which differs by only one amino acid from human insulin. Some people, however, developed an allergic reaction to pig insulin, because it is not identical with human insulin, and was recognized by their immune systems as 'foreign'.

Insulin is now produced in microbes (see Figure 1.15). The human gene has been inserted into bacterial cells using genetic engineering, and these bacteria are able to produce insulin cheaply and in a pure, safe

FIGURE 1.15 How a microbe produces a human protein. (A) A gene is synthesized and inserted into a small circular DNA called a plasmid. (B) Many copies of the plasmid are then inserted into bacterial cells and these cells produce the required human protein.

(A)

Gene

Plasmid

Enzymes

Plasmid containing gene

(B)

Bacterial cell containing lots of plasmids

Protein chemistry

Human protein

form. It is much easier to purify a single protein from bacteria than from mammalian sources.

Fred Sanger

Fred Sanger is the only living person to have won two Nobel Prizes, and the only person to win the Chemistry Prize twice. His 1958 Nobel Prize was for the development of the chemistry of protein sequencing. He chose insulin as his first sequencing project.

 Chapter 9 discusses biological macromolecules and their importance in pharmacy. Very often these compounds (proteins, nucleic acids, lipids) are

targets for drugs. Insulin, however, is not a target, but a drug itself.

 In the *Therapeutics and Human Physiology* book of this series, protein sequencing (Chapter 2), protein structure (Chapter 3), and insulin (Chapter 7) are examined. Also in the same book, the treatment of diabetes is discussed in Chapter 10.

KEY POINT

Plants and microorganisms use chemical reactions to produce a wide range of interesting molecules. Many of these substances can be used as drugs or can be modified to make drugs.

1.4 Glass and plastic test tubes

Quinine, artemisinin, nystatin and penicillin are all found in nature and are called natural products. Morphine, tetracycline and chloramphenicol are also natural products (see Figure 1.16). This is not to say that you buy them in a health food shop in the form of dried leaves. They are drugs that have been tested properly and are administered in carefully regulated forms. Many plants, bacteria and fungi carry out complex chemistry and so produce many (perhaps a third) of the drugs we use today.

There is an increasing demand for genetically engineered protein drugs, produced in microbes or in cell culture. These include insulin, clotting Factor VIII (used to treat haemophilia) and several anti-cancer drugs. Other drugs are synthetic – they are made entirely in the laboratory by chemists. Others still, such as amoxicillin, are semi-synthetic – nature provides a complex starting material and laboratory chemists modify it.

FIGURE 1.16 Morphine, chloramphenicol, tetracycline.

Morphine   Chloramphenicol   Tetracycline

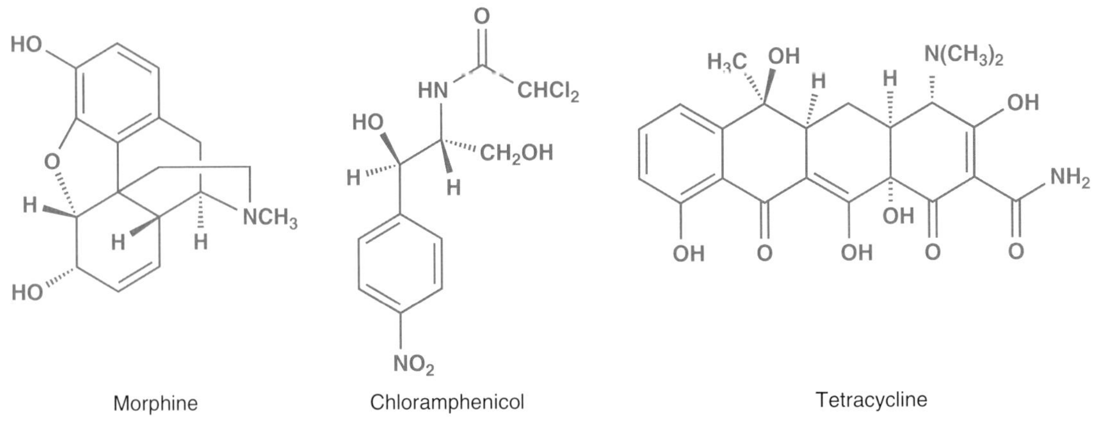

FIGURE 1.17   Do you know this molecule?

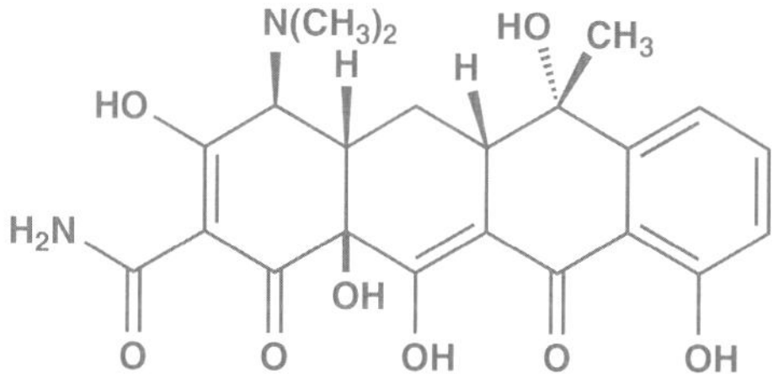

SELF CHECK 1.10

Look at Figure 1.17. What molecule is this?

Semi-synthetic drugs

Semi-synthetic drugs are very common. Microbes or plants put together a complex structure with medicinal activity. Chemists in laboratories then modify the drug to make new molecules that are more active drugs, or drugs that are better tolerated by patients.

Figure 1.18 shows the structure of erythromycin A, a very important antibiotic isolated from a soil bacterium. This molecule is made in the same way as nystatin, except that propionate units, rather than acetate units, are bolted together. Erythromycin A is very effective, but has a few problems. For example, it is very acid-sensitive and rather hydrophilic. Together,

FIGURE 1.18   The main steps of the elaboration of erythromycin A to produce the blockbuster drug azithromycin.

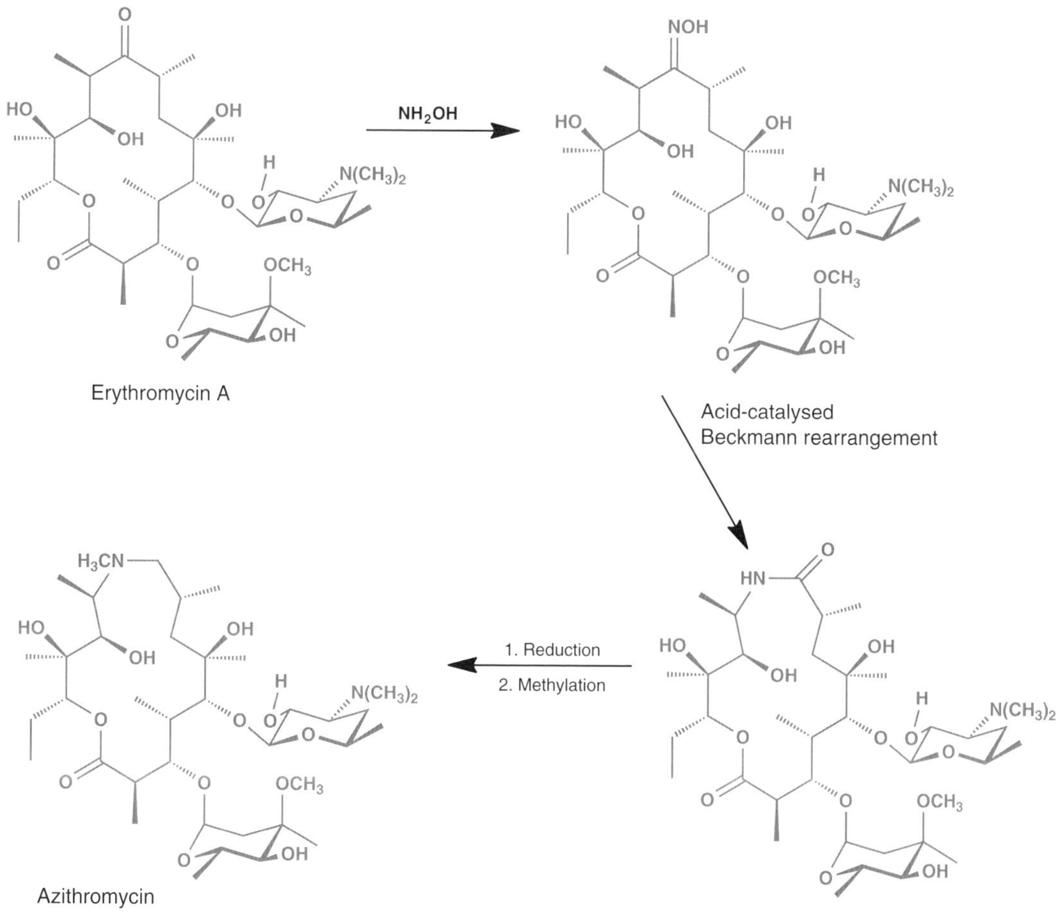

these two factors mean that erythromycin typically needs to be taken four times per day for seven days because the drug is quickly degraded and eliminated in the body. The acid-sensitivity also means that erythromycin A cannot be used to treat infections of the stomach, such as *Helicobacter pylori*, the bacterium that causes stomach ulcers.

A Croatian research group, led by Dr. Slobodan Djokic, developed azithromycin in 1980. Erythromycin A is used as the starting material, so all the chiral centres (18 of them) and the sugars are already present. Erythromycin A is very cheap, because the soil bacteria can make tonnes of it. Figure 1.18 shows the simple chemistry by which erythromycin A can be modified to make azithromycin (it is simplified slightly but all the concepts are present).

Azithromycin was a blockbuster drug with sales of more than 1 billion US dollars per year. It can be given once daily for just three days, and is not acid-sensitive.

Synthetic drugs

We have considered some examples of natural products that have been made in the laboratory. Penicillin, quinine and even erythromycin A have been synthesized, but these syntheses are not economical sources of the drugs. There are, however, a few drugs that are now made purely in the laboratory, even though they were originally isolated from nature. These include the antibiotic chloramphenicol (see Figure 1.19), which was originally isolated from the soil bacterium *Streptomyces venezuelae*, but is now synthesized in the lab. Chloramphenicol is a good choice for laboratory

FIGURE 1.19 Chloramphenicol.

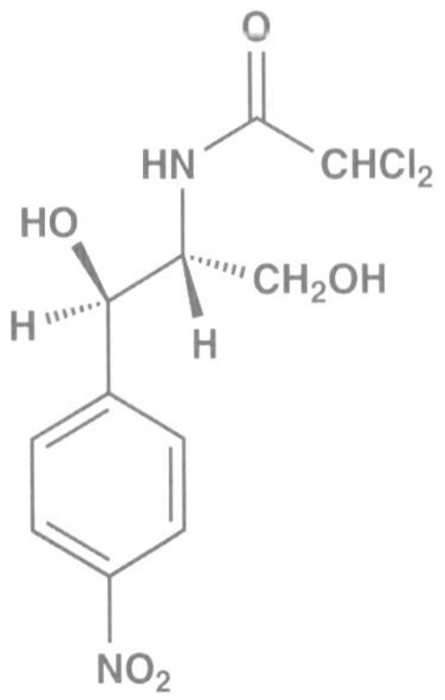

synthesis because it has very few chiral centres, and it is therefore not too difficult to make the correct isomer (see Figure 1.19).

SELF CHECK 1.11

How many chiral centres does chloramphenicol have?

About half of the drugs in use today are not found in nature, but are made in laboratories. If you have not already made aspirin yourself, you probably will, by treating salicylic acid with acetic anhydride, as shown in Figure 1.20.

SELF CHECK 1.12

If you store aspirin in your bathroom cabinet for some months, it may smell of vinegar. Why is that?

Ibuprofen is a painkiller first developed by the Boots company in 1960, and strongly recommended by dentists (among others); it is extremely effective and has fewer side-effects than other common painkillers (see Integration Box 1.1). The original synthesis was eight steps and generated quite a lot of waste.

INTEGRATION BOX 1.1

Common painkillers

When you work in a pharmacy, you will almost certainly have to advise about painkillers. Aspirin is not recommended for children because it is associated with Reyes Syndrome, a disease that causes multiple organ failure and which can be fatal. In some adults, aspirin can cause gastric bleeding. Paracetamol is therefore often preferred, but the tablets are huge (500 mg) and difficult to swallow for many people. In addition, it is relatively easy to overdose on paracetamol, and an overdose can lead to a slow, painful death from liver failure. Ibuprofen has a similar mechanism of action to aspirin, but the side-effects are comparatively rare. An effective dose is normally a single 200 mg tablet. For severe, but non-dangerous, pain (dental or post-operative pain), ibuprofen and paracetamol can both be taken safely because their modes of action are different. Do not take aspirin and ibuprofen at the same time though!

FIGURE 1.20 The synthesis of aspirin from salicylic acid.

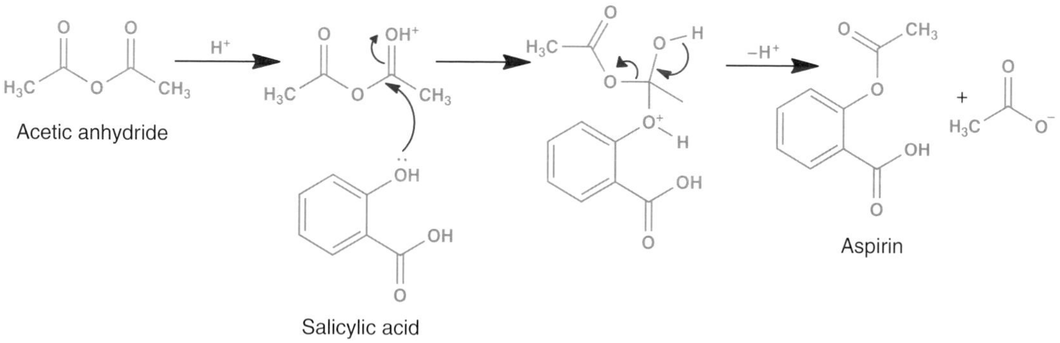

Acetic anhydride

Salicylic acid

Aspirin

FIGURE 1.21 Green synthesis of ibuprofen. The three catalysts can be recovered, so there is very little waste of material.

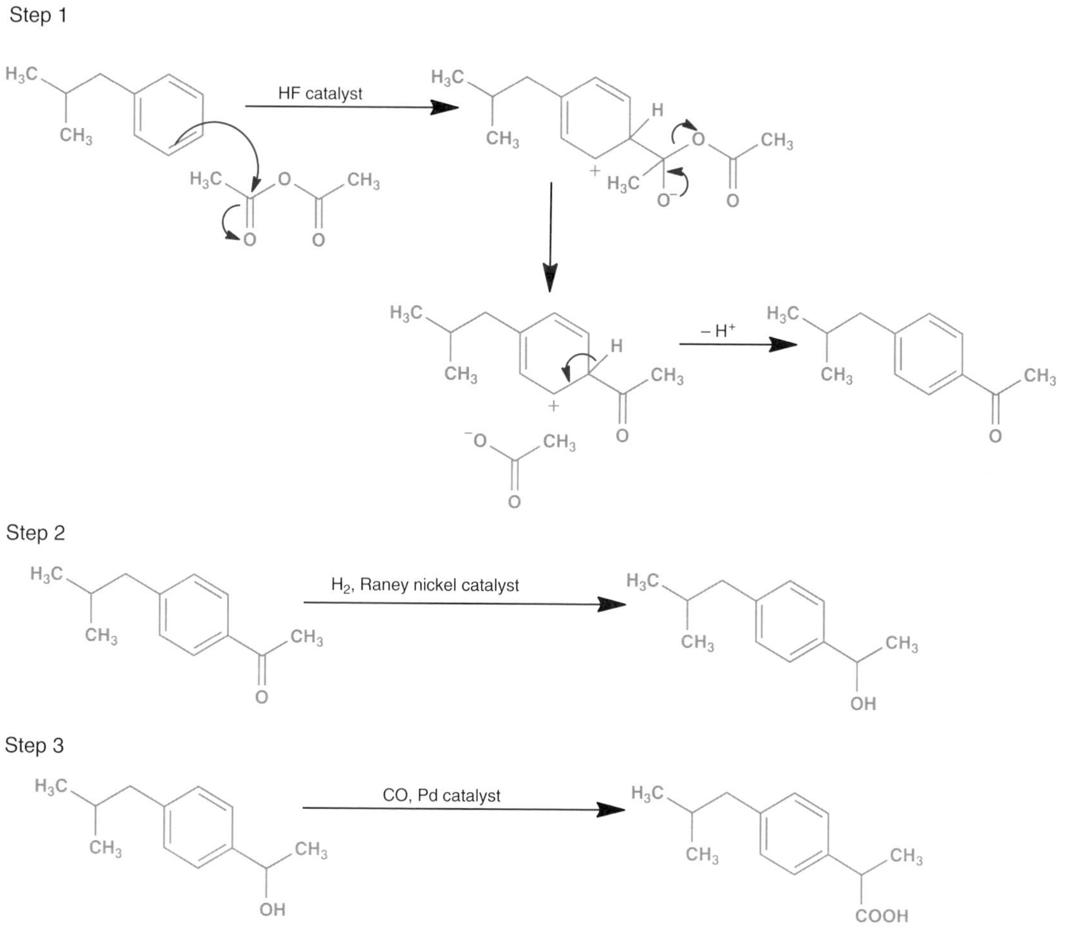

Step 1

HF catalyst

Step 2

H₂, Raney nickel catalyst

Step 3

CO, Pd catalyst

In the mid-1980s a new company (BHC) developed a new 'green' synthesis of ibuprofen, as shown in Figure 1.21.

A **'green' synthesis** aims to minimize its impact on the environment, for example, through reducing the use of hazardous materials such as organic solvents, and reducing by-products, thereby achieving a lower wastage of carbon or other materials. Green synthesis is increasingly important in the 21st century, with global temperatures rising and the population increasing. Green chemistry is chemistry that has a minimal impact on the environment.

SELF CHECK 1.13

There is one small waste molecule in the synthesis of ibuprofen. What is it?

Atorvastatin (Lipitor)

In 2008, atorvastatin (Lipitor) was the world's most profitable drug with sales of $12.4bn. It is a statin used for lowering blood cholesterol. Some medical professionals recommend statins for everyone over about 55 years old (sometimes 50 for men, 65 for women) to reduce the risk of heart attack and stroke. Others feel that statins are already over-used. These drugs make a lot of money for drug companies because they are taken by people who are not sick, every day for years or decades. Contrast this with antibiotics which are taken for a few days until the patient gets well or dies.

 Statins are discussed in Chapter 4 'Introduction to drug action' of *Therapeutics and Human Physiology* and in Chapter 8 'Pharmaceutical care' of *Pharmacy Practice*, both books from this series.

Statins work by inhibiting 3-hydroxy-3-methylglutaryl-coenzyme A reductase (HMGR), a key enzyme in the production of cholesterol. The synthesis of atorvastatin illustrates a major advantage of the glass test tube over the microbial test tube. You can see, in Figure 1.22, how lots of different compounds can be made from the same route. The red part of the molecule is essential and very few variations can be made, but each of the other four substituents can be changed independently. For example, the fluoro-substituted aromatic ring (shown in green) needs to be a hydrophobic group (otherwise the drug does not bind to the target), but there are lots of possibilities. Twenty different **analogues** were tested before the fluoro-substituted aromatic ring was chosen. The isopropyl group (shown in purple) was chosen similarly, followed by the two remaining substituents. The analogues that can be made starting from a natural product are usually much more limited. (Note that chemists cannot always predict whether an analogue is likely to improve a drug molecule; very often they make lots of analogues and choose the best after testing.)

SELF CHECK 1.14

How many chiral centres does atorvastatin have?

SELF CHECK 1.15

Draw atorvastatin with all its carbon and hydrogen atoms labelled, then work out the molecular formula of the drug.

KEY POINT

Many drug molecules are made by chemists in laboratories. Simple molecules are made wholly in the laboratory, whereas more complex drugs are often made by chemical modification of natural products.

FIGURE 1.22 Atorvastatin. The active part of the drug molecule is shown in red.

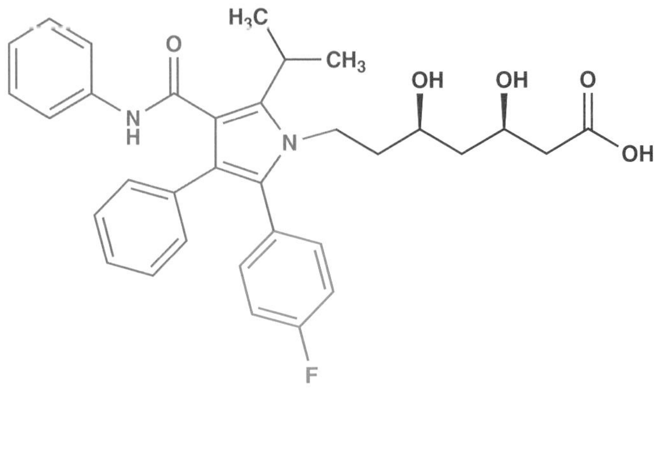

1.5 From test tube to pharmaceutical

Drugs are chemicals and are made using chemistry. Usually, however, a drug has to be converted into a medicine. The pharmacist does not hand out white powders; instead, the drugs are converted into tablets, solutions, suspensions or creams – and sometimes this conversion process relies on chemistry.

Erythromycin A

Erythromycin A is widely used to treat infections in children, because some children cannot tolerate penicillin, and some infections are resistant to penicillin. There is one huge problem with administering erythromycin A to children. It tastes absolutely foul. No amount of sugar can mask the bitter taste, and of course, children cannot swallow the 250 mg tablets that adults take. There is, however, a very clever bit of chemistry that makes erythromycin A tolerable to children.

Figure 1.23 shows a one-step reaction in which erythromycin A is converted to an ester, usually the ethyl succinate shown. Erythromycin esters are

FIGURE 1.23 (A) The preparation of erythromycin A ethyl succinate. (B) The general mechanism for ester formation. X can be Cl or RCOO. Lots of drug molecules are esters.

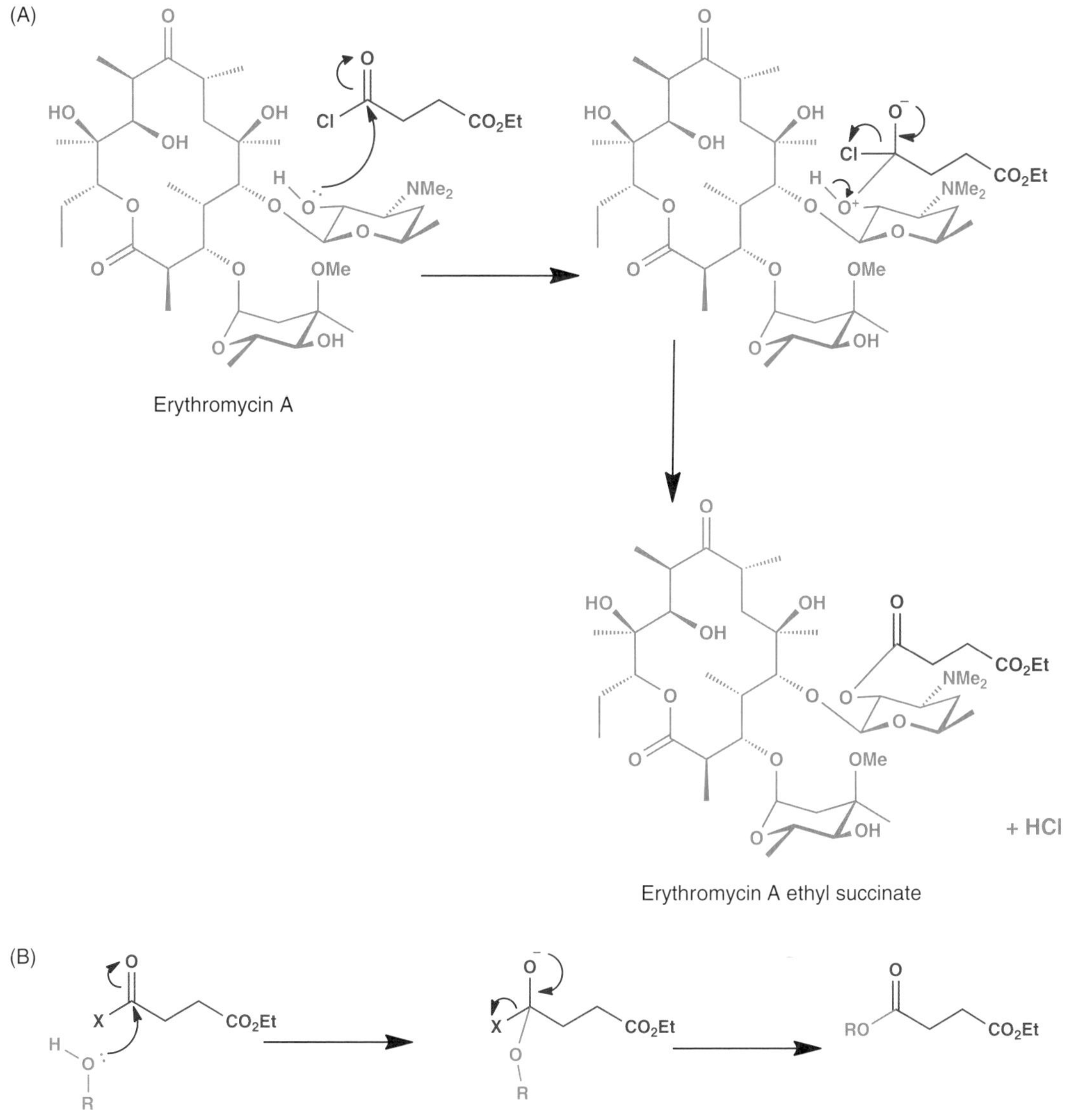

taste-free and poorly soluble in water. The esters have no antibacterial activity and they need to be hydrolysed to erythromycin in the body; they are **prodrugs**. The hydrolysis is base-catalysed and takes place in the intestine and in the blood stream.

There is a problem with this chemistry. The pharmacist adds water to the powdered ester and flavourings to make up the medicine, which is a suspension. The ester then hydrolyses to some extent in the medicine bottle (see Case Study 1.1). This problem has been solved for erythromycin B (see Figure 1.24) by making another derivative (erythromycin B enol ether ethyl succinate), which is converted to erythromycin B

ethyl succinate in acid (such as the stomach). Because erythromycin enol ether succinates are even less soluble than erythromycin esters they do not hydrolyse in the medicine bottle.

SELF CHECK 1.16

Compare Figure 1.20 with Figure 1.23 (B). Can you see that the mechanism for the formation of aspirin is the same as the mechanism for the formation of erythromycin ethyl succinate? What is X in the formation of aspirin?

FIGURE 1.24 The conversion of erythromycin B enol ether ethyl succinate to erythromycin B. Erythromycin B enol ether ethyl succinate is almost completely insoluble in water and so does not hydrolyse to the foul-tasting erythromycin until it is acidified.

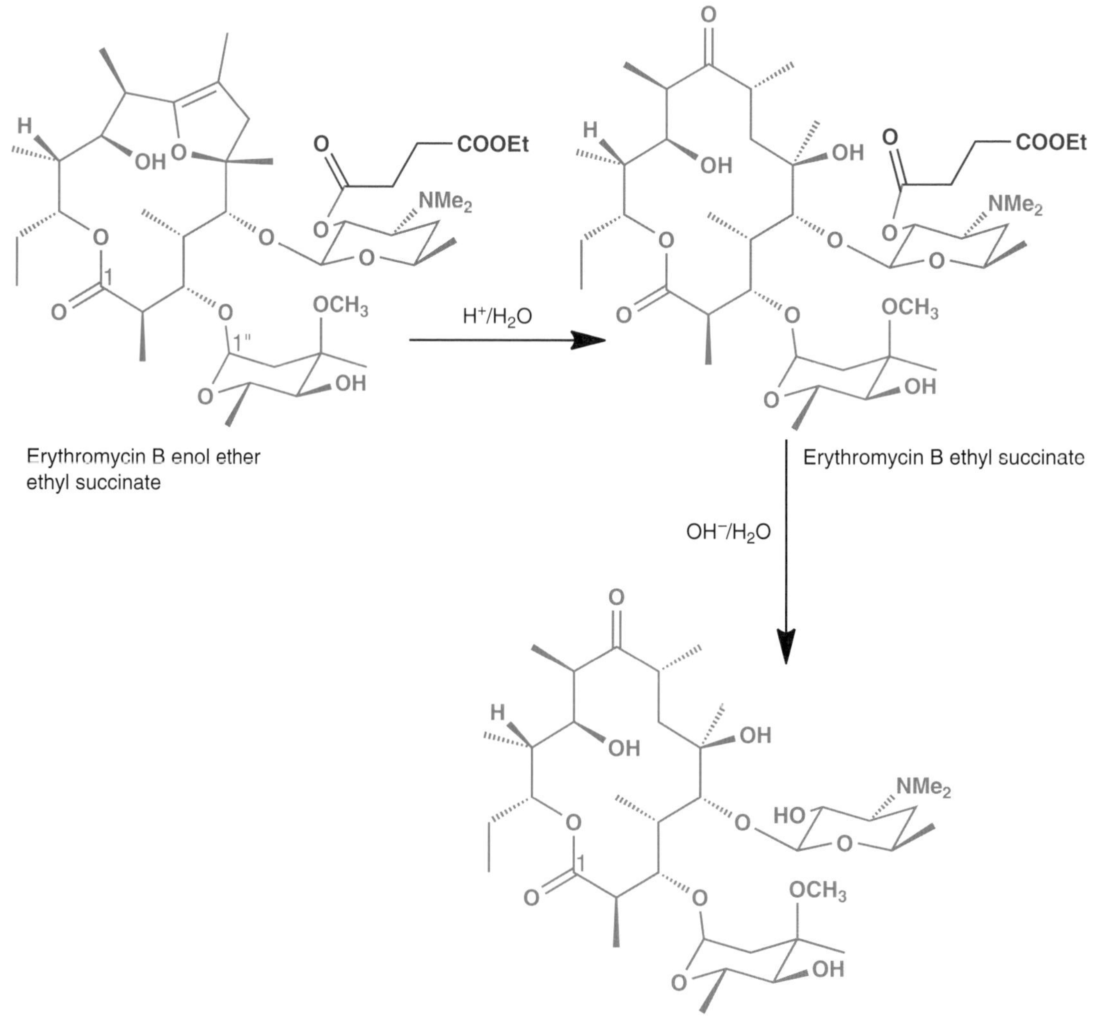

Erythromycin B enol ether ethyl succinate

H^+/H_2O

Erythromycin B ethyl succinate

OH^-/H_2O

Erythromycin B

A parent comes into the pharmacy where you are work-ing and comments that their child does not like taking erythromycin because of the taste. Children's erythro-mycin is an ester that hydrolyses when it gets warm. When this happens, the medicine still works but it leads to the bad taste.

REFLECTION QUESTIONS

1. What would you suggest to this parent?

2. The hydrolysis of erythromycin esters is base-catalysed, so why cannot you make them up in acidic solution to prevent the hydrolysis?

Answers

1 You can give them a fresh suspension and advise them to keep it in the fridge. It will still hydrolyse but much more slowly.

2 Erythromycin is acid-labile, that is it is degraded by acid. It is even broken down by gastric acid, which is why erythromycin is delivered orally as an enteric coated tablet or as a more stable ester compound.

Unfortunately, erythromycin A enol ethyl suc-cinate undergoes quite different chemistry when treated with acid, and although erythromycin B is a good antibiotic, it is much easier for the soil bacteria to make large amounts of erythromycin A, so erythro-mycin B enol ether ethyl succinate is not available in the clinic.

Sunscreen

When the sun shines on pale skin and turns it brown (or red), chemical reactions are taking place. Cells in the skin produce melanin, a pigment that absorbs ultra-violet as well as visible light. By absorbing energetic ultraviolet rays, it protects the skin from DNA damage. Because melanin is coloured, it turns the skin brown or tanned. Sunscreen is designed to reduce the exposure of the skin to damaging ultraviolet rays and may use organic or inorganic chemicals, or both. You may have seen cricketers wearing inorganic sunscreen on their faces. The zinc oxide or titanium dioxide reflects both ultraviolet and visible light so that the skin appears bright white. It is now possible to make inorganic sun-screen clear by using smaller particles, which are not nearly so bright.

Organic sunscreens work by absorbing ultraviolet light and converting it to harmless infrared radiation. To do this, an organic molecule requires a **chromo-phore** that absorbs over the right range of wavelengths. Organic molecules that absorb light in the ultraviolet or visible range have several adjacent double bonds and/or lone pairs of electrons; the larger the number of dou-ble bonds present, the higher the wavelength of most efficient absorption.

Para-aminobenzoic acid is one compound able to absorb harmful UV-B rays, and it is used in organic sunscreens. Figure 1.25 shows the structure of *para*-aminobenzoic acid and its alternate double bonds. However, Figure 1.25 also shows that *para*-aminoben-zoic acid does not absorb over the whole ultraviolet range. If this were the only compound in sunscreen, people would still get burned. So a sunscreen contains lots of different compounds, all absorbing at different wavelengths and between them covering the whole UV-B range (280–320 nm), and often the UV-A range (320–400 nm) as well.

FIGURE 1.25 The ultraviolet spectrum of *para*-ami-nobenzoic acid, a constituent of organic sunscreens.

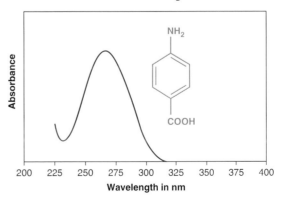

Quality control

Chapter 11 'Introduction to pharmaceutical analysis' describes in detail the methods used to confirm that a medicine really is what it is supposed to be. There was a time, before health and safety regulations, when a chemist could sniff a medicinal preparation and tell you pretty accurately what was in it. Quality control for erythromycin A ethyl succinate was done by taste. A bad preparation would taste strongly of erythromycin, and a good chemist could estimate how much erythromycin was present by tasting it.

Nowadays you can spend a million pounds on an NMR spectrometer (which delivers very high quality information), several hundred thousand pounds on a mass spectrometer (which is incredibly sensitive), and hundreds or thousands of pounds on less informative instrumentation. These instruments can be used to visualize a drug molecule in solution, how a drug interacts with a protein, or what happens to a drug in the body. When you want to know simply whether a medicine (such as tablet) contains the right drug in the right quantity, you will probably use a simple chemical analysis, costing very little.

These analyses are described in the British Pharmacopoeia. Here are the instructions for quality control of aspirin tablets:

Weigh and powder 20 tablets. To a quantity of the powder containing 0.5 g of Aspirin add 30 ml of 0.5 M sodium hydroxide solution, boil gently for 10 minutes and titrate the excess of alkali with 0.5 M hydrochloric acid using phenol red solution as indicator. Repeat the operation without the substance being examined. The difference between the titrations represents the amount of sodium hydroxide required. Each ml of 0.5 M sodium hydroxide solution is equivalent to 45.04 mg of $C_9H_8O_4$.

Source: British Pharmacopoeia

This is simple old-fashioned chemistry (hydrolysis of an ester followed by acid-base reaction) that helps keep medicines safe.

KEY POINT

Drug substances themselves are of little use if they cannot be delivered appropriately to the patient in an appropriate form. This means that drugs are converted into appropriate dosage forms, thus becoming a medicine.

CHAPTER SUMMARY

This chapter is an overview and most of the material will be covered in more detail elsewhere in the book. You should now understand that

➤ Pharmacy is the science of drugs and all drugs are chemicals.

➤ Chemicals, whether in a body, in a plant or microbe or in a test tube, do chemistry, not magic.

➤ Chemical structures can be drawn without labelling carbons and without drawing in hydrogens.

➤ Drugs are made by plants and microbes, by chemists, or by a mixture of the two.

➤ Chemistry can contribute to converting a drug into a medicine.

➤ Chemistry is used in quality control of drugs and medicines.

At this stage of your career, you should be finding out a little bit about every drug you encounter. Find out what medicines your grandparents are taking and ask:

➤ What is it for?

➤ How does it work?

➤ What is its chemical structure?

If you watch medical dramas on the television, ask yourself the same questions about the drugs that are mentioned. You will find most of the answers in the British National Formulary at http://bnf.org/bnf or in Wikipedia at http://en.wikipedia.org/wiki. Wikipedia is written by members of the public and is not systematically peer-reviewed like a scientific journal, but it is *much* more reliable and impartial than most web pages. Use it to find out the basics about non-controversial subjects, including diseases and drugs. (Do not, however, use it as a major source for your final year project!) If you want to know more about the chemistry described here, try a good organic chemistry textbook, such as 'Organic Chemistry' by Clayden, Greeves, and Warren.

You should also just read. Pharmacy is where science meets communication, and reading really helps effective communication. Dorothy L. Sayers' novel 'The Documents in the Case' will help you understand the difference between a synthetic drug and a natural product. If you do not like fiction, try reading Ben Goldacre's 'Bad Science' or Atul Gawande's 'Complications'.

Psalm 139 vs 13 and Psalm 71 vs 18.
Both attributed to King David although many Biblical scholars think that they were written by someone else.

Clayden, J., Greeves, N., and Warren, S., *Organic Chemistry*, 2nd edn. Oxford University Press, 2012.

Penniston, K.L., Nakada, S.Y., Holmes, R.P., and Assimos, D.G. (2008). Quantitative assessment of citric acid in lemon juice, lime juice, and commercially-available fruit juice products. *J. Endourology* 2008; 22(3): 567–70.

Ro, D.K., Paradise, E.M., Ouellet, M., Fisher, K.J., Newman, K.L., Ndungu, J.M., Ho, K.A., Eachus, R.A., Ham, T.S., Kirby, J., Chang, M.C., Withers, S.T., Shiba, Y., Sarpong, R., and Keasling, J.D. Production of the antimalarial drug precursor artemisinic acid in engineered yeast. *Nature* 2006; 440: 940–3.

Djokic, S., Kobrehel, G., Lopotar, N., Kamenar, B., Nagl, A. and Mrvos, D. Erythromycin series. Part 13. Synthesis and structure elucidation of 10-dihydro-10-deoxo-11-methyl-11-azaerythromycin A. *J. Chem. Res. (S)* 1988; 152–3.

Ibuprofen, a case study in Green Chemistry. Royal Society of Chemistry. http://intechemistry.files.wordpress.com/2010/09/ibuprofen-rsc-booklet.pdf, accessed 28.01.13.
This is a resource produced by the Royal Society of Chemistry which you may like to work through, perhaps when you are revising for examinations.

Bhadra, P.K., Morris, G.A., Barber J. Design, synthesis, and evaluation of stable and taste-free erythromycin proprodrugs. *J. Med. Chem.* 2005; 48: 3878–84.

British Pharmacopoeia online, aspirin tablets.
The British Pharmacopoeia online is likely to be available via your university library's electronic resources.

2

Organic structure and bonding

ALASTAIR MANN

The modern periodic table now shows the existence of over one hundred elements and while each of them is unique in its own way, it is fair to say that carbon is the most unique of them all. Indeed, one of the traditional branches of chemistry – organic chemistry – is essentially the chemistry of compounds based on this one element.

In this chapter, we will explore how carbon is able to become a part of such a diverse range of organic compounds. We will consider how we can rationalize the bonding, structure and, with particular relevance to pharmacy, the shapes of organic molecules. We will also examine the types of forces and interactions that occur *between* molecules, as opposed to the covalent bonds which hold them together. Finally, we will look at the idea of reaction mechanisms, which allow us to explain and even predict how such compounds will react in the presence of others.

Learning objectives

Having read this chapter you are expected to be able to:

➤ explain the structure and bonding of organic molecules

➤ recognize the hybridization state of carbon atoms in molecules of pharmaceutical interest and describe how this hybridization influences molecular shape

➤ describe the types of intermolecular forces that operate between organic molecules

➤ show how curly arrows may be used to describe mechanisms in organic chemistry.

2.1 What is organic chemistry?

Today, most scientists have little doubt about what they mean by the term 'organic chemistry'. The term 'organic' is used to refer to carbon-based compounds such as alcohols, amines, esters and so on, in which carbon is covalently combined with other elements, particularly hydrogen, oxygen, nitrogen and sulfur. Two hundred years ago, however, the picture was different.

The idea of there being organic compounds and inorganic compounds has its roots in the theories of

Aristotle: 384–322 BC

Jöns Jacob Berzelius
1779–1848

Friedrich Wöhler
1800–1882

the early 19th century Swedish chemist, Berzelius. Berzelius was in turn reflecting the division of the kingdoms, by the ancient Greek philosopher Aristotle, into mineral (inorganic) and into animal and vegetable (organic). Berzelius felt that there was some kind of 'vital force' which must exist in organic compounds, making them distinct from inorganic compounds. This concept of vitalism was strong and can be found in both oriental and western cultures. Indeed it was so strong that it was believed that organic compounds could not be synthesized from inorganic compounds – an idea which, if true, would have profound implications in the manufacturing sector of the modern pharmaceutical industry!

The idea started to be eroded in 1828 by the German chemist Friedrich Wöhler, who succeeded in synthesizing the organic compound, urea, from the inorganic substance, ammonium cyanate, shown in Figure 2.1.

As is so often the case in science, **serendipity** played a part. Wöhler did not set out to undermine the theory of vitalism, even though that was the eventual effect

of his work. In fact, this wasn't the first time Wöhler had converted an inorganic compound to an organic compound. Four years earlier, in 1824, he had converted the inorganic substance cyanogen $(CN)_2$ to the organic compound, oxalic acid. This toxic compound with the formula HO_2CCO_2H is found in wood sorrel (belonging to the genus *Oxalis*, hence the name of the compound), rhubarb and black tea. Although the production of urea from ammonium cyanate is more famous, the synthesis of oxalic acid from cyanogen is generally regarded as the first synthesis of an organic compound from an inorganic compound.

Many other examples followed, notably the synthesis of ethanoic (acetic) acid from carbon disulphide (CS_2) by Kolbe in 1845, and the idea of a vital force in organic chemistry faded away. We are left with the modern idea that organic chemistry is the chemistry of the compounds of carbon, and typically involves covalent bonding of carbon to H, O, N, S, P and the halogens, with important contributions from many other elements in the periodic table too.

FIGURE 2.1 Wöhler's synthesis of urea via ammonium cyanate.

$$Pb(NCO)_2 + 2NH_3 + 2H_2O \longrightarrow Pb(OH)_2 + 2NH_4(NCO)$$

Lead cyanate   Ammonium cyanate

$$NH_4(NCO) \longrightarrow [NH_3 + HNCO] \longrightarrow (NH_2)_2CO$$

Ammonia and isocyanic acid   Urea

What is so special about carbon?

There are a number of features that make carbon such an important and unique element, which can be summarized as follows:

- It forms a huge variety of compounds. Alkanes and alkenes (see Chapter 4), alcohols and amines (see Chapter 5), carboxylic acids (see Chapter 6) are just some examples.

- It is able to bond with many other elements in the periodic table, including s block elements such as lithium and magnesium, p block elements such as nitrogen, oxygen and the halogens, d block elements such as iron, copper and zinc and even f block elements such as cerium and uranium.

- It can 'catenate'. In other words, it can form long chains of carbon atoms in a way that very few other elements can. (Sulfur also forms chains of atoms, as do selenium, tellurium and silicon to some extent.)

- It can take part in different types of **homonuclear** bond; carbon can form single, double or triple bonds with a neighbouring carbon atom. This leads to different shapes of molecules, which becomes of considerable importance when designing medicines and thinking about how they interact with other molecules present in the body. Carbon nearly always takes part in covalent, rather than ionic, bonding.

Allotropes of carbon

Although this chapter is much more concerned about the *compounds* that carbon forms, we should briefly note that even as an element, carbon is not as straightforward as it might first appear. Like many other elements, it can exist as a number of different *allotropes*, or different forms of the element where the atoms are bonded together in different arrangements. For a long time, the only known forms, or allotropes, of carbon were diamond and graphite.

In 1985, a third allotrope was discovered in which sixty carbon atoms were found to be arranged in the shape of a football. For the discovery of these so-called buckminsterfullerenes (often nicknamed 'buckyballs'), Harry Kroto and his co-workers were awarded the Nobel Prize for chemistry in 1996. Variations on these structures, where carbon atoms are arranged in long tubes (nicknamed 'buckytubes') have since been reported. More recent developments in material science have led to the reporting of another allotrope of carbon. This consists of single sheets of carbon atoms and is known as graphene. These can even be produced by physically separating the layers of atoms that make up graphite with adhesive tape, leaving the bonding *within* the sheet itself, which is now one atom thick, intact. In 2010, the Nobel Prize for physics was awarded to Andre Geim and Kostya Novoselov for this ground-breaking work.

2.2 The shape of molecules

The shape of a molecule is very important. In pharmacy, it may govern the way in which a substrate binds to an enzyme and in turn, the way in which a molecule may act as a drug. That a molecule can act as a drug when given to a patient is of huge importance. It relies on a correct fit between the drug and the receptor binding sites in the patient. Most receptor sites involve proteins, and drugs interact with specific amino acids in these proteins. These interactions trigger a series of reactions in the body which lead to the beneficial effect of the drug (see Integration Box 2.1).

 This is discussed further in Chapters 1 'The scientific basis of therapeutics' and 4 'Introduction to drug action' of the ***Therapeutics and Human Physiology*** book of this series.

Changes in the three-dimensional arrangement of atoms in a molecule can cause significant differences in its properties in the human body. Arranged correctly, ibuprofen is a nonsteroidal anti-inflammatory drug; the mirror image form, however, has no useful effect at all. Similarly, the mirror image form of *S*-penicillamine, a drug used to treat rheumatoid arthritis, would not improve the patient's condition. Much worse, in fact, it would have a toxic effect on the body. The structures of ibuprofen and penicillamine are shown in Figure 2.3.

INTEGRATION BOX 2.1

The importance of the shape of drug molecules

The hormone adrenaline will trigger **bronchodilation** during an asthma attack, so you might think it would be a good treatment for this condition. However, when used as a drug, it has unwanted side effects such as increasing the heart rate. The synthetic compound, salbutamol, was first introduced by Allen and Hanburys (part of GlaxoSmithKline), in 1968, as an alternative treatment for the symptoms of asthma and is still used today. Because the shapes of the two molecules – adrenaline and salbutamol – have similarities (see Figure 2.2), salbutamol will fit into the adrenaline receptors in the muscles in the bronchiole walls, alleviating the well-known feelings of shortness of breath in the asthma sufferer, yet without producing many of the unwanted side effects.

FIGURE 2.2 The structures of adrenaline and salbutamol.

Adrenaline

Salbutamol

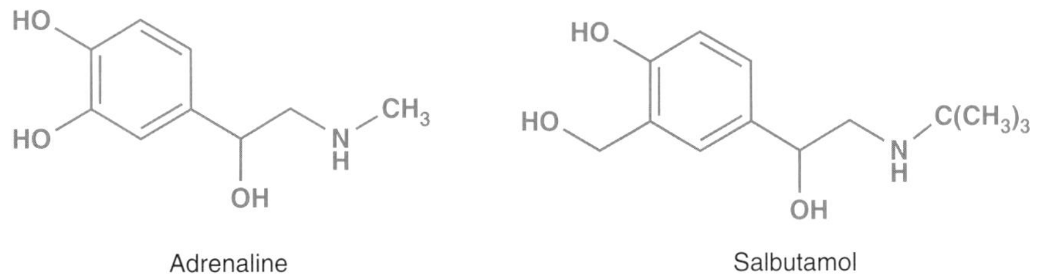

FIGURE 2.3 The structures and properties of ibuprofen and penicillamine.

R-ibuprofen
(inactive form)

S-ibuprofen
(active form)

R-penicillamine
(toxic form)

S-penicillamine
(active form)

Mirror line

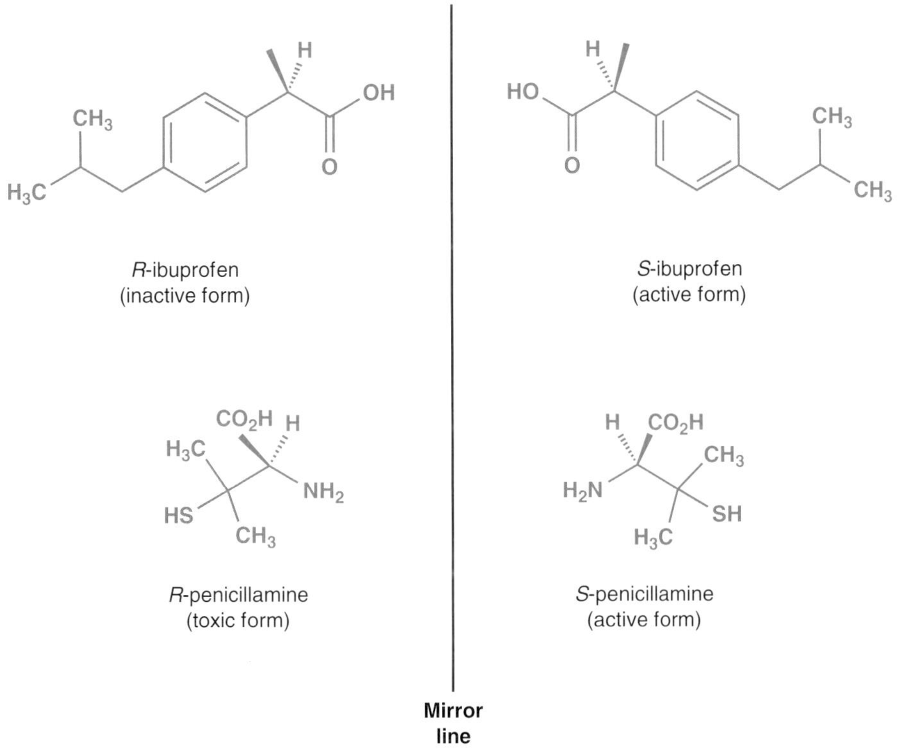

2.3 **The electronic configuration of carbon**

Carbon is the sixth element in the periodic table (atomic number Z = 6) and therefore an isolated carbon atom contains six electrons. We can use a combination of the *aufbau principle* and *Hund's rule* to predict how these electrons will be accommodated in the available atomic orbitals. The *aufbau principle* tells us that we start with the lowest energy **orbital** (the 1s orbital) and fill it with electrons first, before moving onto the next lowest energy orbital (the 2s orbital) and repeating the process. Once the three 2p orbitals (identified as $2p_x$, $2p_y$ and $2p_z$) are reached, we need to apply *Hund's rule* and place electrons into each of these orbitals separately, rather than pairing up two electrons in any single 2p orbital. As a result, we can write the ground state electronic configuration of carbon as $1s^2\ 2s^2\ 2p^2$ just based on the aufbau principle, or more fully as $1s^2\ 2s^2\ 2p_x^1\ 2p_y^1$, applying Hund's rule as well.

The third 2p orbital, $2p_z$ is empty and by convention, we don't write $2p_z^0$; we leave out any empty orbitals.

Alternatively, it may be illustrated as a series of boxes, shown either horizontally or, as in Figure 2.4, vertically in order of increasing energy. Arrows are then used, pointing either up or down to represent the opposite spins of two electrons occupying the same orbital (in the case of the 1s and 2s orbitals). In the singly occupied 2p orbitals, the fifth and sixth electrons have parallel spins and so are shown pointing in the same direction in the lowest energy, or so-called *ground state*.

At this point, we should describe the shapes of the atomic orbitals that have already been referred to in this section. The 1s and 2s orbitals are spheres (they are spherically symmetric about the nucleus). The 2s orbital is bigger than the 1s orbital and the bulk of the electron density is therefore further from the nucleus. The shape of the 2p orbitals is sometimes described as a *dumbbell*, or perhaps more simply, a solid figure of eight. All three 2p orbitals are identical and are the same distance from the nucleus. They all have the same energy and therefore are said to be **degenerate**. As a result of mutual repulsion between them, the three 2p orbitals are arranged mutually perpendicular to each other (all at right angles), lying along the *x*, *y* and *z* axes that originate at the nucleus. The shapes of these are shown in Figure 2.5 below.

FIGURE 2.4 The ground state electronic configuration of carbon. Note that the $2p_x$ and $2p_y$ orbitals each contain one electron, but the $2p_z$ orbital is empty.

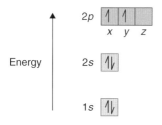

FIGURE 2.5 The shapes of s and p orbitals. Note that the shapes of 1s and 2s orbitals are the same but the 2s orbitals are bigger.

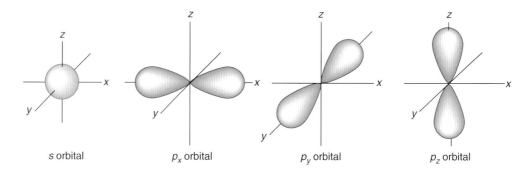

2.4 The shape of organic molecules

The ideas outlined in Section 2.3 are not immediately consistent with experimental observations of organic molecules. For example, the various representations of the ground state electronic configuration of atomic carbon all indicate that there are only *two* unpaired electrons, suggesting that carbon might only be **divalent**, yet we know that it is normally **tetravalent**. (In fact, there are some species known as carbenes in which carbon does behave as if it is divalent, though these are very unusual).

Furthermore, if we look at the relative orientation of the 2p orbitals, it is clear that these are arranged at right angles to each other, so we might expect to find 90° bond angles in carbon compounds. With the possible very rare exception of the so-called cubanes, where eight carbon atoms are arranged at the corners of a cube, these are never observed. Angles of around 109°, 120° or 180° are much more common.

A third difficulty arising from the $1s^2\ 2s^2\ 2p_x^1\ 2p_y^1$ orbital model concerns the observed bond lengths. How can we explain why all four C–H bonds in methane are identical in terms of bond length and bond strength? We would also find it difficult to explain why the C–H and C–C bonds in alkanes, alkenes and alkynes vary in terms of their length and strength. Variability in the length and strength of bonds between carbon and other elements, such as oxygen, nitrogen and sulfur, is also seen.

This variability in bond angle and bond length can be illustrated by looking at three simple hydrocarbons: ethane, ethylene (systematic name, ethene) and acetylene (systematic name, ethyne), shown in Figure 2.6.

Excited states and hybridization

Our understanding of how electrons are accommodated in an isolated carbon atom and the known shapes of molecules, such as the hydrocarbons shown in Figure 2.6, do not match. To overcome these differences between theory and observation, we need to introduce two further ideas.

FIGURE 2.6 Bond angles and bond lengths in ethane, ethene and ethyne.

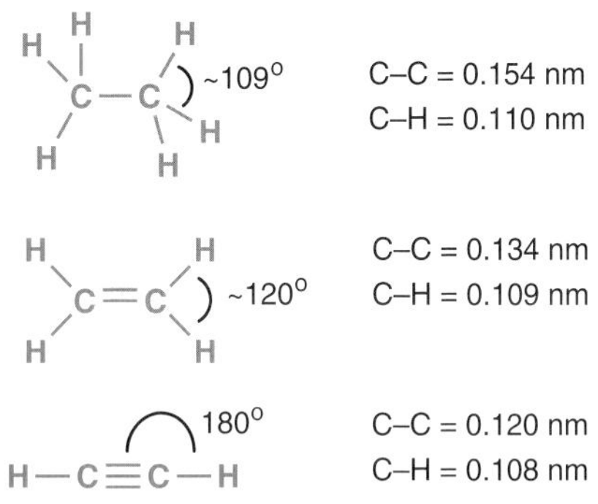

~109°	C–C = 0.154 nm
	C–H = 0.110 nm
~120°	C–C = 0.134 nm
	C–H = 0.109 nm
180°	C–C = 0.120 nm
	C–H = 0.108 nm

The first idea is the *excited state* of atomic carbon. In the excited state, energy has been supplied and, as a result of this, one of the 2s electrons has been promoted (raised to a higher orbital) and now sits in the previously unoccupied $2p_z$ orbital. This is shown in Figure 2.7.

This excited state now contains four unpaired electrons, which are available for bonding to other elements. So carbon now has the potential to be tetravalent, which is consistent with observation. However, if these electrons were used to form bonds while occupying the three 2p orbitals we could still not account for the known shapes of organic molecules. In such a scenario, three of the hydrogen atoms in a compound such as methane (CH_4) would be attached to the carbon atom via a 2p orbital and so would be arranged at right angles to each other, while the fourth, attached to a spherical 2s orbital, would have no preferred orientation. To make matters worse, because of the different energies of the 2s

FIGURE 2.7 Formation of an excited state of carbon.

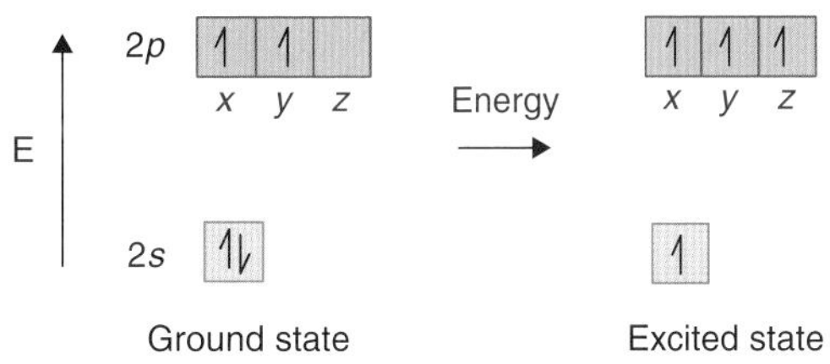

and 2p orbitals and their different distances from the nucleus, the four C–H bonds in methane would not all be of equal length, in contrast to what is well established by physical measurements.

So although the production of an excited state is a useful start, something further is needed. This is the concept of hybridization of the atomic orbitals to produce new hybrid orbitals.

Hybridization is effectively a mixing process, in which combinations of atomic orbitals are taken and blended together to produce the same number of new, hybrid orbitals. These display properties that are somewhere between those of the original atomic orbitals used to produce them. Hybridization is used to rationalize the observed shapes and structures of real molecules, allowing us to overcome the difficulties that we encountered when we tried to explain these observed properties in terms of the atomic orbitals available.

At a more advanced level, hybridization can be described using some sophisticated mathematics. However, a less intimidating, qualitative description of hybridization was first put forward by the American chemist, Linus Pauling, in the 1930s and works perfectly adequately here.

There are three possible ways that the 2s and 2p atomic orbitals in carbon can be combined, or hybridized, differing only in how many 2p orbitals are involved. These are described as:

- **sp^3 hybridization**: here all four orbitals (2s and three 2p) are mixed to produce four new hybrid orbitals, which are labelled sp^3. These hybrid orbitals are used to rationalize the observed shapes of saturated organic molecules such as CH_4, C_2H_6 and many others.

- **sp^2 hybridization**: here only three orbitals (2s and two of the 2p orbitals) are mixed to produce three new hybrid orbitals, which are labelled sp^2. The third 2p orbital remains unhybridized, although it still plays a part in bonding. These hybrid orbitals are used to rationalize the observed shapes of unsaturated organic molecules such as the alkene, ethene (ethylene).

- **sp hybridization**: here only two orbitals (2s and just one of the 2p orbitals) are mixed to produce two new hybrid orbitals, which are labelled sp

(there is no need to put a superscript 1 on this; the correct notation is sp, not sp^1). The other two 2p orbitals remain unhybridized, although they still play a part in bonding. These hybrid orbitals are used to rationalize the observed shapes of unsaturated organic molecules such as the alkyne, ethyne (acetylene).

Let us now consider the three types of hybridization in more detail, in all cases starting from the excited state of carbon illustrated in Figure 2.7. In each case, the following properties of the hybrid orbitals have to be considered:

- the *energy* of the hybrid orbitals with respect to the starting atomic orbitals
- the *shape* of the hybrid orbitals
- the *relative orientation* of the hybrid orbitals with respect to each other and also with respect to any unhybridized orbitals that may be present
- their *use in bonding* in organic compounds

sp^3 hybridization

The way in which the 2s and three 2p orbitals in the excited state of atomic carbon are combined to give four sp^3 hybrid orbitals is shown in Figure 2.8. (The 1s orbital is not shown in this diagram as it is much lower in energy and so is not directly involved in bonding.) The result is the production of four new, degenerate orbitals, each one of which accommodates a single electron from carbon. This makes carbon tetravalent and allows it to form four bonds to itself or other elements. Because these new orbitals are three parts p and one part s, they lie somewhere between the original 2s and 2p orbitals in terms of energy, though rather closer to the original 2p level. Looking ahead, we can imagine that the sp^2 and sp hybrids will progressively drop in energy, as the percentage of s character increases each time. (It is 25% in sp^3, 33% in sp^2 and 50% in sp hybridization.)

The shape of these hybrids is rather harder to explain in detail without adopting a fuller, more mathematical approach. However we can adopt a simpler, more pictorial approach, seeing what would happen in just one dimension when an s orbital is combined

FIGURE 2.8 sp³ hybridization of carbon.

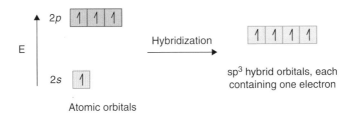

FIGURE 2.9 Shape of a hybrid orbital: same coloured regions add together and different coloured regions cancel.

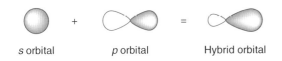

s orbital   p orbital   Hybrid orbital

with a p orbital. This is illustrated schematically in Figure 2.9, where the effect of adding a 2s orbital to a 2p orbital is shown. As a result, the size of the hybrid orbital is reduced on the left of the nucleus, as shown in the diagram, and is increased on the right. The resulting shape is sometimes described as looking a bit like a tadpole.

Again looking ahead, we can predict that as the relative contribution of the s orbital increases as we go to sp² and finally to sp hybridization, the resulting hybrid orbital will be a broadly similar shape, though it will become slightly shorter and slightly rounder each time. We will see later that this shortening of the orbital is the reason that C–H bonds become progressively shorter as we go from alkanes to alkenes and finally, to the alkynes (see Figure 2.6 for details).

The relative orientation of the four sp³ hybrid orbitals is determined by mutual repulsion, which maximizes the separation between four identical and similarly charged objects in space. The optimum way of doing this is to have them pointing towards the corner of a tetrahedron, which results in an angle between them of 109.5°.

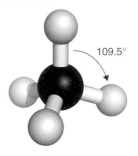

109.5°

This is the angle observed between the C–H bonds in methane illustrated here, or between the C–Cl bonds in tetrachloromethane. However, if the four groups or atoms attached to the central carbon atom are not identical, there are small deviations from the perfect tetrahedral angle to allow for the difference in sizes and electronegativities of the substituent groups surrounding the carbon atom. These deviations are typically only about one or two degrees.

We learn more about bond angles in organohalogen compounds such as tetracholoromethane in Box 2.1.

BOX 2.1

Halothane, an anaesthetic

Halothane (2-bromo-2-chloro-1,1,1-trifluoroethane) is an organohalogen compound, first synthesized by Charles Suckling in 1951 while he was working for ICI (Imperial Chemical Industries). It is trademarked under the name *Fluothane* and was used as a general anaesthetic from the mid-1950s onwards, administered by inhalation. Although it is still used in veterinary medicine and in some parts of the developing world, in many developed countries it has now been superseded by other compounds. These often also contain CF_3 groups, though they tend also to contain ether functions. Halothane and tetrachloromethane are illustrated in Figure 2.10, though without representing the full, three-dimensional aspects of their structures.

SELF CHECK 2.1

In tetrachloromethane, the Cl–C–Cl bond angle is 109.5°. In halothane, the F–C–F bond angles are in the range 107.5° to 108.5°, while the Br–C–Cl bond angle is nearly 112°. Can you explain these differences?

FIGURE 2.10 Halothane and tetrachloromethane.

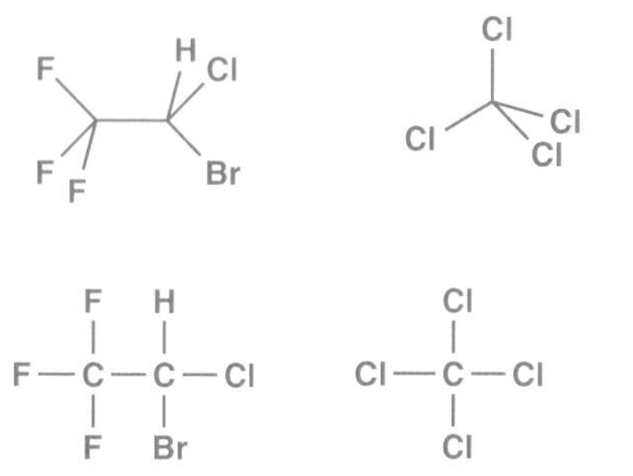

Halothane   Tetrachloromethane

The four sp³ hybrid orbitals are used to form σ-(sigma) bonds (often synonymous with the term 'single bond') by an 'end-on' or 'head-on' overlap of one of these orbitals with another suitable orbital in a second atom. The term 'end-on' implies that these orbitals are pointing straight at each other, along the internuclear axis. Sigma (σ) bonds are the strongest type of covalent chemical bonds and are radially symmetric about the internuclear bond axis – in other words, if you sliced through a σ bond at right angles to the internuclear axis, you would 'see' a circle.

In methane, each sp³ hybridized orbital on carbon overlaps with a 1s orbital on a hydrogen atom, which also contains a single electron. This leads to the for-

mation of four conventional covalent σ-bonds, sometimes more rigorously described as sp³–s σ-bonds to reflect the contributing orbitals. This is illustrated in Figure 2.11. In ethane, three of the sp³ hybrid orbitals on carbon overlap with 1s orbitals on hydrogen, while the fourth overlaps (again end-on) with the sp³ hybrid on a second carbon atom. This gives a sp³–sp³ C–C σ-bond, or more simply, a carbon-carbon single bond.

Alkanes (and many other molecules), employ exclusively single, σ-bonds. Simple alcohols and amines, such as ethanol and ethylamine, also employ only single, σ-bonds. C–O and C–N bonds are formed by end-on overlap of a sp³ hybrid orbital on carbon with an appropriate orbital on either oxygen or nitrogen. However, carboxylic acids, ketones, alkenes, nitriles, and alkynes, require the other two types of hybridization – sp² and sp – to account fully for their bonding and structures.

There is generally completely free rotation about σ-bonds, even when there are different (sometimes quite bulky) groups attached to the atoms making the bond. (An exception to this is in most cyclic compounds, as is discussed in Chapter 3.) We will see in the following section, 'sp² hybridization', that free rotation is prevented when other types of bonds, termed pi (π) bonds, are formed.

FIGURE 2.11 Bonding in methane (A) and ethane (B).

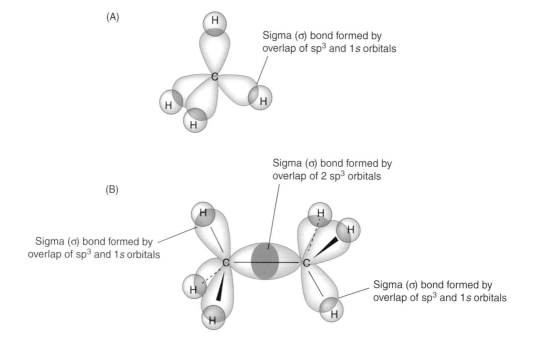

(A)

Sigma (σ) bond formed by overlap of sp³ and 1s orbitals

(B)

Sigma (σ) bond formed by overlap of 2 sp³ orbitals

Sigma (σ) bond formed by overlap of sp³ and 1s orbitals

Sigma (σ) bond formed by overlap of sp³ and 1s orbitals

sp² hybridization

The way in which the 2s and *two* of the three 2p orbitals in the excited state of atomic carbon are combined, to give three sp² hybrid orbitals, is shown in Figure 2.12 (again, the 1s orbital is not shown in this diagram as it is much lower in energy and so is not involved in bonding). The result is the formation of three new, degenerate orbitals, in addition to the unhybridized 2p orbital. Each one of these orbitals accommodates one of the four valence electrons, so again carbon is tetravalent and able to form four bonds to other elements.

Given that these new hybrid orbitals are two thirds p and one third s, they again lie somewhere between the 2s and 2p orbitals in terms of energy. They are still closer to the original 2p level, though this time slightly lower in energy than the sp³ hybrids we saw above (because the orbitals now have 33% s character).

The shape of these hybrids is broadly similar to the sp³ hybrids described earlier. However, they are slightly shorter and rounder than the sp³ hybrids, reflecting the increased contribution from the 2s atomic orbital. As was suggested previously and shown in Figure 2.6, the C–H bond in an alkene is slightly shorter than in an alkane because the sp² hybrid orbital is shorter than an sp³ hybrid orbital.

The relative orientation of the three sp² hybrid orbitals and the unhybridized 2s orbital is again governed by mutual repulsion, with their orientations maximizing their separation in space. However this time the argument is not quite as simple, because the four objects we need to separate are not all identical. This time, the three sp² hybrids point towards the corners of an equilateral triangle (in a so-called *trigonal planar* arrangement), with the unhybridized 2p orbital placed perpendicular to these. This results in an angle of 120° between the three sp² hybrids, which of course is approximately the angle observed between C–H bonds in an alkene such as ethylene. In fact, this H–C–H angle is closer to 118° because of the greater bulk of the rest of the molecule compared with the two hydrogen atoms, so the angles have to adjust slightly to accommodate this (see Figure 2.6).

The three sp² hybrid orbitals are able to form σ-bonds to other elements in exactly the same way as sp³ hybrid orbitals can. A σ-bond is formed by end-on overlap of the appropriate orbitals, and the C–H bonds may be described this time as sp²–s, while the C–C bonds are described as sp²–sp².

The unhybridized 2p orbital on a carbon atom can also become involved in bonding between two adjacent carbon atoms. However, end-on overlap is not possible because the orbitals are pointing the wrong way. Instead, a weaker, 'side-on' overlap between the two parallel orbitals is all that is possible. This leads to the formation of a π-bond between the two carbon atoms, where the electron density is now located above and below the internuclear axis, rather than along it. This time a slice through the internuclear axis would reveal two areas of electron density, one above and one below the axis, rather than the radial distribution seen in the case of a σ-bond. This overall bonding scheme is shown in Figure 2.13.

Side-on overlap is not as effective as end-on overlap, so the π-bond in a so-called double bond – which comprises one σ-bond and one π-bond – is weaker than a σ-bond. This is often reflected in the chemistry of an alkene. Chapter 4 introduces several reactions in which the π bond is lost, perhaps by addition across it, while the σ-bond remains intact. Critically, however, the π-bond is strong enough to prevent free rotation about the C–C axis, whereas this rotation readily occurs when there is only a σ-bond holding

FIGURE 2.12   sp² hybridization of carbon.

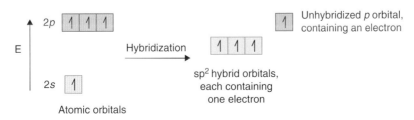

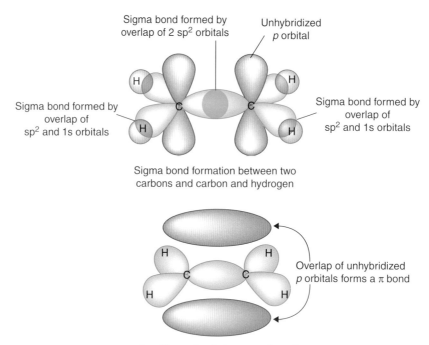

FIGURE 2.13   Bonding in ethene.

Sigma bond formed by overlap of 2 sp² orbitals

Unhybridized *p* orbital

Sigma bond formed by overlap of sp² and 1s orbitals

Sigma bond formed by overlap of sp² and 1s orbitals

Sigma bond formation between two carbons and carbon and hydrogen

Overlap of unhybridized *p* orbitals forms a π bond

π bond formation between carbons by overlap of unhybridized *p* orbitals

the two atoms together. This is what permits the formation of geometric, or *cis* and *trans* isomers in alkenes, and is discussed more in Chapters 3 and 4.

So far, we have only considered the formation of double bonds between two carbon atoms, though in fact a similar description can be used for the double bonds observed between carbon and oxygen in a carbonyl group, C=O, (for example, in a ketone), or in the C=N bond, for example, in an **imine**. In these cases, the carbon atom is sp² hybridized (as is the heteroatom, O or N), with angles of about 120° between any other atoms attached to it.

sp hybridization

The way in which the 2s and just *one* of the three 2p orbitals in the excited state of atomic carbon are combined to give two sp hybrid orbitals, is shown in Figure 2.14. (Once again, the 1s orbital is not involved in this process and so is not shown in the diagram.) The result is the formation of two new, degenerate orbitals, in addition to the two unhybridized 2p orbitals. Each orbital accommodates one of the four valence electrons in carbon, so carbon is again tetravalent and is

able to form four bonds to other elements. Given that these new hybrid orbitals are now made up of equal parts s and p, they lie half way between the 2s and 2p orbitals in terms of energy, below both the sp³ and sp² hybrids seen in the last two sections.

The shape of these sp hybrids is broadly similar to the sp³ and sp² hybrids already described (and the same reasons are used to account for their shapes). However, they are slightly shorter and rounder than the sp³ and sp² hybrids, reflecting the increased contribution from the 2s atomic orbital. As was discussed earlier and shown in Figure 2.6, the C–H bond in an alkyne is slightly shorter than the C–H bonds in either an alkane or an alkene, because it is an sp hybrid orbital that is involved in the formation of the C–H bond.

Once again, the relative orientation of the two sp hybrid orbitals and the two unhybridized 2p orbitals is governed by their mutual repulsion, to maximize their separation in space, and, once again, the four objects being separated are not identical. The optimum arrangement is to have the two sp hybrids pointing away from each other with a bond angle of 180° between them. The two unhybridized 2p orbitals are then placed mutually perpendicular to these (see

FIGURE 2.14 sp hybridization of carbon.

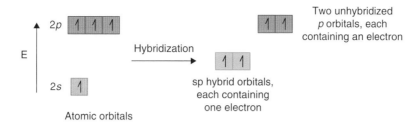

Figure 2.15). Having the two sp hybrids 180° apart is consistent with the shape of a simple alkyne like acetylene (ethyne), where the four H–C–C–H atoms are found in a linear arrangement (see Figure 2.6).

The two sp hybrid orbitals are able to form σ-bonds to other elements in exactly the same way as sp³ and sp² hybrid orbitals, with a σ-bond being formed by end-on overlap of the appropriate orbitals. The C–H bonds are described this time as sp–s, while the C–C bonds are described as sp–sp.

The two unhybridized 2p orbitals on a carbon atom also become involved in bonding between two adjacent carbon atoms and, again, weaker side-on overlap occurs because end-on overlap is not possible. This leads to formation of two π-bonds between the two carbon atoms, which lie in planes arranged at right angles to each other. The electron density associated with these is now located in front and behind the internuclear axis, as well as above and below it. This time, a slice through the internuclear axis would reveal four areas of electron density associated with the π-bonds, in addition to the radial distribution around the axis arising from the C–C σ-bond. This overall bonding scheme is shown in Figure 2.15.

As we saw with bonding in the alkenes in the section discussing 'sp² hybridization', side-on overlap is not as effective as end-on overlap, because the orbitals cannot get close enough to maximize it. As a result, the second and third bonds (the π-bonds) in a so-called triple bond are both weaker than the σ-bond, which influences the chemistry of the alkynes.

SELF CHECK 2.2

What kind of chemical reaction would occur between acetylene (ethyne) and bromine? Draw the structure of the compound that would be produced when ethyne reacts with two moles of Br_2. The names acetylene and ethyne are both acceptable; see Box 2.2.

So far we have only considered the formation of triple bonds between two carbon atoms, though in fact a similar description can be used for the triple bonds observed between carbon and nitrogen in a

FIGURE 2.15 Bonding in ethyne (acetylene).

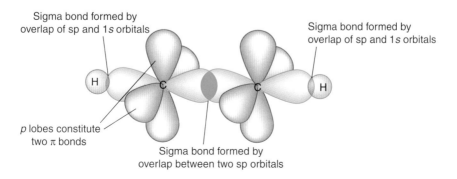

Formation of two π bonds between the carbons by overlap of two pairs of unhybridized p orbitals

Systematic nomenclature and trivial names

By the time you have studied this book you will be familiar with both systematic nomenclature and the trivial names of many common organic molecules. In the early chapters, we will normally give both. Later, we will use whichever is more appropriate in the context.

nitrile (C≡N) group. In this case, the carbon atom is still regarded as being sp hybridized (as is the nitrogen atom), with angles of 180° forming between any other atoms or groups attached to it.

SELF CHECK 2.3

Why does carbon, generally, not form triple bonds to oxygen?

Recognizing hybridization states and the shapes of molecules

The previous sections have described how we can rationalize the experimentally observed bonding and structure of organic molecules by invoking different patterns of hybridization of the valence shell orbitals, 2s and 2p. However, it is also important to think about this the other way round: to be able to look at the structure of a molecule, recognize the hybridization states of its carbon atoms and hence begin to visualize the three-dimensional shape of the compound.

While such visualization may seem difficult initially, there are some simple guidelines that will help

you to do this reliably. You should look at an individual carbon atom and count the number of atoms it is bonded to. This is a reflection of the number of σ-bonds that have been formed to it and hence the number of hybrid orbitals that were required to make this bonding possible. For example, in methane or ethane, we can see that carbon is bonded to four separate atoms and so it needs to be sp³ hybridized. The bond angles between these various attached atoms, or groups, will be about 109°. Likewise we saw that there are three separate atoms attached to the central carbon in an alkene and that sp² hybridization was required to rationalize this. The corresponding bond angles around the carbon atom are seen to be about 120°. Finally, there are only two separate atoms attached to each carbon atom in an alkyne and so sp hybridization is required to account for this. The associated bond angles are 180°.

We also extended these ideas, which were discussed in most detail for the hydrocarbons, to include other atoms, for example O and N, but equally S, P and the halogens too. In other words, it does not matter what other element carbon is attached to when we come to assess its hybridization state in this way.

These guidelines are summarized in Table 2.1.

Some compounds of interest and relevance in pharmacy are shown in Figure 2.16. The three drug structures shown here all contain carbon atoms in all three possible hybridization states!

SELF CHECK 2.4

Now draw all the structures in Figure 2.16 as 'stick' structures (See Chapter 1 if you are not sure how to do this).

TABLE 2.1 Hybridization state and bond angles around carbon atoms

Number of separate atoms attached to a carbon atom	Hybridization state of the carbon atom	Approximate bond angles around the carbon atom
4	sp³	109°
3	sp²	120°
2	sp	180°

FIGURE 2.16 Structures of drugs and hybridization states of carbon in them.

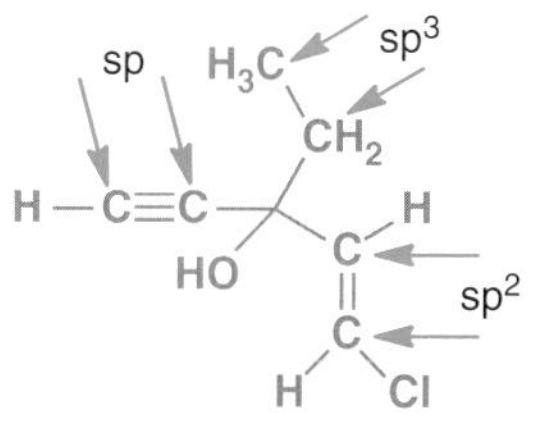

Ethchlorvynol – a sedative

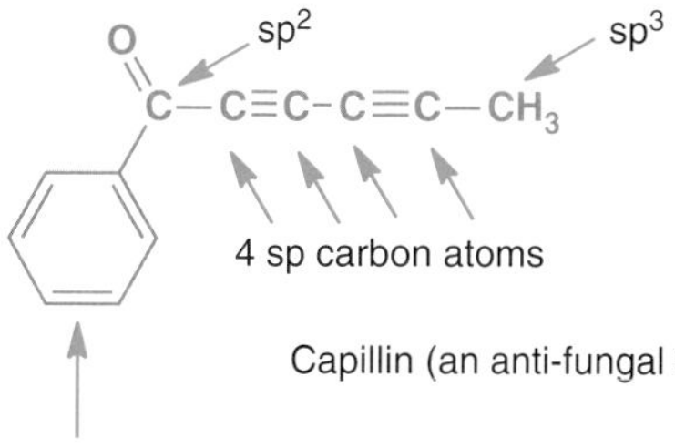

All 6 ring carbons are sp²

4 sp carbon atoms

Capillin (an anti-fungal agent)

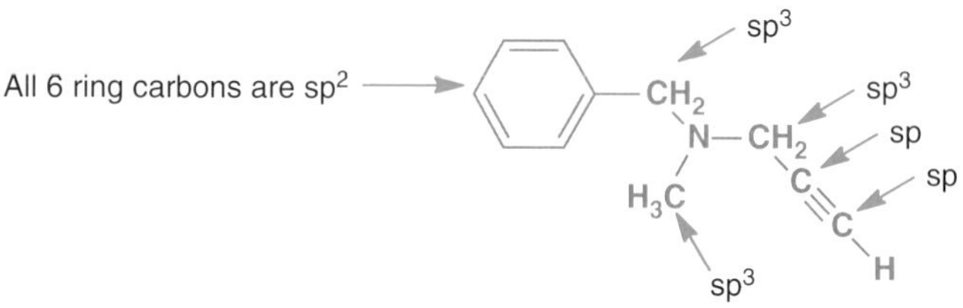

All 6 ring carbons are sp² →

Pargyline (an anti-hypertensive sold as Eudatin and Supirdyl)

FIGURE 2.17 Structures of paracetamol, ethynylestradiol, oblivon.

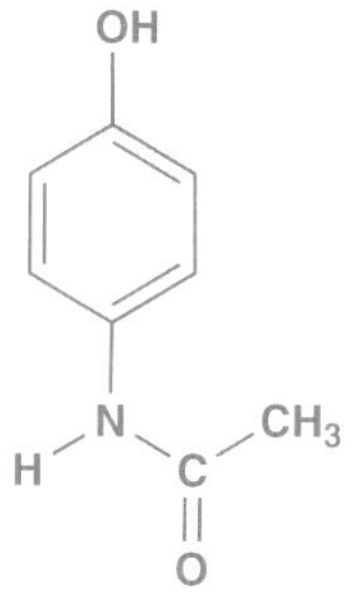

Paracetamol (an analgesic)

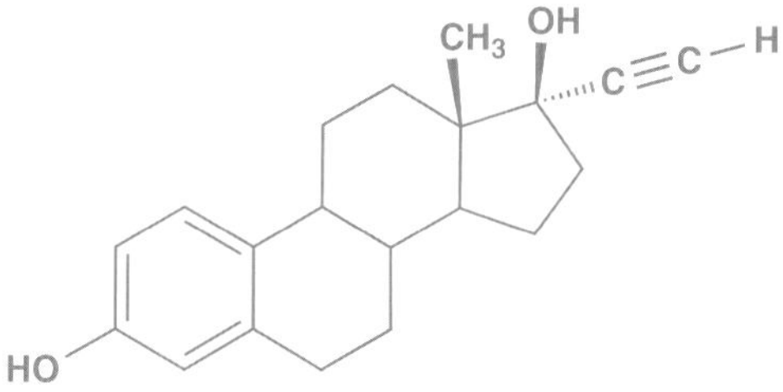

Ethynylestradiol (oral contraceptive)

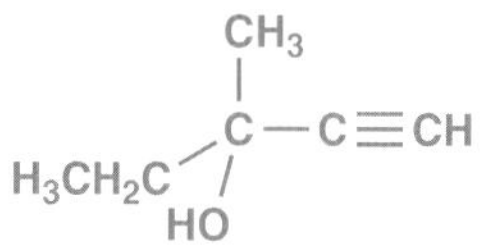

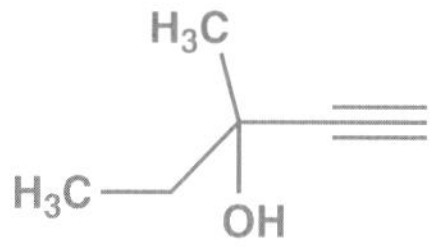

Oblivon (a sedative) alternative representations

SELF CHECK 2.5

Figure 2.17 shows the structures of three more drugs. What are the hybridization states of each of the carbon atoms in them? What are the bond angles around each of the carbon atoms? Note that in two of the structures, paracetamol and oblivon, only two of the three possible hybridization states are observed. However the other example, ethynylestradiol, is more complicated and all three hybridization states of the carbon atom are present.

KEY POINT

Carbon utilizes hybrid electronic orbitals to bond covalently to other atoms. These orbitals determine the overall molecular shape. The shape of a molecule is very important with respect to its chemical and biological activities.

2.5 Intermolecular forces

So far we have only considered the covalent bonds *within* a molecule that hold it together, in other words, the *intra*molecular forces. Before going on to look at how we can describe the reactions of these molecules, we need to be aware of the forces that occur *between* molecules, holding one molecule to the next – in other words, the *inter*molecular forces.

The importance of these forces is shown by thinking about the physical properties of one of the substances most familiar to us – water. For such a small molecule, water has an unexpectedly high boiling point. It is approximately 200° higher than might be predicted by comparison to the boiling points of the other hydrides in the same group of the periodic table: H_2Te (−2.2 °C), H_2Se (−41.5 °C) and H_2S (−60.7 °C). Simple extrapolation of this trend might suggest that the boiling point of water would be around −100 °C, not +100 °C, which we know it to be. Its melting point is also considerably higher than would be expected from comparison to other, related substances.

Relatively strong intermolecular forces exist between water molecules in all of its physical states (solid, liquid and gas), causing its apparently anomalous properties, and without these unusual properties, life could never have evolved in the way that is familiar to us. Although we have used water as an extreme example, in fact intermolecular forces of varying types and strength must exist between the atoms and molecules of *all* substances – otherwise we would never be able to observe them in the liquid or solid states.

There are a number of different types of intermolecular force that can act between covalently bonded molecules, holding them in contact with one another. All are very much weaker than the covalent forces that operate in a σ-bond or π-bond, or the ionic forces that operate in a substance such as sodium chloride. Those that we will examine, and an indication of their strength, are given in Table 2.2 (values for covalent bonds and ionic lattice energies are also given as a comparison, but you can see that these are at least an order of magnitude stronger).

TABLE 2.2 Strengths of intermolecular and intramolecular forces

Type of intermolecular force	Approximate strength (kJ mol^{-1})
London (van der Waals) forces	<2
Dipole-dipole interactions	1–3
Hydrogen bonding	10–30
Intramolecular forces for comparison	
Covalent bonds	300
Ionic bonds (lattice energy)	~760 for NaCl

London (van der Waals) forces

These are the weakest forces that occur between molecules. They are referred to both as London dispersion forces (after Fritz London, a German-American physicist), or as van der Waals forces, although, strictly speaking, van der Waals forces also include other types of intermolecular interaction. London forces occur in substances where there is no permanent dipole. Because the electrons in a molecule are moving continuously, even the electrons in a non-polar molecule can become unevenly distributed within the molecule, causing it to have a temporary dipole. This temporary polarization in a molecule causes neighbouring molecules to become temporarily **polarized** as well, and so the molecules can be weakly attracted to each other. These temporary dipoles are induced, and, therefore, the interactions are also known as induced dipole-induced dipole interactions.

Dipole-dipole interactions

The second type of intermolecular interaction we need to consider are dipole-dipole interactions, which are described as electrostatic interactions. These affect molecules in which there is a *permanent* dipole, as opposed to the temporary ones described in the previous section. The polarization in a bond, and hence in a molecule,

FIGURE 2.18 Dipole-dipole interactions.

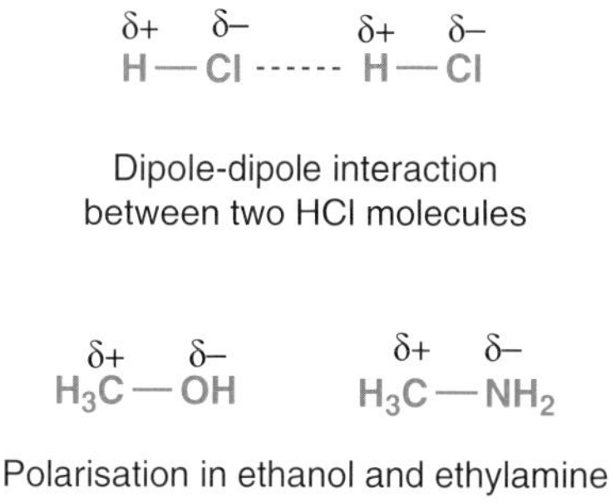

$$\overset{\delta+}{H}\!-\!\overset{\delta-}{Cl}\ \cdots\cdots\ \overset{\delta+}{H}\!-\!\overset{\delta-}{Cl}$$

Dipole-dipole interaction
between two HCl molecules

$$\overset{\delta+}{H_3C}\!-\!\overset{\delta-}{OH}\qquad\overset{\delta+}{H_3C}\!-\!\overset{\delta-}{NH_2}$$

Polarisation in ethanol and ethylamine

FIGURE 2.19 Hydrogen bonding between water molecules and ethanol molecules.

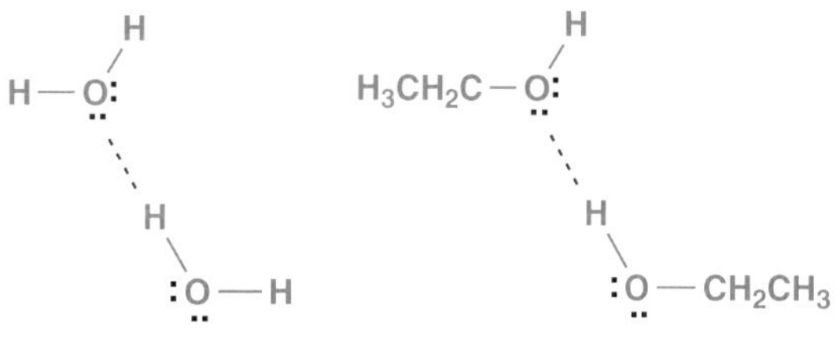

is the result of differences in the electronegativity between the two atoms in a covalent bond, leaving partial positive ($\delta+$) and negative ($\delta-$) charges on the two atoms concerned. Molecules with permanent dipoles are attracted to each other because the molecules can align themselves in such a way that the positive end of one dipole is close to the negative end of another dipole, resulting in an overall attractive force between two adjacent molecules, as shown in Figure 2.18.

Hydrogen bonding

A hydrogen bond is a particularly strong type of electrostatic interaction, or dipole-dipole interaction, which occurs specifically between an H–X bond (where X is an electronegative atom, normally F, O or N) and another electronegative atom, X′. X′ is not only highly electronegative but possesses one or more lone pairs of electrons. In practice, this means that X′ is also nitrogen, oxygen or fluorine. The interaction is typically about one tenth the strength of a covalent bond, and is strong enough to make the interatomic distance less than the sum of the van der Waals radii. (In other words, the two atoms are closer together than they should be, based on their diameters, which implies there is some degree of covalent character in these bonds.)

Figure 2.19 illustrates the hydrogen bonding present between two adjacent water molecules and between two adjacent ethanol molecules. In the liquid phase, a molecule of water, or ethanol, may take part in more than one hydrogen bond by involving both lone pairs of electrons on the oxygen atoms or both hydrogen atoms on the water molecule (though only the single H atom on the OH group of the ethanol). Extensive hydrogen bonding is the reason water has such a high boiling point (100 °C) compared with

other hydrides in the same group of the periodic table. Sulfur, for example, is much less electronegative than oxygen and cannot form strong hydrogen bonds. Consequently, H_2S, a pungent-smelling gas with an aroma of rotten eggs, has a boiling point of −60.7 °C. Several other important properties of water, including its unusual expansion on freezing, may also be attributed to hydrogen bonding.

For more information about the significance of hydrogen bonding in alcohols, see Chapter 5.

SELF CHECK 2.6

Consider the two structural isomers with the formula C_2H_6O. One of these has a boiling point of 78 °C and the other a boiling point of −24 °C. Give the names and structural formulae of both isomers and explain the difference in their boiling points.

Hydrogen bonding turns out to be extremely important in organic molecules, many of which contain N–H and O–H bonds (amines, alcohols and carboxylic acids, for example). Even molecules without N–H or O–H bonds may become involved in hydrogen bonding, as the C=O group in a ketone, for example, contains an oxygen atom with two lone pairs of electrons that may become hydrogen bonded to the O–H or N–H group in a second molecule. The importance of these types of electrostatic bonds in describing how drugs may become attached to receptor sites in the body (or in the holding together of the two strands in a nucleic acid) is immense.

Hydrogen bonding is discussed in Chapter 3 'The biochemistry of cells' and Chapter 4 'Introduction to drug action' of the *Therapeutics and Human Physiology* book in this series.

We can also use Figure 2.19 to illustrate two important ideas – those of a *hydrogen bond donor* and a *hydrogen bond acceptor*. In the case of water, the O–H bond is regarded as the hydrogen bond *donor* (it donates the hydrogen), while the oxygen atom with the lone pair is the hydrogen bond *acceptor* (it accepts the hydrogen). In the case of amines, the N–H bond is the *donor* and the N atom with the lone pair is the *acceptor*.

SELF CHECK 2.7

The structure of the anaesthetic, halothane, was shown in Figure 2.10. Can you identify any part of this structure that might act as a hydrogen bond acceptor?

SELF CHECK 2.8

The structures of water and ethanol are both shown in Figure 2.19. Can you show how hydrogen bonding can occur between a molecule of ethanol and a molecule of water; how might this account for the high solubility of ethanol in water, and vice versa? This ability of an organic molecule to dissolve in water is very important in the distribution of a drug around the body.

SELF CHECK 2.9

The structures of diethyl ether and triethylamine are shown here. How might these two compounds hydrogen bond to water? Why can they not hydrogen bond to each other?

$H_3CH_2C-\ddot{O}\colon$ with CH_2CH_3

Diethyl ether

Triethylamine

KEY POINT

Whereas atoms in organic molecules are held together by covalent bonds, the intermolecular forces between molecules can take a variety of forms, depending on the nature of the molecules. Intermolecular forces comprise van der Waals forces, dipole-dipole attraction and hydrogen bonds.

2.6 Reaction types and the making and breaking of bonds

Chemical reactions involving organic compounds can be classified into four types. These are listed below and illustrated in Figure 2.20.

1. *substitution* – where one group is replaced by another; an example of this is the formation of propanenitrile in the reaction of bromoethane with potassium cyanide

2. *elimination* – where two groups are lost, to create a double (or perhaps triple) bond; an example of this is the formation of propene in the reaction of 1-bromopropane with a base

3. *addition* – the reverse of elimination, where two groups are added across a double (or perhaps triple) bond; an example of this is the formation of 2-bromocyclohexanol when cyclohexene is shaken with bromine water

4. *rearrangement* – where scrambling of the structure occurs and the atoms are now bonded to each other in a different order; an example of this is the Fries rearrangement in which an aromatic ester rearranges in the presence of an aluminium trichloride catalyst to give a mixture of hydroxy aromatic ketones.

These classifications are useful, but they only describe what has happened, rather than *how* it has happened. The latter consideration of 'how' is the stuff of organic reaction mechanisms and is introduced in Section 2.7.

FIGURE 2.20 Examples of the four classes of organic reactions.

Substitution:

$$CH_3CH_2Br + KCN \longrightarrow CH_3CH_2CN + KBr$$

Elimination:

$$CH_3CH_2CH_2Br + KOH \longrightarrow CH_3CH{=}CH_2 + H_2O + KBr$$

Addition:

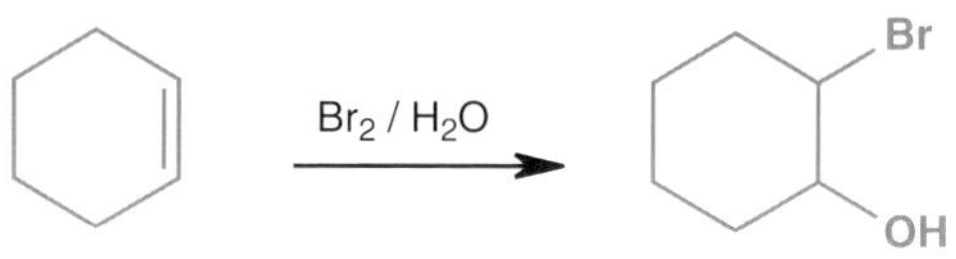

Rearrangement:

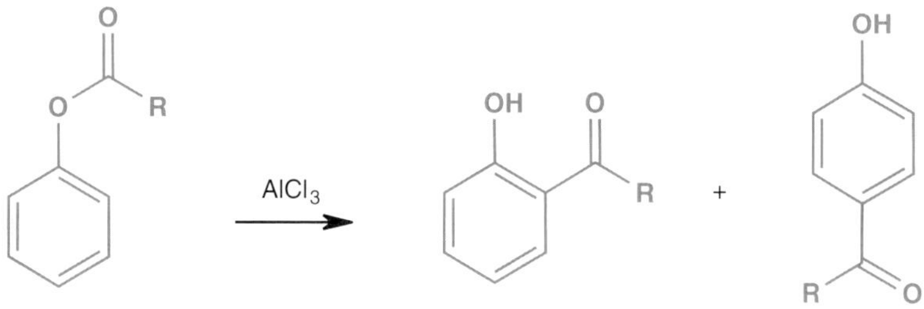

All reactions involve some sort of bond cleavage (breakage) and bond formation. If bond cleavage results in the formation of anions and cations, it is described as heterolytic cleavage. In heterolytic cleavage, the two electrons in the σ-bond between the two atoms move as a *pair* and end up on either one or other of the two atoms involved. Of course with an example represented just as X–Y, this could be illustrated as occurring in two possible ways – both electrons going to X or both to Y. However with 'real' molecules, the two electrons will always go to the more electronegative of the two atoms, which is where the negative charge will reside at the end of the process. Thus, for example, HCl will undergo heterolytic cleavage to give H^+ and Cl^-, rather than H^- and Cl^+. (You can think of this as arising from the partial polarization, represented as δ+ and δ– in Figure 2.18, being taken to an extreme, resulting in a complete separation of the charges into + and –).

On the other hand, in homolytic cleavage, the two electrons in the σ-bond separate and one goes onto each of the two atoms that were originally involved in the bond. The result is the formation of two neutral species called free radicals (or just radicals), each with an unpaired electron.

Simple illustrations of what is meant by the terms *heterolytic cleavage* and *homolytic cleavage* are shown in Figure 2.21.

FIGURE 2.21 Homolytic and heterolytic cleavage.

$X{-}Y \longrightarrow X^{\bullet} + Y^{\bullet}$ (Homolytic cleavage)

$X{-}Y \longrightarrow X^{+} + Y^{-}$ (Heterolytic cleavage)

$X{-}Y \longrightarrow X^{-} + Y^{+}$ (Heterolytic cleavage)

$H{-}Cl \longrightarrow H^{+} + Cl^{-}$ (Heterolytic cleavage, controlled by electronegativities)

There are also some more complex, but very important processes – particularly in some of the rearrangement reactions – where bonds do not appear to break in either of these ways, but nevertheless migration of atoms and bonds has occurred. These are grouped together as the so-called pericyclic reactions, though they are not discussed further here.

2.7 The principles of organic reaction mechanisms

In the introduction to this chapter, the reasons why carbon might be regarded as a unique element were outlined. Perhaps we can now add a further reason. Our understanding of the way in which carbon-based (organic) compounds react, using the ideas of mechanisms to explain them, is particularly well developed – certainly more so than for the chemistry of any other single element. By understanding and using mechanisms, organic chemistry ceases to be an endless list of unrelated reactions – which of course it is definitely not, though it can sometimes be perceived in that way! Instead, it becomes something that can be rationalized and which very often has quite predictable outcomes.

The vast majority of reactions encountered in the early stages of any organic chemist's, or indeed pharmacist's, career involve the motion of electron pairs. However, radical processes, involving the motion of single electrons are extremely important in biological systems (see, for example Integration Box 2.2). They are often oxidation and reduction (redox) processes involving some of the transition metals that have variable oxidation states.

Twice in the preceding paragraph, the word *motion* has been used in the context of electrons in reaction mechanisms. When describing, or illustrating, reaction mechanisms, this motion has to be represented somehow. Conventionally this is done using so-called 'curly arrows' and their use in this way gives rise to the often-used expressions 'curly arrow theory' or 'curly arrow description' of mechanism.

The correct use of these curly arrows is explained below, but before that, it is useful to review *all* the types of arrow that are commonly encountered in chemistry. These are shown together in Figure 2.22.

The most important point to come out of the last two arrows shown in Figure 2.22 is quite simply that they illustrate the motion of electrons, either singly or in pairs. Therefore they start at the place where the electrons *are* and they point towards where the electrons *end up* (in that way showing the direction of motion of the electrons). It is a simple idea, but one that is all too often confused. However once they are understood, mechanisms become a very powerful tool in explaining and even predicting organic chemical reactions.

Examples of 'curly arrows' in ionic and radical processes

The purpose of these examples is *not* to describe in detail the major classes of organic reactions – for example, the S_N1 and S_N2 substitution processes observed in the haloalkanes, the electrophilic addition across a double bond in an alkene, or the electrophilic aromatic substitution of an aromatic compound such as benzene. Rather, it is to show *how arrows work* and what they can do; the specific reactions cited above will all be explained in later chapters of this book. Our purpose here is only to establish the ground rules for using these arrows to describe reactions. We need to begin by restating that where ionic reactions and electron pairs are involved, we use *double headed* arrows to

FIGURE 2.22 **The different types of arrow used in chemistry.**

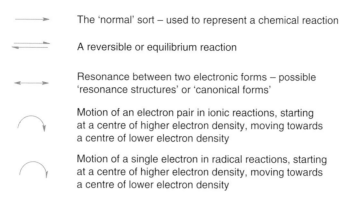

The 'normal' sort – used to represent a chemical reaction

A reversible or equilibrium reaction

Resonance between two electronic forms – possible 'resonance structures' or 'canonical forms'

Motion of an electron pair in ionic reactions, starting at a centre of higher electron density, moving towards a centre of lower electron density

Motion of a single electron in radical reactions, starting at a centre of higher electron density, moving towards a centre of lower electron density

---

INTEGRATION BOX 2.2

The hydroxyl radical

One of the most significant radical species in the body is the hydroxyl radical (HO·). It is one of a group of species referred to as *reactive oxygen species* (ROS), which includes hydrogen peroxide (H_2O_2) and the superoxide ion (O_2^-). Of these, the hydroxyl radical is the shortest lived and most reactive species and is the one that is capable of causing the most damage to organisms. It can be formed by the breakdown of peroxides in the body (sometimes mediated by the presence of iron or other transition metals), but also, importantly, by the action of ionizing radiation on water. It is this reaction that leads to much of the tissue damage that is observed as a result of exposure to ionizing radiation, particularly X-rays or gamma rays.

The hydroxyl radical cannot be eliminated from the body by any enzymatic reaction and it will react quickly and destructively with all types of macromolecules found in living organisms – for example, carbohydrates, lipids and nucleic acids.

It may lead to breaking of nucleic acid strands, or to the alteration of the bases within them so that the nucleic acids will no longer correctly replicate. At a chemical level, this may either involve the hydroxyl radical removing a hydrogen atom from the structure, or alternatively the substitution of an aromatic ring with an OH group.

Lipids and carbohydrates are also attacked, generally by removal of a hydrogen atom from the macromolecule, followed by subsequent attack by O_2 to give a hydroperoxyl (ROO·) radical. It is the reactions that occur with this peroxyl radical that lead to the breakdown of the macromolecule.

There is, therefore, a lot of interest in the role of antioxidants in the body, which can quench the effect of free radicals and interrupt the destructive chemical processes that would otherwise take place. These antioxidants include compounds such as vitamin C, vitamin E and more generally, the polyphenols found in a wide variety of foodstuffs (including red wine).

 More information can be found in Chapter 5 'Alcohols, phenols, ethers, organic halogen compounds and amines' of this book.

The role of radical reactions in the ageing process is another fascinating area, but mostly lies outside the scope of this chapter.

---

describe them and where radical reactions and single electrons are involved, we use *single headed* arrows to describe them.

Neutralization of an acid and a base

One of the simplest (and also one of the fastest) of all chemical reactions is the neutralization of an acid by a base and this provides an excellent starting point for showing how curly arrows work. Figure 2.23 illustrates this reaction, showing the arrow starting at the region of high electron density (the negatively charged hydroxide ion) and moving towards the region of low electron density (the positively charged proton), exactly as the electrons would do in this reaction. In this figure, the lone pairs of electrons around the oxygen atom have been

While waiting for her prescription in Dilip's pharmacy, a customer is browsing the cosmetics counter. Upon collecting her prescription she remarks that most of the 'anti-ageing' creams contain vitamin E and asks Dilip why this is.

REFLECTION QUESTIONS

1. What possible role might vitamin E possess in anti-ageing preparations?

2. Are there any side-effects related to an overdose of vitamin E?

Answers

1 Vitamin E (α-tocopherol) acts as an antioxidant. It operates as a free radical scavenger, halting the propagation of damaging free radical chain reactions. These free radical reactions can damage collagen, potentially leading to the appearance of wrinkles. However, eating the appropriate foods in a normal diet will usually provide sufficient vitamin E without the need for supplementation.

2 No adverse effects from consuming vitamin E in a healthy diet have been identified. Although there are potential risks associated with high dose vitamin E supplements, such levels are unlikely to be achieved by the use of anti-ageing creams.

FIGURE 2.23   Mechanism of neutralization.

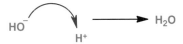

omitted and the arrow is just shown starting from the negatively charged oxygen atom. It is not wrong to draw in the lone pairs, but they do not participate in this reaction, so they are not normally drawn.

Conversion of bromomethane to methanol by hydroxide

This is a simple example of a substitution reaction that serves to illustrate how a reaction can be described using more than one curly arrow. In this case, shown in Figure 2.24, the C–Br bond breaks at the same time as the C–O bond forms, and *both* processes need to be shown. The arrow coming from the hydroxide ion is acting in exactly the same way we saw in Figure 2.23 – electrons flow towards an area of low electron density. Note how the second arrow, showing the heterolytic cleavage of the C–Br bond, starts in the middle of the bond and moves towards the Br. The two bonding electrons leave the C–Br bond and move to (become *localized onto*) the bromine atom, converting it to a bromide ion.

FIGURE 2.24   Mechanism of nucleophilic substitution.

Since the mid 19th century, diethyl ether ($CH_3CH_2OCH_2CH_3$) has been used as an inhalation anaesthetic. It can be prepared by the Williamson synthesis, which involves the reaction of the ethoxide ion ($CH_3CH_2O^-$), used as the sodium salt, sodium ethoxide, and bromoethane (CH_3CH_2Br). Draw the curly arrow mechanism of the substitution reaction involved in the formation of this ether.

Methoxypropane ($CH_3OCH_2CH_2CH_3$) is an isomer of diethyl ether and has also been used as an anaesthetic. It too could be synthesized from a suitable combination of an alkoxide ion and a haloalkane. Can you suggest what the starting materials might be here, and, again, propose a mechanism for the reaction?

Electrophilic addition to an alkene

The electrophilic addition reactions that occur with the alkenes are described in more detail in Chapter 4. However Figure 2.25 shows the first step in such a reaction – in this case, the addition of a proton to the double bond, which is typically what happens when an alkene is hydrated under acid conditions to give an alcohol. The point here is to clarify an apparent contradiction between the idea of 'electrophilic attack *on*

FIGURE 2.25 First step in the mechanism of electrophilic addition of water to an alkene.

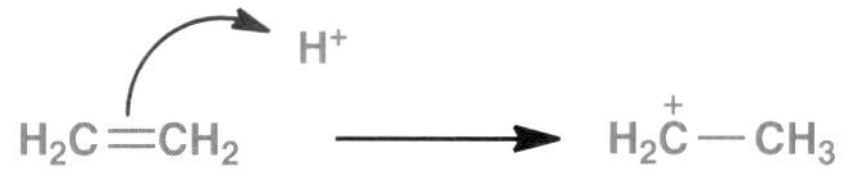

an alkene', yet the arrow is going *away* from it. These ideas are surprisingly often confused, yet there is no contradiction here. The attack is electrophilic, in other words, the proton is attracted to a region of high electron density associated with the π-bond between the two carbon atoms, while the curly arrow is doing precisely what it is meant to do, showing the flow of electrons *away* from the alkene, towards the positively charged proton. Both parts of the description of this reaction are therefore perfectly correct.

SELF CHECK 2.12

Figure 2.25 shows how addition of a proton to an alkene gives an ethyl carbocation. If this was then attacked by water, the product would be $CH_3CH_2OH_2^+$, which is a protonated form of the final product, ethanol. Show how this might be produced, using a curly arrow to indicate the motion of electrons involved.

SELF CHECK 2.13

If the ethyl carbocation in Figure 2.25 was attacked by a bromide ion rather than water, what would the final product be? Again, use a curly arrow to show the motion of the electrons involved.

SELF CHECK 2.14

Why do you think water or a bromide ion cannot attack the C=C bond directly?

Delocalization of electrons in a carboxylate ion

The three previous examples have shown curly arrows representing the reaction between two separate species. However, they may also be used to represent the motion, or delocalization, of electrons within a single species. The carboxylate anion is a good example of this.

The conventional drawing of this carboxylate ion might imply that the two carbon–oxygen bonds are different – with one shorter, stronger double bond and one longer, weaker single bond. However we know from spectroscopic and crystallographic studies that these two bonds are in fact identical.

Acetate (ethanoate) is usually written as shown in Figure 2.26A with a complete octet of eight electrons around each oxygen. We could also place the negative charge on the other oxygen to give the structure shown in Figure 2.26B. However, neither of these structures represents the compound completely accurately. These are referred to as **resonance structures** or **canonical forms**. The actual structure of the compound, or intermediate, is an average of these and is referred to as a **resonance hybrid**. In the resonance hybrid, electrons are delocalized within the CO_2^- part of the structure, producing bonds that are intermediate in character between single and double bonds, and with lengths that are intermediate too.

SELF CHECK 2.15

Give another example of a resonance hybrid structure. Hint: see Chapter 1.

Once again, the curly arrows show the flow of the electrons and, where appropriate, they start in the middle of the bond that is being broken, in this case a π-bond. In this example, you can also see the use

FIGURE 2.26 Mechanism of delocalization in a carboxylate anion.

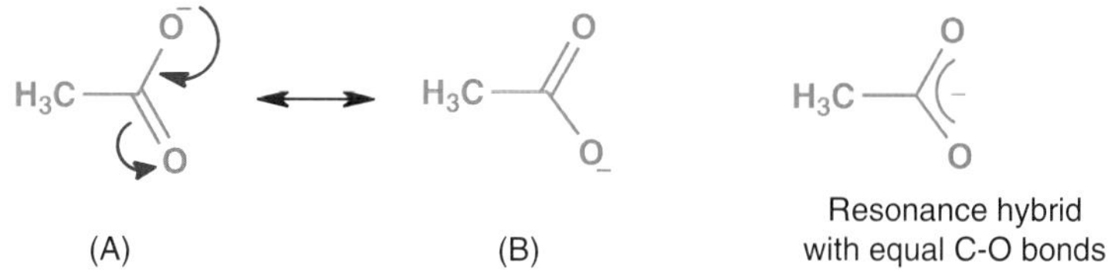

(A)   (B)   Resonance hybrid with equal C-O bonds

45

of the double ended arrow to indicate the resonance between the two forms.

Draw the curly arrows needed to describe how the second resonance structure in Figure 2.26 is converted back to the first one.

Addition of a bromine atom to an alkene

It is possible to add HBr to an asymmetric alkene such as propene ($CH_3CH=CH_2$) in two different ways. Under ionic conditions, Markovnikov's Rule (see Integration Box 2.3) is obeyed and in the major product, the H goes onto the CH_2 group, while the Br goes onto the

FIGURE 2.27 Mechanism of radical addition to an alkene.

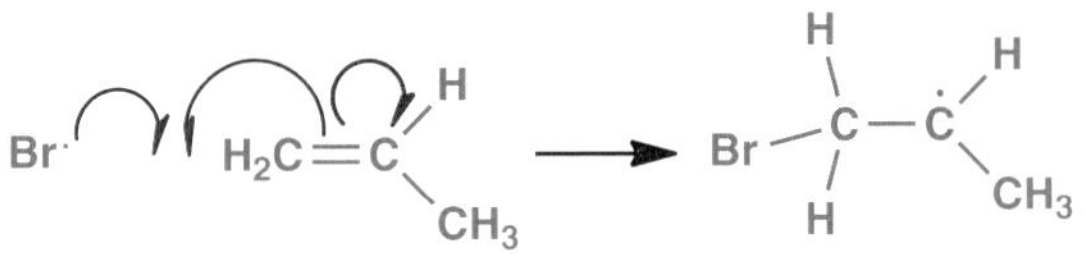

middle carbon atom. The product is 2-bromopropane. However, under free radical conditions, the H and the Br add the opposite way round and the product is 1-bromopropane and the product is described as the anti-Markovnikov product.

While we won't enter into a full explanation of this difference, the first step of the anti-Markovnikov reaction – the addition of a bromine atom to propene – is shown in Figure 2.27. This shows how curly arrows may be used to show the flow of electrons in a *free*

Markovnikov's rule – a reminder

The expression 'Markovnikov's rule' refers to the conclusions derived from a series of empirical observations made by the Russian chemist, Vladimir Markovnikov, which were published around 1870. They allow us to predict the outcome of a chemical reaction when a compound of general formula HX (for example HCl, HBr or even H_2O, which we might regard as H-OH) is added to an asymmetric alkene such as propene.

When we look at the double bond in propene, we can see that it has one H attached to one end and two H atoms attached to the other end of it. There are various ways in which Markovnikov's rule can be expressed, but one form is: in the addition of HX to an asymmetric alkene, the H from the HX goes onto the carbon atom in the double bond which already has the greater number of H atoms.

In fact a *mixture* of products is generally produced, but Markovnikov's rule allows us to identify the *major* product. In this case, it is 2-bromopropane, with 1-bromopropane present as the minor product. It is now perfectly possible to rationalize these original empirical observations in terms of the stabilities of the carbocation intermediates in the reaction, and this is done in Chapter 4.

While Markovnikov's rule works well under ionic conditions, it is possible to vary the reaction conditions so that the major and minor products are reversed – a process often referred to as anti-Markovnikov addition. Typically this will occur if the reaction is carried out under free radical conditions, though, again, the outcome may be rationalized by considering the stability of the radical intermediates that are formed.

The two processes are summarized in Figure 2.28.

FIGURE 2.28 Markovnikov and anti-Markovnikov addition of HX to an alkene.

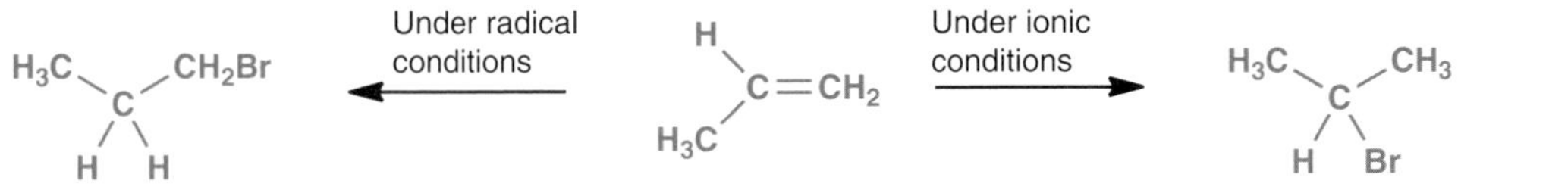

1-bromopropane
anti-Markovnikov product

2-bromopropane
Markovnikov product

radical process. Three electrons are involved – one from the bromine atom and the two in the carbon-carbon π-bond – so this time we need three, single-headed curly arrows to show the process in full. However their sense is exactly the same as we have seen previously, and, again, all they are doing is showing the direction of motion of the electrons.

KEY POINT

The way in which organic molecules react has been extensively studied and detailed reaction mechanisms are widely used and accepted. The movement of electrons during these reaction mechanisms are represented by curly arrows.

CHAPTER SUMMARY

We close this section by emphasizing the importance of the basic, chemical ideas presented in this chapter. You have been provided with a toolkit which will allow you to understand how and why molecules have some of the physical and chemical properties that they have. This toolkit is necessary to allow you to understand how drugs can be synthesized, how they might interact with receptor sites in the body, how they act as medicines, how they might be metabolized, and, maybe eventually, how they might be improved. By mastering the material here, you now have the basics to enable you to go on confidently and tackle the rest of this book, and, finally, to develop a full understanding of the very things you will be dealing with in your professional lives.

FURTHER READING

Clayden, J., Greeves, N., and Warren, S., *Organic Chemistry*, 2nd edn. Oxford University Press, 2012.

Stereochemistry and drug action

ROSALEEN J. ANDERSON AND ADAM TODD

The common analgesic ibuprofen, which you met in Chapter 1, actually consists of two forms. One of these forms is used to treat headaches, the other is completely inactive. The two forms have mostly identical physical properties, but their three-dimensional shapes are different – they are mirror images of one another. The three-dimensional shape of biological molecules and drugs is profoundly important for their action, as we will explore in this chapter.

This chapter introduces the concepts of isomerism, structure, and shape, particularly with reference to the activity of drugs, and relates strongly to the hybridization and bonding you studied in Chapter 2.

Learning objectives

Having read this chapter you are expected to be able to:

➤ recognize, differentiate, and discuss the key features of constitutional isomers, conformational isomers and stereoisomers (geometric and optical isomers);

➤ assign the stereochemistry of asymmetric alkenes and chiral molecules using the IUPAC nomenclature;

➤ explain how molecular shape, size and stereochemistry affect the activity and pharmaceutical action of drugs.

3.1 Introduction

The term '*isomerism*' is used to describe the ways in which molecules can have identical compositions in terms of carbon, hydrogen, nitrogen, etc, but differences in their patterns of bonding or **conformation** that may lead to dramatically different three-dimensional shapes. There are many drugs used in medicine that exhibit isomerism. For example, it is essential to have a good understanding of isomerism to explain why:

• the selective serotonin reuptake inhibitors (SSRIs) escitalopram and citalopram are used at different doses to treat depression, or

• the antibiotic metronidazole interferes with the anticoagulant agent warfarin (and thus increases the risk of a patient having a haemorrhage), or

• thalidomide – an anti-angiogenesis drug used in the treatment of multiple myeloma – is contraindicated in pregnancy.

You will find explanations of these first two points in the Online Resource Centre (www.oxfordtext-books.co.uk/orc/ifp) and you will learn about the third point, relating to thalidomide, during this chapter.

The structures of citalopram, warfarin and thalidomide are shown in Figure 3.1. It is their stereoisomerism that imparts their particular pharmacological properties; we will consider this property in more detail in this chapter.

FIGURE 3.1 The chemical structures of citalopram, warfarin and thalidomide; all of these molecules have one thing in common – a chiral (or stereogenic) carbon.

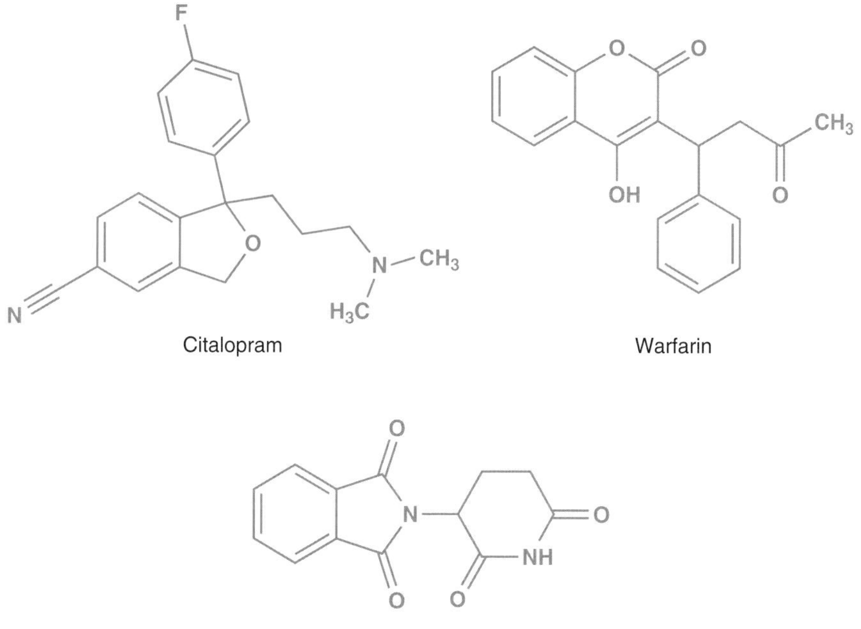

Citalopram

Warfarin

Thalidomide

By the time you have read through this chapter, you should realize the impact these concepts have on the everyday practice of a pharmacist and how important it is to have an understanding of isomerism. We will also explore how different isomers can display dramatically different physical, chemical and biological properties, discussing these as examples throughout this chapter.

As we look at the effects of isomerism on pharmaceutical science, we will consider how isomers can act differently on **receptors**, which are physiological species such as proteins and nucleic acids, and can be activated by natural molecules and drugs to cause specific effects. Molecules that activate a receptor to produce a specific response are called **agonists**; molecules that bind to a receptor, but do not trigger the response (and block the binding of an agonist at the same receptor in doing so) are called **antagonists**. You will meet these concepts in greater detail when you study pharmacology.

 More information can be found in Chapter 4 'Introduction to drug action' in the *Therapeutics and Human Physiology* book of this series.

3.2 **Constitutional isomerism**

Isomers are molecules that contain the same atoms but which are bonded together differently. If the atoms in the isomers are connected in different ways (i.e. their *connectivity* is different), then they are called constitutional isomers (sometimes called structural isomers). Consider the structure of paracetamol, $C_8H_9NO_2$ **A**, shown in Figure 3.2. Paracetamol is an **analgesic**, which also has **antipyretic** properties and is therefore often given to children after immunizations to reduce fever.

If we swap the position of the acetyl (ethanoyl) group from the nitrogen atom to the phenol oxygen

of paracetamol (that is the oxygen bonded to the benzene ring), we obtain another compound, as shown in Figure 3.2 (structure **B**). This compound has the same molecular formula as paracetamol ($C_8H_9NO_2$), but is not paracetamol; it is an example of a **constitutional isomer**. In fact, there are many constitutional isomers for paracetamol, some of which are shown in Figure 3.3. You will meet the different functional groups in the other chapters in this book; here, our focus is that these constitutional isomers may have different physical and chemical properties from paracetamol and probably will not treat your headache.

Constitutional isomers are sometimes formed as by-products of organic synthesis and may even be toxic to the patient. The British Pharmacopoeia (BP) is responsible for setting assays and standards for the purity of pharmaceuticals and has limit tests for such by-products.

FIGURE 3.2 Paracetamol (*para*-acetylaminophenol) (**A**) and a constitutional isomer (**B**).

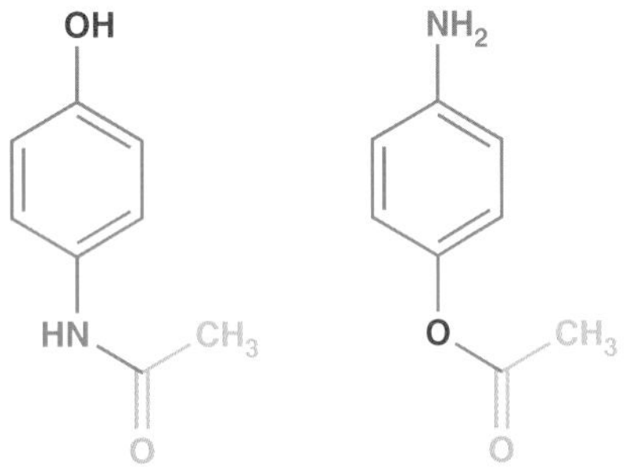

FIGURE 3.3 Some examples of constitutional isomers of paracetamol, $C_8H_9NO_2$.

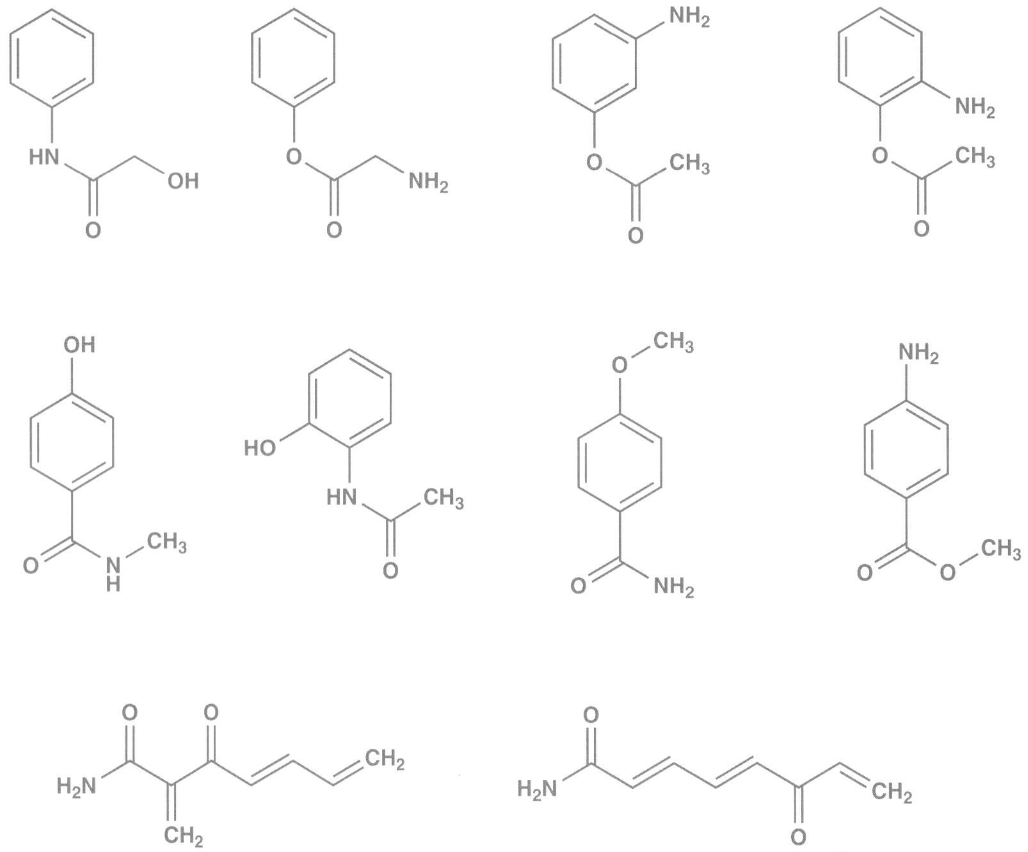

More information on these assays can be found in Chapter 11 'Introduction to pharmaceutical analysis' of this book.

SELF CHECK 3.1

Figure 3.3 shows ten different constitutional isomers of paracetamol. There are many more; can you draw some more examples?

3.3 Conformational isomerism

Conformational isomerism arises because rotation is possible about σ-bonds, allowing molecules to adopt a variety of shapes. Conformational isomers are just different shapes adopted by the same molecule. Rotation about σ-bonds allows a molecule to find its most stable conformation or to adopt a shape that can bind to an enzyme or receptor. Conformational isomerism is therefore very important for pharmaceuticals and their medicinal action. Restricting conformation is an important strategy in drug design and discovery — by using structures and groups that have a limited range of conformations available to them, a desired shape for binding to a particular medicinal target can be promoted, with the resulting prize of strong interactions with the receptor and greater potency of the drug.

KEY POINT

Conformational isomers can interconvert by the rotation of bonds and movement of atoms, without the need to break any bonds. This makes conformational isomerism distinct from other forms of isomerism.

Conformations of linear molecules

Let us look, firstly, at a relatively simple molecule; we can apply the principles we consider here to larger molecules. Nabumetone is a pro-drug that is activated by liver enzymes to a non-steroidal anti-inflammatory drug (NSAID), which acts to reduce pain and inflammation in osteoarthritis and rheumatoid arthritis.

There are two methylene (CH_2) carbon atoms (C2 and C3) in nabumetone, C2 with a ketone substituent, $COCH_3$, and C3 with a substituted aromatic naphthyl

 You have already met one NSAID, ibuprofen, at the beginning of this chapter and will meet many more in Chapter 7 'Communication systems in the body – autocoids and hormones' of the *Therapeutics and Human Physiology* book in this series.

group. We can draw the structure as shown, with an extended (spread out) CH_2-CH_2 central unit; but does this represent the actual conformation? How are the ketone and aromatic groups arranged around this central unit?

Nabumetone

We can represent and view this structure in a number of ways; the two most popular, when considering conformation, are the 'sawhorse' and Newman projections, Table 3.1. The sawhorse form views the molecule from one end of the C2–C3 bond, slightly from the side, whereas the Newman projection looks directly along the C3–C2 bond, with C2 behind C3. The circle in the Newman projection represents C2 and the point at which the bonds from 'naph' and the two H atoms meet is C3.

When the atoms or groups on the front carbon atom are lined up with the atoms or groups on the back carbon atom (i.e. on the same 'side' of the carbon–carbon bond), they are said to be **eclipsed**. When the atoms or groups on the two carbon atoms lie on opposite sides of the carbon–carbon bond, they

TABLE 3.1 Sawhorse and Newman projections for two conformations of nabumetone (naph represents the substituted naphthyl unit)

Linear structure	Sawhorse projection	Newman projection
naph ‑‑‑‑ COCH₃ Staggered	naph H H / COCH₃ / H H	naph / H H / H COCH₃
naph ‑‑‑ COCH₃ Eclipsed	COCH₃ / naph / H H / H H	CH₃OC naph / H H H H

are said to be **staggered**; this type of conformation is generally lower in energy than the eclipsed forms, as there is less repulsion of spatially close atoms and bonding electrons (which cause *steric* and *torsional* strain, respectively) in the molecule. Other nomenclature is sometimes used to describe conformations of linear molecules; you can read about them on the support website (http://www.oxfordtextbooks. co.uk/orc/ifp).

We can use the Newman projection to look along the C3–C2 bond and consider the possible conformations nabumetone can adopt. If we start with the two substituents arranged one behind the other, we have our starting point, conformation **A**; then, keeping the front carbon and its attached atoms/groups stationary and rotating the rear carbon atom clockwise around the C–C bond, we can trace the energy changes in this molecule as it passes through several high and low energy conformations (see Figures 3.4A and 3.4B).

The most stable conformations for simple linear molecules are generally those that involve a staggered arrangement of substituents on the central C–C bond,

most often in the antiperiplanar form; however, other features in some molecules, such as hydrogen bonding, can result in alternative conformations becoming more stable.

 The example of cysteamine, $NH_2CH_2CH_2SH$, is discussed in the Online Resource Centre.

Conformations of cyclic molecules

Of course, cyclic molecules do not have total freedom of rotation around their bonds – they are tethered by the cyclic structure. However, if at all possible, these molecules minimize their potential energy by adopting the lowest energy conformation available to them. Aromatic molecules are cyclic and planar, so they cannot readily change conformation. Aliphatic cyclic compounds, by comparison, can adopt different conformations, within the restrictions imposed by their cyclic nature. As far as possible, the energy in these molecules is minimized by adopting conformations in which the substituents are staggered, rather than eclipsed, and the bond angles are as normal as possible for the hybridization of the atoms. In this chapter, we are concentrating on 6-membered rings, particularly cyclohexane and its related structures; however, there are many examples of pharmaceuticals containing aliphatic cyclic systems with 7, 5, 4, and 3 atoms in the ring. You will find some of these examples on the support website.

SELF CHECK 3.2

Study Figures 3.4A and B and explain why each of conformations **A** to **F** gives rise to an energy maximum or minimum. Why is **A** the highest energy conformation, while **D** is the minimum energy conformation?

FIGURE 3.4A   Key conformations around a substituted two carbon unit; in this case, nabumetone.
          B   Plot of energy changes as a function of dihedral angle calculated at every whole degree between 0–360° for nabumetone. *[Calculations performed by Dr. Peter Dawson using Gaussian software.]* The dihedral angle between two substituents, X and Y, is the angle between the two planes in which the substituents lie and is described by θ (theta).

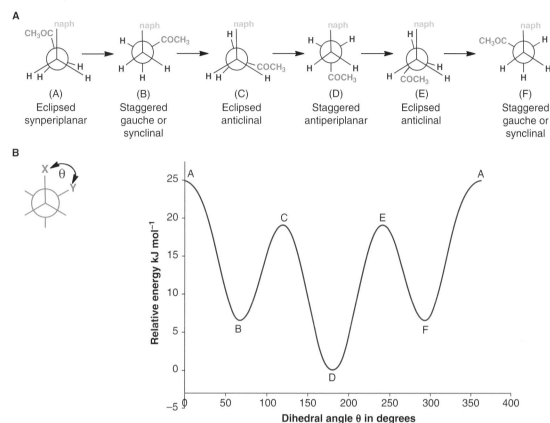

Let's consider a cyclohexane ring. We often draw it in the same way as a benzene ring, as a hexagon, but without the π-system. However, a regular hexagon has internal angles of 120°, whereas an sp³ hybridized carbon atom has a preferred bond angle of 109.5°. In addition, the two H atoms on each cyclohexane carbon would be eclipsed with those on the adjacent carbon atoms, so this is obviously not the best conformation. If you make a model of cyclohexane with a molecular model kit, you find that it naturally adopts a non-planar structure to achieve carbon bond angles of about 109.5°. There are two main conformations adopted by a cyclohexane ring: a *chair* form and a *boat* form, as shown in Figure 3.5.

 More information on hybridization can be found in Chapter 2 'Organic structure and bonding' of this book.

In the chair form of cyclohexane, each carbon bears one hydrogen vertically up or down from the ring. These are termed the axial bonds, because of their position on the axes, as seen in Figure 3.6, structure A. The second hydrogen atom on each carbon is sited around the plane of the ring on the equatorial bonds, structure B. To draw the equatorial bonds, note that they are parallel to the ring bond on the adjacent carbon atom: in C they are shown in pink, in D as green, and in E as turquoise.

In general, substituents in an **equatorial** position are favoured over the **axial** position, as there is less steric strain. In Figure 3.7: **A** views cyclohexane from the same plane as the ring; **B** shows how two axial substituents can be co-planar and close in space, even though they are on non-adjacent ring atoms; **C** shows how equatorial substituents on adjacent ring atoms do not lie in the same plane, which reduces steric strain.

FIGURE 3.5 Representations of cyclohexane (the hydrogen atoms are omitted for clarity in the last three structures; see Figure 3.6 for structures with H atoms included).

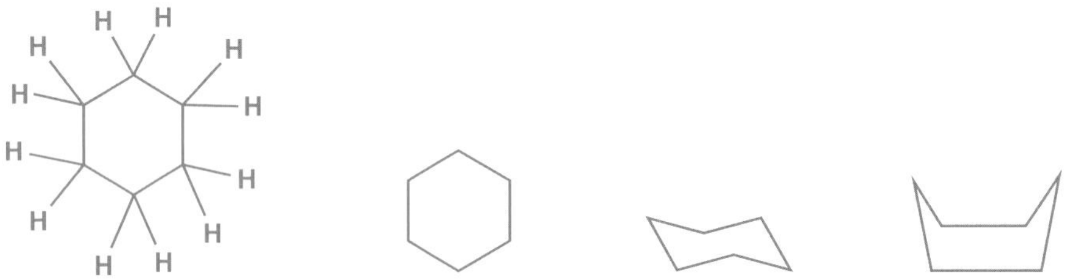

Cyclohexane

Cyclohexane skeleton

Cyclohexane chair form

Cyclohexane boat form

FIGURE 3.6 The position of the bonds on a six-membered ring in the chair conformation.

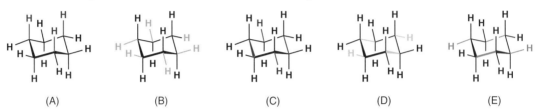

(A)   (B)   (C)   (D)   (E)

Because of the possibility of *ring flip*, which interconverts the two chair forms, the axial and equatorial positions are often interchangeable, allowing the most stable conformation to be adopted. This is very important in pharmaceuticals, as it makes a significant difference to the shape of the molecule.

 This is a broad generalization, as there are several factors to consider when studying the conformations of cyclic molecules. You will find a more detailed discussion of the factors affecting the preferred conformation of cyclic molecules on the support website, along with an explanation of ring flip.

An interesting example is provided by pethidine, an opioid and potent analgesic (see Figure 3.8). In this molecule, the six-membered ring is a piperidine, with two substituents on the same carbon atom. Does the molecule prefer the phenyl ring to be equatorial, or the ester group? It was argued that the larger phenyl ring would cause greater steric strain than the ester group if it was axial, which agreed with the crystal structure of pethidine showing the phenyl ring in the preferred equatorial position (see Figure 3.8), which also puts the *N*-methyl group into the preferred equatorial position.

Six-membered rings with a chair conformation are common in biological systems and in pharmaceuticals. Examples include the anticancer agents, doxorubicin and daunomycin; many steroids, such as cholesterol; and the aminoglycosides, an important class of antibiotics. Figure 3.9 shows two aminoglycosides, gentamicin (a natural product) and amikacin (a

FIGURE 3.7 Arrangement of bonds around the chair conformation of a six-membered ring.

      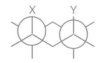   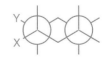

(A) Chair conformation: all bonds on adjacent carbon atoms are staggered

(B) X and Y both axial sited in the same plane, they are close together

(C) X and Y both equatorial in different planes, they are further away in space

FIGURE 3.8 The two chair conformations of pethidine: the crystal structure shows that the second conformation, with the phenyl ring in the equatorial position, is preferred.

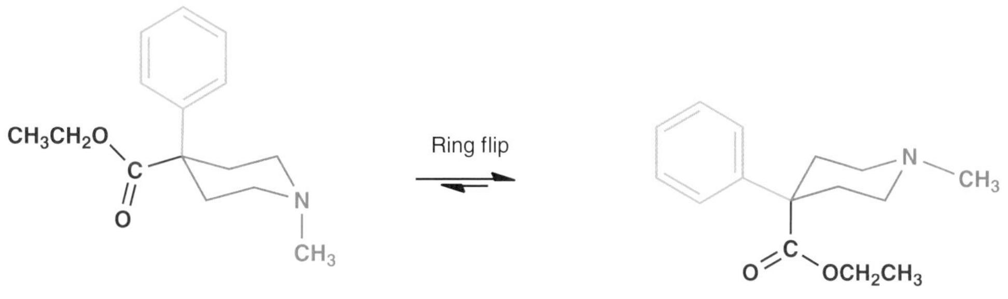

Axial phenyl, equatorial ester

Axial ester, equatorial phenyl

FIGURE 3.9 Chair conformations of cyclohexane and pyranose rings in two examples of aminoglycoside antibiotics, gentamicin and amikacin.

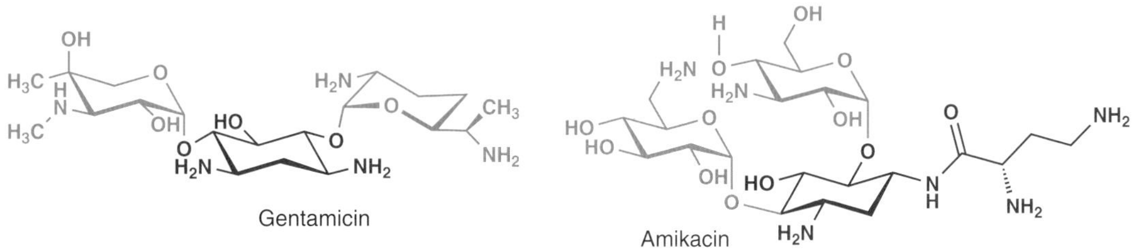

Gentamicin

Amikacin

semi-synthetic drug consisting of a three-ring microbial product to which a side chain is added synthetically). In both of these antibiotic examples, the central cyclohexane ring adopts a chair conformation that puts all of the substituents into equatorial positions. The pyranose rings also adopt chair conformations that put as many as possible of their substituents into equatorial positions.

The boat conformation for six-membered rings is much less common due to adverse strain and interactions (see Figure 3.10). The *geminal* H atoms on one end of the structure are coloured to show their spatial relationship. (Geminal H atoms are bonded to the same central atom; a CH_2 group has two geminal H atoms attached to the same carbon atom. The term comes from the Latin word *Gemini*, meaning twins). As you can see in the 'end view' conformation, this results in several bonds lying in the same plane, giving rise to torsional strain, and the 'flagpole' atoms are very close in space, causing steric strain, particularly if one or both of these atoms is not H. In fact, the boat conformation is usually only seen when there are few or no conformational options.

FIGURE 3.10 The boat conformation of a six-membered ring suffers from steric and torsional strain.

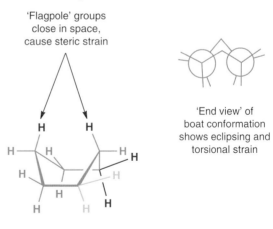

'Flagpole' groups close in space, cause steric strain

'End view' of boat conformation shows eclipsing and torsional strain

The importance of conformation on shape can be seen in the case of donepezil, one of the leading treatments for Alzheimer's disease (marketed as Aricept®). This drug inhibits the enzyme acetylcholinesterase by binding snugly into a defined cavity or 'pocket' in the enzyme – the active site. (The active site of an enzyme is the region of the enzyme that carries out the catalytic activity; it usually has a defined

FIGURE 3.11 The two chair conformations of donepezil: the most stable chair conformation of the piperidine ring puts the two substituents into **equatorial** positions (shown on the left).

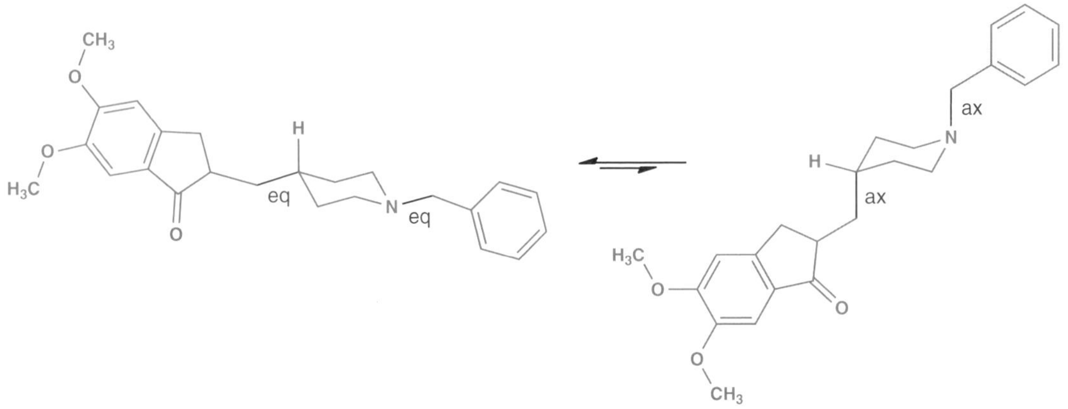

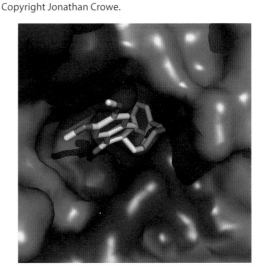

three-dimensional size and shape to which the substrate binds with strong affinity.) Fortunately, this involves the most stable chair conformation for the central piperidine ring with the large substituents in the equatorial positions (see Figure 3.11). If the diaxial conformation had been a better fit in the active site, donepezil is unlikely to have been a good inhibitor, as it would have to adopt a less favourable, higher energy, conformation to bind.

Figure 3.12 shows how donepezil binds to the acetylcholinesterase enzyme. Look at this figure, and notice how donepezil fits snugly into the active site, whose shape is emphasized by the surface plot of the enzyme depicted here. The favoured conformation with the substituents in the equatorial positions is relatively horizontal and has good complementarity to the size and shape of the active site of this enzyme. The diaxial substituted conformation would not be able to fit into this site.

FIGURE 3.12 Complementarity of the stable chair conformation of donepezil to the active site of acetylcholinesterase.
Copyright Jonathan Crowe.

 Other cyclic systems, based on 5-membered and 4-membered rings, are important in medicine. They also have defined preferred conformations, which you can study on the support website.

3.4 Stereoisomerism

Many drugs used in medicine exhibit stereoisomerism, so it is important that, as pharmacists/pharmaceutical scientists, we are familiar with this subject. In this part of the chapter, we will meet some more concepts that relate to the three-dimensional (3D) shape of molecules and the way this affects their properties, including their pharmacological actions.

Stereoisomers of a molecule have the same connectivity of atoms to each other, but these can be arranged differently in three dimensions. Unlike conformational isomers, stereoisomers cannot usually interconvert easily. In this section, we will consider geometric isomers and optical isomers; for both cases, we will consider the effects of this type of isomerism on shape and pharmaceutical action.

Geometric isomerism

Geometric isomers occur when groups of atoms are arranged asymmetrically across a double bond. We will use the drug, tamoxifen, an oestrogen receptor antagonist used in the treatment of breast cancer, as an example; the structures of two isomers of this molecule – we call them the Z and E isomers – are shown in Figure 3.13. (These structures are actually an over-simplification, as tamoxifen exerts the majority of the antioestrogen effect via its metabolite, (Z)-4-hydroxytamoxifen.)

 More discussion on geometric isomers can be found in Chapter 4 'Introduction to drug action' in the *Therapeutics and Human Physiology* book of this series.

As you can see, tamoxifen contains an alkene group. In Chapter 2, you learnt that the alkene group has two kinds of bond – a **strong** σ-bond and a **weaker** π-bond. The weaker π-bond has an area of electron density both above and below the σ-bond and, as a result, rotation is not usually possible about a double bond (in other words, the double bond is 'locked' in place). Look at (Z)-tamoxifen again, and notice how the two unsubstituted phenyl rings are on either side of the double bond, diagonal to each other. If we want

the two phenyl rings to be on the same side of the double bond, as in (E)-tamoxifen, we have to break and reform the π-bond, which requires a lot of energy. (Z)-Tamoxifen and (E)-tamoxifen cannot easily interconvert and are therefore classed as configurational isomers, specifically **geometric isomers**.

But what do the letters Z and E actually tell us? They relate to the Cahn-Ingold-Prelog system, which is used to name or classify geometric isomers; it was published in 1966 and adopted later by IUPAC. This system enables scientists to communicate about stereochemical isomers without ambiguity. The Cahn-Ingold-Prelog system is explained in more detail later in this chapter when we discuss enantiomers and chirality. These assignments come from German: Z stands for zusammen, which (roughly translated) means together, while E stands for entgegen, which means opposite.

To use the Cahn-Ingold-Prelog system, we must first assign a priority to the groups at either end of the alkene based upon the atomic number of each of the atoms. The atom with highest atomic number is given top priority. If the two highest priority groups are on the *same* side of the alkene, the alkene is assigned as Z (= together = on the same side); if the groups given top priority are on *different* sides, the alkene is assigned as E (= opposite = on different sides).

Let's take the structure of tamoxifen as an example.

1. Firstly, draw an imaginary line through the centre of the double bond (we now have two parts to the molecule – shown in red and blue). Now we need to assign priority to each atom at either side of the double bond, based on the atomic number.

Draw an imaginary line down the centre of the double bond

2. Let's concentrate on the red half of the molecule first. We have two carbon atoms bonded directly to the alkene, one atom is part of the ethyl group and the other is part of the benzene ring. Fortunately, there is a way of differentiating between them.

The carbon atom which forms part of the benzene ring is bonded to another two carbon atoms, while the carbon atom which forms part of the ethyl group is bonded to two hydrogen atoms and only one carbon atom. The carbon atom with more atoms of higher atomic number attached to it takes higher priority; the carbon atom in the benzene ring therefore takes priority 1 and the carbon atom in the ethyl group takes priority 2.

3. We now follow the same principles with the blue half of the molecule: both carbon atoms attached to the alkene are part of aromatic rings, and the 2 and 3 positions of the aromatic rings are identical. The oxygen atom on carbon-4 (attached to the top benzene ring) eventually makes the difference. Oxygen on the substituted aromatic ring has a higher atomic number than hydrogen in the unsubstituted ring, so the top ring (as drawn) is given priority 1.

4. We can now decide whether this molecule is E or Z. In the case of tamoxifen, the two groups given priority 1 are on the same side of the alkene, so it is assigned as the Z isomer.

If you have previously met *cis* and *trans* nomenclature, you might name (Z)-tamoxifen as *trans*-tamoxifen because the two identical phenyl rings are on opposite faces at either end of the double bond (these terms come from the Latin translation; *cis* means 'on the same side' and *trans* means 'on the other side'). You would be correct! The *cis* and *trans* system predates the (Z) and (E) system, and is limited to alkenes in which one of the substituents on each alkene C is the same. (This is a rare example in the pharmaceutical world, when *cis* does not correlate with (Z) or *trans* with (E).) Although you will occasionally meet *cis* and *trans* nomenclature, alkene groups in pharmaceuticals are assigned explicitly using the (Z) and (E) system adopted by IUPAC. It is essential that there is no doubt that the correct form of a drug is being used.

Let's get back to the world of medicine and see how geometric isomerism can have an impact on the patient. The 3D shape of a drug molecule is crucial to the desired therapeutic response in the body. To re-cap, tamoxifen is an oestogen receptor antagonist and is given as the *Z* isomer. In Figure 3.14, you can see tamoxifen drawn in three dimensions, as a 'stick' diagram and with an added surface. The *E* isomer is shown in the same way. The two isomers differ dramatically in 3D shape. In fact, they differ so much that the *E* isomer cannot bind to the oestrogen receptor. It is quite useless in the treatment of breast cancer.

 When considering aliphatic cyclic systems, the *cis*- and *trans*- nomenclature is still used to describe the relationship of ring substituents; as the bonds in the ring are tethered and unable to rotate, the relative orientation of the substituents is fixed. This is discussed more, along with several examples of such molecules, on the support website.

FIGURE 3.14 (A) Stick structures of (Z)-tamoxifen (left) and (E)-tamoxifen (right); (B) surface representations of (Z)-tamoxifen (left) and (E)-tamoxifen (right) showing the dramatic difference to shape imparted by geometric isomerism. The colour represents partial charge on individual atoms: at one extreme, blue atoms have a partial negative charge, while, at the opposite extreme, pink represents atoms with a partial positive charge. The shades of purple represent atoms that are close to being neutral.

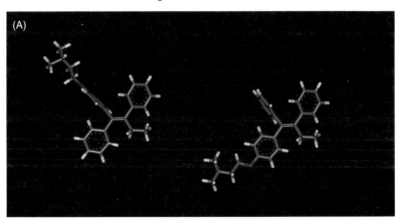

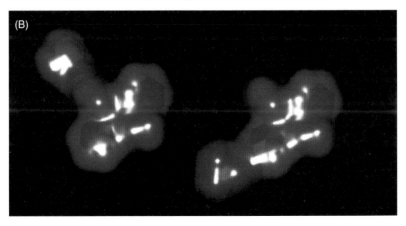

Optical isomers

Optical isomers are so-called because of the different way the isomers interact with plane polarized light. We will look at this property later in this section; first, we will consider how these isomers arise structurally.

Let us consider the structure of a natural amino acid, 2-aminopropanoic acid (alanine), whose structure is shown in Figure 3.15, to illustrate how chirality can arise.

In Chapter 2, we saw that an sp³ hybridized carbon atom has a tetrahedral shape. To represent the tetrahedral 3D shape in the 2D medium of paper, we use hashed line bonds (these represent the bonds going away from you) and bold wedge bonds (these represent the bonds coming towards you), alongside the bonds in the plane of the paper.

These compounds both have the molecular formula $C_3H_7NO_2$ and the atoms are connected together in the same order; however, you cannot arrange these molecules side by side so that they look identical. If you do not believe us, make models of the two compounds and try to superimpose them! The two structures are actually non-superimposable mirror images of each other (similar to the way in which your hands are mirror images of one another — providing you are not wearing any jewellery!) and are called **enantiomers** (see Figure 3.16). Again, if we use our hands as an example, each hand would be an enantiomer. One simple test for chirality considers whether two mirror images are superimposable or not; if the mirror images are superimposable (the pair can be overlaid exactly onto each other, like most socks), then those articles or compounds are **achiral** (meaning, not chiral) — while non-superimposable mirror images (like most shoes) are enantiomers of a chiral species.

Note: In this book, we mostly use the terms 'chiral' and 'achiral', these being currently more common. A **chiral** molecule exhibits 'handedness' – it exists as

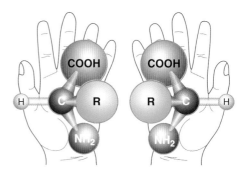

FIGURE 3.16   The chirality of hands and asymmetric molecules.

optical isomers that, like our hands, are non-superimposable mirror images. An **achiral** molecule does not exhibit handedness. The terms **stereogenic** and **non-stereogenic**, respectively, may be used instead.

Enantiomers are non-superimposable and are called chiral because they have **no internal plane of symmetry**. An equimolar mixture of enantiomers is called a **racemate**, or a **racemic mixture**. **Racemization** occurs when a sample of 100% of one enantiomer is converted into the racemate. Any carbon atom bonded to four different groups has no plane of symmetry and is therefore chiral – we say the molecule has a chiral or stereogenic carbon. A molecule containing one or more chiral atoms is usually chiral, except for examples with an internal plane of symmetry (for example, see the section on 'Molecules with more than one chiral centre'). A molecule with a plane of symmetry is achiral and therefore **cannot** exist as enantiomers.

Molecules with chiral carbon atoms are the most common form of chirality observed in medicine, but other forms of chiral centre do exist. For example, on the support website you will meet omeprazole, a 'proton pump' inhibitor that is used to prevent and treat stomach and duodenal ulcers, which has a chiral sulfur atom; and cyclophosphamide, an anticancer agent with a chiral phosphorus atom.

As we have seen previously, many drugs work by interacting with receptors that are formed from building blocks that are chiral (such as a protein or nucleic acid), or by binding to and inhibiting enzymes; the 3D shape of a molecule is therefore crucial if the molecule is going to 'fit' into the enzyme active site or receptor and elicit a therapeutic response. To illustrate this, let us consider morphine – an opioid analgesic extracted from the poppy, *Papaver somniferum,* which you will

FIGURE 3.15   2-Aminopropanoic acid (alanine) – a simple molecule that displays chirality.

meet again in Chapter 10. *Papaver somniferum* (sometimes called the opium poppy) is a source of opium from which many clinically useful alkaloids are obtained, including: morphine, codeine, papaverine, and noscapine. Codeine and morphine are analgesics; papaverine is used for the treatment of erectile dysfunction; and, noscapine acts as a cough suppressant. Morphine has a very complex structure, including several chiral centres (see Figure 3.17); it binds to opioid receptors in the brain where it elicits an analgesic (pain killing) effect.

For morphine to bind to the opioid receptor in the brain, it has to have a specific shape, illustrated in Figure 3.18A (left). If we invert all of the chiral centres in morphine, we change its shape (see Figure 3.18A (right)); the two 'stick' structures of morphine look quite similar, but, if we put a surface on the molecules

FIGURE 3.17 The chemical structure of morphine.

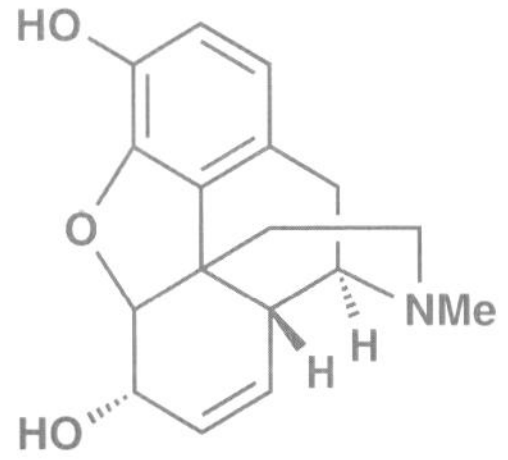

(see Figure 3.18B), you can see how the charge (shown in different colours) and 3D shape varies between the two structures.

SELF CHECK 3.3

How many chiral carbon atoms can you find in the structure of morphine?

SELF CHECK 3.4

We have been looking at molecules with a chiral centre that are drawn in 3D, where it is easier to identify which is the chiral carbon. Of course, molecules are not always drawn in 3D, so it is important to be able to identify which are the chiral carbons in stick diagrams. Look at the examples below and identify the chiral carbon atoms in each structure; some may have more than one.

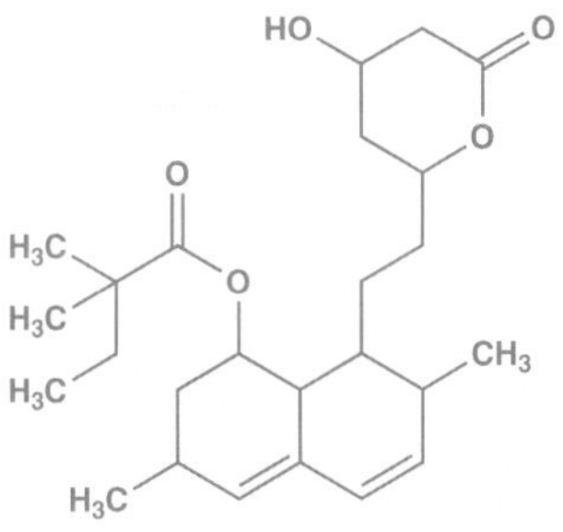

Simvastatin

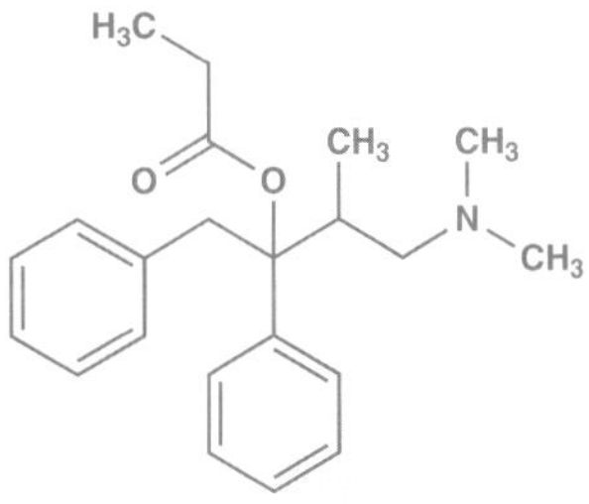

Dextropropoxyphene

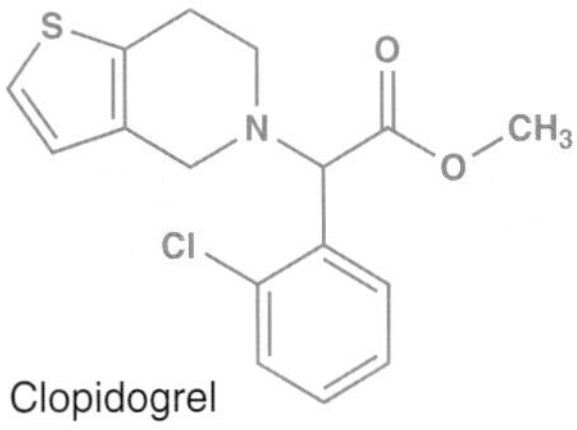

Clopidogrel

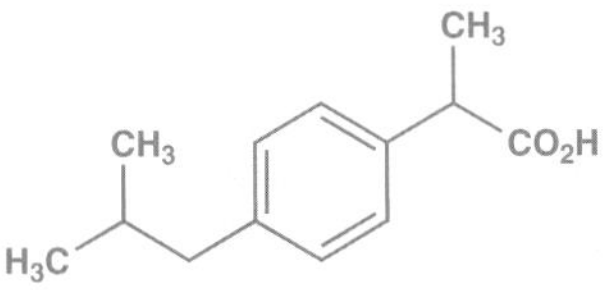

Ibuprofen

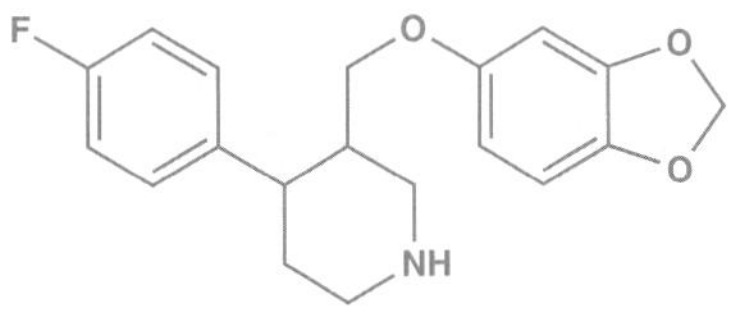

Paroxetine

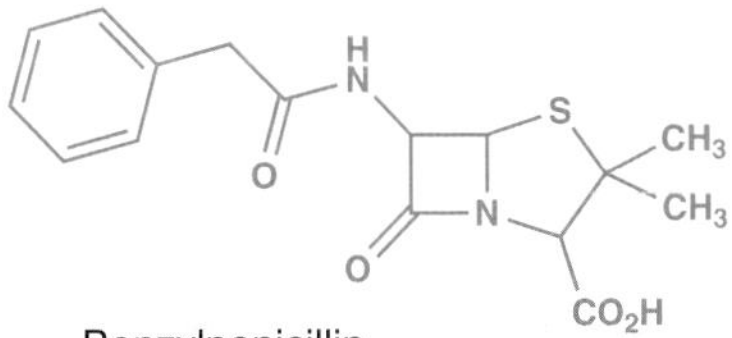

Benzylpenicillin

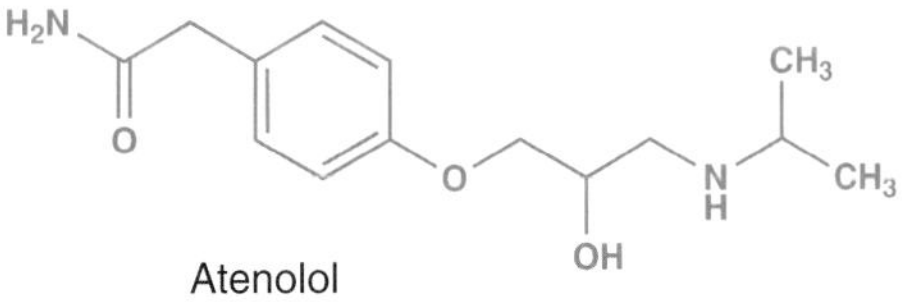

Atenolol

FIGURE 3.18 The structure of morphine ((A) left) and with a surface to show molecular shape ((B) left); the structures on the right in (A) and (B) belong to the mirror image form of morphine in which the chiral centres have been inverted. Again, the atoms are coloured according to their partial charge: blue/cyan represents partial negative charge and red/pink atoms have a partial positive charge.

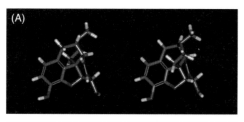
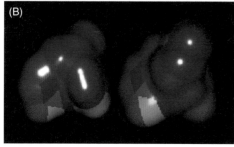

We learn more about the importance of understanding the impact of enantiomers on the pharmaceutical industry in Integration Box 3.1.

Stereochemical nomenclature of chiral compounds

Now you can identify molecules with a chiral carbon atom and can see that one chiral carbon in a molecule leads to two, non-superimposable, mirror image isomers called enantiomers, we need a method to communicate which enantiomer is under consideration at any particular time – we need some standard nomenclature.

IUPAC nomenclature is used internationally by the majority of scientists, most of the time (one specific exception will be discussed later); for stereochemistry, we use the same Cahn-Ingold-Prelog system of nomenclature that we introduced in the earlier section on geometrical isomers. So how do we go about using this system in the context of chiral molecules? First, we must represent a **three-dimensional chiral sp³ carbon** in two dimensions on paper. On page 60,

we met alanine drawn in 3D, with two adjacent bonds in the plane of the paper, a hashed bond behind the plane, and a wedge bond in front of the plane.

SELF CHECK 3.5

There are many ways to draw a tetrahedral shape correctly; two are shown (see Figure 3.19), next to an incorrect structure. Why is the third drawing incorrect? [Hint: examine a tetrahedral model!]

FIGURE 3.19 Two different ways of correctly representing a chiral atom in three dimensions.

With four atoms or groups attached to the chiral carbon drawn in a correct tetrahedral format, we can follow a procedure to identify the stereochemistry:

1. Assign a priority to each atom attached directly to the chiral carbon, based upon the atomic number Z of each of the atoms. The atom with highest atomic number has top priority (number 1), the atom with the second highest atomic number gets number 2 priority, the third highest atomic number gets number 3 priority, and the atom with the lowest atomic number is given the lowest priority (number 4).

2. Once identified, the number 4 priority atom or group is put to the back of the tetrahedral structure. This is the hashed wedge bond on our stick diagram, leaving the other three atoms or groups in the foreground (like the steering wheel of a car) – the assignment of stereochemistry is based upon these three atoms or groups.

3. If the order of the atoms proceeds 1,2,3 in a clockwise direction, the nomenclature *R* (from the Latin, rectus, meaning 'right-handed') is used, as in the example here; if the atoms prioritized as 1,2,3 follow an anticlockwise direction, then the nomenclature *S* (from the Latin, sinister, meaning 'left-handed') is used. It is convention to use italicized capital *R* and *S* for this nomenclature.

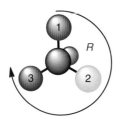

Worked example

In our example, alanine, there are four atoms or groups arranged around the chiral carbon (indicated with a *): the methyl (CH_3) group and the carboxylate (CO_2^-) group are both in the plane of the paper, while the ionized amine (NH_3^+) group is at the front and the H atom is at the back. We can see that the chiral C has four different atoms/groups attached to it: CH_3, CO_2^-, NH_3^+ and H.

1. To prioritize them, we need the atomic numbers *Z*, which you can obtain from a periodic table: **N**, *Z*=7; **C**, *Z*=6; **H**, *Z*=1. (It is tempting to assign stereochemistry using the relative atomic masses of each element to prioritize the atoms; however, be aware that doing this can result in the wrong priorities being assigned. The only way to be correct with certainty is to use the atomic number, Z.)

2. We can immediately see that **N** takes priority 1 and **H** priority 4. Adding the priority numbers to the structure, we find that the number 4 priority atom is already at the back, on the hashed wedge bond.

3. To assign priorities 2 and 3, we need to distinguish between the two carbon atoms, by taking account of the atoms that are attached to each one. Once again, we consider the atomic numbers of the next attached atoms to identify which has highest

priority. In our example, one C atom has three hydrogen atoms attached (**H** *Z*=1), while the other has oxygen atoms (**O** *Z*=8); we therefore assign the carboxylate group (CO_2^-) as priority 2 and the CH_3 as priority 3. **An atom with multiple bonds, such as a carbonyl group (C=O), is given the same priority as multiple single bonds to the same atom or group**; in the case of a carbonyl, the carbon is equivalent to a carbon having two bonds to oxygen (O-C-O).

4. We can now ignore the H at the back, and look at the direction of the 1,2,3 priority groups. In this molecule, they go anticlockwise from NH_3^+ to CO_2^- to CH_3, so the nomenclature *S* is assigned. We can distinguish this enantiomer from its mirror image by naming it (2*S*)-2-aminopropanoic acid. This conveys the exact formula and connectivity, along with the correct stereochemistry around the chiral carbon.

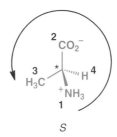

A more complex example is provided by vigabatrin, an anti-convulsant drug. Only one of the two enantiomers is credited with its anti-convulsant activity.

NH₃⁺ ... (structure)

Active enantiomer of vigabatrin

First, we need to identify the chiral carbon atom (C4); it is often highlighted by the stereochemistry shown around it. In this case, we need to add a H on a hashed wedge bond that is not currently shown on the structure. Then we can prioritize the atoms attached to C4, the chiral carbon.

There is a N atom, two C atoms and a H atom, with atomic numbers Z of 7, 6 and 1, respectively. We can immediately assign the N as priority 1 and the H as priority 4, but we need to differentiate between the two carbons atoms (C3 and C5) to decide which has higher priority.

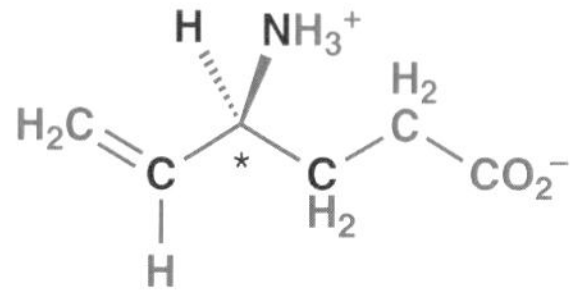

The priority 4 group (H) is at the back, so we can now ignore it and look at the order of the 1,2,3 priority groups: the sequence goes anticlockwise, so this has the nomenclature S.

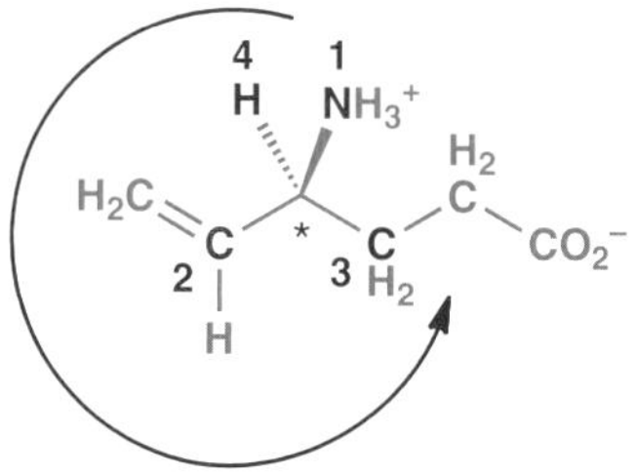

The methine alkene carbon (C5) has a double bond to the terminal CH_2 group, equivalent to two single bonds to carbon, plus one bond to an H atom. The other carbon atom C3, shown to the right of the chiral carbon, has one single bond to another carbon atom (C2) and two bonds to H atoms. We can summarize the bonding to each of C3 and C5 to help us assign the priority:

- C3 has 1C and 2H bonded to it,

- C5 has the equivalent of two single bonds to C and 1H bonded to it (remember we consider a multiple bond, such as the alkene to C5, as multiple single bonds to the same atom or group, so that we consider C5 as having two bonds to C);

  therefore,

- C5 has higher priority than C3, as it has more carbon atoms (higher atomic number) attached to it, so

- C5 gets priority 2, with C3 getting priority 3.

The S enantiomer of vigabatrin irreversibly inhibits gamma-aminobutyric acid (GABA) transaminase and is the active drug with the desired pharmacological properties.

Sometimes, a chiral molecule is drawn with the number 4 priority group at the front, instead of at the back. In this case, you are seeing the structure reversed, so you can reverse the stereochemistry to get the correct assignment. For example, if you were given the structure of fluoxetine shown on the left in Figure 3.20, with the H at the front, it only needs the whole molecule to be turned around through 180° to move the H to the back as in the structure on the right. The stereochemistry looks like it will be R on the structure shown on the left, but this is viewing it from the wrong side. When the whole molecule has been rotated, the correct stereochemistry, as shown on the structure on the right, is S.

FIGURE 3.20   (S)-Fluoxetine viewed from front and back faces.

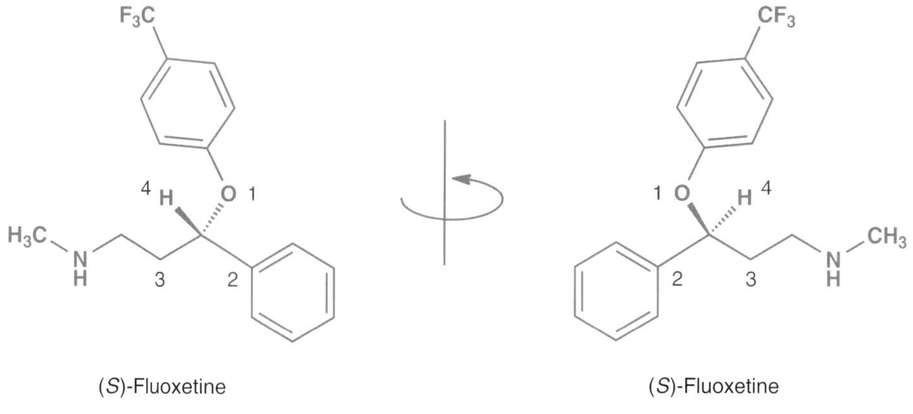

(S)-Fluoxetine

(S)-Fluoxetine

When a chiral molecule is drawn so that the number 4 priority group is in the plane of the paper, you can imagine looking along the bond from the chiral carbon to the number 4 priority group (the C–H bond in this example). If you find this challenging, you could try redrawing the molecule, but this requires great care so that you do not invert it while redrawing.

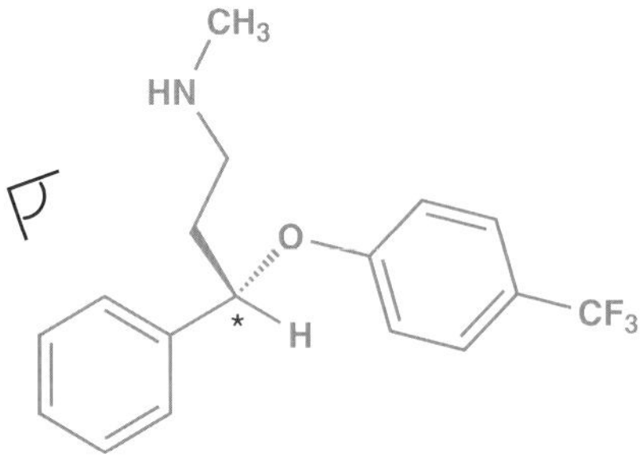

Fluoxetine

If you want to try this approach, you will find fluoxetine as a worked example on the support website. Other practice examples, including the rules for assigning the stereochemistry of cyclic compounds can also be found on the support website.

Stereoisomerism in helices

Helical structures are commonly found in biological systems and are a source of chirality on the macromolecular scale; notable examples include the double helix of DNA and the α-helix structural motif that is commonly found in proteins.

There are many examples of helical stereoisomerism beyond biology: a spiral staircase is a helix, as is a corkscrew. If you are familiar with the use of a corkscrew, you will know that they are designed for use by

SELF CHECK 3.6

Asthma is a condition that causes the airways to become inflamed, and, in the UK alone, there are over 5 million people receiving treatment for asthma – this equates to a staggering 1 in 12 adults and 1 in 11 children. Thankfully, there are many drug treatments available for managing asthma, some of which are shown below. You will notice that each drug contains a chiral centre – can you assign them as either *R* or *S*? In practice, some of these drugs are given as an equimolar mixture of both enantiomers (the racemate).

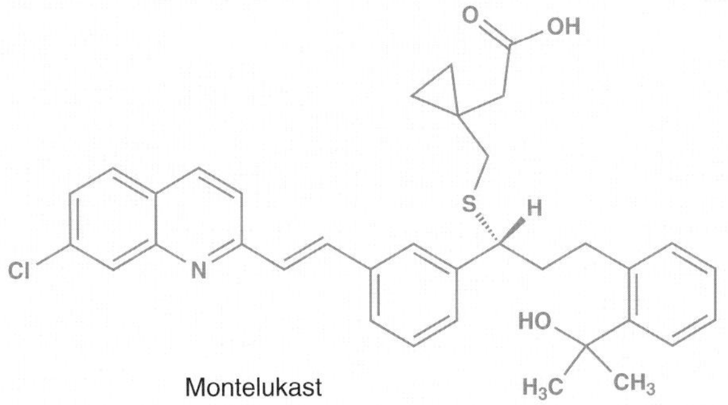

Montelukast

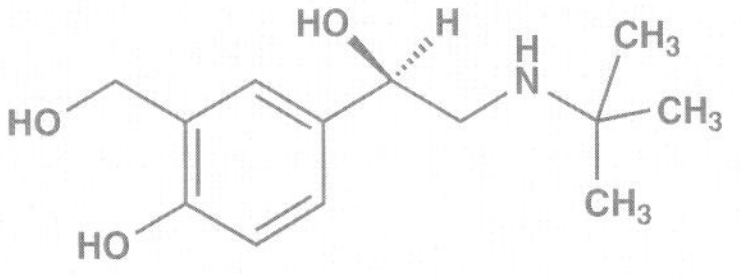

Salbutamol

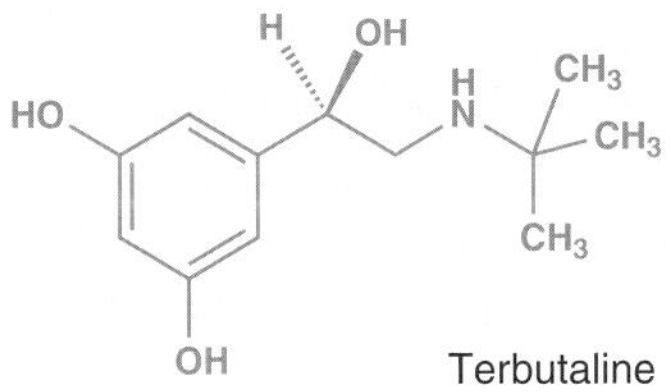

Terbutaline

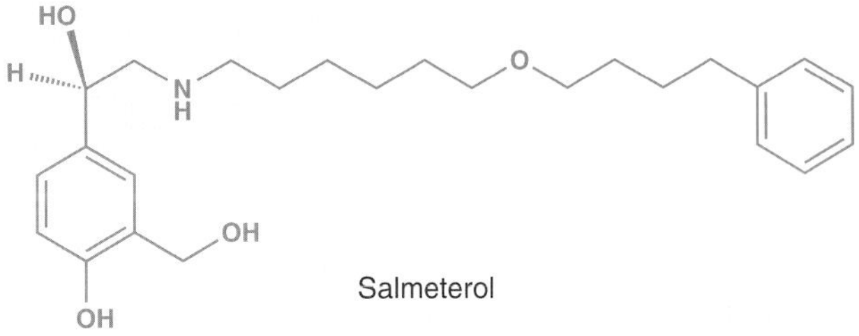

Salmeterol

The pharmaceutical impact of stereochemistry – Thalidomide

You may be asking yourself why it should matter whether or not a drug has any chiral centres. That is a good question and one that arose as a consequence of a drug approved for human use, and then later withdrawn because it caused horrendous adverse effects. You may have heard of thalidomide and that it caused severe deformities in children born to mothers who had taken this drug early in their pregnancy, but you may not be aware that it was a turning point in drug development and regulation worldwide.

Thalidomide was first synthesized and patented in 1953 by the pharmaceutical company Chemie Grünenthal, then licensed as a hypnotic sedative in 1957. It was thought to be so safe that it was released in Germany as an 'over the counter' agent in 1960 and promoted as an entirely safe remedy for morning sickness and restless sleep in pregnancy. More than 40 other countries worldwide introduced this new wonder-drug with many indications, including asthma, hypertension and migraine, as well as morning sickness. In the USA, regulation was more strict, following a tragedy involving a toxic taste-masking agent, and the Food and Drug Administration (FDA) refused to license thalidomide.

Over the next two years, there were numerous reports of deformities in children born to mothers who had taken thalidomide. It is estimated that between 10,000 and 20,000 foetuses were affected. Thalidomide was taken off the market worldwide in 1961–62.

At that time, the concept that the enantiomers of a drug could have different activities in the body, or that their effects could be different in a foetus to the responses in an adult, was not recognized. After thalidomide was withdrawn from the UK market in 1962, an enquiry was initiated to investigate how such a huge mistake could have been made. It was eventually found that

the (R)-enantiomer of thalidomide possessed the antiemetic, anxiolytic properties that promoted sleep and benefitted morning sickness, while the (S)-enantiomer had teratogenic properties, which meant it resulted in severe adverse effects to a foetus.

Although it is easy to think of rescuing the situation by marketing only the (R)-thalidomide, it was shown that each enantiomer of thalidomide was racemized *in vivo*, so that administration of the pure (R)-enantiomer leads to production of some of the teratogenic (S)-thalidomide.

As a result of this tragedy, the yellow card reporting system of adverse drug reactions (ADR) was introduced in the UK in 1964 and the Medicines Act of 1968 established a Medicines Commission and strict standards on the efficacy and safety of drugs. Many years later, in the 1990s, the regulatory bodies[1] with control of the registration of new drugs introduced new regulations that require the absolute stereochemistry of the active form(s) of a chiral agent to be identified early in drug development and justification for the use of a racemic mixture over a pure enantiomer. Such requirements usually entail stereoselective synthesis of each single enantiomer, or sometimes a pair of diastereoisomers, and the pharmacological properties of each individual chiral species to be established independently.

Thalidomide has now been reintroduced for the treatment of multiple myeloma and certain skin diseases, such as leprosy. Care is taken to ensure no patients on thalidomide are, or may become, pregnant.

[1] The regulatory body for the UK is the MHRA – the Medicines and Healthcare products Regulatory Agency, while the corresponding body in Europe is the EMA – the European Medicines Agency – and in the USA is the FDA – the Food and Drug Administration.

the right hand; in fact, if you are left-handed, you may have already searched for a 'left-handed corkscrew' (yes, they do exist!). Right-handed and left-handed corkscrews are mirror images of each other and are non-superimposable: they meet the criteria we used

earlier to identify chiral molecules and are, in fact, stereoisomers. The double helix usually found in DNA is right-handed, although a left-handed form is known; the α-helix found in proteins is also right-handed (see Figure 3.21).

FIGURE 3.21 Left and right-handed helices

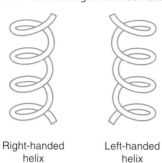

Right-handed
helix

Left-handed
helix

Fischer's nomenclature: D and L

When you study macromolecules, such as sugars, peptides and proteins (see Chapter 9), you will learn that they are based on chiral building blocks; the way in which substrates, inhibitors or receptors bind to them is crucial to the biochemical processes that go on in our bodies. Biologists often talk about L-amino acids or D-sugars, not mentioning *R* or *S*; you may wonder why these molecules use a different system and why you have to learn about two sets of nomenclature for chiral molecules.

It is true that using the single IUPAC system streamlines the learning process, but the 'old' (or 'trivial') nomenclature system is retained because it simplifies the names of biological molecules; see Figure 3.22 for some examples.

The system of D and L nomenclature is based upon the work of Emil Fischer, who studied sugars, such as glucose, mannose and fructose, between 1884 and 1894. His careful work showed that they are isomers of each other and he established their stereochemistry. To achieve this aim, he synthesized each sugar from a known molecule with established stereochemistry, D-glyceraldehyde, using a reaction that partly bears his name – the Kiliani-Fischer reaction; C2 of D-glyceraldehyde becomes C5 in each sugar.

Comparison of the stereochemistry of natural glucose, mannose and fructose showed they had the same stereochemistry at C5 as observed at C2 of D-glyceraldehyde; by carrying out many reactions with each sugar and characterizing the products, he used the results to solve the absolute stereochemistry at each chiral centre. This was a seriously impressive achievement, particularly when you remember that he did it without the use of modern characterization techniques, like IR and NMR spectroscopy, mass spectrometry and X-ray crystallography! (Fischer also studied the chemistry of purines and the nature and structure of peptides and proteins, discovering that the amino acid building blocks were held together by amide (peptide) bonds. He was awarded the Nobel Prize in 1902 for his contributions to chemistry.)

Note that the convention is to use small capitals for the Fischer nomenclature: D and L rather than D and L. However, we use 'full size' capitals for IUPAC Z/E and *R/S* nomenclature.

FIGURE 3.22 Carbohydrate examples of IUPAC and Fischer nomenclature. Although glucose, fructose and mannose are represented as acyclic molecules here, in aqueous solution the cyclic form of each predominates: the hemiacetal forms of glucose and mannose and the hemiketal form of fructose. You will meet these structures in more detail in Chapter 9.

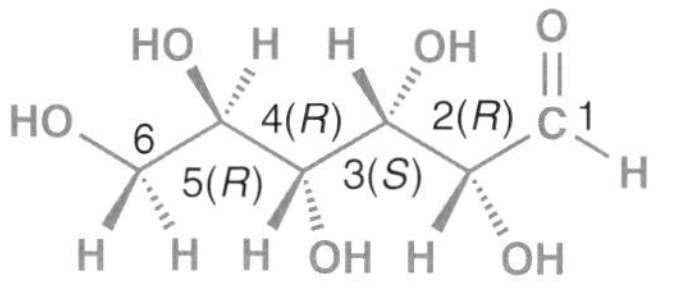

(2*R*, 3*S*, 4*R*, 5*R*)-2,3,4,5,6-pentahydroxyhexanal
[D-glucose]

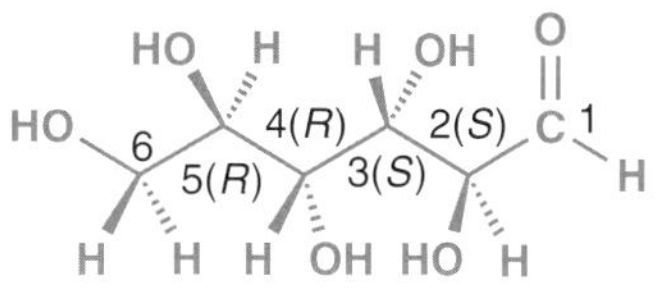

(2*S*, 3*S*, 4*R*, 5*R*)-2,3,4,5,6-pentahydroxyhexanal
[D-mannose]

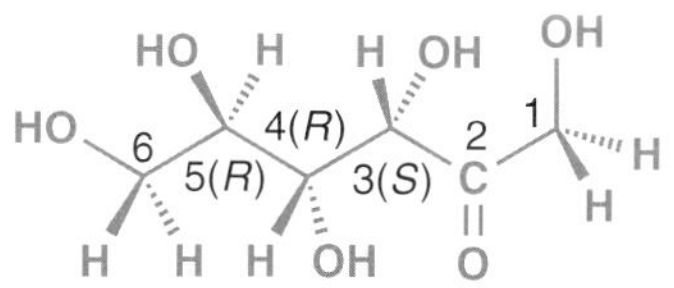

(3*S*, 4*R*, 5*R*)-1,3,4,5,6-pentahydroxy
hexan-2-one [D-fructose]

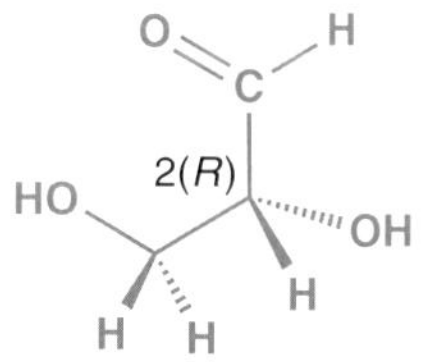

(2*R*)-2,3-dihydroxypropanal
[D-glyceraldehyde]

You will study a range of biological molecules and their chemistry in Chapter 9 'The chemistry of biologically important macromolecules', and in Chapter 11 'Introduction to pharmaceutical analysis', you will learn more about IR and NMR spectroscopy, and mass spectrometry.

In Figure 3.22, the upper chemical name under each structure is the IUPAC name, while the lower name is the old name. You can see that there are sometimes advantages to using the trivial name – sugars are very well known molecules and D-glucose is much easier to use, and type, than $(2R, 3S, 4R, 5R)$-2,3,4,5,6-pentahydroxyhexanal!

What do D and L stand for and how do we decide which to use? Fischer devised a way of representing sugars to make their stereochemistry clear and unambiguous. He placed the most oxidized carbon atom, numbered C1, at the top of a page and the chain of carbon atoms vertically down. As a result, the two other groups on every carbon atom stuck up away from the paper and towards the viewer. With a light held above the structure, the 2D shadow of the 3D molecule was cast on the paper below, illustrated for D-glyceraldehyde in Figure 3.23, and is called a **Fischer projection**. When drawn according to Fischer's rules, the shadow (and so the drawing) of a particular chiral molecule will always be the same.

This system is most usually applied to sugars and amino acids, which have either aldehyde and hydroxyl (OH), or carboxylic acid and amino groups (NH_2) in common. **If the OH or NH_2 on the *highest numbered chiral carbon* (the one nearest the bottom of the page) is on the right of the vertical carbon chain, then the molecule is designated D, if it is on the left of the vertical carbon chain, then it is designated as L.** The D and L forms of a compound are **enantiomers**, that is, they are mirror images. The notation comes from the Latin; D for dextro, meaning 'right', and L meaning laevo (or levo) for 'left'.

You can try this for yourself using a molecular model kit and a torch (or the torch application on your phone), using Figure 3.23 as a guide.

FIGURE 3.23 The origin of the Fischer system of nomenclature for chiral biological building blocks, such as sugars and amino acids, demonstrated for D-glyceraldehyde.

(2R)-2,3-dihydroxypropanal
[D-glyceraldehyde]

D-glyceraldehyde

Fischer projection
of D-glyceraldehyde
(shadow)

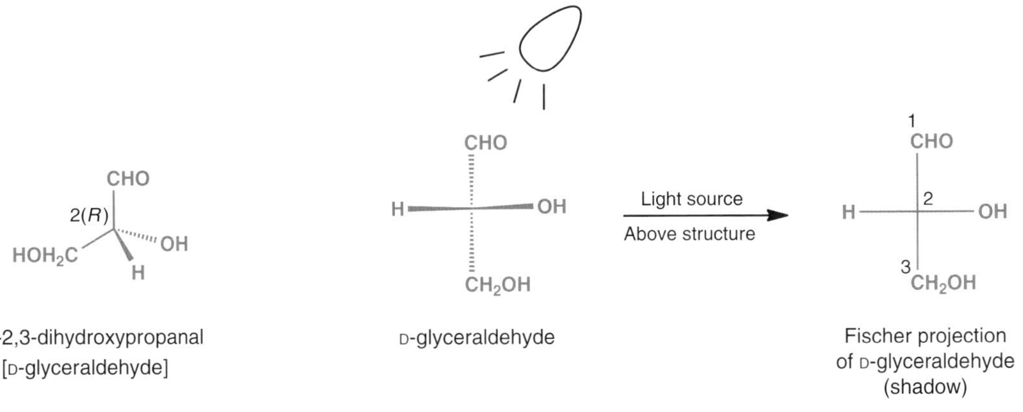

SELF CHECK 3.7

The same rules as those described above are applied to glucose, any other sugar, and a range of other molecules, such as amino acids. The two enantiomers of glucose are shown opposite; can you use Fischer's rules to decide which is the natural D-glucose and which is the L-form. Two natural amino acids, alanine and serine, are also shown in Figure 3.24; use Fischer's rules to decide if the D- or L-form is the naturally occurring form of each.

FIGURE 3.24 The enantiomers of glucose and two natural amino acids as Fischer projections.

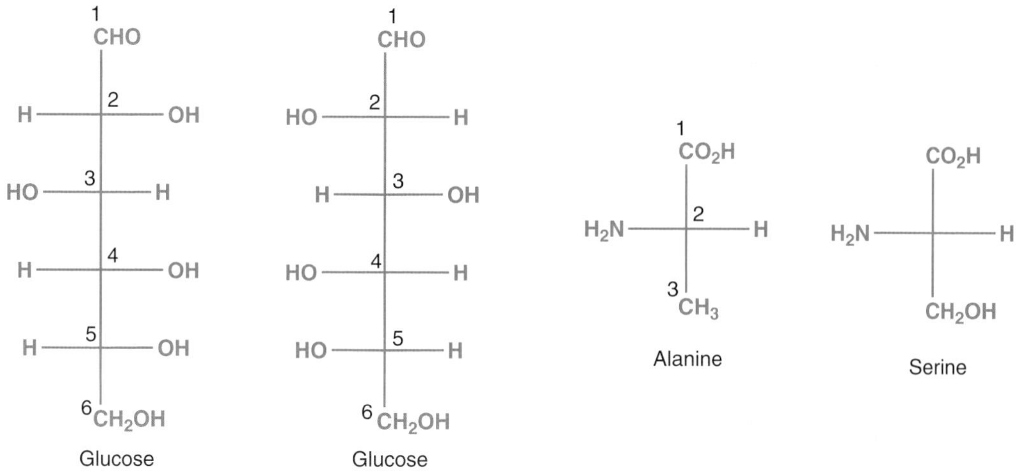

Glucose   Glucose   Alanine   Serine

Manipulating Fischer projections

There are some rules to using Fischer projections that must be adhered to for accuracy: if you turn an entire Fischer projection through 90° on paper, then the stereochemistry inverts. You can check this by making models of each structure below (remembering the horizontal bonds always stick up out of the paper towards you) and putting each into the Fischer projection with the most oxidized carbon at the top and the carbon chain vertically down, then check where the OH group is – left (L) or right (D) (see Figure 3.25).

The stereochemistry also inverts if you swap the positions of any two groups; in the examples shown in Figure 3.26 the two groups that have been swapped are highlighted. Again, you can convince yourself by making the models.

Properties of enantiomers

Constitutional isomers and geometric isomers have different physical and chemical properties: their melting points, boiling points, etc., are different. In contrast, a pair of enantiomers has identical physical

FIGURE 3.25 Illustration of inversion of Fischer projection nomenclature by rotation of the structure.

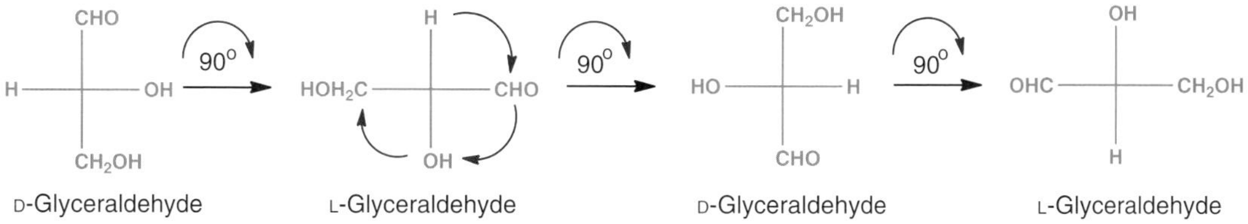

D-Glyceraldehyde   L-Glyceraldehyde   D-Glyceraldehyde   L-Glyceraldehyde

FIGURE 3.26 Each reversal of two adjacent groups inverts Fischer stereochemical assignment.

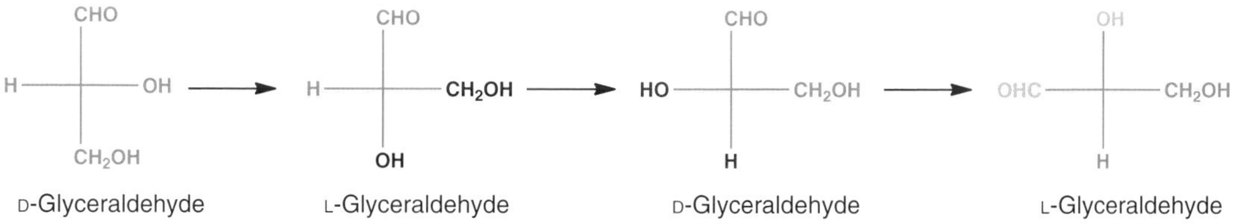

D-Glyceraldehyde   L-Glyceraldehyde   D-Glyceraldehyde   L-Glyceraldehyde

and chemical properties: they have the same melting points, boiling points, solubilities, and even have the same NMR spectra. There are only two key differences in their dynamic properties; we will look at these distinguishing features in some detail in this section.

In Table 3.2, some physical properties of the enantiomers of ibuprofen, whose structures are shown in Figure 3.27, are listed.

Notice how the melting points of the enantiomers are the same, whereas the melting point of the racemate is different. It is common for the melting point of the racemate to be higher than either of the enantiomers from which it is formed; the enantiomers combine in a 1:1 ratio with high affinity to form an ordered crystalline solid, requiring greater energy (a higher temperature) to separate the enantiomers. The tight crystal packing can also contribute to lower solubility of the racemate. You may wonder why the racemate of ibuprofen is used, when the activity resides largely in the (S)-enantiomer. There are three main reasons:

1. Economics: the racemic mixture is cheaper to synthesize.

2. Toxicology: the (R)-enantiomer does not have any significant adverse effects.

3. Pharmacokinetics: although the standard 400 mg dose contains only 200 mg of the active (S)-

TABLE 3.2 Selected physical properties of the enantiomers and racemic mixture of ibuprofen

Molecule	Mol. wt. (g/mol)	Melting point (°C)	Aq. Solubility at pH 1.5 (mg/100 mL)	pK_a
(R)-Ibuprofen	206.28	52	9.5	4.43
(S)-Ibuprofen	206.28	52	9.5	4.43
Racemic ibuprofen (1:1 R/S)	206.28	76	4.6	4.43

enantiomer, about 50–60% of the (R)-enantiomer is slowly converted to the active (S)-form, resulting in an overall total dose of 300–320 mg of (S)-ibuprofen being delivered.

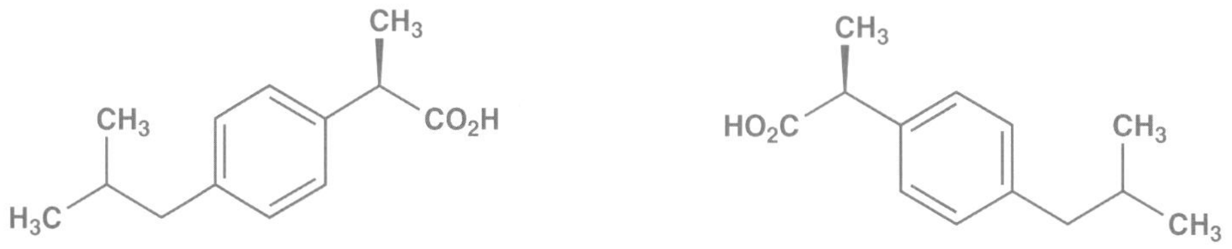

Distinguishing properties of enantiomers

The properties of a pair of enantiomers only differ in:

- the way each enantiomer interacts with plane polarized light (PPL);
- the way in which each enantiomer reacts or interacts with another chiral species.

Because biological molecules are chiral, this latter difference is most relevant to us in the world of pharmaceutical science and it helps to explain the 'thalidomide disaster' (see Integration Box 3.1).

Interaction with plane polarized light

Light from the sun, or from a normal light bulb, radiates out from the source in all directions and planes. Plane polarized light has been passed through a filter which only allows light to pass through in the plane of the filter (see Figure 3.28). This is the principle upon which polarized sunglasses and car windscreens are based; the filter cuts the glare that arises from bright sunlight.

In 1815, Jean Baptiste Biot discovered that plane polarized light could be rotated by some naturally occurring compounds. Many years later, the action was defined: the enantiomers of a compound rotate plane polarized light by the same amount, but in opposite directions. One enantiomer rotates the light in a *clockwise* direction by X degrees, while the other enantiomer rotates the light *anticlockwise*, also by X degrees. An isomer that rotates plane polarized light in a clockwise direction was classified as dextrorotatory

FIGURE 3.27 The enantiomers of ibuprofen.

FIGURE 3.28 The origin of plane polarized light.

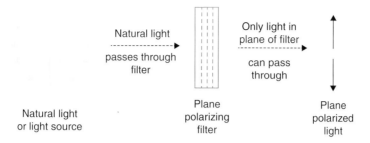

(d) and an isomer that rotates plane polarized light in an anticlockwise direction was labelled as laevorotatory (l); however, these lower case letters were often confused with the Fischer nomenclature of D and L, so dextrorotatory is now labelled as (+) and laevorotatory as (–) (see Figure 3.29).

Interaction with chiral molecules

The other property that distinguishes enantiomers is the way in which they interact with another chiral molecule; a chiral target molecule will often preferentially interact with one of a pair of enantiomers. There are many examples of this 'recognition' in our everyday life – perhaps not altogether surprising, given that

we have already seen that our hands are chiral. The left-handed readers are at this point nodding sagely – we know all about this! Those of us who are left-

FIGURE 3.29 Dextrorotatory and laevorotatory illustrated.

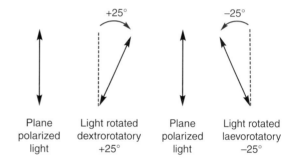

The observed rotation of a solution of a mixture of isomers of known specific rotation can be used to calculate the % optical purity of the major component. A worked example is provided on the support website, along with an example calculation using % concentration units (commonly used in pharmaceutical analysis).

(a) Using the information about specific rotation, calculate the specific rotation of chloramphenicol in ethanol at 27 °C, when a solution of 48.6 mg/mL gives an observed rotation of 0.904° using a cell of length 10 cm. (b) Calculate the observed rotation you would expect to see from a 50 mg/mL solution of (S)-naproxen in chloroform (cell length 1 dm).

Stereochemical nomenclature – make sure you know which is which!

Note that R and S correspond neither to D and L, nor to + or −. (Likewise, D and L do not correspond to + and −. It is important not to get these different assignment systems confused.) In the case of ibuprofen, we have (S)-(+)-ibuprofen and (R)-(−)-ibuprofen. The anti-inflammatory activity of the related molecule, naproxen, resides in the (S)-(+)-enantiomer, which has a specific rotation $[\alpha]_D$ of +66°, yet the $[\alpha]_D$ of (S)-naproxen sodium salt is −11°, showing that the exact structure of a molecule can have an acute effect upon its physical properties. Both naproxen and naproxen sodium are used clinically as analgesics and are marketed as Naprosyn® and Synflex®, respectively.

The solvent in which a compound is dissolved can also have a profound effect upon the specific rotation; the example of (−)-ephedrine can be found on the support website.

(S)-Naproxen
$[\alpha]_D = + 66° (CHCl_3)$

(S)-Naproxen, sodium salt
$[\alpha]_D = − 11° (H_2O)$

handed will, at some stage, have struggled with a pair of scissors, a can opener, a golf club, or a corkscrew, which have been designed for a right-handed person.

Feet, like hands, are chiral; your left foot does not superimpose on your right foot. Socks, however, are (normally) achiral and we do not have to check each morning that we are putting the right sock on the right foot. Our feet interact with socks in much the same way as a chiral drug interacts with an achiral molecule, such as a solvent.

Not so with our shoes! From being a child, we have been taught that only one of our shoes will fit on the left foot and the other shoe will only fit on the right foot –

indeed, getting it wrong is uncomfortable. Continuing the analogy, the chiral shoes recognize the chirality in our feet and each enantiomeric shoe interacts differently with our enantiomeric feet. This recognition feature is found extensively throughout nature, even down to the molecular level, where a chiral molecule, such as a receptor, recognizes the enantiomers of another chiral molecule, for example a drug, and usually selects (fits) one better than the other.

We experience selective recognition events daily, not only in the way biological molecules and pharmaceuticals interact in our cells, but also in the way external molecules affect us. For example, the olfactory and

FIGURE 3.30 The enantiomers of carvone and tryptophan.

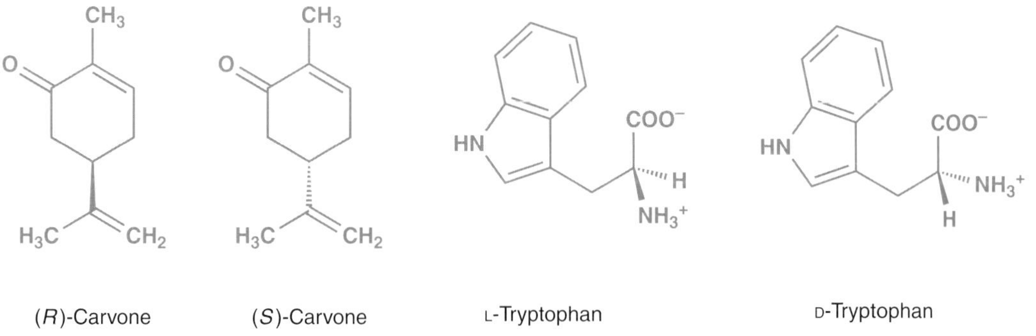

(R)-Carvone   (S)-Carvone   L-Tryptophan   D-Tryptophan

gustatory systems of smell and taste, respectively, rely on the way in which molecules interact with receptors in the nose and on the tongue. The enantiomers of some chiral molecules have an entirely different individual smell or taste, because each enantiomer binds to the chiral receptors in a different way. Two examples are carvone, a natural terpenoid, and tryptophan, a natural amino acid. (R)-Carvone smells (and tastes) of spearmint, while its (S)-enantiomer smells (and tastes) of caraway; L-tryptophan (the natural form found in proteins) has a bitter taste, but its unnatural enantiomer, D-tryptophan, tastes sweet, about 35 times sweeter than sucrose (sugar) (see Figure 3.30).

SELF CHECK 3.9

The chirality of tryptophan has been given, as usual for amino acids, using the Fischer nomenclature – can you correctly assign each structure in Figure 3.30 as R or S?

Stereochemical consequences of two or more chiral centres

We saw in Chapter 1 that many pharmaceuticals have several chiral carbon atoms and we now consider the effect of two or more chiral centres. We have already seen some examples of drugs with more than one chiral carbon in this chapter. In fact, chirality is hugely important in the worldwide multi-billion pound pharmaceutical market: more than 65% of newly approved drugs have at least one chiral centre, a large proportion of these have more than one and about 40% of the pharmaceutical market consists of single enantiomers (single enantiomer drugs are marketed

as the active enantiomer alone). In 2008, the three biggest sellers in the pharmaceutical world were all single enantiomer drugs: atorvastatin (used to lower blood cholesterol levels), esomeprazole (a 'proton pump inhibitor' used to treat gastric and duodenal ulcers), and clopidogrel (used in the prevention of atherothrombotic events).

The industrial scale synthesis of one single enantiomer of a chiral drug, especially for those with more than one chiral centre, involves complex and challenging chemistry, which is generally reflected in their higher cost (see also Integration Box 3.2).

Molecules with more than one chiral centre

For any molecule with n chiral carbon atoms, there are 2^n possible stereochemical isomers. We can see why when we consider an example. Chloramphenicol is a broad spectrum antibiotic, in which you should find two chiral carbon atoms: C1 and C2.

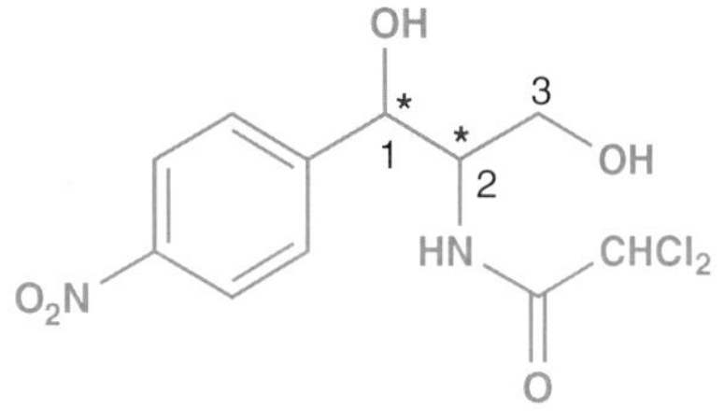

Chloramphenicol

If we draw out all of the possible stereoisomers, we find that there are four: (1R, 2R), (1S, 2S), (1S, 2R) and (1R, 2S), because each chiral carbon atom can take either R or S stereochemistry. The relationship between stereoisomers is illustrated in Figure 3.31:

Chiral switching

When a patent on a drug expires, there is no longer any legal protection, allowing anyone to make a generic form and sell it for the same indications. Chiral switching is a term used to describe the actions of the pharmaceutical industry when this happens. The original company re-patents the drug in its enantiomerically pure form, giving the owners another term of many years exclusivity. To re-register the active enantiomer, the company has to prove

that it has therapeutic advantages over the racemic form, which is sometimes more difficult than it sounds! In the case of ofloxacin, a fluoroquinolone antibacterial agent, the greater aqueous solubility of the (S)-enantiomer resulted in an increased antibacterial action of 8–125 times better than the racemate, depending upon the target bacterial species, and the added bonus of decreased toxicity, result-ing in the approval of a patent for levofloxacin (Tavanic®).

FIGURE 3.31 The four possible stereoisomers of chloramphenicol.

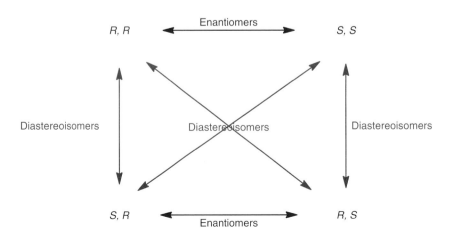

enantiomers are non-superimposable mirror images and diastereoisomers are non-superimposable, non-mirror image stereoisomers. The term 'diastereoisomer' is often written as 'diastereomer'.

The antibacterial activity resides in only *one* of these stereoisomers: the (1*R*, 2*R*) isomer, which tells us that the stereochemistry makes a big difference to the biological activity. We will look at this phenomenon in more detail when we look at properties of diastereoisomers. Some limitations to the use of chloramphenicol are outlined in Integration Box 3.3.

KEY POINT

For any molecule with *n* chiral carbon atoms, there are 2^n possible stereochemical isomers.

INTEGRATION BOX 3.3

Chloramphenicol

Chloramphenicol is a broad spectrum antibiotic with valuable activity against both Gram positive and Gram negative bacteria, used mostly for the topical treatment of bacterial infections, but used systemically sometimes to treat life threatening infections, such as meningitis caused by *Haemophilus influenzae*. It acts to inhibit bacterial protein synthesis, but is not entirely selective for the bacterial process. When used systemically, careful monitoring of blood levels is required to avoid adverse events, such as aplastic anaemia and 'grey baby syndrome', caused by its inhibition of mammalian mitochondrial protein synthesis, particularly in bone marrow cells, when used at high concentrations.

Molecules with two identical chiral carbon atoms

Consider the example of DMSA, **dim**ercapto**s**uccinic **a**cid, which has two chiral carbon atoms. Having read the last section, you would probably expect there to be four stereoisomers of DMSA and yet there are only three. To find out why, we need to draw out all of the possible structures (see Figure 3.32).

As before, there is a pair of enantiomers, the *R,R* and *S,S* pair, but the *R,S* and *S,R* pair are identical to each other. We can see why if we rotate the whole *S,R* stereoisomer through 180° as it becomes the mirror image *R,S* form, indicating that it is superimposable (see Figure 3.33); this is the same structure drawn in different ways. It is called the *meso-form*.

FIGURE 3.32 The three possible stereoisomers of dimercaptosuccinic acid (DMSA).

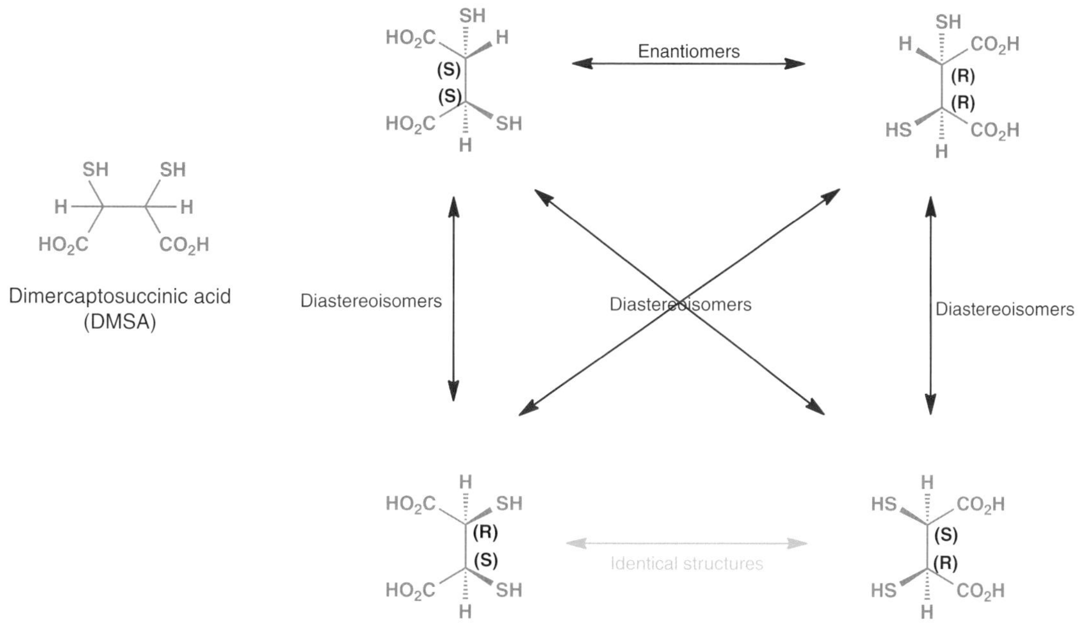

FIGURE 3.33 The *meso*-form of DMSA has a mirror plane through the middle of the structure.

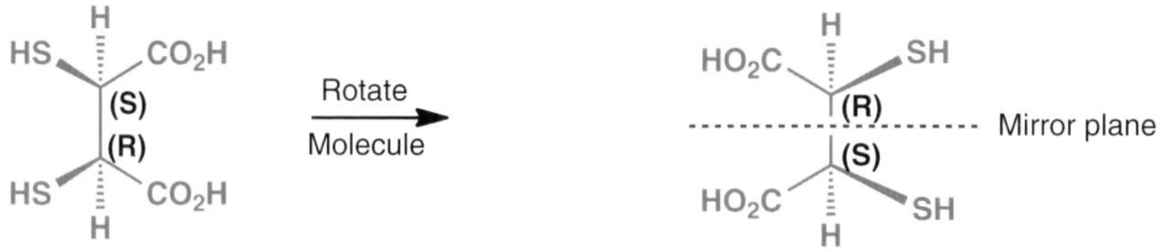

A *meso*-form results when a molecule with two or more chiral carbon atoms has identical substituents on the chiral carbon atoms and a mirror plane through the centre of the molecule.

Meso-DMSA can be used medically to reduce the toxic effects of ingested heavy metal ions, such as lead, mercury, arsenic and cadmium. The two thiol (SH) groups must be co-planar to complex with (chelate) the metal ion, illustrated for lead (Pb^{2+}). Lead poisoning leads to irreversible central nervous system (CNS) damage in children at a concentration of 20–25 mg Pb^{2+}/100 mL blood.

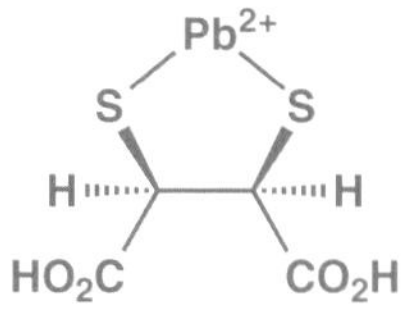

SELF CHECK 3.10

Study the structures of ephedrine and pseudoephedrine. They both have a long history of use in Traditional Chinese Medicine, sourced from the herb Ma Huang (*Ephedra sinica*), and are also components of the British National Formulary. Assign the stereochemistry of each chiral carbon and decide on the relationship of the two molecules: are they identical, enantiomers or diastereoisomers?

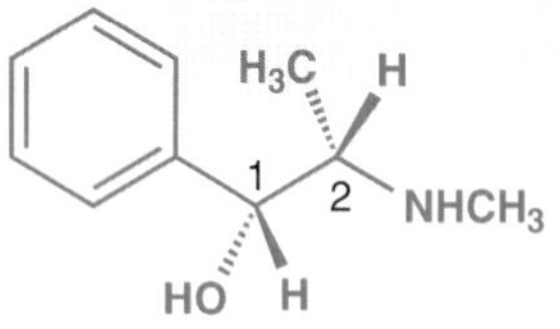

Ephedrine

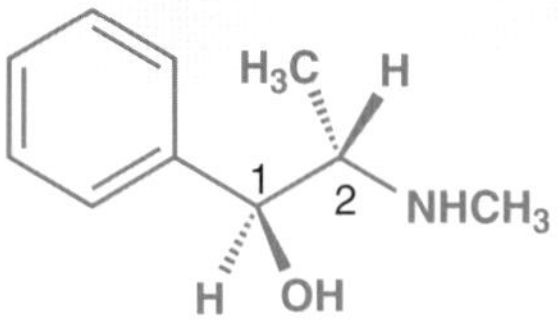

Pseudoephedrine

Properties of diastereoisomers

Pairs of enantiomers have identical physical and chemical properties, with the two exceptions noted before (interactions with plane polarized light and with other chiral species). However, diastereoisomers often have very different physical properties despite having similar chemical properties (owing to having the same functional groups).

In the 1920s, two compounds were isolated from the Chinese herb Ma Huang; you have just met these compounds in self-check 3.10 above. One, named ephedrine, is a potent bronchodilator. Pseudoephedrine, a nasal decongestant, was originally thought to be unrelated to ephedrine, because its physical and pharmacological properties are quite different (see Table 3.3), but it was later found to be a diastereoisomer. These compounds are related to amphetamine and are banned substances in sport.

The diastereoisomeric relationship of ephedrine to pseudoephedrine, with different physical properties, allows them to be separated fairly easily. Amines are often more easily crystallized as salts; in the 1920s, when the properties of these compounds were being studied, it was discovered that they could be separated by crystallization from water. The salt formed from the acid–base reaction of (–)-ephedrine with oxalic acid is poorly soluble in cold water, whereas (+)-pseudoephedrine oxalate is very soluble in cold water.

TABLE 3.3   Selected properties of the stereoisomers of ephedrine

Stereoisomer	Alternative name	$[\alpha]_D$ (25 °C) (free amine* in EtOH)	Melting point of free amine (°C)	Source; pharmacological activity
1R,2R-ephedrine	(–)-pseudoephedrine	−52	119	Not naturally occurring; inactive
1S,2S-ephedrine	(+)-pseudoephedrine	+52	119	Natural plant metabolite; sympathomimetic, nasal decongestant
Racemic pseudoephedrine	(±)-pseudoephedrine	0	118	Synthetic product; minimal activity
1R,2S-ephedrine	(–)-ephedrine	−6.3	35–40	Natural plant metabolite; anti-asthmatic, potent bronchodilator
1S,2R-ephedrine	(+)-ephedrine	+6.3	35–40	Not naturally occurring; inactive
Racemic ephedrine	(±)-ephedrine	0	76–78	Synthetic product; minimal activity

*The specific rotation values of the hydrochloride salts are different to those of the free amines; likewise, other salts of the amines, such as sulfate salts, also have different specific rotation values.

 You will find similar information about tartaric acid, along with self check questions, on the support website.

Separation of stereoisomers: the resolution of enantiomers

Enantiomers cannot be readily separated because of their identical physical properties, but they *can* be separated if converted into diastereoisomers with *different* physical properties. The process of separating enantiomers through the preparation of appropriate diastereoisomeric derivatives is called **resolution** and is a classical method of separation.

To convert an enantiomeric (R, S) pair into diastereoisomers, we can react them with a *single enantiomer* of another chiral molecule (e.g. an S enantiomer) to form a diastereoisomeric pair – the (R, S)-diastereoisomer and the (S, S)-diastereoisomer. We now have diastereoisomers with different physical properties, allowing their separation. For example, to resolve the enantiomers of ibuprofen, which has a carboxylic acid group, the racemic mixture can be reacted with one enantiomer of a chiral base, such as (S)-phenylethylamine to form the diastereoisomeric salts, the (R)-acid:(S)-base and the (S)-acid:(S)-base. The (S, S)-diastereoisomer is poorly soluble in aqueous solution and can be separated by filtration from the mixture, leaving the relatively pure (R, S)-diastereoisomer in solution (see Figure 3.34).

Once the diastereoisomers are in separate vessels, they can be acidified to release the ibuprofen carboxylic acid enantiomers as poorly soluble solids. The chiral base can also be recovered. The chiral molecule used to prepare a diastereoisomeric pair for resolution often comes from a natural source, and there are quite a number to choose from, including sugars and amino acids. These molecules make up a group often referred to as the 'chiral pool' – easily accessible chiral molecules of high enantiomeric purity.

KEY POINT

Drugs which can exist as optical isomers are increasingly being required to be tested and supplied as single enantiomers. If the synthesis of the drug produces a racemic mixture, as is often the case, the enantiomers can be separated (resolved) by formation of diastereoisomers; however, chiral synthesis resulting in the pure active isomer is preferred over wasting 50% of the material.

FIGURE 3.34 Separation of ibuprofen enantiomers through the formation of diastereoisomeric salts.

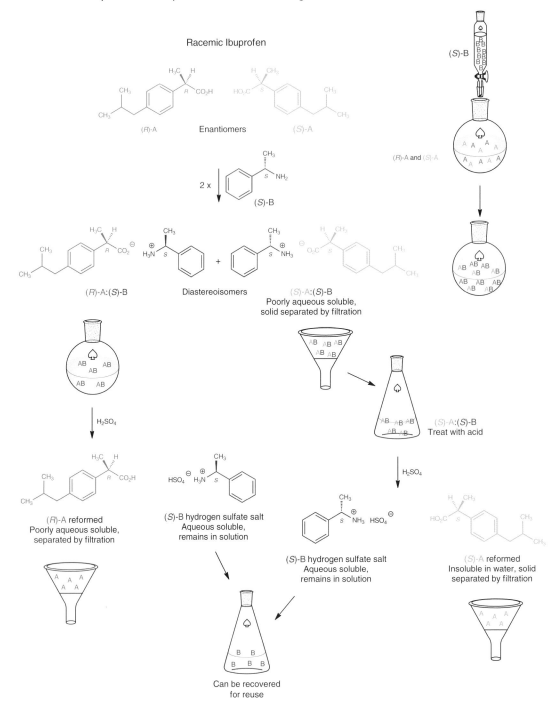

Racemic Ibuprofen

(R)-A   Enantiomers   (S)-A

(S)-B

(R)-A and (S)-A

2 x

(S)-B

(R)-A:(S)-B   Diastereoisomers   (S)-A:(S)-B
Poorly aqueous soluble,
solid separated by filtration

H₂SO₄

(S)-A:(S)-B
Treat with acid

H₂SO₄

(R)-A reformed   (S)-B hydrogen sulfate salt   (S)-B hydrogen sulfate salt   (S)-A reformed
Poorly aqueous soluble,   Aqueous soluble,   Aqueous soluble,   Insoluble in water, solid
separated by filtration   remains in solution   remains in solution   separated by filtration

Can be recovered
for reuse

Stereochemistry, warfarin and pharmacogenetics

Warfarin is widely used as an anticoagulant – to prevent a patient's blood from clotting too easily. Without this treatment, some patients are at risk of thrombosis (a blood clot). If a thrombosis forms in a heart blood vessel, it could cause a heart attack.

Warfarin

Warfarin acts by inhibiting vitamin K epoxide reductase cofactor subunit 1 (VKORC1), an essential enzyme in blood clotting. Unusually, the enantiomers of warfarin have similar activity on the target enzyme (VKORC1). Their metabolism is, however, very different. The (R)-enantiomer is metabolised by at least 3 different cytochrome P_{450} (CYP) enzymes and is quickly removed from circulation and excreted. The (S)-enantiomer, by contrast, is metabolised by only one enzyme, called CYP 2C9; (S)-warfarin is mostly responsible for the anticoagulant effect.

Approximately 10% of people have a small change in the gene encoding CYP 2C9 and produce an almost inactive form of the enzyme. Another small group have extra copies of the CYP 2C9 gene and produce more CYP 2C9 than usual.

Patient variability in response to a drug caused by genetic differences is called pharmacogenetics.

Because of these pharmacogenetic variations, the dose of warfarin has to be identified for each patient individually and varies from 3 to 9 mg per day. You may hear pharmacists and doctors talking about the INR – international normalised ratio – of a patient taking warfarin. The INR is used to determine how quickly a patient's blood coagulates (or clots). Generally, for patients taking warfarin, an INR range of 2.5 – 3.5 is recommended.

REFLECTION QUESTIONS

1. Identify the chiral carbon in warfarin. Draw each stereoisomer and assign each structure as R or S stereochemistry.

2. Why do you think the enantiomers are metabolised by different enzymes?

3. What will be the effect on a patient receiving warfarin if they produce an inactive form of CYP 2C9, or if they produce more CYP 2C9 than usual?

Answers

1. There is one chiral carbon atom in warfarin – it is marked with a red star in the first structure. The priority of the groups are shown below; remember to reverse the assignment when the priority 4 group (H, in this case) is at the front on the solid wedge bond.

Warfarin

(R)-Warfarin

(S)-Warfarin

2. Substrates for an enzyme tend to 'fit' neatly into the active site. Non-bonding interactions – such as hydrogen bonds and lipophilic interactions – 'hold' the substrate within the active site of the enzyme. The 3-dimensional structures of the two enantiomers are different and they fit preferentially into the active sites of different enzymes.

3. If a patient produces more CYP 2C9 than usual, the (S)-warfarin enantiomer is metabolised quickly and they have less (S)-warfarin in the blood to exert an anticoagulant effect than someone who metabolises it slowly. On the other hand, if a patient produces an inactive form of CYP 2C9, the (S)-warfarin is metabolised more slowly. If the dose of warfarin is not reduced to account for this, the patient could have a haemorrhage. This is why monitoring the effects of warfarin by checking a patient's internationalised normalised ratio (INR) is so very important.

3.5 **Protein folding diseases**

So far, we have explained how the shape of molecules is very important, especially in pharmacy, and we have used common drugs, including tamoxifen, morphine and donepezil, to illustrate this point. Molecular shape also has an impact upon the function of large macromolecules found in our bodies and, to help illustrate our point, we will use proteins as an example.

Proteins are made up of amino acids, linked together by peptide (or amide) bonds. In our bodies, chains of amino acids are folded into three-dimensional shapes to form proteins; this 3D-shape is essential to allow a protein to exert a specific function. In some cases, however, proteins can 'mis-fold', which changes their 3D-shape and ultimately leads to a change, or even a loss, of their function. The mis-folding of proteins can, therefore, lead to the development of disease. Examples of protein folding diseases include bovine spongiform encephalopathy (or 'Mad Cow' disease), Creutzfeldt-Jacob disease (the human equivalent of Mad Cow disease), Alzheimer's disease, cystic fibrosis, and even some cancers.

CHAPTER SUMMARY

New chiral drugs are now often marketed as the single active enantiomer, even when they are made synthetically (chiral natural products usually exist as single enantiomers); however, there are still many examples of new racemic drugs. In some cases, the inactive forms cause no major adverse effects and, in others, the inactive form can cause serious adverse effects or can even augment the action of the therapeutically active enantiomer. The FDA and MHRA both require evidence that a mixture of stereoisomers has no adverse effects and, through these more careful considerations of stereochemistry and its influences upon interactions with human biochemistry, we can feel more confident that another thalidomide-type disaster will not occur.

As the 'expert on medicines', pharmacists have a crucial role to play in understanding how stereochemistry affects drug action, identifying potential problems, responding to changes, and supporting other healthcare professionals in executing their roles. Medicinal chemists contribute to the design and synthesis of chiral drugs and pharmacologists evaluate and rationalize their biological actions; together the pharmaceutical science team ensure that each patient is treated by the safest and most effective medicine. We hope this chapter has helped you to achieve a sound understanding, not only of stereochemistry, but of all forms of isomerism, which you can build upon as you progress through your studies.

FURTHER READING

Cahn, R.S., Ingold, C.K., and Prelog, V. Specification of molecular chirality. Angewandte Chemie Int. Ed. Engl. 1966; 5: 385–415.

For more about the involvement of stereochemistry in drug design:

Nogrady, T. and Weaver, D.F. *Medicinal chemistry: A molecular and biochemical approach.* 3rd edn. Oxford University Press, 2005.

Properties of aliphatic hydrocarbons

ANDREW J. HALL

In Chapter 2, you learnt about the ways in which atoms are bonded together and the remarkable ability of carbon to form bonds with other carbon atoms. This means that molecules can be produced that have short or long chains, or more complicated arrangements of carbons atoms. Hydrocarbons are compounds made solely of carbon and hydrogen atoms, in which each carbon forms four covalent bonds.

Hydrocarbons are extremely important in our everyday lives. They are a source of energy to heat and light our homes, and power our vehicles, and they provide us with many of the plastic items we use. Hydrocarbons are also abundant in nature – for example, as flavours or fragrances produced by plants, as insect pheromones, and as natural rubber and lipids.

In this chapter, we will explore the properties of the **aliphatic** hydrocarbons: the **alkanes**, **alkenes** and **alkynes**. In Chapter 2, you saw that the types of bonds formed between atoms affect the shape of the molecules formed. Further, the differences in strength between the first (σ) bond and second/third (π) carbon-carbon bonds enable us to anticipate the different reactivity of the different classes of aliphatic hydrocarbon.

As we proceed, you will learn how to name aliphatic hydrocarbons and discover their physical properties. You will also learn where they come from, how they are made and how they react. You should come to realize the importance of aliphatic hydrocarbons in pharmaceutical and biological systems.

The properties of the aromatic hydrocarbons are altogether another story and will be covered in Chapter 7.

Learning objectives

Having read this chapter you are expected to be able to:

➤ name any given hydrocarbon following the rules that chemists use

➤ relate the physical properties of aliphatic hydrocarbons to their behaviour

➤ give examples of the reactions of aliphatic hydrocarbons

➤ give examples of the importance of aliphatic hydrocarbons in pharmacy.

4.1 Nomenclature

Before we go any further, we need to sort out how we name organic compounds. This is a little like learning a foreign language and is absolutely necessary for you to be able to navigate your way through the world of organic chemistry.

Alkanes

The best place to start is with the alkanes, as their names form the basis for naming the majority of organic molecules. These molecules contain *only* **single bonds** between the carbons and hydrogens.

Straight-chain alkanes

In **straight-chain alkanes**, the carbons form a continuous, non-branched chain. The first four members of the series are methane, ethane, propane and butane (see Figure 4.1).

You should notice that the alkanes have the general formula C_nH_{2n+2}, where n is a whole number and that, as the series continues, the next member differs from the previous one by a $-CH_2-$ unit. Such a family of molecules is called a **homologous series**. Beyond butane, Latin or Greek prefixes are used for the number of carbons in the chain, followed by the 'ane' suffix. The series continues with pentane (C_5), hexane (C_6), heptane (C_7), octane (C_8), nonane (C_9), decane (C_{10}), undecane (C_{11}), dodecane (C_{12}), etc.

SELF CHECK 4.1

Complete Table 4.1 by drawing 'stick' structures (see Chapters 1 and 2) for ethane, propane and butane. Can you see why we do not normally draw a stick structure for methane?

TABLE 4.1 Structural representations of alkanes

Alkane	Structure	Stick Structure
Ethane	H_3C-CH_3	
Propane	 $H_3C\diagup\diagdown CH_3$	
Butane	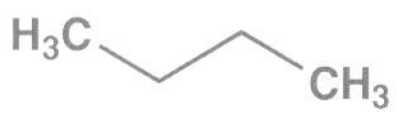 $H_3C\diagdown\diagup\diagdown CH_3$	
Pentane	 $H_3C\diagup\diagdown\diagup\diagdown CH_3$	

Branched-chain alkanes

Alkanes with more than three carbons can have branched chains. A hydrogen atom from a carbon in the middle of the chain is replaced by an **alkyl** group. For example, two possible structures fit the formula C_4H_{10}, the linear butane molecule and its branched isomer, isobutane (2-methylpropane). As the number of carbon atoms increases, so does the number of isomers. While there are just three isomers of pentane C_5H_{12}, (see Figure 4.2), there are 75 isomers of decane, $C_{10}H_{22}$, and more than 300,000 possible isomers of eicosane, $C_{20}H_{42}$!

SELF CHECK 4.2

Draw structures for decane and eicosane (just the straight chain forms).

In Figure 4.2, one of the isomers of pentane has also been given its common name (isopentane), but as we move to larger molecules with differing numbers of branches or, as you will see later, different functional groups, then we need a more systematic naming system to avoid confusion. The rules of this system are laid out below.

FIGURE 4.1 Straight-chain alkanes (note that these are drawn as projections for clarity).

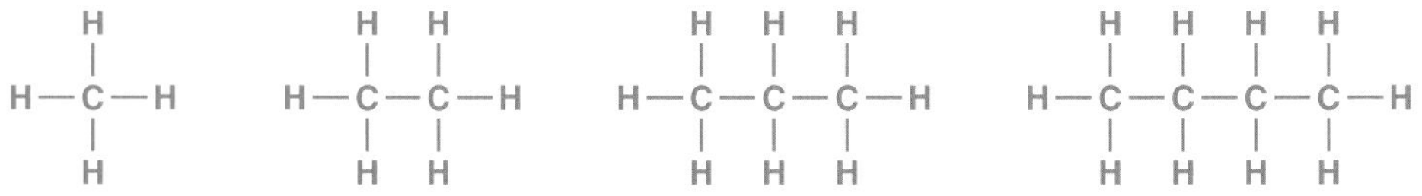

Methane CH_4 Ethane C_2H_6 Propane C_3H_8 Butane C_4H_{10}

FIGURE 4.2 Isomers of pentane.

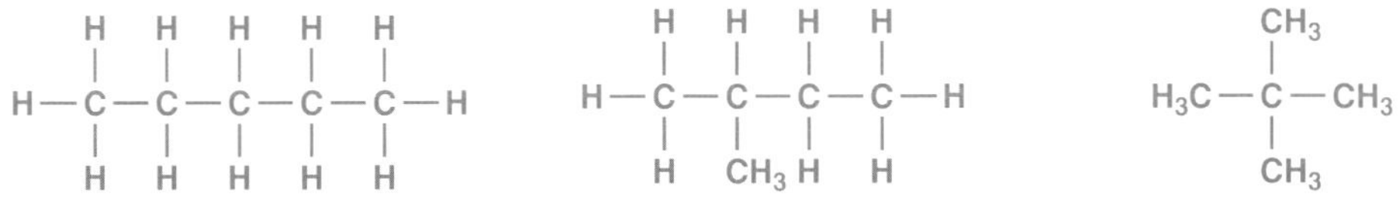

Pentane C_5H_{12} 　　　　2-Methylbutane (isopentane) 　　　　2,2-Dimethylpropane C_5H_{12}
　　　　　　　　　　　　　　　　　C_5H_{12}

1. Identify the longest continuous carbon chain.

2. Identify the branching points on that chain and give the groups attached at these branching points names corresponding to the number of carbons they contain: methyl (1 carbon), ethyl (2 carbons), propyl (3 carbons) and so on. Note that these are called **alkyl groups**.

3. Using the lowest numbers possible, number the carbon atoms on the longest chain to describe the positions of the branching groups.

4. Write the branches in alphabetical order.

5. If you find more than one branch with the same name, use the prefixes di-, tri-, tetra-, etc.

The examples below should help you to visualize these rules.

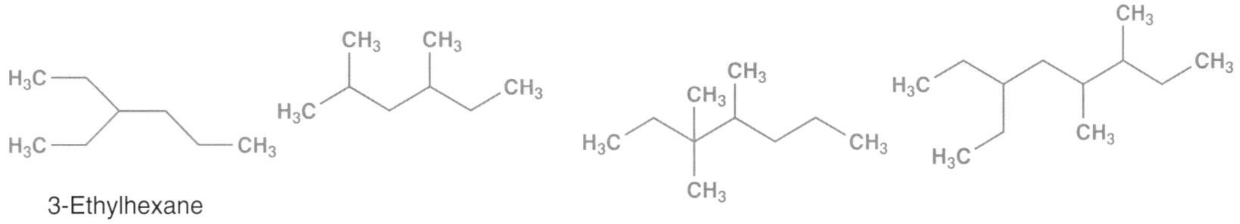

3-Ethylhexane

2,4-Dimethylhexane

3,3,4-Trimethylheptane

6-Ethyl-3,4-dimethyloctane

SELF CHECK 4.3

Draw structures for 2-methylhexane, 3-ethyl-4-methyl-hexane, 3-ethyl-2,5-dimethylheptane and 3,5-diethyl-2,8-dimethyldecane, then name the compounds shown below.

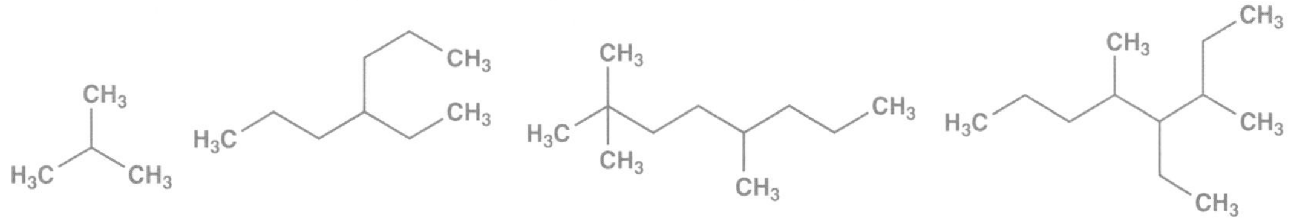

Cycloalkanes

Not all alkanes contain a straight or branched chain of carbon atoms. **Cycloalkanes** have their carbon atoms arranged in a ring. As a result, a cycloalkane has two fewer hydrogen atoms compared with the acyclic alkane with the same number of carbons, leading to a general formula of C_nH_{2n} for cycloalkanes. Some common cycloalkanes are shown in Figure 4.3 in skeletal form, which is how they are usually drawn.

These molecules are named simply by inserting the prefix **cyclo** before the name describing the number of carbons in the ring. The rules for naming cycloalkanes follow those used for their acyclic counterparts.

1. For cycloalkanes with an alkyl substituent, the ring is the parent hydrocarbon.

2. If the ring has more than one substituent, write them in alphabetical order and give position number 1 to the first substituent.

FIGURE 4.3 Common cycloalkanes.

Cyclopentane　　　　Cyclohexane　　　　Cycloheptane　　　　Cyclooctane

The examples below should help you to visualize these rules.

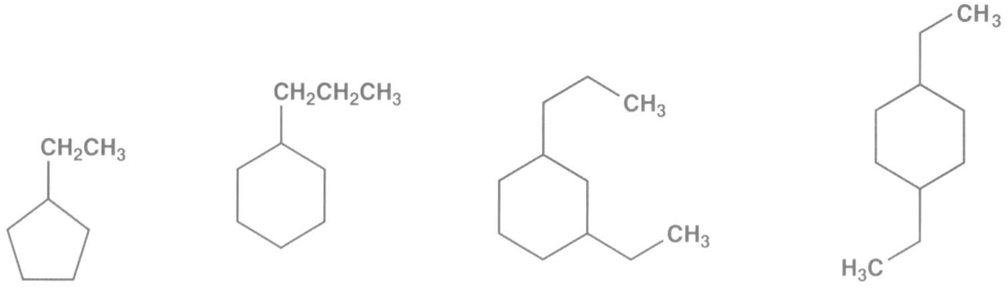

CH₂CH₃

Ethylcyclopentane　　Propylcyclohexane　　1-Ethyl-3-propylcyclohexane　　1,4-Diethylcyclohexane

SELF CHECK 4.4

Draw skeletal structures for butylcyclobutane, 1,4-diethyl-cyclooctane and 1-ethyl-3-methyl-5-propylcyclohexane, then name the compounds shown below.

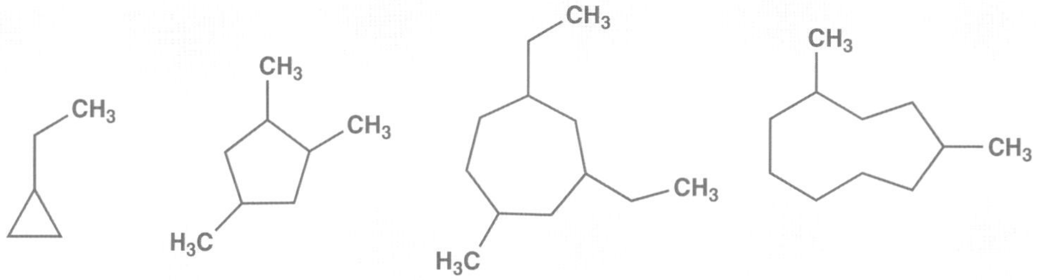

Alkenes

Alkenes, with the general formula C_nH_n contain carbon–carbon double bonds (see Figure 4.4).

You get the systematic name for an alkene by replacing the 'ane' suffix at the end of the parent hydrocarbon's name with the suffix 'ene'. After this, we use the following rules:

1. Determine the longest continuous chain of carbon atoms **containing the double bond**.

2. Number the carbon atoms in the chain from the end nearest the double bond.

3. Pick the carbon with the lowest number to describe the position of the double bond.

4. If the chain has two double bonds, use the suffix 'diene'.

5. Place substituent names in front of the name of the longest continuous carbon chain.

6. List multiple substituents alphabetically, as before.

FIGURE 4.4 Common alkenes.

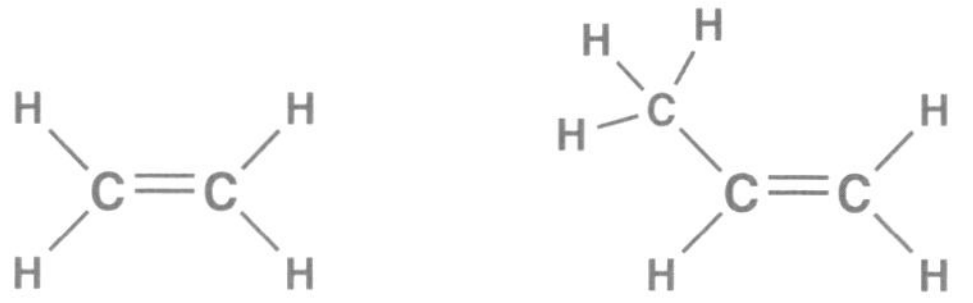

Ethene　　　　　　　　Propene

The examples below should help you to visualize these rules.

But-1-ene   But-2-ene (*E* isomer)   Hexa-2,4-diene (*E*, *E*-isomer)   3-Methylpent-2-ene (*E*-isomer)

3, 5-Dimethyloct-4-ene (*E*-isomer)

(We will deal with the term *E*-isomer below.)

Cycloalkenes

For cycloalkenes, which have the general formula C_nH_{2n-2} there are two more rules:

1. The carbons in the double bond always take numbers 1 and 2, so such a molecule without substituents bears no numbers. The position of substituents takes the lowest number counting around the ring, bearing this in mind.

2. If the ring contains more than one double bond, numbers are used to define their positions.

The examples below should help you to visualize these rules.

Cyclopentene   Cyclohexene

2-Methylcyclopenta-1,3-diene

3-Ethylcyclohexa-1,4-diene

Isomerism in alkenes

If we have an alkene with the structural formula, C_4H_8, there are three possible structural arrangements. However, there are in fact four isomers corresponding to this molecular formula, as there are two isomers of but-2-ene: the methyl groups can be on the same side of the planar double bond or on opposite sides of it (see Figure 4.5). These two arrangements are termed *cis* ('on the side') and *trans* ('across').

As there is no free rotation around the carbon–carbon double bond, the two molecules have atoms oriented differently in space and are, therefore, physically and chemically different. They are **geometric isomers**. If one of the carbons in the double bond has two identical substituents, then there is only one possible structure for the compound and so there are no *cis* and *trans*-isomers (see Figure 4.5).

Interconversion between *cis* and *trans*-isomers is possible, but only if the molecule absorbs enough energy, in the form of heat or light, to cause the π-bond to break (see Figure 4.6). Once this bond has broken, free rotation around the remaining σ-bond can occur.

One important example of this sort of interconversion is the light-catalysed conversion of *cis*-retinal to *trans*-retinal, which is the central chemistry required in vision (See Figure 4.7). This is discussed in more detail in Chapter 1.

If each of the carbon atoms in the double bond has only one substituent, then we can use the *cis/trans*-nomenclature. However, this falls down when the number of substituents increases, e.g. 1-bromo-2-chloropropene.

FIGURE 4.5 Isomerism in alkenes.

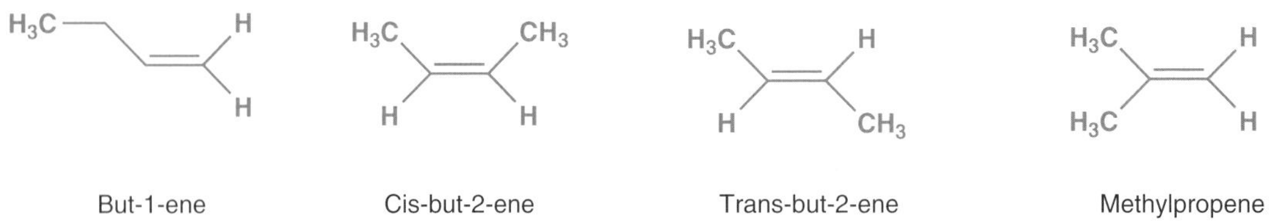

But-1-ene   Cis-but-2-ene   Trans-but-2-ene   Methylpropene

FIGURE 4.6 Interconversion between *cis* and *trans*-isomers.

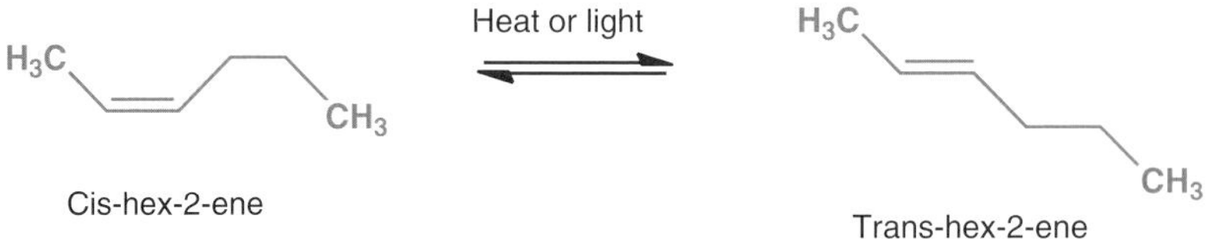

Heat or light

Cis-hex-2-ene

Trans-hex-2-ene

We need a new naming system to differentiate these two molecules – it is called the *E,Z* **system** (see Figure 4.8).

This system is described in detail in Chapter 3, but it is sufficiently important that we will briefly revisit it here. We first assign priorities (following the Cahn-Ingold-Prelog convention) to the substituents on the carbons of the double bond. If the higher priority groups are on the **same side**, then we have the *Z*-isomer (from the German *zusammen* meaning 'together'). If the higher priority groups are on **opposite sides**, we have the *E*-isomer (from the German *entgegen* mean-ing 'opposite'). Note that this nomenclature can be used for ALL alkenes, not just those with three or four substituents.

 A more detailed discussion of the Cahn-Ingold-Prelog convention is given in Chapter 3 'Stereochemistry and drug action'

Priorities are determined by the atomic numbers of the atoms bonded directly to the carbons of the double bond. The greater the atomic number, the higher the

FIGURE 4.7 The conversion of *cis*-retinal imine to *trans*-retinal imine in vision.

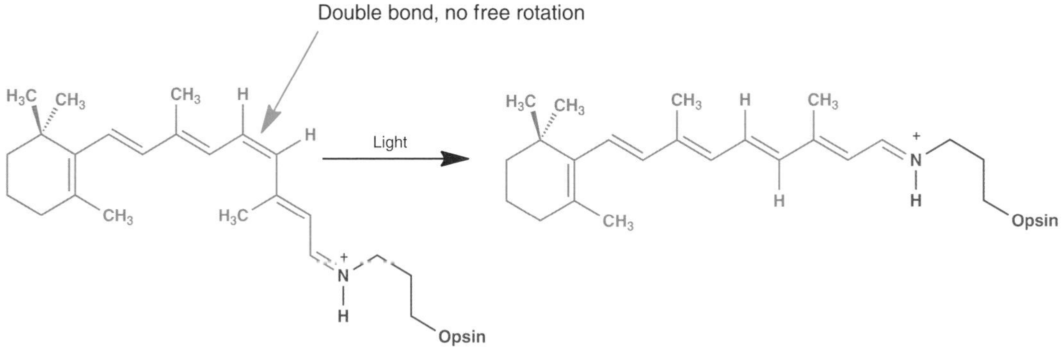

Double bond, no free rotation

Light

Opsin

Opsin

FIGURE 4.8 The E,Z system for naming molecules.

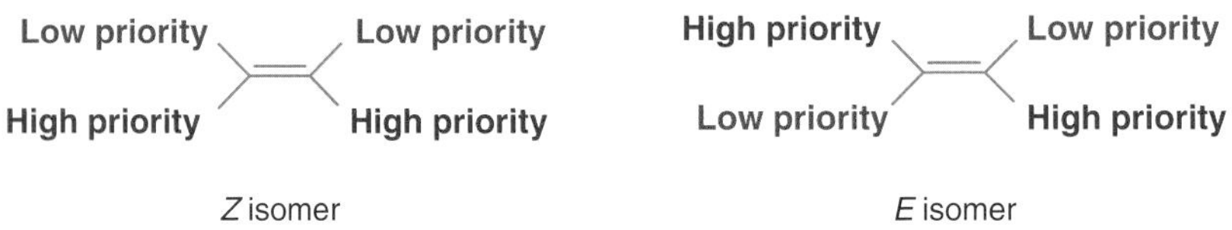

Low priority   Low priority
High priority   High priority

Z isomer

High priority   Low priority
Low priority   High priority

E isomer

FIGURE 4.9 Examples of *E, Z* nomenclature.

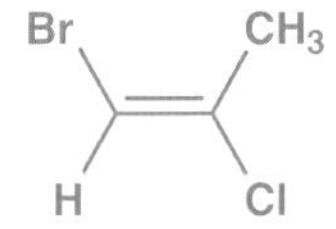

(*E*)-1-bromo-2-chloropropene

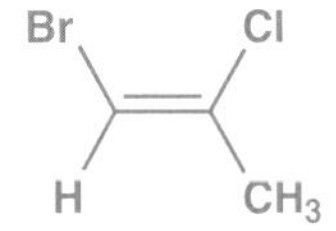

(*Z*)-1-bromo-2-chloropropene

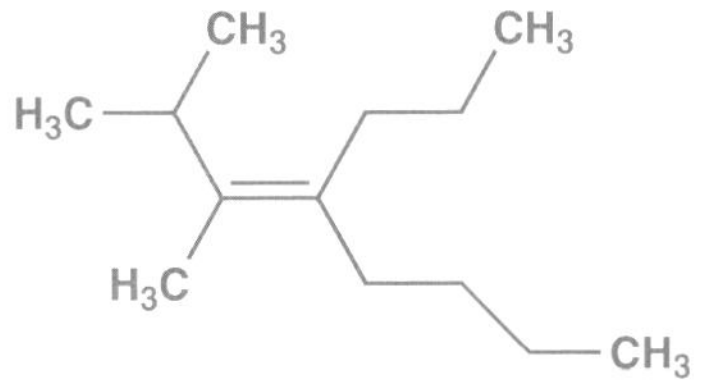

(*E*)-2,3-dimethyl-4-propyloct-3-ene

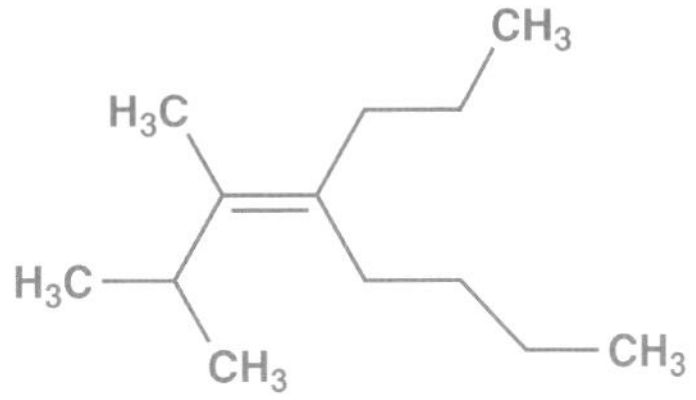

(*Z*)-2,3-dimethyl-4-propyloct-3-ene

priority. If two groups start with the same atom, we move to the next atom attached to the 'tied' atoms. The final 'rule' in this system concerns the case where the atom attached to a carbon of the double bond is doubly bonded to another atom. The priority system treats it as if it were bonded to two of these atoms (see Figure 4.9).

SELF CHECK 4.5

Draw structures for the *E*- and *Z*- isomers of 1-chloro-3-ethylhept-3-ene and 2-bromopent-2-ene, then identify the isomers below.

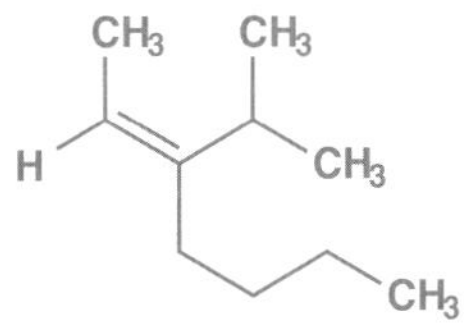

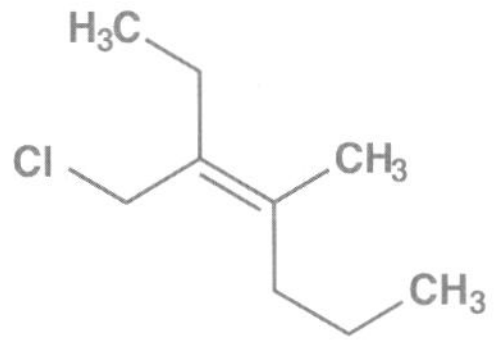

 A more detailed guide to alkene stereochemistry and nomenclature is given in Chapter 3 'Stereochemistry and drug action'

KEY POINT

Alkenes can exhibit geometrical isomerism. Geometric isomers have different physical, chemical and biological properties.

Alkynes

To obtain the systematic name of an alkyne, we take the alkane name and substitute the 'ane' ending with 'yne'. Again, the longest continuous chain containing the carbon–carbon triple bond is numbered to give the alkyne functional group the lowest possible number.

The triple bond can be within the chain (**internal** alkynes) or at the end of the chain (**terminal** alkynes). In some circumstances, counting along the chain from either direction can lead to the alkyne having the same number. In such a case, the correct systematic name is the one where the substituents carry the lowest numbers. See the examples 1-bromo-5-methylhex-3-yne and 2-bromo-3-chlorooct-4-yne.

Ethyne (acetylene)

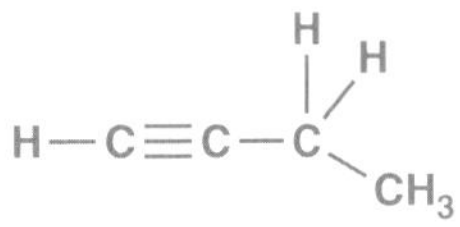

But-1-yne

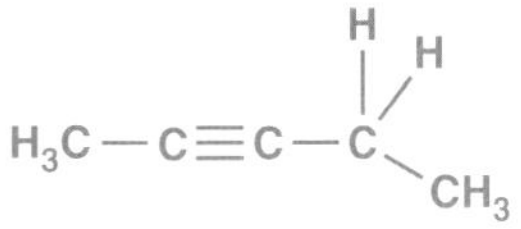

Pent-2-yne

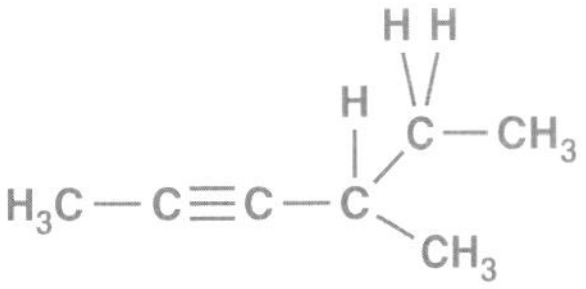

4-Methylhex-2-yne

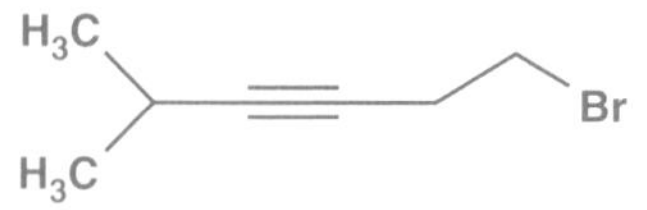

1-Bromo-5-methylhex-3-yne

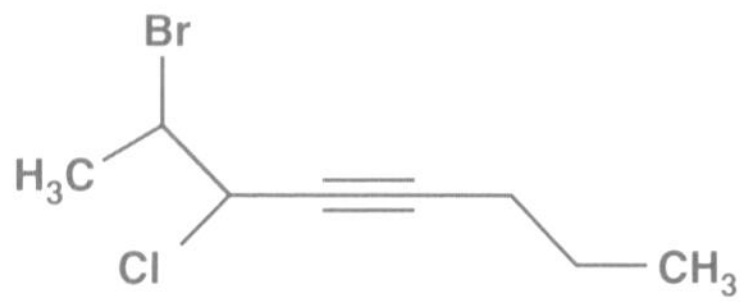

2-Bromo-3-chlorooct-4-yne

SELF CHECK 4.6

Draw structures for 4-methoxyhex-2-yne and but-3-yn-2-ol, then name the structures below.

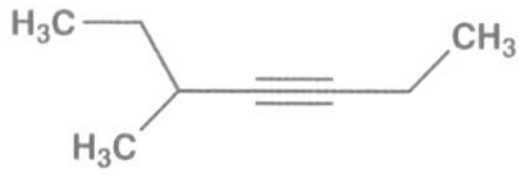

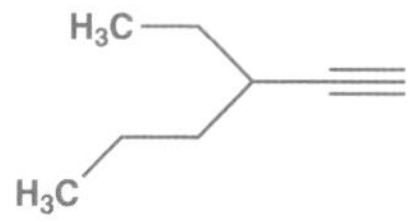

KEY POINT

Alkanes, alkenes, and alkynes can exist as straight-chain, branched-chain, or cyclic structures.

4.2 Physical properties of aliphatic hydrocarbons

In this section, we will examine the properties that are responsible for the physical characteristics of the aliphatic hydrocarbons. These properties include their intermolecular forces, their polarity (or rather lack of it) and, perhaps surprisingly, their acidity.

Boiling and melting points

The small difference between the electronegativities of carbon and hydrogen means that the bond between them is only very weakly polar – and the aliphatic hydrocarbons themselves are non-polar. The only intermolecular forces holding one molecule in contact with another are weak van der Waals' forces. This helps to explain why smaller alkanes, alkenes and alkynes are very volatile; it takes little energy to overcome the interactions between molecules.

As the molecular weight increases, so does the size of the molecules. This leads to greater contact between molecules and a consequent increase in the van der Waals' forces between them. This is why the boiling points increase as each series progresses, as shown in Table 4.2.

Chain branching causes a decrease in the area of contact. So, if two alkanes have the same molecular weight, the more highly branched one will have the lower boiling point. Alkynes have stronger van der Waals' interactions than alkenes and, consequently, higher boiling points, because of their more linear structures.

The melting points of aliphatic hydrocarbons also increase with size, but in a less regular manner. Under standard conditions, alkanes with 1–4 carbons are gases, from 5–17 carbons are liquids and from 18 carbons onwards are waxes and solids.

Solubility and density

As the hydrocarbons are non-polar, they tend to be very insoluble in water and other very polar solvents. Instead, they prefer to dissolve in non-polar solvents, such as benzene and diethyl ether. Thus, hydrocarbons can be described as hydrophobic (literally 'water-hating') or lipophilic (literally, 'fat loving'). The lipophilicity of hydrocarbon groups is especially important when considering how drug molecules get distributed in the human body after administration; it also impacts on their mode of action, as we see in Integration Box 4.1.

The hydrocarbons are less dense than water, meaning that they float on the surface of water, as seen to disastrous effect during the various oil spills that have occurred around the world.

TABLE 4.2 Physical properties of selected aliphatic hydrocarbons

	Number of carbons	Name	Boiling point (°C)	Melting point (°C)	Density at 25 °C (g/mL)
Alkanes	1	methane	−167.7	−182.5	–
	2	ethane	−88.6	−183.3	–
	3	propane	−42.1	−187.7	–
	4	butane	−0.5	−138.3	0.579 (20 °C)
	5	pentane	36.1	−129.8	0.626
	6	hexane	68.7	−95.3	0.659
Cycloalkanes	3	cyclopropane	−33	−128	–
	4	cyclobutane	12.5	−91	–
	5	cyclopentane	49	−94	0.751
	6	cyclohexane	80.7	4–7	0.779
	7	cycloheptane	118.5	−12	0.811
	8	cyclooctane	151	10–13	0.834
Alkenes	2	ethene	−103.7	−169	–
	3	propene	−47.6	−185	–
	4	but-1-ene	−6.1	−138	–
	4	*cis*-but-2-ene	3.7	−139	–
	4	*trans*-but-2-ene	0.9	−105	–
	5	pent-1-ene	30.2	−165	0.641
	5	*cis*-pent-2-ene	36	−180	0.65
	5	*trans*-pent-2-ene	37	−135	0.649
Alkynes	2	ethyne	−84	–	–
	3	propyne	−23.2	−102.7	–
	4	but-1-yne	8.1	−125.7	–
	4	but-2-yne	27.0	−32	0.691
	5	pent-1-yne	40.2	−105	0.691
	5	pent-2-yne	57	−109	0.71

INTEGRATION BOX 4.1

Polyene macrolide antibiotics

Polyene macrolide antibiotics play a major role in the treatment of systematic and topical fungal infections. Their lipophilic structure enables them to insert into fungal cell membranes, making them 'leaky'. This leads to the loss of small molecules from the cells, which eventually die.

(Continued)

Nystatin

Amphotericin B

Nystatin

Phospholipid

Typical examples are nystatin, used to treat topical *Candida* infections, and Amphotericin B, which is also used to treat systemic infections. The latter drug suffers from toxicity issues, such as lowering of blood pressure and even kidney damage, but it remains the drug of choice for the treatment of life threatening fungal infections.

Acidity of aliphatic hydrocarbons

Of all the species you will encounter in organic chemistry, alkanes are without doubt the weakest **Brønsted acids**. In fact, they are so weakly acidic that it is difficult to measure just how acidic they are! In general, their pK_a values are very high. For example, the pK_a of ethane is greater than 50 (see Integration Box 4.2).

INTEGRATION BOX 4.2

What is pK_a?

You are probably rather surprised to find out that pK_a can be defined for a hydrocarbon or that a proton can dissociate from a C–H bond! Let's take a step back to define what pK_a actually represents. The general reaction of an acid with water is:

$$HA + H_2O \rightleftharpoons H_3O^+ + A^- \quad K_a = \frac{[H_3O^+][A^-]}{[HA]} \quad pK_a = -\log_{10} K_a$$

The acid dissociation constant, K_a, indicates the relative strength of the acid. If K_a is large, it means that the reaction tends to the right hand side (the forward reaction is favoured), indicating a strong acid. Given the relationship of pK_a to K_a, this means that a strong acid will have a low pK_a value.

Let's take methane as an example of a typical hydrocarbon. The dissociation reaction would be:

$$CH_4 + H_2O \rightleftharpoons H_3O^+ + H_3C^-$$

The forward reaction is not very favourable as the methyl anion, H_3C^-, is very reactive. Consequently, the back reaction (which favours the reactants) happens to a greater extent. This means that K_a will be very small and, consequently, the pK_a will be very large, which is indeed the case.

This will be discussed in more detail in Chapter 6 'Acids and Bases' of the *Pharmaceutics* book in this series

Moving to alkenes, there is an increase in the s-character of the carbon–hydrogen bond. (Remember that the carbon of the double bond is sp^2 hybridized (33% 's') rather than sp^3 hybridized (25% 's') as in alkanes.) This increased s-character leads to an increase in acidity. For example, the pK_a of ethene is estimated to be 44, meaning that it is at least six orders of magnitude more acidic than ethane.

$$H_3C — CH_3 \qquad H_2C = CH_2$$

Ethane   Ethene (ethylene)

pKa > 50   pKa = 44

$$HC \equiv CH$$

Ethyne (acetylene)

pKa = 25

In terminal alkynes, the s-character of the carbon–hydrogen bond is further increased (to 50%), leading to another increase in acidity. The pK_a of acetylene (ethyne) is 25. This means that it is 10^{19} times stronger as an acid than ethene. The relative acidity of terminal alkynes is useful in the synthesis of higher alkynes, as you will see in Section 4.5 'Alkynes – preparation and reactions'.

A word of caution: although we refer to the terminal alkyne proton as acidic, this is only a relative term. While terminal alkynes are more acidic than other hydrocarbons, they are still very weak acids. To put this into perspective, compare the pK_a value of ethyne to that of water (15.7), itself a very weak acid, and to acetic (ethanoic) acid (4.7), and hydrochloric acid (–7)!

4.3 Alkanes – preparation and chemical properties

Alkanes are saturated hydrocarbons and are generally quite unreactive. Combustion of alkanes is, however, a very important reaction.

Isolation and preparation of alkanes

Alkanes are obtained primarily from fossil fuels. Natural gas contains 60–90% methane, the first molecule in the alkane series, with smaller amounts of other small alkanes such as ethane, propane and butane. Crude oil is a mixture of alkanes and cycloalkanes (together with aromatic hydrocarbons). Untreated, this would not be useful, but fractional distillation allows separation of the compounds into groups (fractions) of similar chain length.

Most industrial chemicals, including plastics and pharmaceuticals, originate from the alkanes obtained from crude oil. Small alkanes are obtained from the less useful long chain alkanes by a process called cracking. Cracking can be achieved thermally by heating the alkanes at high temperatures, thus generating smaller alkanes and small alkenes. Catalytic cracking involves heating the alkanes in the presence of a silica–alumina catalyst, leading to shorter alkanes suitable for use in petrol. In another type of process, catalytic reforming, linear alkanes are broken down and then reassembled into branched alkanes. The process can be used to convert low-grade petrol (low octane number) into higher grades.

Reactions of alkanes

Alkanes are exceptionally stable substances. They do not react with ionic or polar substances and they are resistant to the action of acids and bases. There is little surprise, then, that one of the earlier names for these compounds was paraffins, from the Latin for 'little affinity'. The alkanes are saturated molecules; other atoms cannot (easily) add to them. Consequently, alkanes exhibit little chemical reactivity.

SCHEME 4.1

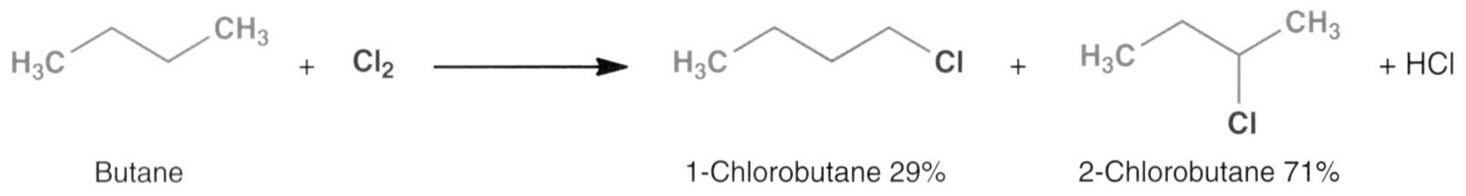

$$H_3C\diagup\diagdown^{CH_3} + Cl_2 \longrightarrow H_3C\diagup\diagdown\diagdown^{Cl} + H_3C\diagup^{CH_3}_{Cl} + HCl$$

Butane   1-Chlorobutane 29%   2-Chlorobutane 71%

Combustion of alkanes

One important reaction of alkanes is combustion, that is, they burn in a plentiful supply of oxygen. This is an oxidative process and follows a free-radical mechanism. The products of this reaction are water and the greenhouse gas carbon dioxide. This is a very exothermic reaction, which explains the use of alkanes as fuels. For example, the change in enthalpy (ΔH) for the combustion of methane is $-890\,kJ\,mol^{-1}$. Each additional $-CH_2-$ group adds $630-670\,kJ\,mol^{-1}$ to the energy released.

$$CH_4(g) + 2O_2(g) \rightarrow CO_2(g) + 2H_2O(l)$$
$$C_{11}H_{24}(g) + 17O_2(g) \rightarrow 11CO_2(g) + 12H_2O(l)$$

The lower alkanes are efficiently oxidized, but heavier alkanes can burn with a sooty flame. This is because there is insufficient oxygen to convert all the alkane to CO_2 (complete combustion), leading to the formation of some carbon and some carbon monoxide (incomplete combustion).

Alkanes and alkyl groups are also oxidized in the body by metabolizing enzymes. These enzymes act to render the lipophilic alkane more hydrophilic and suitable for excretion, as discussed in Section 4.6 'Hydrocarbons in pharmacy'.

Halogenation of alkanes

Another major reaction of alkanes is halogenation, which occurs very rapidly in the presence of ultraviolet (UV) light, following a free-radical mechanism. The amounts of halogen and alkane present determine what products are formed. For example, the chlorination of methane can lead to chloromethane, dichloromethane, trichloromethane (chloroform) and tetrachloromethane (carbon tetrachloride).

With higher alkanes, mixtures of different isomers are generally formed. The relative amounts of each isomer formed are related to the stability of the free radicals involved in the reaction. In the example above, 2-chlorobutane is the major product because it is formed via a more stable secondary radical (see Scheme 4.1).

These reactions produce mixtures of isomers and mixtures of products with different degrees of substitution, so they are not very suitable for the preparation of alkyl halides in the laboratory. They are used on an industrial scale, however, when it is economically worthwhile to separate the individual compounds.

KEY POINT

Compared to other hydrocarbons, alkanes are relatively unreactive. They burn, but do very little else.

 You will learn more free-radical processes and about radical stability in Section 4.4, 'Alkenes'.

4.4 Alkenes – preparation and chemical properties

In this section we will examine the preparation and reactions of alkenes. Unlike alkanes, in the previous section, alkenes undergo a wide range of reactions, largely because of the presence of the π bond.

Isolation and preparation of alkenes

As mentioned earlier, smaller alkenes are obtained from the cracking of alkanes. On an industrial scale, alkenes are also prepared by the dehydrogenation of alkanes (literally, the removal of hydrogens). As these methods tend to produce mixtures of compounds, they are not well suited for synthesizing alkenes in the laboratory where a particular target product may be sought.

Preparation by elimination reactions

There are a number of methods for producing alkenes in the laboratory, the most common being the elimination of HX (X = halogen atom) from alkyl halides or the elimination of water from alcohols. Alkenes can also be prepared via the hydrogenation of alkynes, as you will see later in this chapter, and using the Wittig reaction (you can find details of this reaction in an organic chemistry textbook).

 The elimination reactions of alkyl halides and alcohols are described in more detail in Chapter 5 'Alcohols, phenols, ethers, organic halogen compounds, and amines'.

Dehydrohalogenation (elimination of HX) can take place via either an E1 or an E2 mechanistic pathway.

(The numbers '1' and '2' indicate how many species are involved in the reaction's rate-limiting step – the step that determines how quickly the reaction proceeds. The E1 pathway is a two-step reaction where one species is involved in the rate-limiting step; the E2 pathway is a one-step reaction where two species are involved in the rate-limiting step.)

The E2 (Elimination Type 2) pathway is more useful for synthetic purposes because competing reactions are less important. It requires the use of a moderately strong base, for example KOH, and, sometimes, heat. This mechanism is **concerted** – there is no formation of an intermediate carbocation (hence it is a one-step reaction). Instead, the movement of electrons represented by the two curly arrows take place at the same time. Unhindered haloalkanes undergo this reaction more readily than haloalkanes in which the approach of the base is hindered by bulky groups (see Figure 4.10).

The dehydration of alcohols requires the presence of an acid catalyst, commonly sulfuric acid; sometimes, the reaction mixture must also be heated. The reaction proceeds by an E1 mechanism, with the acid protonating the hydroxyl group, converting it to a good leaving group. Water leaves and the intermediate carbocation is formed (this is the first step). Loss of a proton (step two) then gives the alkene (see Scheme 4.2).

FIGURE 4.10 E2 elimination mechanism.

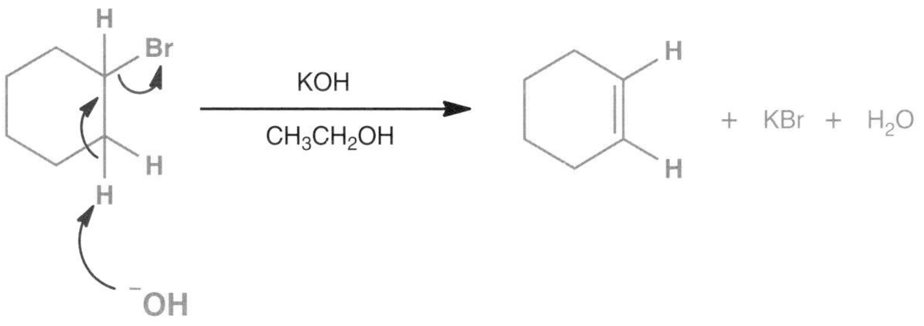

E1 elimination mechanism

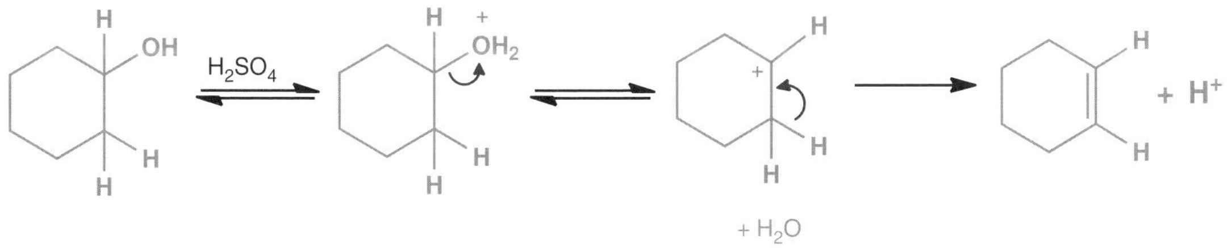

The reaction is an equilibrium reaction and the equilibrium constant is not particularly high. To ensure the forward reaction is favoured, one of the products needs to be removed from the mixture. Commonly, the alkene product has a lower boiling point than either water or the starting alcohol and can be distilled out of the mixture as it is formed.

E1 reactions are favoured according to the following series: tertiary alcohol favoured more than secondary; secondary alcohol favoured more than primary. This is because the intermediate carbocation is stabilized by σ-**conjugation**. Sigma conjugation occurs when electron density in a C–H or C–C bond interacts with a region of low electron density (in this case, an empty p-orbital) and helps to stabilize it. A tertiary carbocation is stabilized by three interactions at any particular time, a secondary carbocation is stabilized by two, and a primary carbocation by a single interaction.

We will meet sigma conjugation again later in this chapter and in Chapter 7 'Introduction to aromatic chemistry'.

BOX 4.1

Alkenes in nature

Alkenes have many important roles in biology and are therefore quite abundant in nature. Even the simplest alkene, ethene (also known as ethylene) is vital; it is a plant hormone that controls various stages of plant growth, including seed germination and fruit ripening.

The two enantiomers of the alkene limonene have different aromas: the (+)-isomer is responsible for the smell of oranges, while the (–)-isomer smells more of lemon (though less generous noses may say turpentine!). α-phellandrene can be found in the oil of eucalyptus.

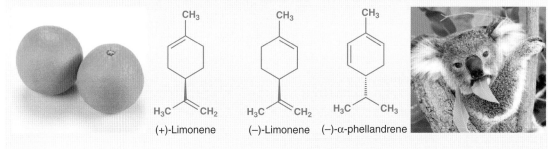

(+)-Limonene   (–)-Limonene   (–)-α-phellandrene

Source: Koala: David Coleman; Oranges: Mark Mason

Pheromones are chemical substances released by insects that can be detected by other insects of the same species. Many of the sex and alarm pheromones of insects are alkenes. For example, bombykol and multifidene are the sex pheromones of the silk moth and brown algae, respectively.

Bombykol
(10E, 12Z)-hexadeca-10,12-dien-1-ol

Multifidene

Relative stabilities and reactivities of alkenes

Almost all the hydrocarbons that we will discuss are stable, almost indefinitely, at room temperature. Alkenes do not generally decompose spontaneously! However, some alkenes are more readily formed than others (their ground state energy is lower) and these tend also to be less reactive than other alkenes.

When chemists wish to prepare alkenes, they often find that the major product is the most stable alkene. Many reaction pathways allow for the rearrangement of carbon–carbon double bonds to the most stable isomer. Therefore, it is useful to know how the structure of an alkene relates to its relative stability.

An alkene with alkyl groups attached to the sp^2 carbons is more stable than one with only hydrogens attached. Indeed, the greater the number of alkyl groups, the more stable the alkene. Evidence for these differences in stability can be obtained from heats of hydrogenation for isomeric alkenes, exemplified using 3-methylbut-1-ene and 2-methyl-2-butene (see Figure 4.11).

The trisubstituted isomer is more stable by 14 kJ mol⁻¹. That alkyl groups have this stabilizing effect on alkenes has been summarized as Zaitsev's rule:

> *More substituted double bonds are usually more stable.*

The alkyl groups stabilize the alkene as they are able to donate electron density, via σ-conjugation to the π-bond. This leads to an extended molecular orbital that stabilizes the system. In addition, the alkene will be more stable (less reactive) if bulky substituents are situated as far apart in space as possible. In alkenes, sp^2-hybridized carbon atoms separate the substituents by 120° (compared to 109.5° for sp^3-hybridized carbons in alkanes), acting to relieve steric effects.

 You will encounter similar effects in Chapter 7 'Introduction to aromatic carbons'.

As you have seen in the nomenclature section, alkenes can exist as geometric isomers, either *cis* or *trans*. Which of these is more stable? Let us take *cis*-pent-2-ene and *trans*-pent-2-ene as an example. Both molecules have two alkyl groups – a methyl and an ethyl group – bonded to the sp^2 carbons, so Zaitsev's rule cannot help us here. We need to look at how the molecules are arranged spatially. The *cis*-isomer has the alkyl substituents close together in space, leading to steric strain. In the *trans*-isomer, the two substituents are much further apart and so this is the more stable of the two isomers. Evidence from the heats of hydrogenation points to *trans*-pent-2-ene being more stable than the *cis*-isomer by 4 kJ mol⁻¹.

Reactions of alkenes

Like alkanes, alkenes burn in a plentiful supply of oxygen, forming carbon dioxide and water. As with alkane combustion, lots of heat energy is given off. For the combustion of ethene (ethylene), shown below, $\Delta H = -1410$ kJ mol⁻¹

$$C_2H_4(g) + 3O_2(g) \rightarrow 2CO_2(g) + 2H_2O(l)$$

FIGURE 4.11   The heats of hydrogenation for isomeric alkenes.

$\Delta H° = -127$ kJmol⁻¹

$\Delta H° = -113$ kJmol⁻¹

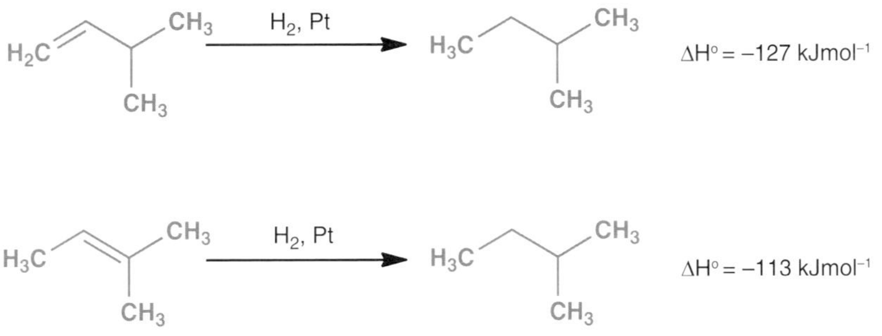

However, this is where the similarity in their reactivity ends. Alkenes have their own set of reactions, similar to one another but different from the reactions of alkanes. There is little mystery in this. Recall the structure of alkenes. You should remember that they are planar molecules with a region of electron density above and below the plane of the carbon–carbon double bond. This is due to the π-bond formed between the unhybridized p orbitals of each carbon atom. The π electrons are relatively loosely held and are attracted to electrophiles (electron-loving species). Thus, the first step in alkene reactions is the addition of an electrophile (E) to one of the sp^2 carbons; this is followed by addition of a nucleophile (Nu$^-$) to the remaining sp^2 carbon. The overall result is the breaking of the π-bond and the formation of two new σ-bonds, which is energetically favourable. This characteristic reaction of alkenes is called an **electrophilic addition reaction**, as the first species to add is the electrophile (see Scheme 4.2).

Alkenes also undergo electrophilic addition reactions mediated by free radicals. These are especially important in the synthesis of a variety of everyday polymers, such as polystyrene and poly(methyl methacrylate) [Plexiglas®]. We will return to polymers later in this chapter.

Addition of hydrogen halides to alkenes

Alkenes engage in electrophilic addition reactions with hydrogen halides (HX), where the proton from the hydrogen halide acts as the electrophile, adding first, and the halide ion is the nucleophile which adds in the second step. The first step is slow and determines the overall reaction rate (see Scheme 4.3).

If the sp^2 carbons in the alkene bear the same substituents, as shown, the product of the reaction is easy to predict. Whichever carbon the electrophile adds to in the first step, the same product will be formed. This is not the case when the sp^2 carbons are not identically substituted. Let us take the addition of HCl to 2-methylpropene as an example. There are two possible products and we need to consider the intermediate in the reaction, the carbocation, and its stability, to determine which one forms. The most stable carbocation is the one fastest to form – and it is the carbocation that determines the final product (see Figure 4.12).

The only product formed in this reaction is 2-chloro-2-methylpropane (*tert*-butyl chloride), as the intermediate *tert*-butyl cation is more stable than the isobutyl cation involved in the alternative pathway. The relative order of stability for carbocations is tertiary > secondary > primary (see Scheme 4.4). Once again, σ-conjugation is at work. Alkyl groups are able to stabilize the positive charge, because overlap between the adjacent sigma bonds and the empty p-orbital spreads the positive charge and stabilizes it. (Some books call this effect hyperconjugation, but sigma conjugation is more descriptive.)

SCHEME 4.2

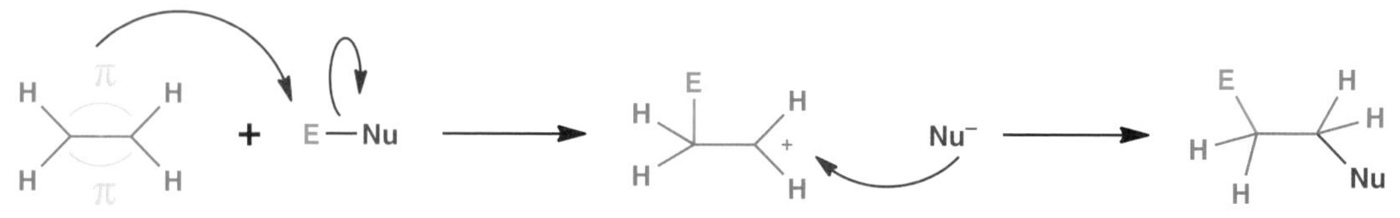

SCHEME 4.3

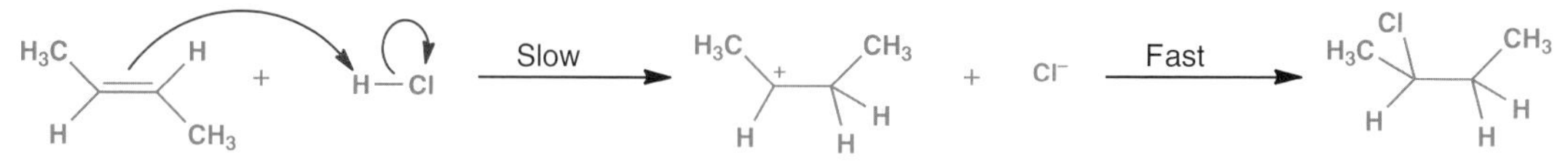

SCHEME 4.4

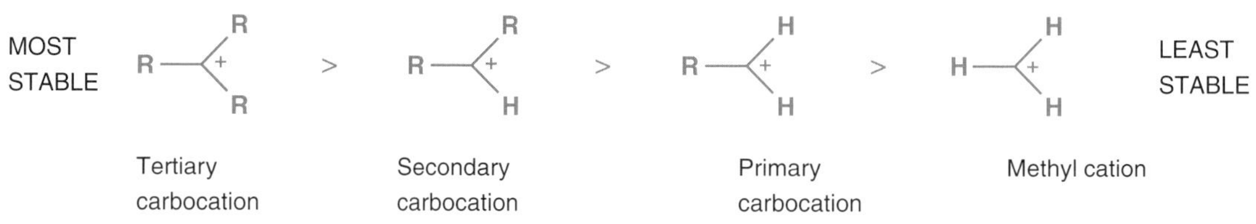

| MOST STABLE | | > | | > | | > | | LEAST STABLE |

Tertiary carbocation   Secondary carbocation   Primary carbocation   Methyl cation

FIGURE 4.12 Addition of HCl to 2-methylpropene.

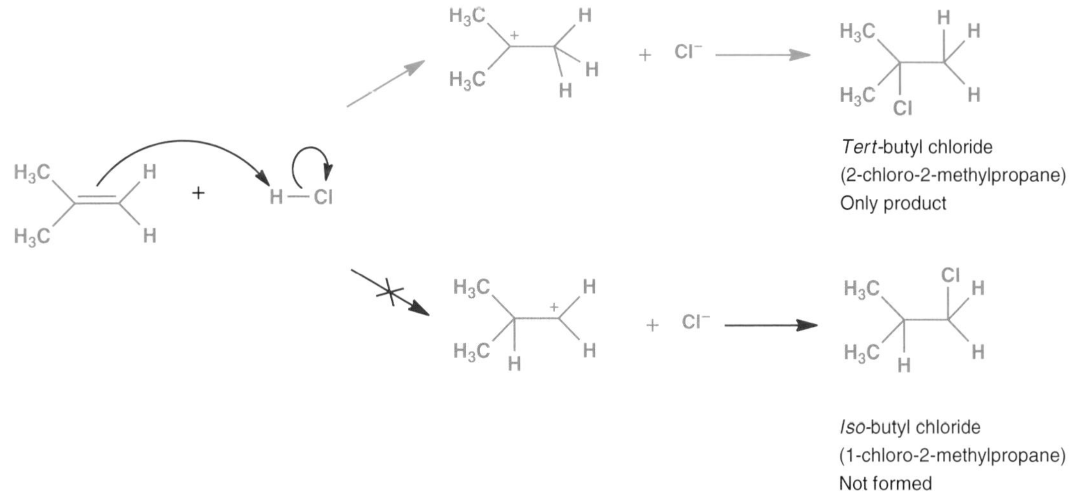

Tert-butyl chloride
(2-chloro-2-methylpropane)
Only product

Iso-butyl chloride
(1-chloro-2-methylpropane)
Not formed

These electrophilic addition reactions are **regioselective** (see Figure 4.13). This term is applied to any reaction where two (or more) isomers are produced, but one predominates. Note that, in many cases, the minor isomer is actually undetectable, but the reaction is still regioselective because the mechanism allows a second isomer to be formed in principle.

When pent-2-ene is reacted with HBr, the reaction is not regioselective, as a secondary carbocation is the intermediate in the pathway to each of the products, 2-bromo- or 3-bromopentane (see Scheme 4.5).

A general rule for determining the products of such reactions is as follows:

The proton of an acid (H-X) adds to the carbon in the double bond that already has the most hydrogens.

This is often referred to as Markovnikov's rule, after the Russian scientist who was the first to recognize

FIGURE 4.13 An example of regioselectivity in electrophilic addition.

1-Methylcyclohexene

1-Bromo-1-methyl
cyclohexane
MAJOR PRODUCT

1-Bromo-2-methyl
cyclohexane
MINOR PRODUCT

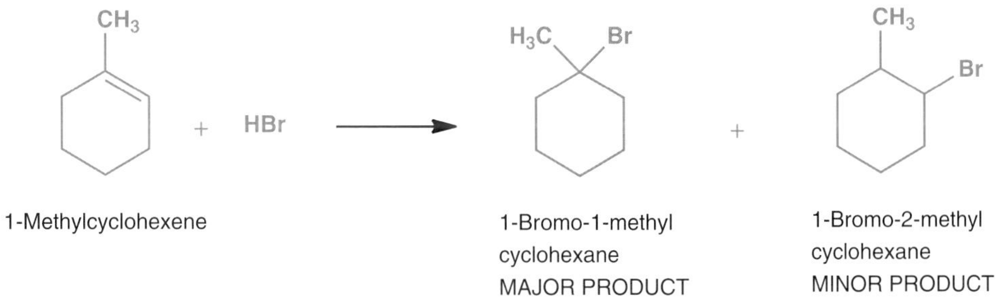

SCHEME 4.5

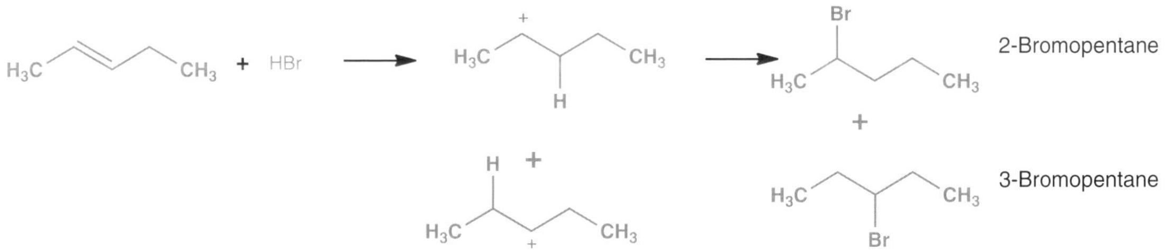

2-Bromopentane

3-Bromopentane

this phenomenon. A more modern and general take on Markovnikov's rule is:

In an electrophilic addition to an alkene, the electrophile (E^+) adds in such a way as to form the most stable intermediate.

This second definition is the more useful.

 Markovnikov's rule is discussed in more detail in Chapter 2 'Organic structure and bonding' of this book.

BOX 4.2

Vladimir Vasilyevich Markovnikov

Vladimir Vasilyevich Markovnikov (1837–1904), was born in Nizhny Novgorod and studied at Kazan University. Although initially an economist, he changed to chemistry. After graduating, he worked at the Universities of Kazan and Saint Petersburg. Later, he was a professor of chemistry at Kazan, Odessa and Moscow Universities.

Although best known for 'Markovnikov's rule', he also contributed to organic chemistry by synthesizing rings containing four and seven carbons, thus disproving the idea that carbon could form only five- and six-membered rings.

Addition of HBr to alkenes in the presence of peroxides

HBr can add to alkenes to give so-called **anti-Markovnikov** products. This type of addition proceeds via the free-radical pathway shown in Figure 4.14. In this case, it is the stability of the radical intermediate that determines which product is formed. Free radical stabilities run in the same order as the carbocation stabilities shown in Scheme 4.4, and for the same reasons.

Note that in the reaction pathway in Figure 4.14, it is the bromine radical which first attacks the double bond. The radical intermediate then takes a proton from HBr to give the product. The mechanism has three distinct phases – initiation, propagation and termination, the first two of which are shown in the reaction scheme. Polymers can also be formed by free-radical reactions, as we shall see later.

Termination involves the reaction of two radicals to give a product which cannot propagate the chain reaction further. Only HBr has the correct bond energy to undergo this free-radical reaction. The HCl bond is too strong, while HI tends to break down to form ions rather than free radicals.

FIGURE 4.14 An example of a free-radical reaction pathway.

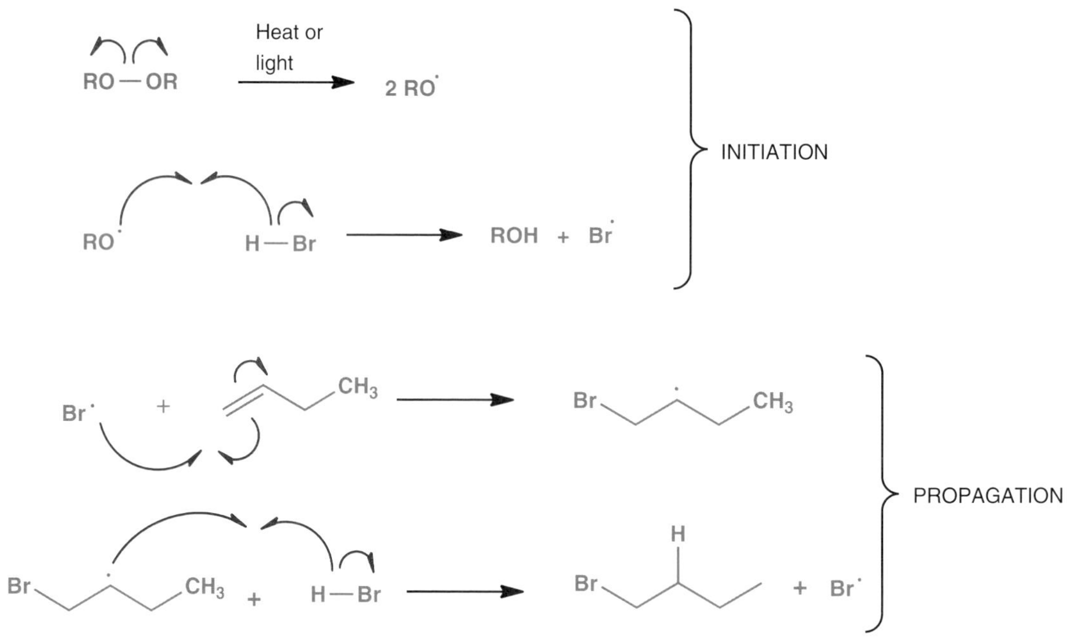

What would be the major products, A, B and C, from the reactions shown?

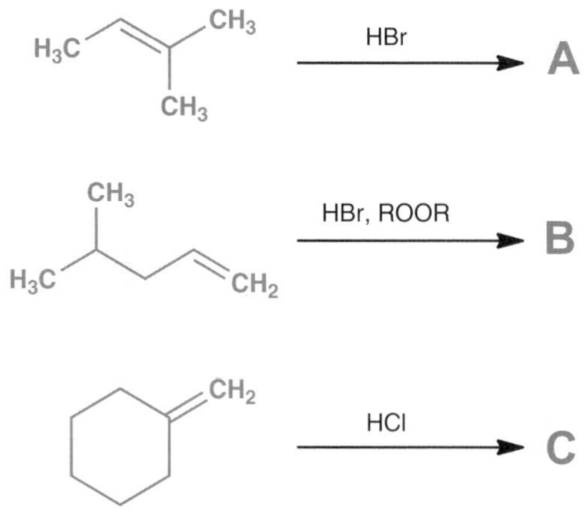

Addition of water and alcohols to alkenes

The O–H bond in water is strong, so water is only a very weak acid and is unable to react with an alkene as an electrophile. Therefore, the addition of water to an alkene (hydration) requires some assistance in the form of an acid catalyst, most often sulfuric acid (H_2SO_4). The presence of the acid now provides the reaction with an electrophile, namely H^+ (as H_3O^+) (see Scheme 4.6).

The product of the hydration of an alkene is an alcohol, and the mechanism for its formation is shown in Scheme 4.7. Note that the Markovnikov product is formed.

SCHEME 4.6

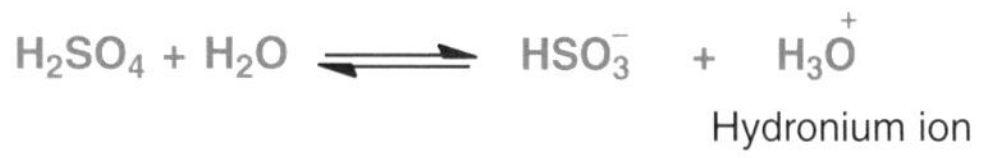

$$H_2SO_4 + H_2O \rightleftharpoons HSO_3^- + H_3\overset{+}{O}$$

Hydronium ion

SCHEME 4.7

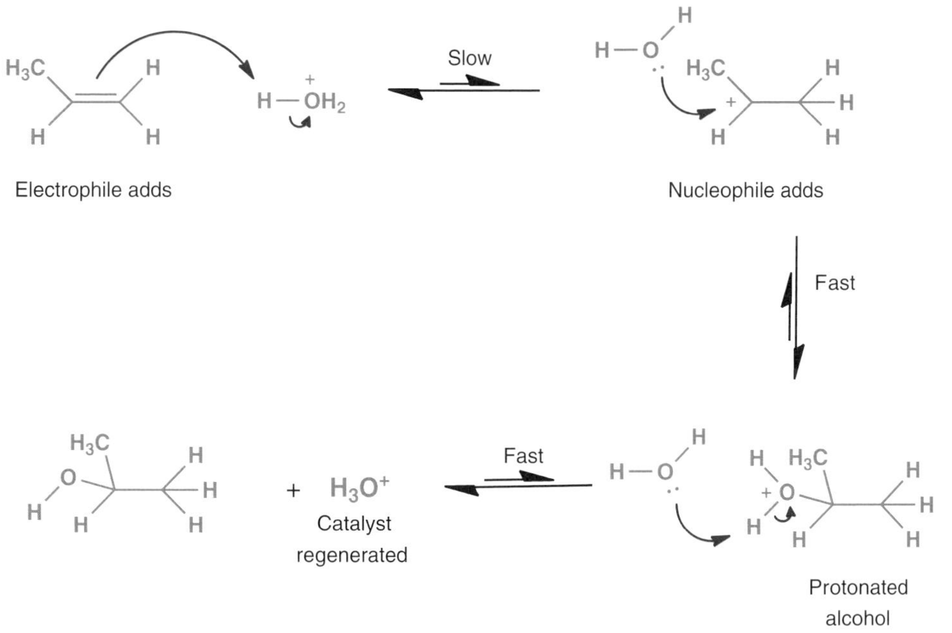

Electrophile adds

Nucleophile adds

Fast

Protonated alcohol

$+$ H_3O^+

Catalyst regenerated

Fast

As you might expect, alcohols react with alkenes in much the same way as water. Again, a strong acid catalyst is required and the products of such reactions are ethers. As you can see in Scheme 4.8, the mechanism is very similar to that for hydration.

SCHEME 4.8

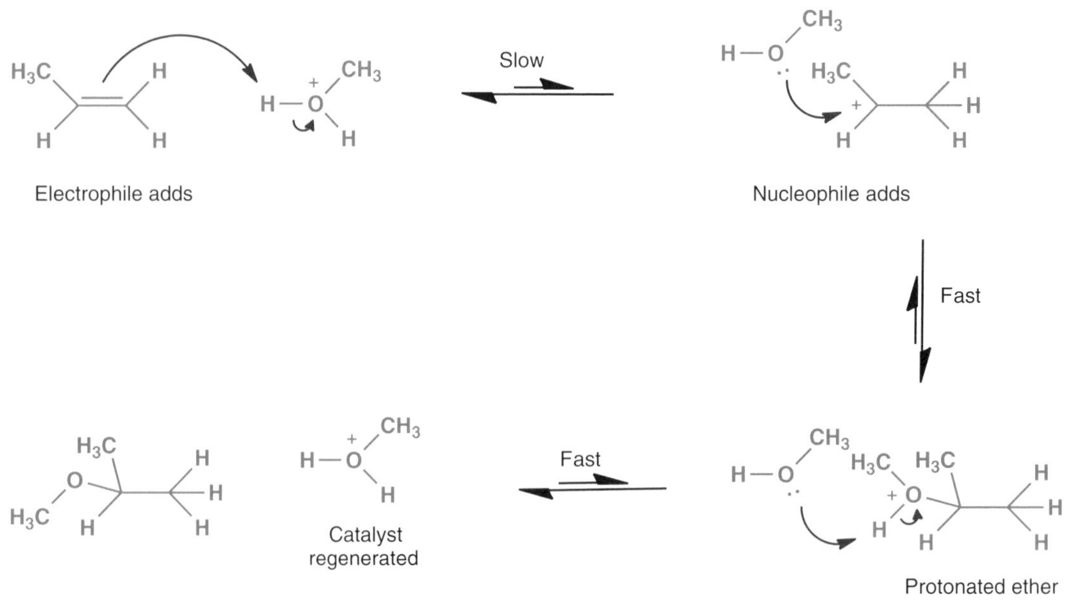

Electrophile adds

Slow

Nucleophile adds

Fast

Catalyst regenerated

Fast

Protonated ether

SELF CHECK 4.8

What alkenes could be used as starting materials to make the compounds A, B, C, and D? What other reagents would be required?

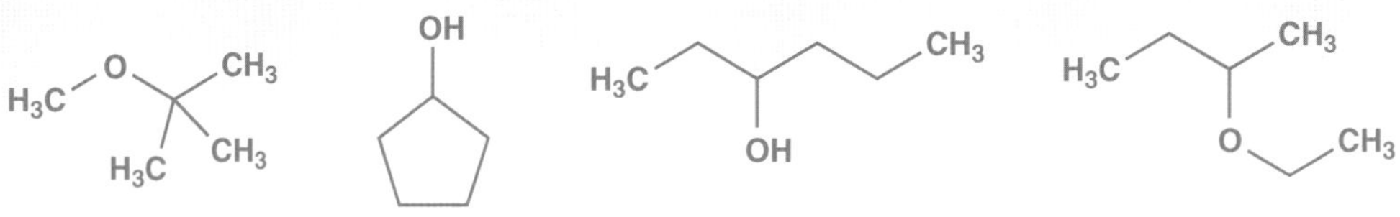

(A) (B) (C) (D)

 The preparation of alcohols is discussed in more detail in Chapter 5 'Alcohols, phenols, ethers, organic halogen compounds, and amines'.

Addition of halogens to alkenes

Chlorine (Cl$_2$), bromine (Br$_2$) and sometimes iodine (I$_2$), add across the alkene carbon–carbon double bond to form **vicinal dihalides**. Indeed, the decolorization of bromine solution is used as a test for the presence of alkenes.

A carbocation mechanism would lead to a mixture of *syn*- and *anti*- addition, but halogenation always involves *anti*-addition. (*Syn* addition refers to addition on the same side of the double bond; *anti*-addition describes addition to opposite sides of the double bond.) The π electrons of the double bond attack to give a cationic intermediate and a bromide ion. The intermediate is not a simple carbocation, but something called a cyclic halonium ion (here a bromonium ion), which is more stable. The bromide ion then attacks the cation from the rear, leading to the vicinal dibromide product (see Scheme 4.9).

SELF CHECK 4.9

Draw the products of the reaction of chlorine with 2-methylpropene and 3-methylbut-1-ene.

SCHEME 4.9

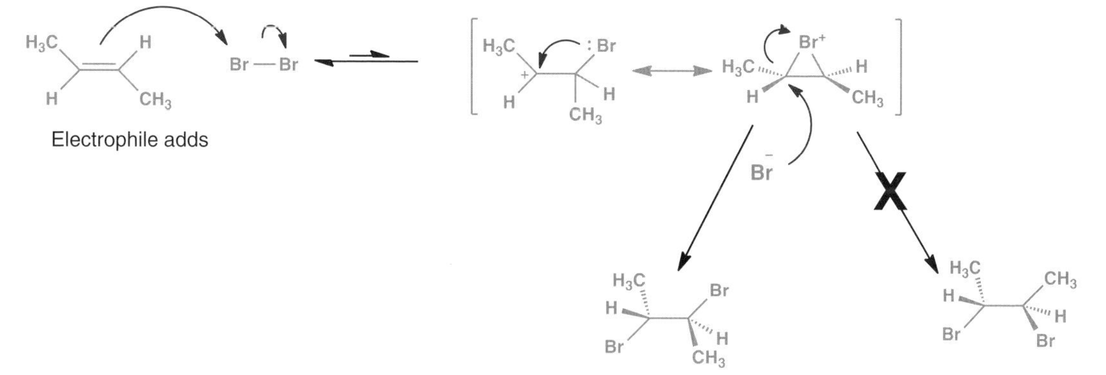

Electrophile adds

Addition of hydrogen to alkenes (reduction)

The addition of hydrogen to an alkene can be achieved in the presence of precious metal catalysts, such as palladium or platinum. This process is called **catalytic hydrogenation** and leads to the production of an alkane. The metals are usually in a finely divided state and are adsorbed onto activated charcoal.

While the overall hydrogenation reaction is highly exothermic, it does not occur in the absence of an appropriate catalyst. Rather than lowering the high activation energy of the original reaction (an energy barrier in the reaction pathway which must be overcome), the catalyst provides an alternative pathway (see Figure 4.15). Remember that a catalyst increases the rate of a reaction without itself changing or becoming part of the product.

A full picture of the mechanism for such reactions does not yet exist, but it is known that both the alkene and the hydrogen must be adsorbed to the metal surface and that all the bond breaking and making events happen on that surface. Because the two hydrogens add from the solid surface, they add with *syn* (same side) stereochemistry. After formation, the alkane diffuses away from the surface as it is less strongly adsorbed.

Oxidation of alkenes

Alkenes are susceptible to combustion, typically to carbon dioxide and water, but also to a number of other oxidative reactions that introduce oxygen into

---

CASE STUDY 4.1

James is 32 and, because of his family history, is at risk of developing coronary heart disease. His GP has advised him to take statins to reduce his blood cholesterol levels but he does not want to take a drug that he will have to take for the rest of his life. Does he have any alternative ways of reducing his cholesterol levels?

REFLECTION QUESTIONS

1. Would modifying his diet help to reduce his blood cholesterol levels?

2. What are trans-fats?

Answers

1 A diet high in trans-fats can raise blood cholesterol levels. If James reduces the amount of trans-fats in his diet, this should help to reduce his blood cholesterol levels. Trans-fats are found naturally in low levels in some foods. However, artificial trans-fats are found in many processed foods such as cakes and biscuits so reducing his intake of processed foods would help.

2 Trans-fats are formed by partial hydrogenation of the double bonds in unsaturated fatty acids. The normal configuration of the double bonds in unsaturated fatty acids is *cis*. Catalytic partial hydrogenation of unsaturated fatty acids converts some of the remaining double bonds to the trans configuration. Trans-fats raise levels of LDL ('bad') cholesterol and lower levels of HDL ('good') cholesterol.

FIGURE 4.15 Hydrogenation of an alkene.

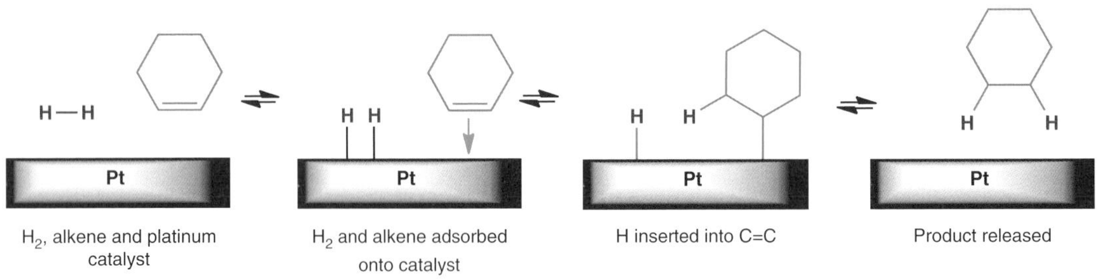

Pt	**Pt**	**Pt**	**Pt**
H₂, alkene and platinum catalyst	H₂ and alkene adsorbed onto catalyst	H inserted into C=C	Product released

molecules. These reactions include epoxidation, hydroxylation and oxidative cleavage.

Epoxides are three-membered cyclic ethers. They are useful synthetic intermediates and can be produced by reacting alkenes with peroxy acids. The reaction mechanism is a one-step concerted process, with several bonds breaking and forming at the same time. This means that there is no chance for rearrangement during the reaction and that the stereochemistry in the starting alkene is preserved in the epoxide (see Scheme 4.10).

The epoxides are generally stable and can be isolated. However, the presence of aqueous acid leads to protonation and then attack of the protonated epoxide by water to give a *trans*-1,2-diol (a glycol). There is no need to isolate the epoxide, and performing the reaction in aqueous peroxy acid leads directly to the diol product (see Scheme 4.11).

To form a *cis*-1,2-diol from an alkene, different reagents are required. Hydroxylation reactions are performed by reacting an alkene with either catalytic quantities of osmium tetroxide in the presence of hydrogen peroxide, or with cold, dilute aqueous potassium permanganate. The former gives higher yields, but osmium tetroxide is expensive, volatile and highly toxic. In both cases, the two C–O bonds are formed simultaneously through esters (osmate or manganate), meaning that they add to the same face of the double bond and give the *cis*-diol.

For more information about epoxides and glycols, Chapter 5 'Alcohols, phenols, ethers, organic halogen compounds, and amines'.

Oxidative cleavage can be achieved using warm or acidic solutions of potassium permanganate, leading to ketones and aldehydes. Any aldehydes formed are quickly oxidized to carboxylic acids (see Scheme 4.12).

Ozone is also capable of cleaving carbon–carbon double bonds. It is a milder oxidizing agent than

SCHEME 4.10

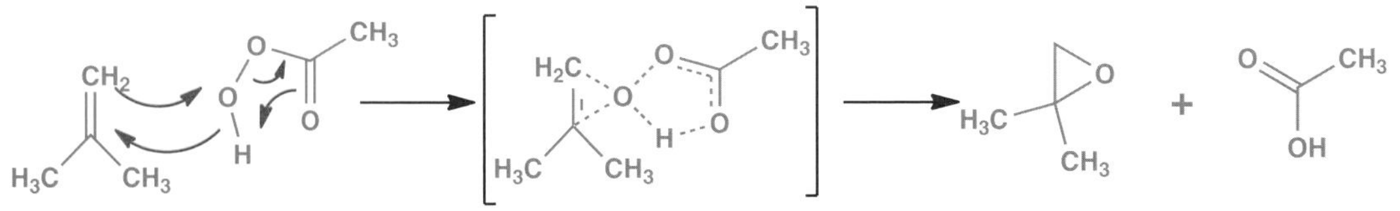

SCHEME 4.11

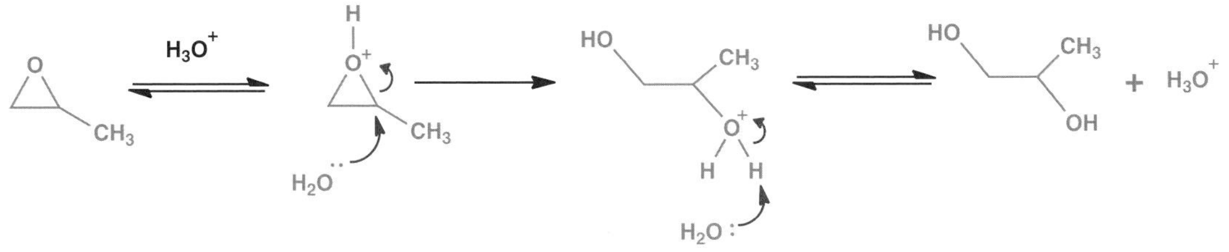

SCHEME 4.12

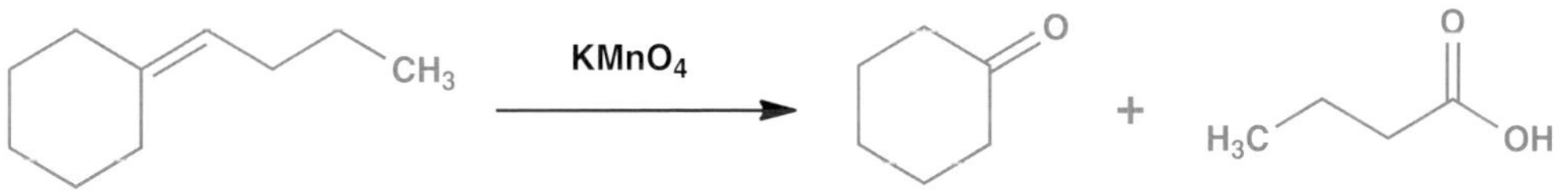

SCHEME 4.13

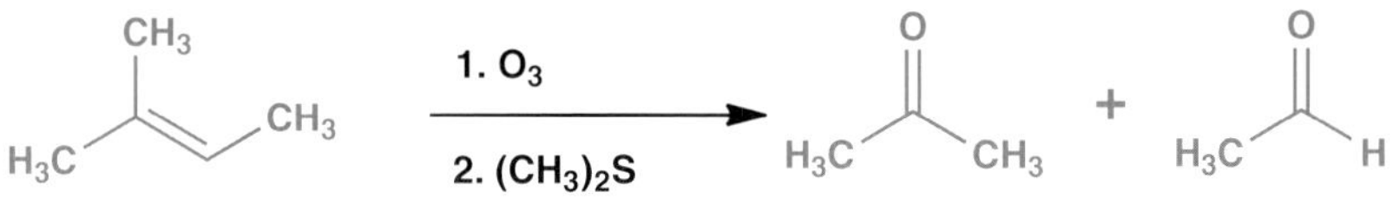

permanganate and allows for isolation of both the ketones and aldehydes produced (see Scheme 4.13).

> **KEY POINT**
>
> Because of the presence of the π bond, alkenes are very reactive. The type of reaction they undergo is electrophilic addition.

Synthetic polymers

We encounter polymers (from the Greek for 'many things') every day of our lives, generally without giving them a second thought. You may have spent the early years of your life in disposable nappies. The absorbent material in these products is the sodium salt of poly(acrylic acid), which can absorb up to 200–300 times its mass in water! As a student, could you really imagine being without food wrap, Superglue, plastic bottles and synthetic fibres? One day you may benefit from artificial joints. All of these are made from polymers. Some common polymers that will have touched your lives are shown in Table 4.3.

Many important polymers (though by no means all) are prepared from substituted alkenes. These are termed **chain-growth polymers** due to the mechanism of their formation. Table 4.3 provides a small selection and should bring home to you the importance of these compounds.

So what is a polymer? Well, essentially, it is a large molecule (**macromolecule**) made by linking together smaller repeating units called **monomers** in a process called **polymerization**. The most common method for making polymers from alkenes is **free-radical**

103

TABLE 4.3   Some common vinyl polymers

Monomer	Repeat unit	Name	Uses
$H_2C{=}CH_2$	$-[CH_2{-}CH_2]-$	polyethene polyethylene	films, toys, bottles, plastic bags
$H_2C{=}CHCl$	$-[CH_2{-}CH(Cl)]-$	poly(vinyl chloride) (PVC)	flooring, pipes, window surrounds, 'squeeze' bottles
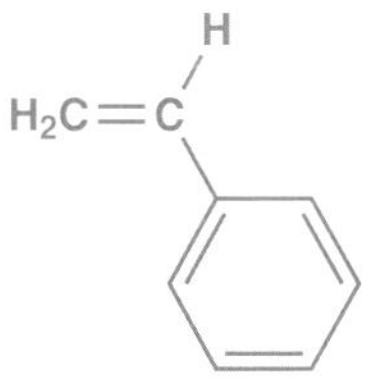 $H_2C{=}C(H)(C_6H_5)$	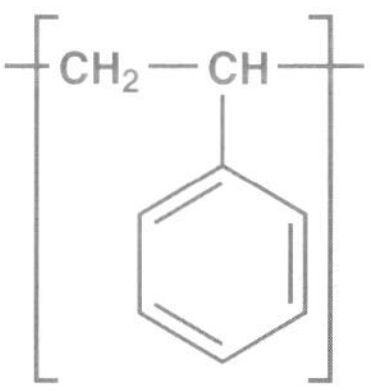 $-[CH_2{-}CH(C_6H_5)]-$	poly(styrene)	toys, egg cartons, packaging, hot drink cups

(continued)

TABLE 4.3 Some common vinyl polymers   (continued)

Monomer	Repeat unit	Name	Uses
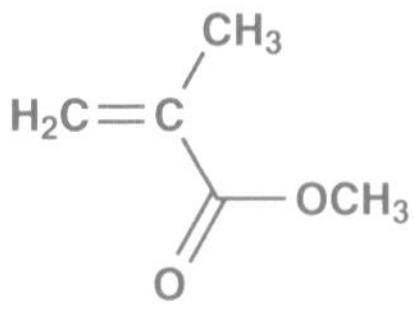 $H_2C=C$ with CH_3 and OCH_3, O	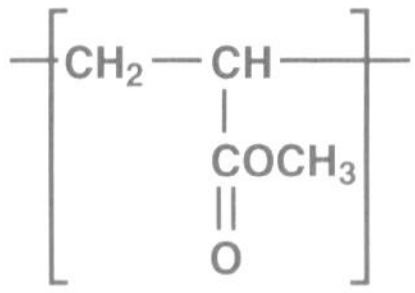 $[CH_2-CH]$ with $COCH_3$, O	poly(methyl methacrylate) (Perspex, Plexiglas)	skylights, signs, lighting fixtures, solar panels
$F_2C=CF_2$	$[CF_2-CF_2]$	poly(tetrafluoroethylene) (Teflon)	non-stick surfaces, cable insulation
$H_2C=CHCOOH$	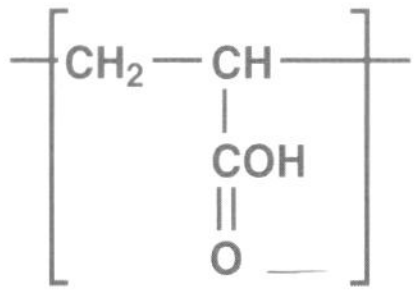 $[CH_2-CH]$ with COH, O	poly(acrylic acid)	disposable nappies

FIGURE 4.16 Initiation, propagation and termination in radical polymerization.

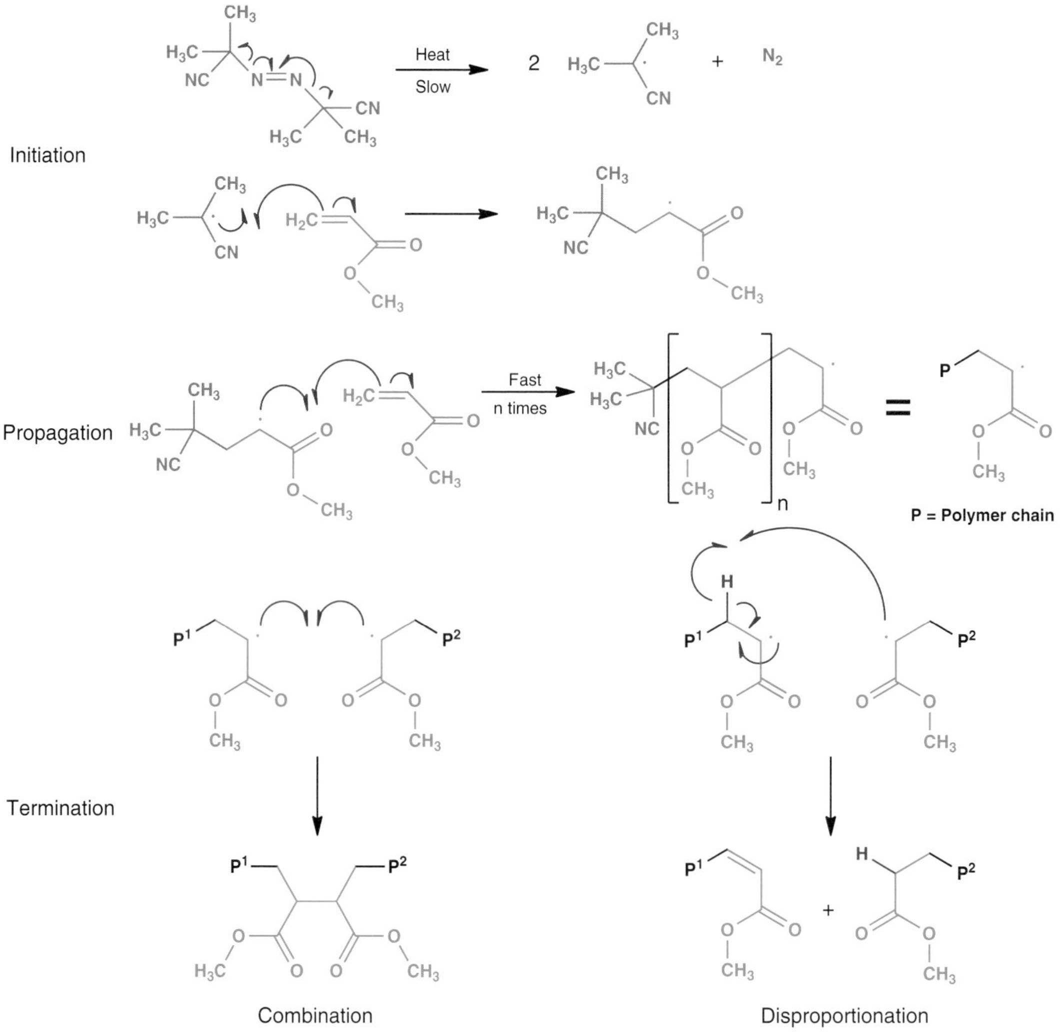

polymerization. You have seen earlier in the chapter that radical reactions have three distinct steps: initiation, propagation and termination.

In radical polymerization, the initiation process is actually two steps, one to create the radicals and the second to form the radical that will propagate the chain reaction. The radical is an electrophile and thus adds to the carbon bearing the greater number of hydrogen atoms. Once the radical is formed, it adds to another monomer molecule, creating another radical and propagating the chain reaction (see Figure 4.16).

Destruction of the propagating site stops the chain reaction and this can happen in a number of ways. Two major pathways are (i) the combination of two free radicals and (ii) the disproportionation of two free radicals. Look at Figure 4.16, and note that the disproportionation reaction leads to an alkene, from which further initiation can occur, and a 'dead' chain (the alkane on the right). The third major pathway (not shown) is through reaction of the propagating chain with impurities.

 It is important to realize that biological polymerizations are not really very different from chemical polymerizations. While the mechanisms for the syntheses of DNA, RNA and proteins are different from that shown above, they also have initiation, propagation and termination steps. The synthesis of nucleic acids and proteins are discussed in Chapter 2 'Molecular cell biology' of the *Therapeutics and Human Physiology* book of this series.

4.5 **Alkynes – preparation and reactions**

In this section, we will deal with the reactions of alkynes. As you might expect, these reactions are very similar to those of alkenes, because of the presence of two π bonds. However, you need to be aware of some significant differences.

Preparation of alkynes

Alkynes are much more reactive than alkanes and alkenes. This means that they are much rarer, and are normally synthesized in the laboratory rather than

Alkynes in medicines

Alkynes are not as prevalent in nature as alkenes, but some plants use them to protect against disease or predators. An example is capillin, which shows fungicidal activity and is produced by the oriental worm-wood plant. The alkyne functional group is also rare in drugs, the most common example being 17-ethynyl estradiol, a common ingredient in birth control tablets.

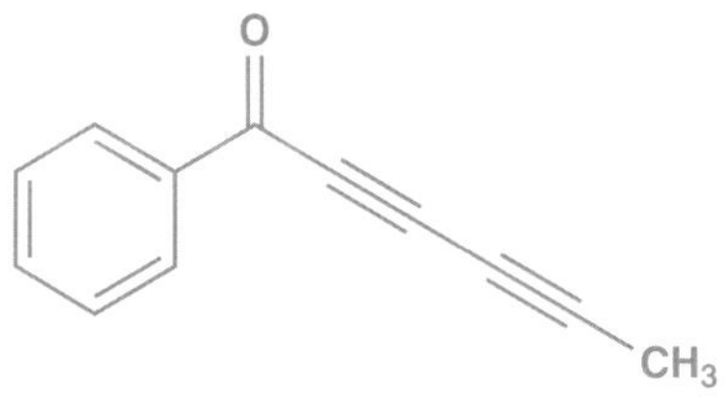

Capillin

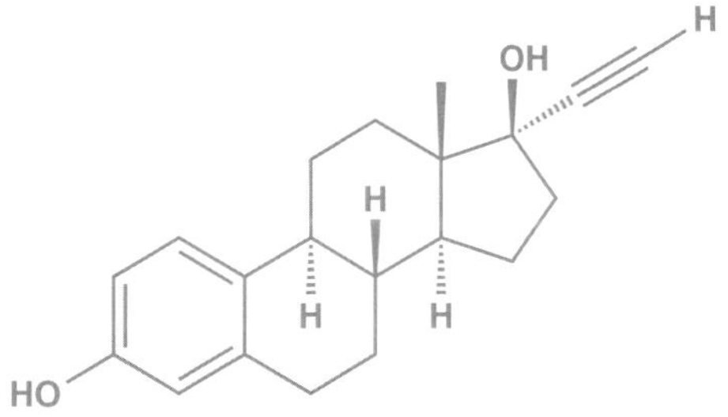

17-Ethynyl estradiol

SCHEME 4.14

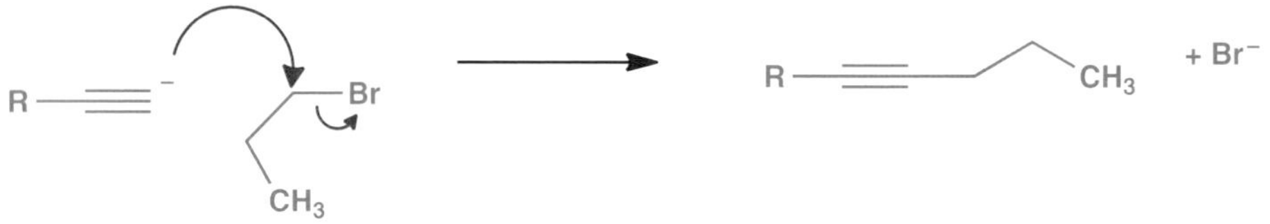

being isolated from natural sources (see, however, Integration Box 4.3).

Acetylide ions as synthetic intermediates

As mentioned in Section 4.1, the proton of a terminal alkyne is *relatively* acidic. Thus, a strong base may abstract this proton to give an acetylide ion. The acetylide ion is a very useful synthetic intermediate.

The formation of carbon–carbon bonds is extremely important in organic chemistry and acetylide ions allow chemists to perform such reactions. For example, reaction of an unhindered alkyl halide with an acetylide ion leads to formation of a carbon–carbon bond to give a larger internal alkyne (see Scheme 4.14).

This alkylation reaction is a bimolecular nucleophilic substitution (S_N2), the mechanism of which is well understood. The negatively charged acetylide anion attacks the carbon bearing the halide, which carries a partial positive charge, kicking out the bromide ion. These reactions are very useful, as they allow terminal alkynes to be turned into internal alkynes of any length we choose, simply by picking the right alkyl halide (see Scheme 4.15).

Acetylide ions also react with carbonyl groups, as found in aldehydes and ketones (see Chapter 6), to give alcohols, after protonation of the alkoxide ion initially formed (see Scheme 4.16).

Addition to formaldehyde gives a primary (1°) alcohol, while addition to higher aldehydes gives secondary (2°) alcohols. Finally, addition of acetylide ions to ketones furnishes us with tertiary (3°) alcohols (see Scheme 4.17).

The preparation of alkynes using elimination reactions

It is possible, in some cases, to prepare alkynes through a double dehydrohalogenation of **vicinal** or **geminal** dihalides. The first loss of HX gives a vinyl halide, which can lose a second HX molecule under extremely basic conditions (see Scheme 4.18).

High temperatures are also required and these rather brutal conditions can lead to side reactions, which lower the yield. The use of KOH tends to give the most stable internal alkyne, as shown in Scheme 4.19.

Reactions of alkynes

The carbon-carbon triple bond in an alkyne is made up of one σ and two π-bonds. These π-bonds are perpendicular to both the plane of the σ-bond and to each other.

SCHEME 4.15

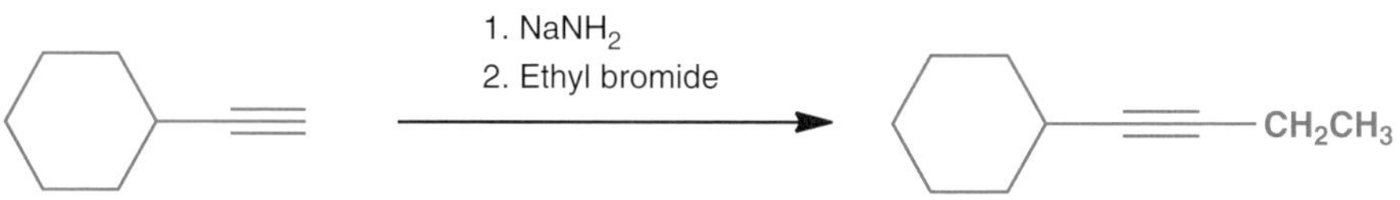

Ethynylcyclohexane

1-Cyclohexylbut-1-yne (70%)

SCHEME 4.16

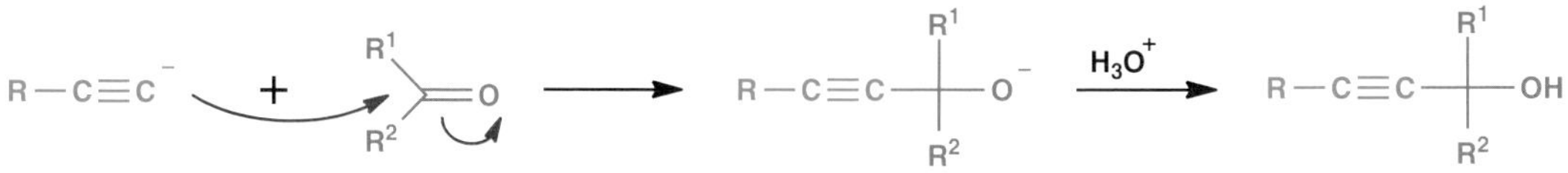

Acetylide anion

Acetylenic alcohol

SELF CHECK 4.10

What would be the major products (A and B) from the reactions shown?

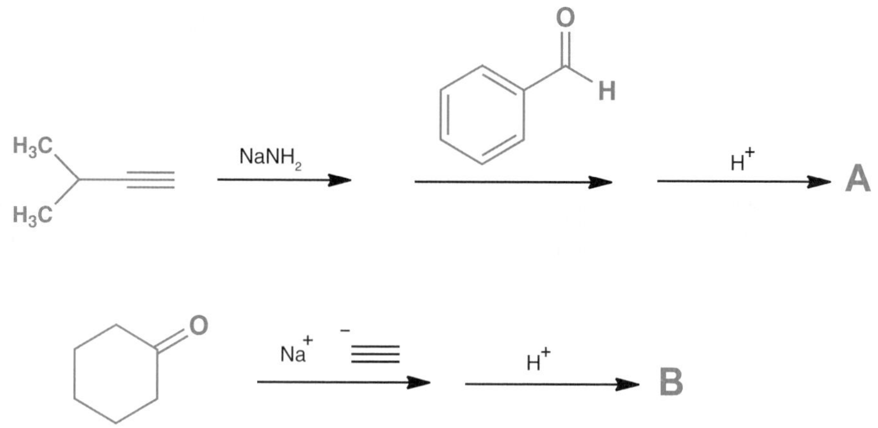

SCHEME 4.17

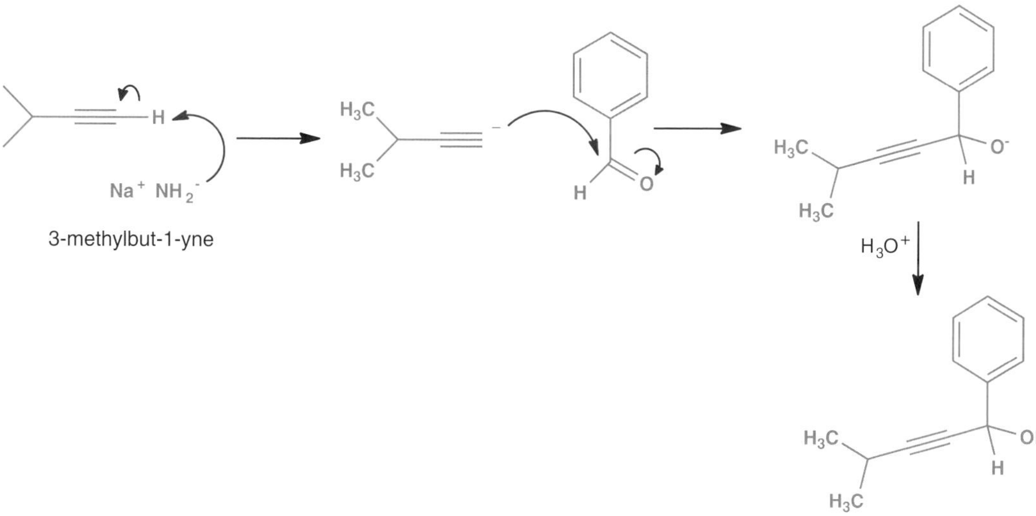

3-methylbut-1-yne

4-Methyl-1-phenylpent-2-yn-1-ol

SCHEME 4.18

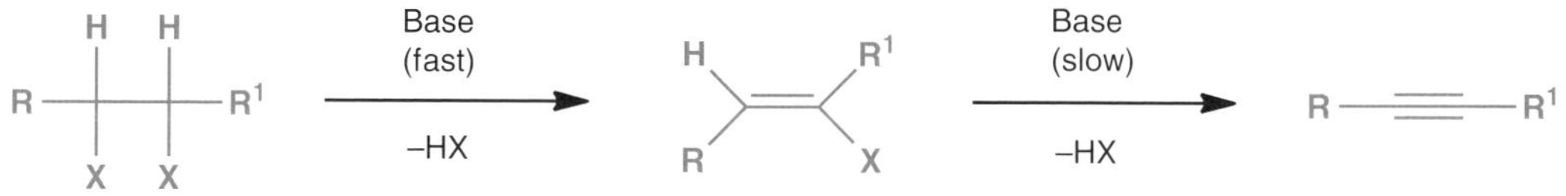

Vicinal dihalide   Vinyl halide   Alkyne

SCHEME 4.19

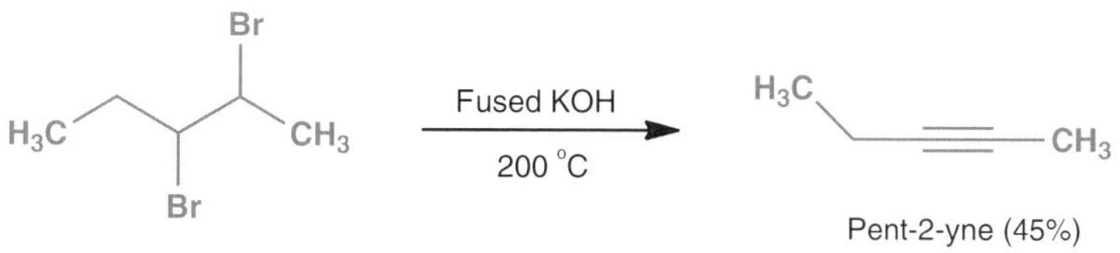

Pent-2-yne (45%)

Synthesis of ethchlorvynol

The synthesis of ethchlorvynol, a drug used to cause drowsiness and induce sleep, involves the addition of acetylide ion to a carbonyl group. The non-polar nature of the drug enhances its distribution into the fatty tissue of the central nervous system.

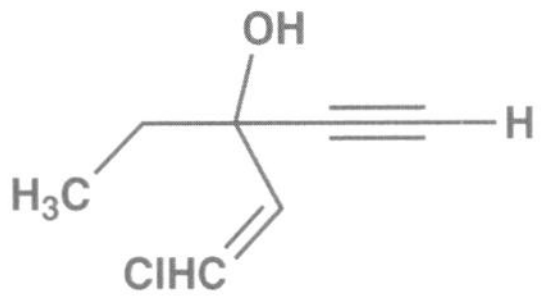

This leads to areas of high electron density both above and below, and in front and at the back of the σ-bond. Because of the shape of these π-bonds, they blend to form a cylinder of electron density which encircles the σ-bond. This makes alkynes electron rich and, therefore, they behave as nucleophiles. Consequently, the reaction profile of alkynes is very similar to that of alkenes, that is, they typically undergo addition reactions across the carbon-carbon triple bond.

Before we consider these addition reactions, we should mention the combustion of alkynes. In air, alkynes burn with a luminous, smoky flame. When they are combined with pure oxygen, the mixtures are explosive and are used in welding (for example, in oxyacety-lene torches). The reaction is extremely exothermic ($\Delta H = -1300$ kJ mol^{-1}) and temperatures reach around 3000 °C (Figure 4.17). Acetylene(ethyne) undergoes complete combustion under these conditions.

$$2C_2H_2(g) + 5O_2(g) \rightarrow 4CO_2(g) + 2H_2O(l)$$

Addition of hydrogen halides and halogens to alkynes

Moving to the addition reactions, let us first consider the addition of hydrogen halides (HX). The mechanism of these reactions is the same as for addition to alkenes. The first, slow step is the breaking of a relatively weak π-bond. The second step, the reaction of the carbocation intermediate with the negatively charged halide anion, is rapid (see Scheme 4.20).

For terminal alkynes, the proton adds to the sp carbon bearing the hydrogen, as this leads to the more stable secondary carbocation. For example, the addition of HBr to pent-1-yne leads only to the Markovnikov product, 2-bromopent-1-ene. In the presence of excess HX, a second addition occurs, again following Markovnikov's rule.

When excess HX is added to internal alkynes, where the sp carbons bear different groups, two products are formed. This is because H⁺ can add with equal ease to either carbon of the triple bond in the first step of the reaction, as shown for the addition of excess HBr to pent-2-yne in Scheme 4.21.

SCHEME 4.20

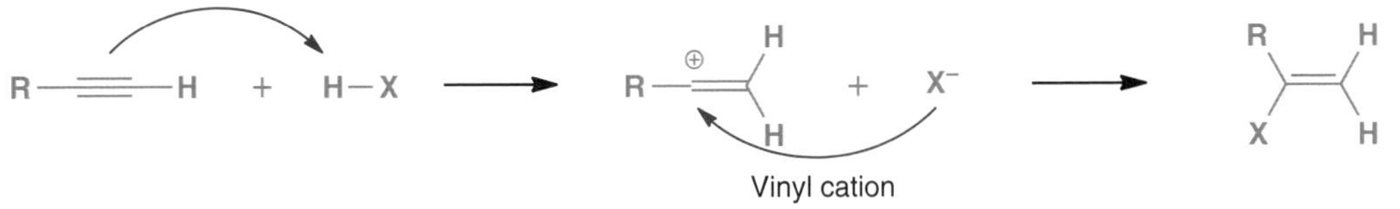

Vinyl cation

SCHEME 4.21

2-Bromopent-2-ene
(E/Z mixture)

3-Bromopent-2-ene
(E/Z mixture)

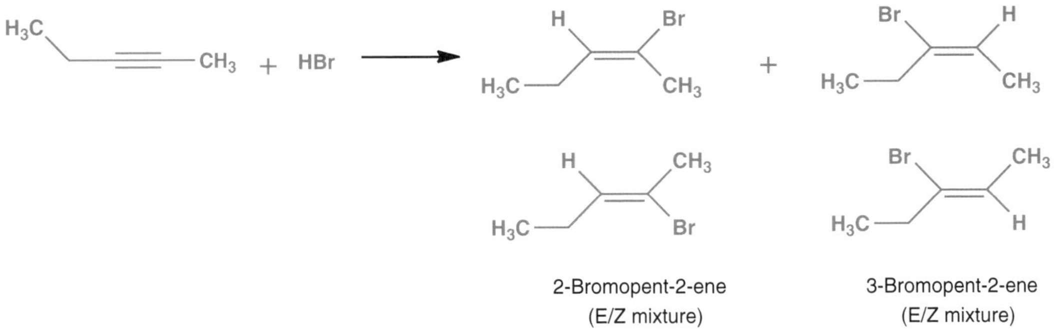

SCHEME 4.22

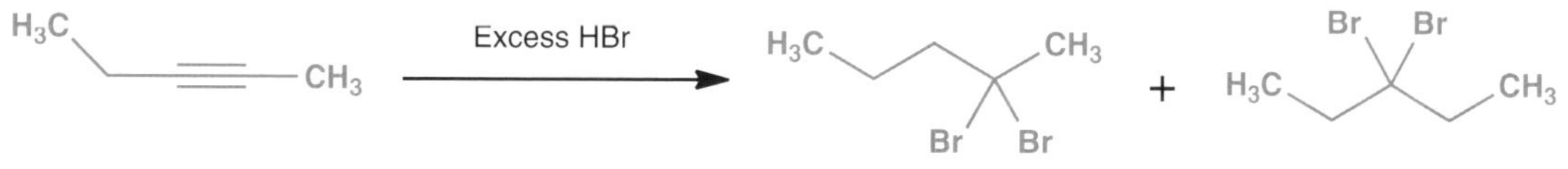

SCHEME 4.23

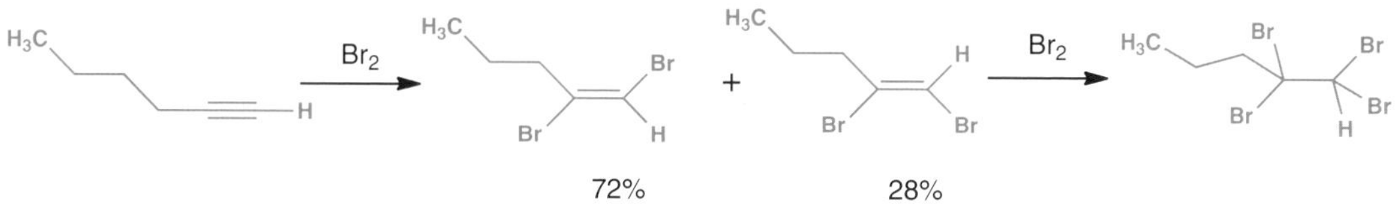

72%   28%

In the presence of excess HBr, the alkene products of Scheme 4.21 react to give two geminal dihalides – 2,2-dibromopentane and 3,3-dibromopentane (see Scheme 4.22). This happens as the second molecule of HBr generally adds with the same orientation as the first.

Addition of HBr to symmetrical alkynes (internal alkynes bearing the same substituents) is much less complex. Reaction of an excess of HBr with hex-3-yne leads only to 3,3-dibromohexane.

Halogens add to alkynes in the same way as to alkenes, generally yielding mixtures of *cis* and *trans*-isomers. In the presence of excess halogen, a second addition reaction can occur to give the tetrahaloalkane (see Scheme 4.23).

SELF CHECK 4.11

Draw the mechanism for the reaction of excess HBr to hex-3-yne.

FIGURE 4.17 Using an oxyacetylene torch to make crème brulee. The high temperatures create the perfect crisp top!
Source: Skip ODonnell/Istock.

SELF CHECK 4.12

Draw the products of the following reactions.

1. But-1-yne + HBr (1 equivalent)

2. 4-Methylhex-2-yne + HCl (1 equivalent)

3. 2-Bromo-but-1-ene + HCl (excess)

4. Oct-4-yne + HBr (excess)

5. Oct-2-yne + HCl (excess)

Addition of water to alkynes

Acid-catalysed addition of water (hydration) to alkynes is also seen, as with alkenes. The initial products of these reactions are **enols**, which rearrange immediately to give ketones (see Chapter 6). The enol and ketone differ only in the location of the double bond and are termed **keto-enol tautomers**. **Tautomers** are isomers that are in rapid equilibrium. The more stable isomer predominates and this is usually the ketone (see Scheme 4.24).

SCHEME 4.24

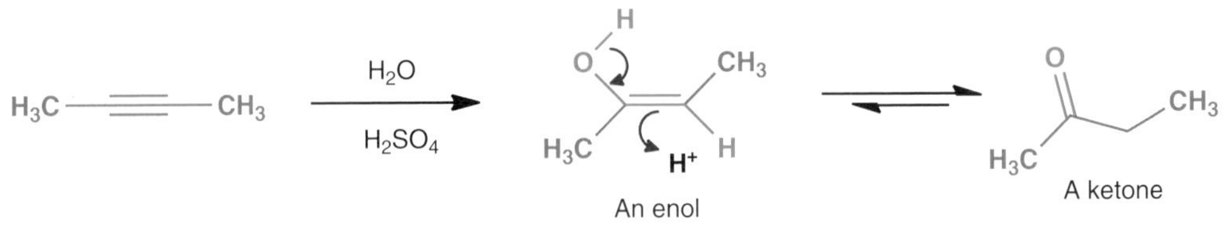

SCHEME 4.25

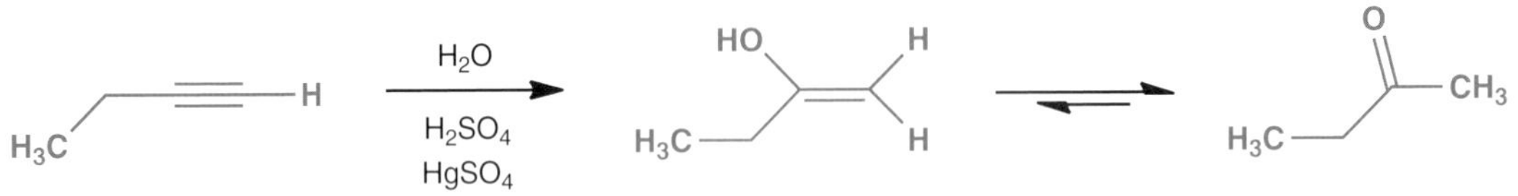

With symmetrical, internal alkynes, a single product results. Hydration of an internal alkyne bearing different substituents on the *sp* carbons may lead to the formation of two products.

SELF CHECK 4.13

What is the product of the hydration of hex-3-yne? There are two possible products of the reaction of pent-2-yne with water/acid. What are they?

Terminal alkynes are less reactive towards the addition of water than their internal alkyne relatives. However, they can be hydrated if mercuric acid (Hg$^+$) is added to the acidic medium. The mercuric acid acts as a catalyst to increase the rate of the reaction (see Scheme 4.25).

Addition of hydrogen to alkynes (reduction)

Hydrogen adds across the alkyne triple bond in the presence of a metal catalyst, typically palladium or platinum, much as with alkenes. The initial product of this hydrogenation is an alkene, but it is difficult to halt the reaction at this point. The efficiency of the catalysts used means that hydrogen will tend to add across the double bond of the alkene formed. Thus, an alkane is generally the result of such a reaction (see Scheme 4.26).

To limit the hydrogenation reaction and allow the preparation of alkenes from alkynes, a 'poisoned' or partially deactivated catalyst is used (see Scheme 4.27). The most common of these is Lindlar's catalyst, which can be prepared by treating the conventional palladium catalyst with, for example, lead acetate or quinoline.

The hydrogens add to the same side of the triple bond – this is termed *cis* (syn) addition – to give a *cis*-alkene. *Trans* (anti) addition of hydrogen to an alkyne, to give a *trans*-alkene, is achieved by reacting the alkyne with either sodium or lithium in liquid ammonia, generally at −78 °C. The reaction proceeds via a radical mechanism.

SCHEME 4.26

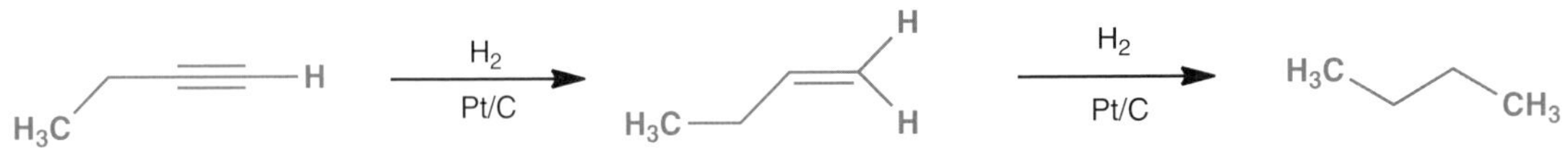

SCHEME 4.27

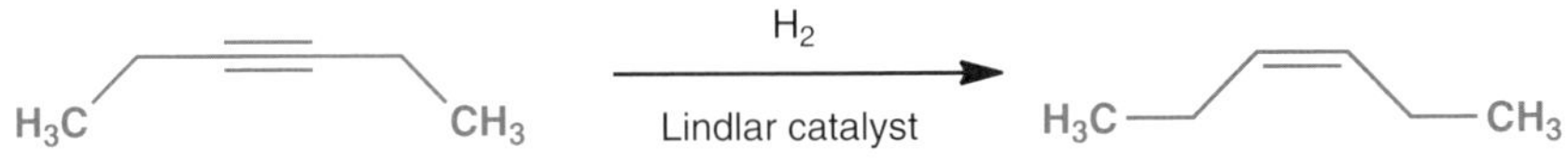

SCHEME 4.28

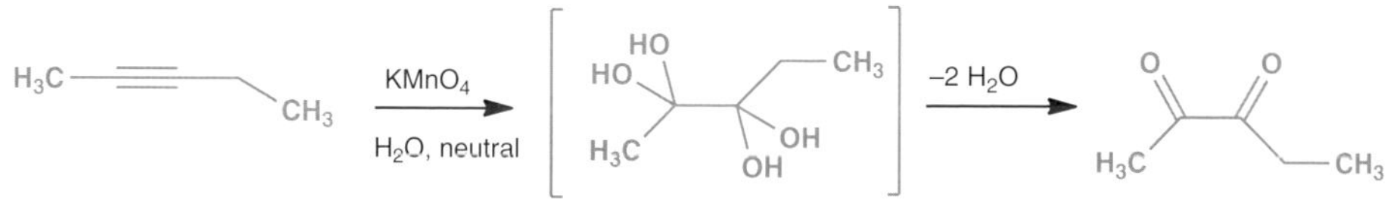

SCHEME 4.29

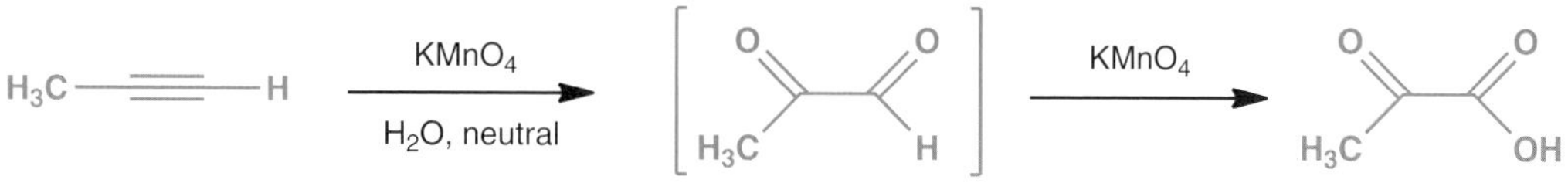

Oxidation of alkynes

The addition of cold, aqueous potassium permanganate to an internal alkyne under near-neutral conditions leads to the formation of an α-diketone. The reaction is actually a double hydroxylation of each of the π bonds of the alkyne, followed by the loss of two molecules of water (see Scheme 4.28).

Under the same conditions, terminal alkynes yield keto-acids, probably via keto-aldehydes which are further oxidized (see Scheme 4.29).

The product in Scheme 4.29 is pyruvic acid, the anion of which (pyruvate) is a key compound in several metabolic pathways. In the laboratory, this reaction is reasonably straightforward, even if the conditions are a little harsh, while nature has to work quite hard to prepare this molecule. In the body, 2 moles of pyruvic acid are made from 1 mole of D-glucose through

the process called glycolysis. This involves a lengthy series of enzyme-catalysed reactions.

 More details about glycolysis can be found in Chapter 3 'The biochemistry of cells' of the *Therapeutics and Human Physiology* book in this series.

When an internal alkyne is treated with warm or basic potassium permanganate, the diketone is oxidatively cleaved to give carboxylic acid salts. These are converted to the free acid by adding dilute mineral acid, for example HCl (see Scheme 4.30).

The cleavage of a terminal alkyne gives a carboxylate ion and the formate ion. The latter is further oxidized to carbonate which, after protonation, yields carbonic acid. Under the aqueous conditions used, carbonic acid decomposes to water and carbon dioxide. You will see that this gives us a carboxylic acid that

SCHEME 4.30

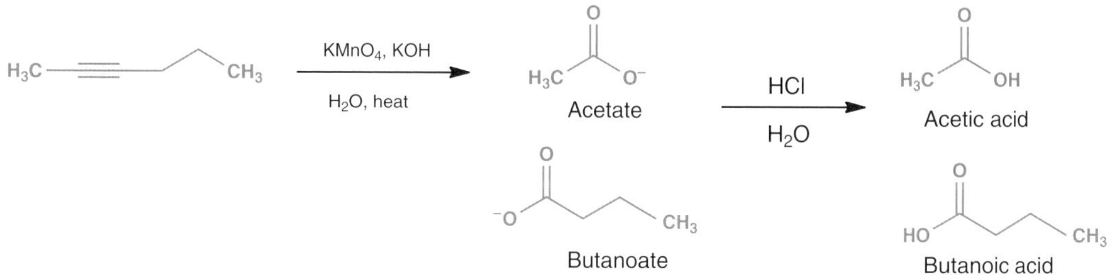

SCHEME 4.31

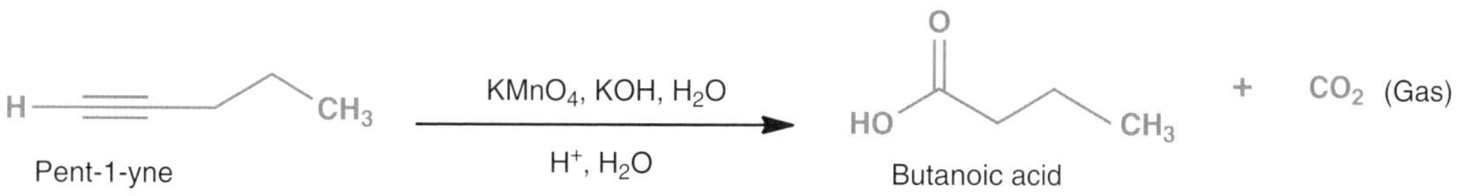

is one carbon shorter than the alkyne we started with (see Scheme 4.31).

Alkynes also undergo ozonolysis, i.e. the addition of ozone, O_3. Hydrolysis of the addition product cleaves the triple bond to give two carboxylic acids. For example, under these conditions 2-pentyne gives acetic (ethanoic) and propionic acids, while 1-pentyne gives butanoic and formic (methanoic) acids. Thus,

the method can be used analytically, to determine the position of the carbon–carbon triple bond.

KEY POINT

Alkynes, like alkenes, also undergo electrophilic addition reactions. Generally, but not always, two molecules of reactant are added.

4.6 Hydrocarbons in pharmacy

In this section we will show how hydrocarbons may be used in pharmacy as drugs, excipients and biologically important molecules.

Active pharmaceutical ingredients (APIs)

No common APIs consist only of aliphatic hydrocarbons, although 'liquid paraffin' is still used occasionally as a laxative, especially in a veterinary setting. The limited use of aliphatic hydrocarbons stems from their lack of polarity: the interactions of drugs with receptors are normally polar. However, hydrophobic groups are very important in drugs – they help drugs cross membranes, so aliphatic hydrocarbons can play a role here. An alkyl chain can also contribute conformational flexibility; longer chains give more conformational flexibility.

An example of the effect of a hydrocarbon chain on drug action comes from the barbiturate family of hypnotic drugs. The parent compound, barbituric acid, has no hypnotic properties, but such properties can be introduced by the addition of substituents at the 5-position. The length and degree of branching in these chains influences both how potent a particular barbiturate is and also its duration of action. For example, secobarbital is slightly more potent than pentobarbital owing to a single extra carbon in one of the 5-substituents. Addition of a methyl group to one of the ring nitrogen atoms leads to methohexital. This compound shows a rapid onset and short duration of action, but also increased levels of side effects.

Generally, any modification that causes an increase in lipophilicity in this series leads to an increase in potency and rate of onset, but also increased excretion and a reduction in the duration of action.

Pentobarbital

Secobarbital

Methohexital

These two cases are examples of the kind of **structure-activity relationship studies** that are generally part of the modern drug discovery process. You will meet this topic in later stages of your pharmacy studies.

More information about drug discovery processes can be found in Chapter 10 'Origins of drug molecules'

Stability of medicines

When considering the stability of a medicinal product, it is not only the active pharmaceutical ingredient (API) or 'drug' that we need to think about. Most of the mass of a tablet actually comes from other ingredients, called excipients. These play diverse roles and include fillers, binding agents, lubricants, plasticizers, antioxidants, colouring and flavouring agents. More detailed information on excipients can be found in the companion book *Pharmaceutics* in this series. In topical formulations, oils of various types are used, e.g. fatty acids, triglycerides, and 'soft' paraffin (petroleum jelly).

Sodium stearate (lubricant)

Sodium oleate (lubricant)

A mixed triglyceride formed from glycerol and three fatty acids (top to bottom): palmitic, oleic and linoleic acids

Many excipients contain hydrocarbons. The primary route through which the hydrocarbon chain degrades is through oxidation. As you saw earlier, there are many ways in which to oxidize hydrocarbons but, in terms of the stability of medicines, the most interesting is spontaneous oxidation in air at ambient temperatures.

Oxygen, although a diradical, is not a good initiator of the process, but light (and especially ultraviolet light) has sufficient energy to initiate oxidation reactions. Among aliphatic hydrocarbons, the most susceptible compounds are those with allylic groups, because the formation of an allylic radical is favoured. Allylic radicals are stabilized by resonance and so are more stable than the tertiary alkyl radicals you saw earlier.

The oxidation of alkenes in unsaturated fatty acids causes fats and oils to taste and smell rancid. The oxidation of linoleic acid is shown in Figure 4.18, as an example.

The hydroperoxide formed in Figure 4.18 can decompose via a number of pathways to give carboxylic acids, acid aldehydes and aldehydes. It is the aldehydic compounds in particular, e.g. 2-nonenal, that lead to the bad smell of oxidized fats.

 SELF CHECK 4.14

Draw a structure for 2-nonenal

Linoleic acid is one of the essential fatty acids required by humans. The body cannot make it and so we get it from our diet, e.g. from vegetable oils. As well as being a cell membrane component, metabolism of linoleic acid in the body generates arachidonic acid and, subsequently, prostaglandins, which are essential to the contraction of smooth muscle.

More details of prostaglandins can be found in Chapter 7 of the *Therapeutics and Human Physiology* book in this series.

FIGURE 4.18 The oxidation of linoleic acid.

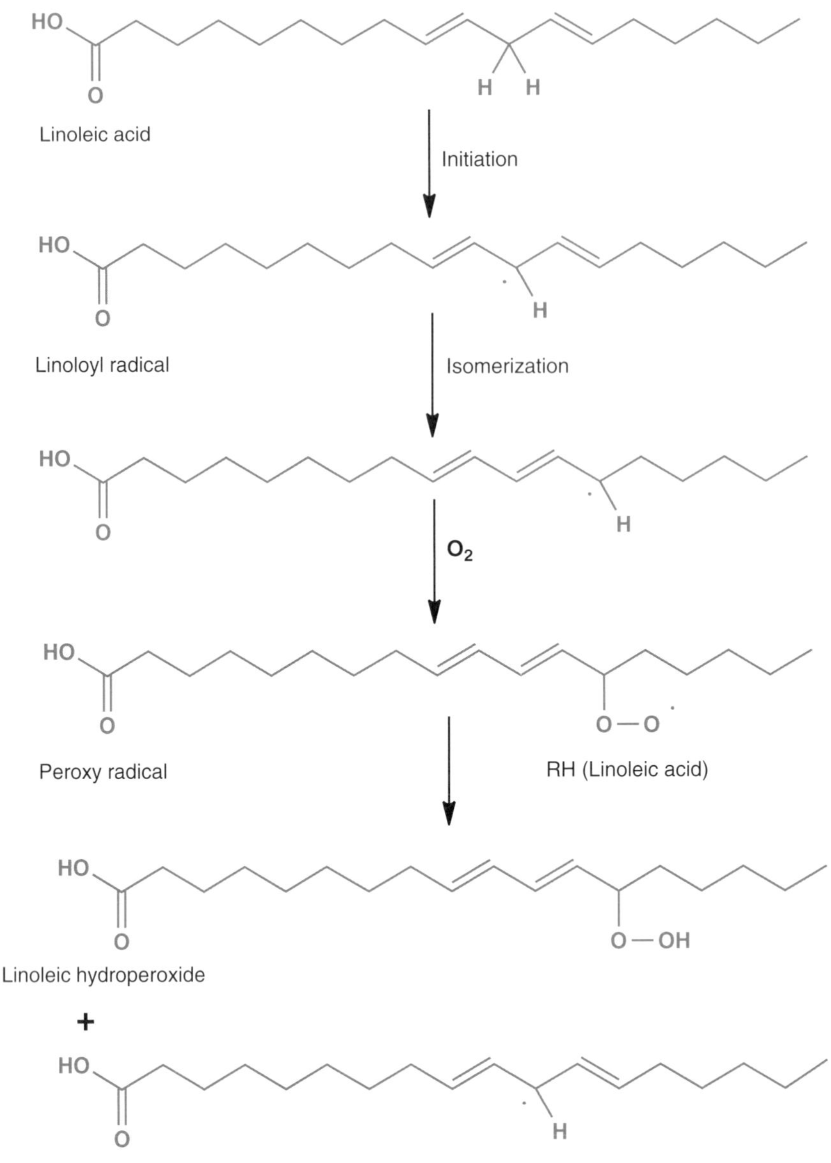

Linoleic acid

Initiation

Linoloyl radical

Isomerization

O_2

Peroxy radical

RH (Linoleic acid)

Linoleic hydroperoxide

+

Linoloyl radical

Absorption, distribution, metabolism and excretion (ADME)

The presence of lipophilic hydrocarbon groups in drugs affects each of these processes. While a wider discussion is beyond the scope of this chapter, we will briefly consider the ADME of hydrocarbons.

 You will find more detailed information on ADME in Chapter 1 'The scientific basis of therapeutics' of the *Therapeutics and Human Physiology* book in this series.

Most drugs are taken orally, at least in the UK. This means that the drug must dissolve in the gastrointestinal tract, an aqueous environment, before it can be absorbed. The presence of hydrocarbon chains in a drug molecule will lower its water solubility; lipophilic groups such as these are, however, necessary for a drug molecule to cross cell membranes, e.g. in the small intestine, and to enter into the systemic circulation. The distribution of a drug within the body depends on its lipophilicity and on its affinity for different environments within the body, such as tissue and plasma proteins.

Pharmaceuticals are, in evolutionary terms, rather recent. Our bodies recognize drugs as foreign substances (xenobiotics) which need to be disposed of, generally in our urine or faeces. However, most drugs need to be altered such that they become more water soluble before this can happen. The body, most especially the liver, has a wide range of enzymes which are capable of performing such reactions. These enzymes have evolved to deal with all manner of endogenous compounds and xenobiotics and are also capable of metabolizing drug substances.

Alkyl groups are mostly inert to metabolism, meaning that such groups are often excreted from the body unchanged. However, there are some occasions when alkyl groups are reactive. For example, hydroxylation (addition of OH) can occur at terminal (omega, ω) methyl groups and at the ($\omega - 1$) position (the carbon next to the end) through the action of mixed-function oxidases. Further metabolic reactions can lead to the generation of carbonyl compounds and to shortening of the carbon chain. As shown for the non-steroidal anti-inflammatory drug ibuprofen, oxidation can occur at either position, leading to either the alcohol (OH) metabolite or, after a second oxidation, the carboxylic acid (COOH) metabolite (see Scheme 4.32). Neither of these metabolites shows any pharmacological activity, which illustrates how important the lipophilic side chain is for the activity of ibuprofen.

The alkene components of drugs and other substances are oxidized to epoxides by cytochrome P450 enzymes, found mainly in the human liver. Epoxides are rather reactive species and are easily hydrolysed by enzymes called epoxide hydrolases, to give diols. The change from alkene to diol greatly increases the aqueous solubility of the substance and the likelihood of its excretion from the body. This is illustrated by the metabolism of carbemazepine, an anticonvulsant and mood stabilizing drug, as shown in Scheme 4.33. While 33 different metabolites of carbemazepine have been identified, the pathway shown predominates. The epoxide metabolite shows anticonvulsant activity comparable to that of carbemazepine. The diol metabolite is excreted either unchanged or in a conjugated form. The diol itself accounts for 30% of carbemazepine metabolites found in a patient's urine.

There are other important metabolic pathways for alkenes and these include hydration, peroxidation and reduction.

Like alkenes, alkynes are oxidized readily, and usually much more quickly. The products of these oxidations depend upon which carbon of the alkyne is attacked. Attack at a terminal alkyne carbon leads to a carboxylic acid metabolite (via a very reactive

SCHEME 4.32

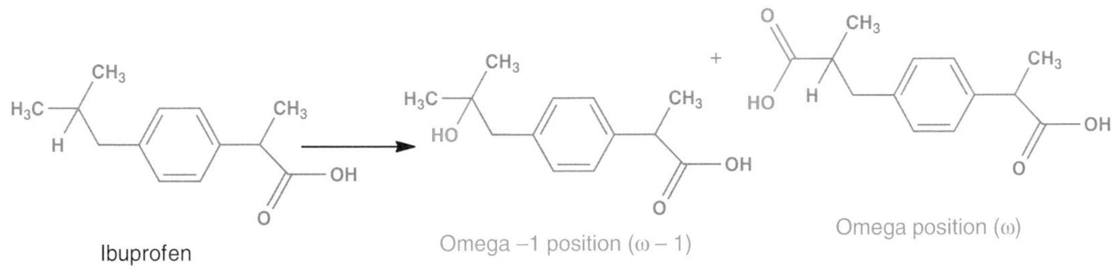

Ibuprofen

Omega −1 position ($\omega - 1$)

Omega position (ω)

SCHEME 4.33

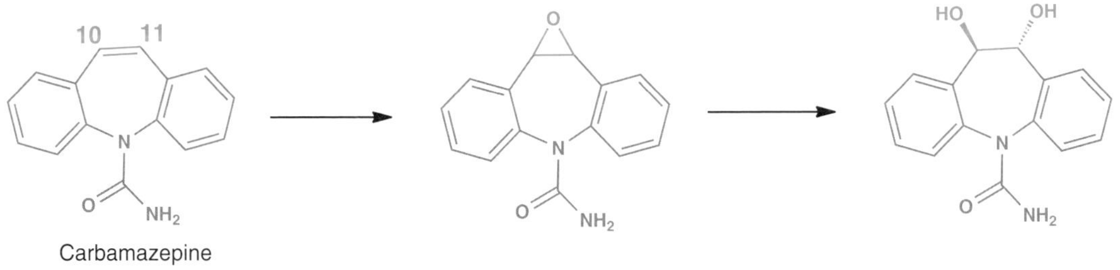

Carbamazepine

substance called a ketene), while attack at the internal alkyne carbon causes the compound to become permanently attached to the enzyme performing the oxidation, inactivating it irreversibly. This latter mechanism has been proposed as one pathway in the metabolism of drugs such as the female contraceptive, 17-α-ethinylestradiol.

17 α-ethinylestradiol

CHAPTER SUMMARY

This chapter contains information on the more important physical properties of the aliphatic hydrocarbons and their chemical reactivity. You should now understand:

➤ how to name aliphatic hydrocarbons

➤ how each class of aliphatic hydrocarbon is made or isolated

➤ the way in which alkanes, alkenes and alkynes react

➤ that aliphatic hydrocarbons can be important features of drugs and excipients.

FURTHER READING

There is an enormous number of organic chemistry texts available. The list below contains some personal favourites! The internet also contains a number of decent sites. Those below may be trusted, but you should always exercise some care when looking for information on a topic for the first time. Gathering information from more than one source is always a good idea.

Clayden, J., Greeves, N., and Warren, S., *Organic Chemistry*, 2nd edn. Oxford University Press, 2012.

Wade, L.G. *Organic Chemistry*, 8th edn. Pearson, 2012.

Bruice, P. *Essential Organic Chemistry International Edition*, 2nd edn. Pearson, 2009.

Baird, C. *Chemistry in your Life*, 2nd edn. W. H. Freeman 2006.

http://www.organic-chemistry.org/.

http://www.chemguide.co.uk/orgmenu.html.

http://www.chemtube3d.com/.

http://www.periodicvideos.com/.

Alcohols, phenols, ethers, organic halogen compounds, and amines

CHRIS ROSTRON

In this chapter we consider a number of **functional groups**. You will be familiar with oxygen (O_2) in the air that you breathe and as a component of water (H_2O), but oxygen atoms are also found in many molecules as parts of functional groups. Functional groups are made, changed, and destroyed in chemical reactions.

Oxygen's value as a component of functional groups stems from its high **electronegativity**, which means that it strongly attracts electrons. As a consequence, it often has a direct and important role in chemical reactions. In the periodic table, oxygen is flanked by nitrogen and fluorine. These too are electronegative elements: nitrogen is somewhat less electronegative than oxygen, but fluorine is the most electronegative element of all. The other halogens, chlorine, bromine and iodine, are also electronegative. Because of the electronegativities of these elements, their functional groups tend to take part in a range of similar chemical reactions, and we therefore discuss selected functional groups containing these elements in this chapter.

The hydroxyl group (–OH) is probably the most biologically significant functional group considered so far, as it is one of the most widely occurring in nature, being present in carbohydrates, proteins and nucleic acids. The properties of carbohydrates, for example, are essentially a combination of hydroxyl (OH) chemistry and the chemistry of aldehydes and ketones. Equally important is the amine functional group (NH_2). This group also occurs widely in nature, often in its protonated form ($-NH_3^+$); it is found in proteins, enzymes and nucleic acids. The amine group is also present in many drugs, often having a vital role in the interaction of the drug with a receptor. Halogens, although rarely found in nature, are often found in drugs because they can have a profound influence on distribution of a molecule throughout the body.

In this chapter, we study hydroxyl-containing compounds, ethers, amines, and halogen-containing compounds, before turning to the carbonyl group ($C=O$) in the next chapter.

Having read this chapter you are expected to be able to:

➤ identify the differences between the primary, secondary and tertiary alcohols and amines

➤ recognize how the presence of a functional group influences the physical properties of molecules

➤ recognize how the presence of a functional group influences the chemical properties of molecules

➤ explain the biological and pharmaceutical significance of these physical and chemical properties.

5.1 The hydroxyl group

The hydroxyl group (–OH) needs very little introduction. People with no background in science are usually aware that water is H_2O or H–OH. Those who have studied chemistry at school are often aware that ethanol (ethyl alcohol) is C_2H_5OH, and many will also have realized that the sugar (sucrose) molecule contains hydroxyl groups. The hydroxyl group features in the molecules of everyday life and every day we encounter more hydroxyl groups than we could ever realize or count.

Some important terminology

Before considering the properties of the OH group, we must introduce some important terminology. There are two pieces of terminology associated with the OH group, which must not be confused. The first of these refers to the number of OH groups present in a molecule. Monohydric, dihydric and trihydric molecules contain one, two and three OH groups, respectively. Often dihydric and trihydric alcohols are referred to as polyhydric alcohols. They have pharmaceutical and biological significance and we will return to these later in this chapter.

The second type of terminology relates to the type of carbon to which the OH group is attached. These different OH groups are referred to as primary (1°), secondary (2°) and tertiary (3°).

Figure 5.1 shows how a primary alcohol has two hydrogens and a single carbon on the carbon to which the OH group is attached, a secondary alcohol has one hydrogen and two carbons, and a tertiary alcohol has

no hydrogens and three carbons attached. This is an important distinction because the three types of OH group have different physical properties, react at different rates and sometimes even undergo different reactions.

Identify the following alcohols as primary, secondary, or tertiary: methanol, propan-1-ol, butan-2-ol, 2-methylpropan-2-ol, cyclohexanol. (Hint: Draw the structures in full before attempting to answer.)

Find the structure of hydrocortisone, a widely-used steroid drug. Identify the OH groups in its structure as primary, secondary, or tertiary.

Physical properties of alcohols

The physical properties of alcohols are quite different from those of the hydrocarbons considered in

FIGURE 5.1 General structures of primary, secondary, and tertiary alcohols.

Primary   Secondary   Tertiary

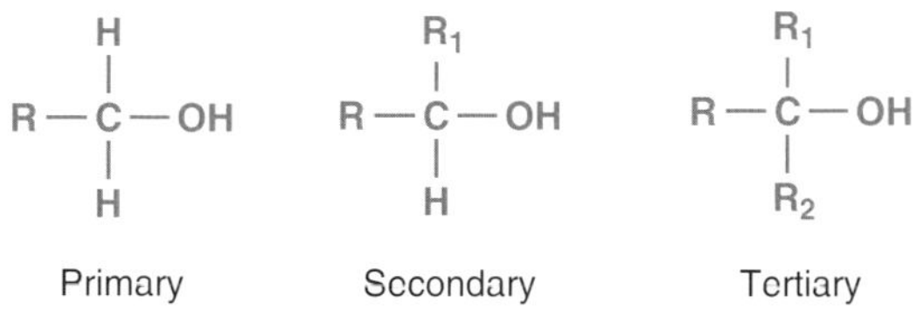

the preceding chapter. The hydroxyl group has a profound influence on boiling point and solubility; these properties in turn affect drug action and drug metabolism.

Boiling points

Methane and butane are, as we know from Chapter 4, gases at room temperature, but the equivalent alcohols, methanol and butan-1-ol, are liquids. Table 5.1 shows how the simpler alcohols are low boiling point liquids, but the boiling points increase quite quickly with increasing molecular mass.

What is responsible for this difference between alcohols and hydrocarbons? The answer lies in the ability of alcohols to form **hydrogen bonds**, which dominate the physical properties of alcohols. Hydrogen bonds are stronger intermolecular forces than **van der Waals interactions** or **dipole-dipole interactions**. Consequently the boiling points of alcohols are higher than those of hydrocarbons (which experience van der Waals intermolecular forces, but do not form hydrogen bonds) and halogenated hydrocarbons (which experience dipole-dipole intermolecular forces, but also lack hydrogen bonds).

If the hydrocarbon chain to which the OH group is attached is branched, the boiling point is lowered. Secondary and tertiary alcohols have lower boiling points than primary alcohols with the same molecular mass, because their branched structures interfere with the formation of intermolecular hydrogen bonds. Table 5.1 shows the boiling points of some representative alcohols, which illustrate these points.

 Look at Chapter 2 'Organic structure and bonding' to remind yourself about intermolecular forces.

TABLE 5.1 Boiling points of selected alcohols

Examples of alcohols	Boiling point °C
Methanol (primary)	65
Butan-1-ol (primary)	118
Decan-1-ol (primary)	233
Butan-2-ol (secondary)	100
2-Methylpropan-2-ol (tertiary)	83

SELF CHECK 5.3

Explain why propane boils at −45 °C, whereas propan-1-ol boils at 97 °C.

Solubility

Unlike hydrocarbons, simple alcohols are generally soluble in water. This property is also explained by the ability of hydroxyl groups to take part in hydrogen bonds. The smaller alcohols are totally soluble in water (**miscible** with water), a property that permits the preparation of screen wash, alcoholic handwash and, of course, alcoholic beverages. Table 5.2 shows how water solubility decreases with increasing size of the hydrocarbon chain, because larger hydrocarbon chains interfere with the hydrogen bond formation.

As the length of the hydrocarbon chain increases, an interesting property of long chain alcohols begins to emerge. At one end of the chain there is a water-soluble OH group; the rest of the molecule is a water-insoluble hydrocarbon chain. Such a molecule will accumulate at an **oil-water interface**, with the OH group residing in the water phase and the hydrocarbon chain in the oil phase. Molecules with both water-soluble and water-insoluble portions are called amphipathic molecules, and can be used as **emulsifying agents**. Emulsifying agents can be used to make water-insoluble and water-soluble substances mix (see Figure 5.2). The water-insoluble molecules cluster into tiny droplets, which are coated by the amphipathic molecules. The hydrocarbon portion of each **amphipathic** molecule interacts with the water-insoluble molecules inside the droplet, while their OH groups interact with the water molecules surrounding the droplet.

TABLE 5.2 Water solubilities of selected alcohols at 20 °C

Examples of alcohols	Solubility in g/100 g
Methanol	Completely miscible
Butan-1-ol	7.9
Hexan-1-ol	0.6

FIGURE 5.2 Formation of an oil-in-water emulsion

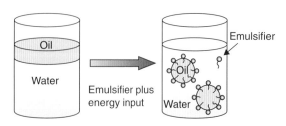

 KEY POINT

The ability to form hydrogen bonds dominates the physical properties of alcohols, both in terms of boiling points and water solubility.

You can learn more about emulsifying agents and emulsions in Chapter 10 'Disperse systems' in the *Pharmaceutics* book of this series.

SELF CHECK 5.4

Arrange the following molecules in order of increasing solubility in water (least soluble first): (a) pentan-1-ol, (b) pentane, (c) propan-1-ol.

The influence of hydroxyl groups on drug metabolism

The human body makes good use of the increased water solubility brought about by the presence of an OH group. One of the key transformations brought about by the **xenobiotic-metabolizing enzymes** in the liver is the introduction of an OH group into a molecule. These transformations increase the water solubility of the molecule, making it more easily excreted from the

body by the kidney. An example of this process can be seen in Figure 5.3. Diazepam is an anxiolytic drug (a drug used to treat anxiety) with a very long duration of action. One of its metabolites is temazepam, which is formed from diazepam by the introduction of an OH group. Temazepam is also used as a drug in its own right when a shorter duration of action is required. The shorter duration of action of temazepam is due to its increased water solubility, and hence more rapid excretion, as a result of the presence of the OH group.

 See Chapter 1 'The importance of pharmaceutical chemistry' of this volume and Chapter 1 'The scientific basis of therapeutics' of the *Therapeutics and Human Physiology* book in this series for more information about drug metabolism.

Chemical properties of OH-containing compounds

The polarity of the C–O–H functional group (see Figure 5.4) allows the formation of the hydrogen bonds that so profoundly influence the physical properties of hydroxyl compounds. This polarity also plays a significant role in determining the chemical properties of OH-containing compounds: both the C–O bond and the O–H bond are readily broken in chemical reactions because of the partial polarization of these bonds.

Breaking the C–O bond

The δ^+ charge on the carbon of the C–O bond (see Figure 5.4) makes this carbon susceptible to nucleophilic attack, leading to substitution reactions. As described

FIGURE 5.3 Metabolic transformation of diazepam.

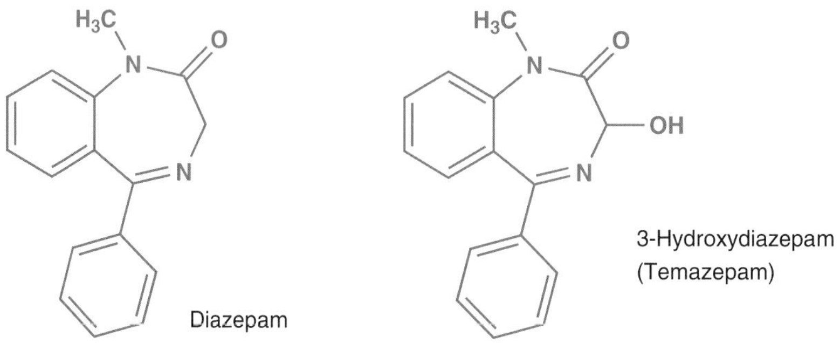

Diazepam

3-Hydroxydiazepam
(Temazepam)

FIGURE 5.4 Partial polarization of the C-O-H group.

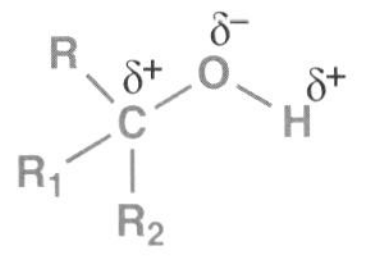

in Chapter 2, however, nucleophilic substitution is often in competition with an elimination reaction. Nucleophilic substitution of alcohols (or their derivatives) is largely of synthetic significance, whereas the competing elimination reaction – the removal of water (dehydration) – is a very important reaction in biological systems as well as in the laboratory.

Elimination reaction of alcohols – dehydration

In the laboratory, dehydration can be achieved by the action of heat on an alcohol, but is usually assisted by the use of a **catalyst**. A strong acid, such as sulfuric acid, is a good example of a catalyst in the elimination of water from an alcohol (see Figure 5.5). We have encountered this reaction in Chapter 4, because it is a method of preparing alkenes.

As can be seen in Figure 5.5, this dehydration reaction involves the formation of a **carbocation**. The relative stabilities of carbocations formed from the different types of alcohols determines the rate of the dehydration reaction. A tertiary alcohol will dehydrate more readily because the reaction involves the formation of a relatively stable tertiary carbocation (see Box 5.1).

BOX 5.1

Relative reaction conditions for dehydration of alcohols

The rate-determining step in the acid-catalysed dehydration of alcohols is the formation of an intermediate carbocation. The relative ease with which the dehydration takes place, mirrors the ease of formation of the carbocation, which in turn mirrors the stability of the carbocation: 3° > 2° > 1°. Therefore, tertiary alcohols are most readily dehydrated (5% H_2SO_4 at 50 °C), followed by secondary alcohols (75% H_2SO_4 at 100 °C) and primary alcohols are very difficult to dehydrate (95% H_2SO_4 at 170 °C).

SELF CHECK 5.5

What is the name of the product of the dehydration reaction in Figure 5.5?

Revise the stability of carbocations in Chapter 4 'Properties of aliphatic hydrocarbons'.

Sometimes dehydration reactions in the laboratory can give rise to unexpected alkene products because of rearrangement of the intermediate carbocation. In the human body, however, dehydration reactions are catalysed by enzymes and so just one specific product is formed (see Figure 5.6).

FIGURE 5.5 Dehydration reaction scheme.

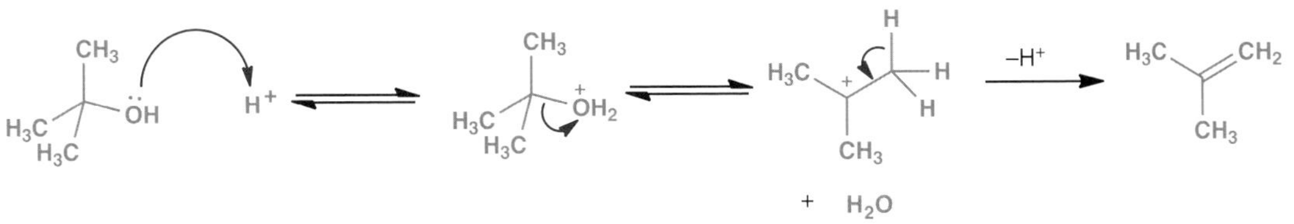

FIGURE 5.6 An example of biological dehydration.

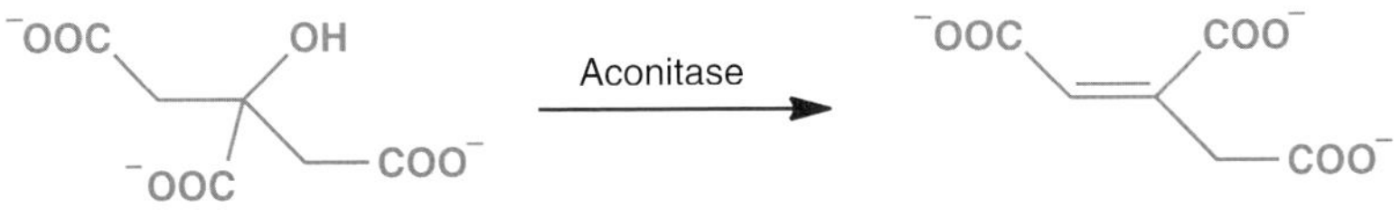

Citrate

Cis-aconitate

SCHEME 5.1

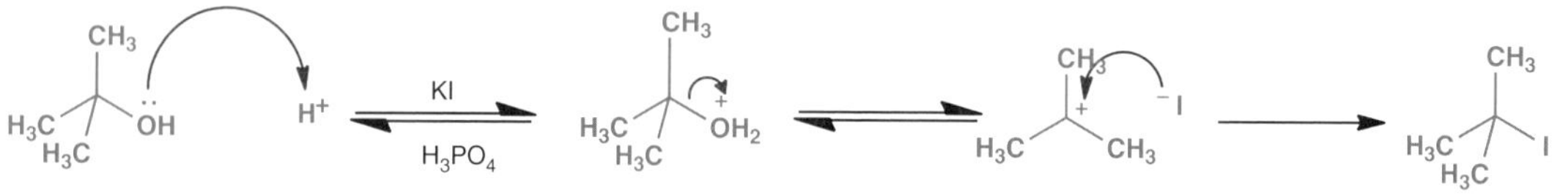

We encountered aconitase, which converts citrate to isocitrate, in Chapter 1 'The importance of pharmaceutical chemistry'. Figure 5.6 shows the intermediate in this reaction: *cis*-aconitate.

SELF CHECK 5.6

Which of the following alcohols would you expect to undergo dehydration in acid most readily: butan-2-ol, butan-1-ol, 2-methylbutan-2-ol?

Substitution reactions of alcohols

If a secondary or tertiary alcohol is treated with acid in the presence of a nucleophile, the intermediate carbocation may react with the nucleophile, rather than losing a proton. In this case there is a substitution reaction, rather than an elimination reaction (see Scheme 5.1).

The formation of the carbocation is once again the slow step, so the ease of acid-catalysed substitution is once again in the order $3° > 2° > 1°$. This reaction is known as an S_N1 reaction: substitution nucleophilic, type 1. We will meet S_N2 reactions later in the chapter.

Breaking the O–H bond

Two reactions of biological and pharmaceutical significance involve the O–H bond being broken. The first of these is esterification, which we explore in detail in the following chapter. The other reaction is **oxidation**. Chemically, oxidation is achieved by reacting an alcohol with an **oxidizing agent** (typically potassium permanganate or potassium dichromate) in the presence of acid.

Figure 5.7 shows how this oxidation reaction is one in which the three types of alcohol behave differently.

- primary alcohols are oxidized to aldehydes;
- secondary alcohols are oxidized to ketones;
- tertiary alcohols are not oxidized at all.

FIGURE 5.7   Oxidation of alcohols.

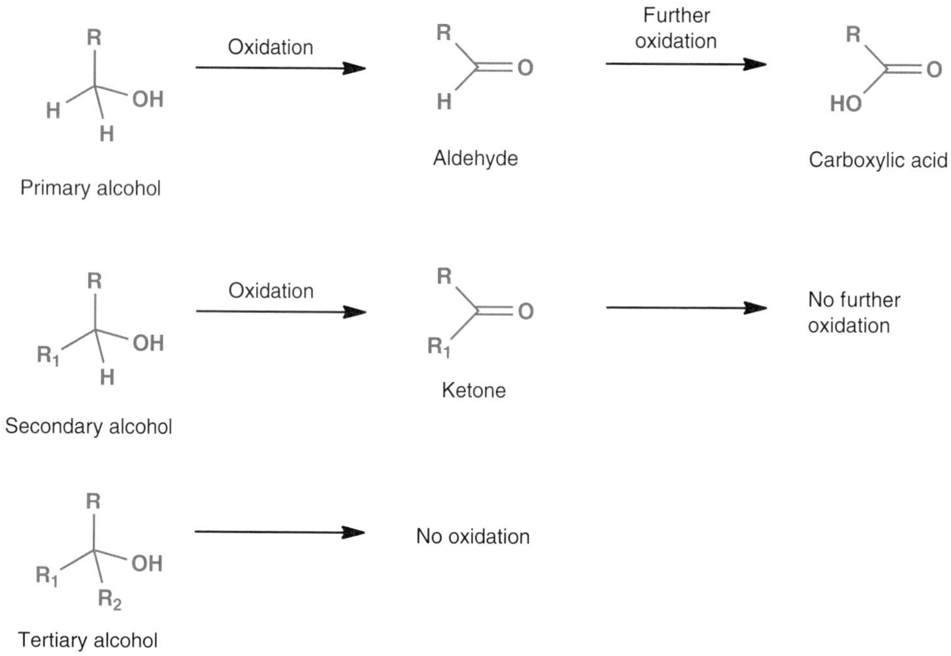

Primary alcohol → Oxidation → Aldehyde → Further oxidation → Carboxylic acid

Secondary alcohol → Oxidation → Ketone → No further oxidation

Tertiary alcohol → No oxidation

Aldehydes and ketones are both types of carbonyl compound; carbonyl compounds are discussed in Chapter 6.

Biological oxidation follows the same pattern as the chemical oxidation described above. Ethanol ($R = CH_3$ in Figure 5.7) is an example of a primary alcohol, and is oxidized in the body to an aldehyde (acetaldehyde or ethanal). The enzyme involved is alcohol dehydrogenase. Of course, when something is oxidized (the alcohol), something else must be reduced. In this case, as with so many biological oxidations, the compound that is reduced is **nicotinamide adenine dinucleotide (NAD⁺)**. This molecule is referred to as a **coenzyme** – it operates in conjunction with the enzyme catalysing the oxidation.

 More details on biological oxidation can be found in Chapter 1 'The importance of pharmaceutical chemistry' of this book and Chapter 3 'The biochemistry of cells' in the *Therapeutics and Human Physiology* book from this series.

Low molecular weight secondary alcohols are oxidized biologically to ketones by the same enzyme (alcohol dehydrogenase). For example, propan-2-ol (isopropyl alcohol, IPA, see Box 5.2) is oxidized to the ketone, propan-2-one (acetone). Acetone cannot be oxidized further and, being volatile and water-soluble, can be excreted via the lungs or in the urine.

SELF CHECK 5.7

Why is IPA used in antiseptic wipes, rather than ethanol?

 **CASE STUDY 5.1**

Sarah Patel had been to the Freshers' party the previous evening and was now feeling quite poorly – she had a hangover! Initially at the party she had felt a little lonely because she didn't really know anyone. Eventually a third year student started chatting to her and he bought her a couple of drinks. After these she felt more relaxed and soon she was enjoying herself on the dance floor and chatting to all sorts of people.

The drinks continued to flow and after a short while Sarah started to feel a little unsteady on her feet. Fortunately one of her flat mates decided to take Sarah back to her flat before any more damage was done. The next thing Sarah knew was the following morning when she woke up feeling quite ill. She knew she had a hangover but as it the first time, she didn't know what to do. She decided to ring her older sister Rachel for advice. Rachel was not particularly sympathetic and said there was very little that could be done. She knew that one of the reasons for feeling bad was that alcohol had been metabolized to acetaldehyde, which is the molecule primarily responsible for the hangover symptoms.

Sarah said that she had only had a few drinks so why did she feel so bad? Rachel did not know, told her to take two paracetamol tablets, drink plenty of water and sleep it off – that had always worked for her. She also said she would not tell Dad.

REFLECTION QUESTIONS

1. Why did Sarah feel more relaxed after a couple of drinks?

2. Why did Sarah start to feel unsteady on her feet after a few more drinks?

3. Why did Sarah have such a bad hangover after having a relatively small amount to drink?

1 A small amount of ethanol in the blood acts as a stimulant. At a blood level of up to 0.1% v/v inhibitions are decreased and anxieties are reduced, causing a feeling of pleasant relaxation.

2 At higher blood alcohol concentrations ethanol acts as a depressant. Increased disruption of the central nervous system leads to a loss of muscle coordination, slurred speech and an inability to concentrate.

3 Had Sarah spoken to Dilip, her dad, he would have been very cross, but as a pharmacist he would have been able to explain why Sarah was so poorly after only a few drinks. Ethanol is metabolized to acetaldehyde (responsible for most of the symptoms). Acetaldehyde is then further metabolized to acetic acid by aldehyde dehydrogenase. This enzyme exists in two forms – the cytosolic and the mitochondrial forms. Unfortunately for Sarah, she is one of the 50% of Asians who are missing the mitochondrial form. This delays the removal of acetaldehyde, prolonging her hangover symptoms. Also unfortunately, the lectures on this topic did not take place until well after the Freshers' party!

Answers

BOX 5.2

Pharmaceutical use of IPA

Isopropyl alcohol (propan-2-ol, IPA) is used as a topical disinfectant. Various concentrations in water are used, ranging between 50 and 95%. Antiseptic wipes are often pieces of gauze or similar material impregnated with IPA.

Tertiary alcohols are not oxidized, chemically or biologically. When the liver metabolizes tertiary alcohols, it normally converts them to water-soluble esters such as sulfates and glucuronides, known as **conjugates**, which can be excreted in the urine.

SELF CHECK 5.8

Why are tertiary alcohols not oxidized? (Hint: Draw the structure of a tertiary alcohol, such as 2-methylpropan-2-ol.)

Metabolic oxidation of primary and secondary alcohols takes place in the liver, as does the conjugation of tertiary alcohols. When a drug is administered orally, it is normally absorbed from the gastrointestinal tract into the hepatic blood circulation, ensuring it passes through the liver before reaching the general circulation. This ensures that any metabolic oxidation or conjugation of the drug will take place before it reaches the general circulation. This process is referred to as first pass metabolism. If a drug is rapidly inactivated by this process, consideration may have to be given to an alternative route of administration so that first pass metabolism can be avoided.

SELF CHECK 5.9

What would be the oxidation products of the alcohols in Self Check 5.4?

KEY POINT

Dehydration is a key reaction for alcohols, both chemically and biologically. Another key reaction is oxidation. In both of these reactions, it is important to recognize the differences between primary, secondary and tertiary alcohols.

 Numerous examples of biological oxidation of alcohols in the citric acid cycle, and the β-oxidation of fatty acids, can be found in Chapter 3 'The biochemistry of cells' in the ***Therapeutics and Human Physiology*** book of this series.

5.2 **Polyhydric alcohols**

We referred earlier in this chapter to polyhydric alcohols. These alcohols have certain properties that are useful in a variety of ways (see Box 5.3). Figure 5.8 shows some well-known polyhydric alcohols.

As you would expect, their numerous OH groups make small polyhydric alcohols miscible with water. In fact, they are so effective at forming hydrogen bonds with water molecules that they disrupt the normal ordered structure of water and depress its freezing point. That is why polyhydric alcohols, such as ethylene glycol (ethane-1,2-diol), are used as anti-freezes in car radiators. Another property that arises from their highly effective hydrogen bonding to water

is **hygroscopicity**. They form hydrogen bonds so effectively with water, that they actually absorb water from the air.

You might expect the boiling points of polyhydric alcohols to be elevated relative to their monohydric counterparts, and you would again be right. The boiling point of propan-1-ol is 97 °C, but the boiling point of glycerol (propan-1,2,3-triol) is 290 °C. Ethanol has a boiling point of 78 °C, but ethylene glycol (ethane-1,2-diol) boils at 197.3 °C.

 Hydrogen bonds are also discussed in Chapter 2 'Organic structure and bonding'.

FIGURE 5.8 Polyhydric alcohols and some of their uses.
Source: Soap: Gareth Boden; Antifreeze: Jane Norton/istock; Cupcake: Chris Leachman/istock.

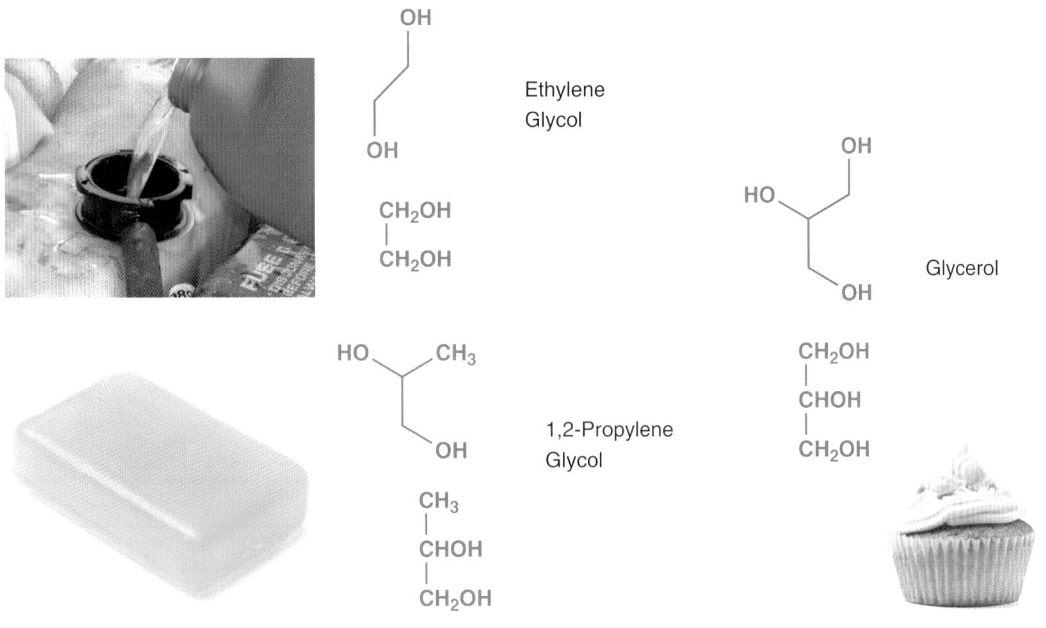

OH

Ethylene Glycol

OH

CH_2OH
CH_2OH

HO

OH

Glycerol

OH

HO CH$_3$

1,2-Propylene Glycol

OH

CH_2OH
CHOH
CH_2OH

CH$_3$
CHOH
CH_2OH

SELF CHECK 5.10

Why might you add glycerol (also known as glycerine) as an ingredient when preparing icing for a cake?

Finally, in this section, the different metabolic fates of simple polyhydric alcohols are of interest. Glycerol is metabolized, by oxidation, to glycerate, which is then converted to 3-phosphoglycerate – a component of the glycolytic pathway; 3-phosphoglycerate poses no problem to the body. On the other hand, ethylene glycol is metabolized by oxidation to oxalic acid. This product is extremely toxic because it interferes with the functioning of the electron transport chain, a process which is vital to the operation of cells (see Figure 5.9).

> More details on glycolysis and the electron transport chain can be found in Chapter 3 'The biochemistry of cells' in the *Therapeutics and Human Physiology* book in this series.

BOX 5.3

Pharmaceutical uses of glycerol

Glycerol is used in pharmaceutical products because it provides lubrication and has the ability to absorb water (it is a humectant). It is found in syrups, elixirs and pastilles used to treat coughs and sore throats, as well as in mouthwashes and skin care products. It is sometimes used as a binding agent in the preparation of tablets, and glycerol suppositories can be used to treat constipation.

BOX 5.4

Treating ethylene glycol poisoning

Strange as it may sound, one of the ways of treating ethylene glycol poisoning is by administering ethanol. Ethanol acts by competing with ethylene glycol for the enzyme, alcohol dehydrogenase. Ethanol has a much higher affinity than ethylene glycol for this enzyme (~ 100×) and effectively blocks the conversion of ethylene glycol to its toxic products.

FIGURE 5.9 Oxidation of some polyhydric alcohols in the body.

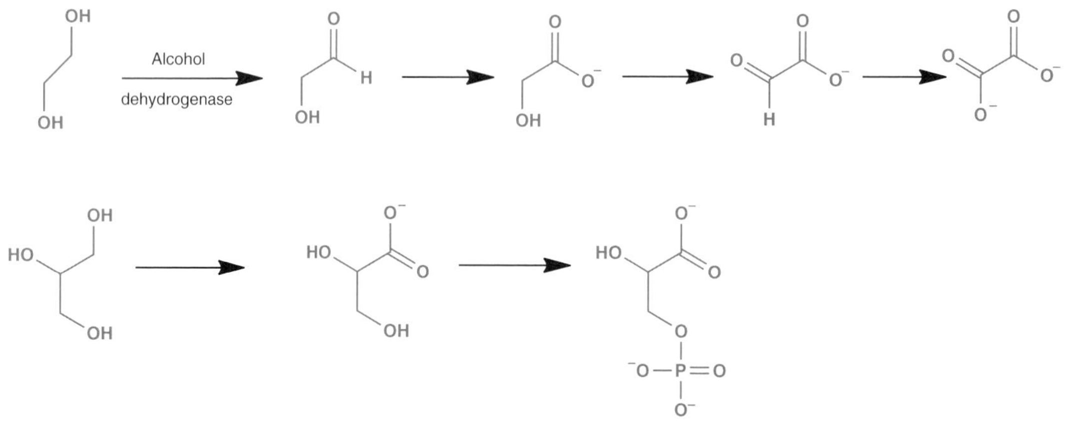

Glycerol   Glycerate   3-Phosphoglycerate

5.3 **Phenols**

There is one more type of OH group that we need to consider: an OH group attached to an aromatic hydrocarbon ring, such as benzene. The aromatic ring dramatically changes the physical and chemical properties of the OH group; indeed, the differences are so great that this type of OH group is regarded as a different functional group and is given a different name. It is not an alcohol but a phenol.

The OH group attached to an aromatic ring is more polarized than in an alcohol, causing stronger intermolecular hydrogen bonds. You might expect this to have a significant effect on boiling/melting points and water solubility. However, the rest of the molecule is hydrocarbon in nature and the resulting properties are as a consequence of a balance between these two factors. Increasing the number

TABLE 5.3 Physical properties of some phenols

Molecule	Structure	Melting point (°C)	Water solubility (g L^{-1})
Phenol	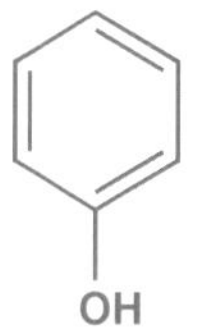	41	83
Benzene-1,3-diol	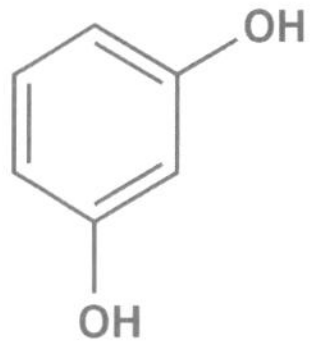	110	1100
Naphthalen-2-ol	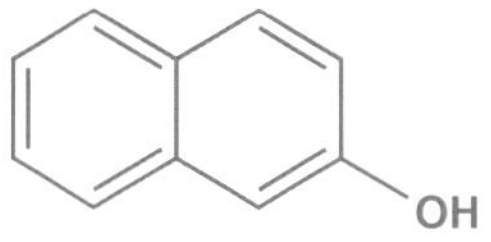	123	7

FIGURE 5.10 Delocalization of the negative charge of phenolate anion.

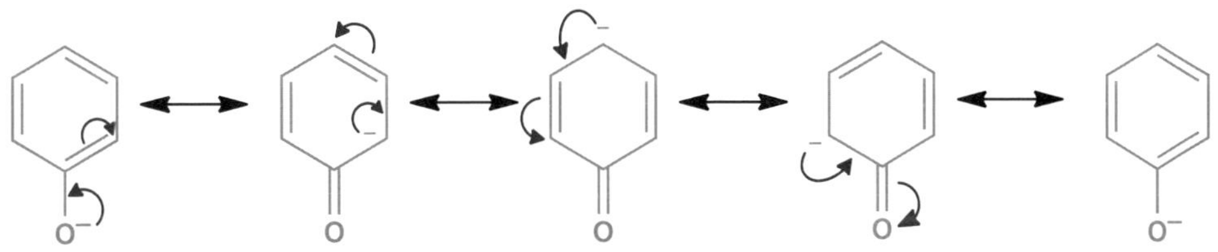

Phenolate

of OH groups, increases the melting point and water solubility, but increasing the hydrocarbon component dramatically reduces the water solubility (see Table 5.3).

Acidity of phenols

Unlike alcohols, which are not acidic (the pK_a of ethanol is about 16), phenols are weak acids (see Box 5.5). They are acidic enough to turn blue litmus paper red, and to react with sodium carbonate to yield carbon dioxide. The acidity of a phenol stems from the way that the hydrogen attached to the oxygen of the OH group is more easily lost as a proton (relative to an alcohol), with a pK_a of about 9.

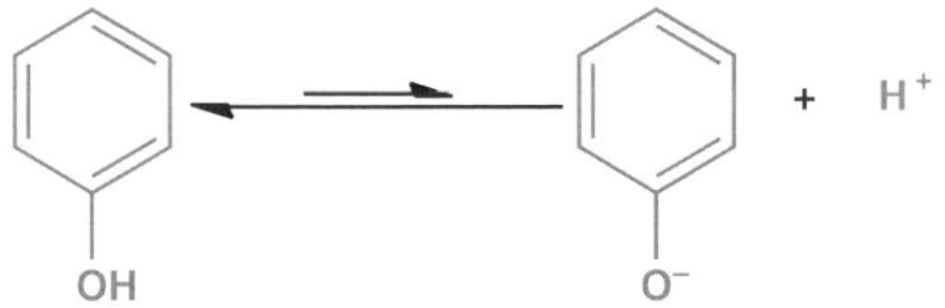

+ H⁺

OH   O⁻

Phenol   Phenolate

The ease with which a proton is lost depends upon the stability of the resulting anion: the more stable the anion, the more easily the proton will be lost. If the negative charge on the anion can be effectively delocalized (shared with other atoms), then the anion will be relatively stable. In phenols, the negative charge on the phenoxide anion can be delocalized into the aromatic ring as illustrated in Figure 5.10, hence the weak acidity observed.

Such delocalization is not possible in alcohols and so they are not normally regarded as acidic ($pK_a \sim 16$). You will see in Chapter 6 that the anion of a carboxylic acid is highly stabilized and hence they are more acidic even than phenols; their pK_as are around 4.

 Chapter 6 'Acids and bases' in the *Pharmaceutics* book in this series, discusses pK_a in more detail.

SELF CHECK 5.11

The side-chain of the amino acid tyrosine is a phenol, with a pK_a of about 9. What percentage of tyrosine residues are ionized in the small intestine at pH 8?

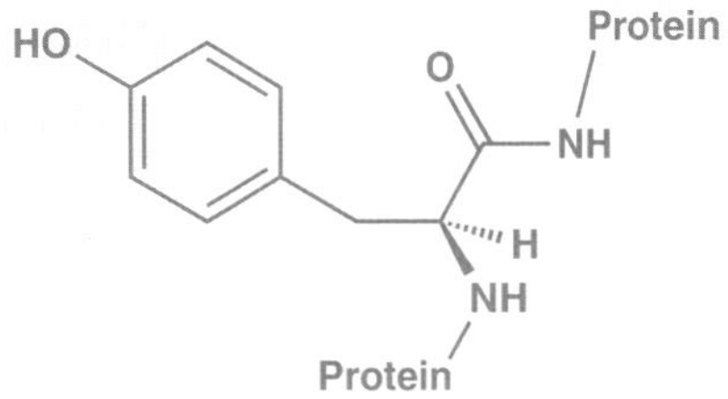

A tyrosine residue in a protein.

BOX 5.5

The Henderson-Hasselbalch equation

The pK_a of ethanol is about 16; this tells us that ethanol is half-ionized at pH 16. In other words, the following equilibrium reaction lies in the middle at pH 16:

The Henderson-Hasselbalch equation tells us that

$$pH = pK_a + \log\frac{[A^-]}{[HA]}$$

So at pH 7, $\log\frac{[A^-]}{[HA]}$ is -9, so $\frac{[A^-]}{[HA]}$ is 10^{-9}.

So, at pH 7, just one molecule of ethanol per billion (10^{-9}) is ionized.

SCHEME 5.2

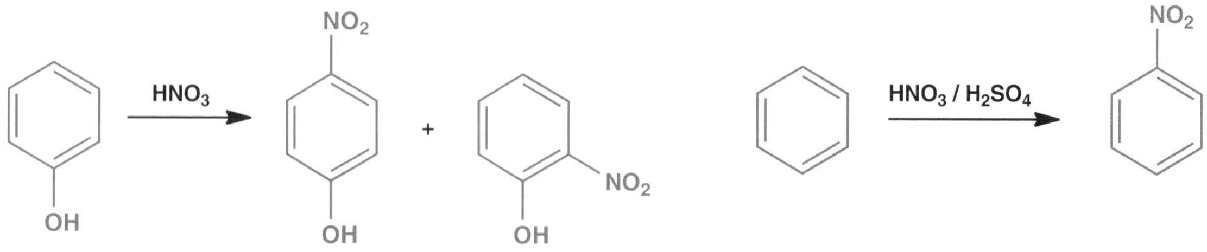

Reactions of phenols

Because alcohols and phenols both possess the OH group, one might expect them to undergo similar reactions. However, whereas alcohols readily undergo nucleophilic substitutions and elimination reactions, phenols undergo neither. Phenyl carbonium ions are very difficult to form because of the geometry of the benzene ring, so S_N1 and elimination reactions are out of the question. The geometry also rules out S_N2 reactions (see section on 'Chemical properties' of 'Haloalkanes and other organic halogen compounds' in this chapter). Instead, phenols undergo electrophilic aromatic substitution and do so very easily because the OH group is strongly ring-activating.

See more about electrophilic aromatic substitution in Chapter 7 'Introduction to aromatic chemistry'.

For example, phenol itself can be nitrated using dilute nitric acid, whereas benzene requires a mixture of concentrated sulfuric and nitric acids (see Scheme 5.2).

Phenols, like alcohols, can be oxidized. Phenols exposed to air for a period of time often become

FIGURE 5.11 Oxidation of benzene-1,4-diol, an example of oxidation of a phenol.

1,4-Dihydroxybenzene   1,4-Benzoquinone

coloured because of the formation of oxidation products. The oxidation of benzene-1,4-diol to 1,4-benzoquinone is actually reversible and this interconversion is an important biological reaction (see Figure 5.11).

Phenols as antioxidants

Phenols are useful as antioxidants, both pharmaceutically and biologically (see Box 5.6). They are known as **free radical scavengers**, that is, they mop up free radicals by reacting with them and rendering them unreactive. Figure 5.12 illustrates this reaction; look at this figure and notice how the hydrogen atom of the phenolic OH is transferred to the free radical species, thus neutralizing it. Of course, this also results in the formation of a phenolic free radical. This is not a problem, however, as phenolic free radicals have a great tendency to react with each other (**dimerize**), rather than with anything else. This dimerization inactivates the free radicals, rendering them harmless.

Free radical oxidation is discussed further in Chapter 12 'Kinetics and drug stability' in the *Pharmaceutics* book of this series.

Nature also utilizes phenols as antioxidants; indeed the antioxidant properties of phenols lies behind the proposed benefits to health of drinking small quantities of red wine, which contains polyphenol antioxidants (see Figure 5.13).

SELF CHECK 5.12

Can you identify the most commonly occurring natural antioxidants?

FIGURE 5.12   Reaction scheme for antioxidant activity of phenols.

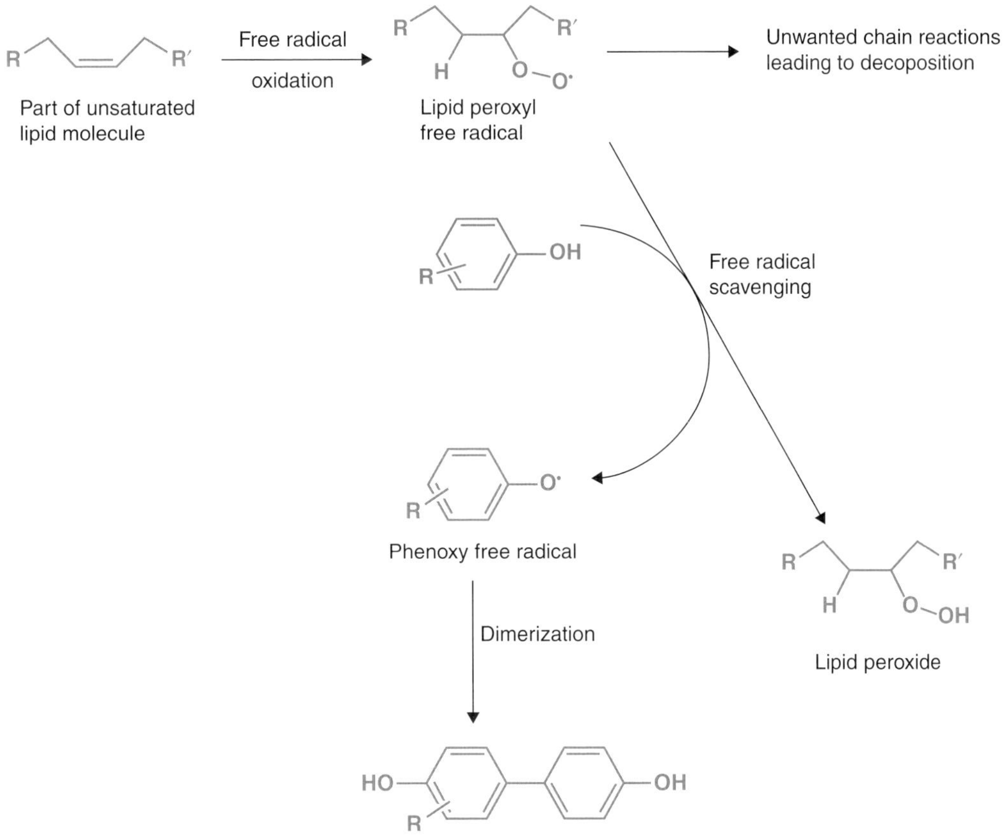

Part of unsaturated lipid molecule

Free radical oxidation

Lipid peroxyl free radical

Unwanted chain reactions leading to decoposition

Free radical scavenging

Phenoxy free radical

Lipid peroxide

Dimerization

BOX 5.6

The biological problem of free radicals

Free radicals are highly reactive and can cause significant damage to biological structures. This damage is often associated with reactive oxygen species, such as superoxide (O_2^-) and the hydroxyl free radical (OH$^.$) and is referred to as oxidative stress. Oxidative stress can cause damage to cells by initiating chain reactions, such as lipid peroxidation or by oxidizing DNA or proteins. Damage to DNA, for example, can cause mutations, which may increase the risk of cancer. Because of the risk of damage, plants and animals possess a variety of antioxidants such as vitamins C and E. These compounds act as free radical scavengers like the phenolic antioxidants, which are now widely added to foodstuffs and pharmaceutical preparations.

 More information can be found in Chapter 4 'Properties of aliphatic hydrocarbons' of this book, and Chapter 12 'Kinetics and drug stability' in the **Pharmaceutics** book in this series.

KEY POINT

Phenols are acidic and alcohols are not. Alcohols (but not phenols) can undergo nucleophilic and elimination reactions. Phenols can undergo electrophilic substitution reactions. Both functional groups can undergo oxidation reactions but the circumstances in which this happens are different.

FIGURE 5.13 Tannic acid is present in red wine.
Source: Wine: Photodisc.

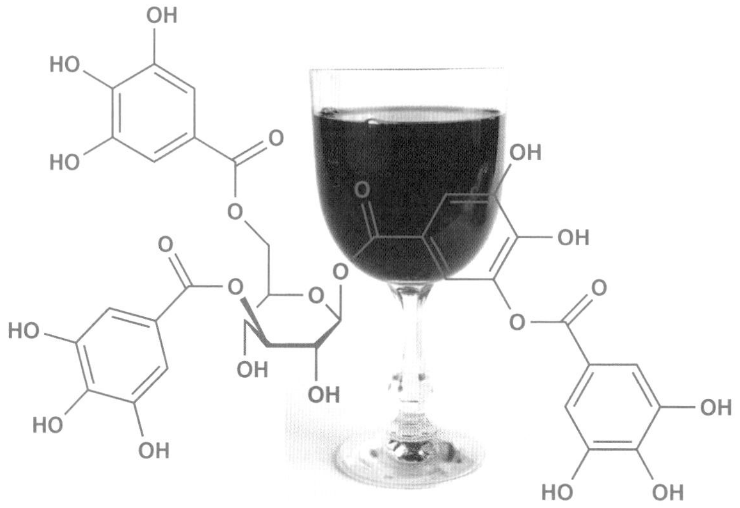

5.4 **Ethers**

Ethers are formed by the replacement of the hydrogen of an alcohol OH group by an alkyl (hydrocarbon) group, which generates an –OR group. Figure 5.14 shows how an ether oxygen can be present in an open chain or as part of a cyclic structure.

The replacement of the hydrogen by an alkyl group results in significant changes to both the physical and chemical properties of ethers, when compared with alcohols.

Physical properties of ethers

Generally speaking, ethers have low boiling points and are immiscible with water (quite unlike alcohols). These differences arise because of the loss of intermolecular hydrogen bonding capacity as a result of there being no hydrogen attached directly to the oxygen. This loss, and the addition of the extra hydrocarbon component, results in the water insolubility.

Although they cannot form intermolecular hydrogen bonds, ethers can self-associate by dipole-dipole attraction. This self-association is not especially effective because the two hydrocarbon groups attached to the oxygen often prevent molecules approaching each other closely. Ethers are, however, good solvents for a wide range of organic molecules. They are widely used as extraction solvents (see Box 5.7) because their volatility makes them easy to remove at the end of the extraction process. There are, however, disadvantages to their use. As one might expect, they are highly flammable and have a tendency to undergo free radical oxidation to generate peroxides, which are highly explosive (see next section).

FIGURE 5.14 Some common ethers.

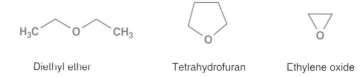

Diethyl ether   Tetrahydrofuran   Ethylene oxide

Ether-water extraction

A hydrophobic compound can be extracted from water into ether. Ether is less dense than water, so forms a layer above the water. The water layer is easily removed, but, although water does not mix with ether, a small amount (1.5 g per 100 mL ether) does dissolve in the ether. Before evaporating the ether, it is necessary to remove the dissolved water (to dry the solvent), usually by mixing it with a solid drying agent, such as magnesium sulfate ($MgSO_4$), then filtering off the drying agent.

The volatility of ethers and their high affinity for lipid tissue also makes them effective general anaesthetics (see Box 5.8).

Diethyl ether is widely used as a solvent in Northern Europe, but not in Australia. Why is this?

General anaesthetics

Anaesthesia is required for three purposes – to achieve analgesia (pain relief), to cause loss of consciousness, and to bring about muscle relaxation. However, until the middle of the nineteenth century, when diethyl ether was first used in a surgical procedure, no general anaesthetics were used in surgery.

A good inhalational anaesthetic should be highly volatile, lipid soluble, chemically stable and non-toxic. Diethyl ether fulfils most of these criteria but is dangerous to use because of its flammability. Nitrous oxide (laughing gas) and chloroform have been used in the past. However, these have all been superseded by halogenated compounds such as halothane and enflurane (see Figure 5.21).

There is still considerable debate about the mechanism by which general anaesthetics bring about their action. A number of theories have been proposed over the years, but they are all variations on a basic theme – that general anaesthetics cause reversible changes in nerve cell membranes.

Explain why propane boils at −45 °C, dimethyl ether boils at −23 °C and ethanol boils at 78 °C.

Chemical properties

Chemically, ethers are unreactive. They, therefore, bear more resemblance to saturated hydrocarbons than to alcohols. This lack of reactivity adds to their usefulness as organic solvents and general anaesthetics. However, their tendency to form **peroxides** renders their use potentially hazardous. The peroxides are formed by free radical reaction with oxygen, as illustrated in Figure 5.15.

In particular, this free radical reaction is catalysed by light, which explains the instruction that ethers should be stored 'in dark, well-filled bottles'. Use of well-filled bottles reduces the amount of oxygen available for reaction and dark glass reduces entry of the ultraviolet component of light which initiates the free radical reaction.

To see what can happen if these restrictions are not observed see the video clip linked to from http://www.oxfordtextbooks.co.uk/orc/ifp for an example of an ether explosion.

The replacement of the hydrogen of the OH group by an alkyl group renders both the physical and chemical properties of ethers significantly different from alcohols.

Epoxides

Generally, cyclic ethers have similar chemical properties to those of acyclic ethers. Three-membered cyclic ethers (also called epoxides) are quite different, however. The structures of some epoxides are shown in Figure 5.16.

Epoxides (more rarely called oxiranes) are very reactive. This is because a three-membered ring is very strained and any reaction that will relieve that strain by breaking the ring is favoured. This reactivity is the basis for the use of ethylene oxide for cold gaseous sterilization, as explained in Box 5.9.

FIGURE 5.15 Peroxide formation from diethyl ether.

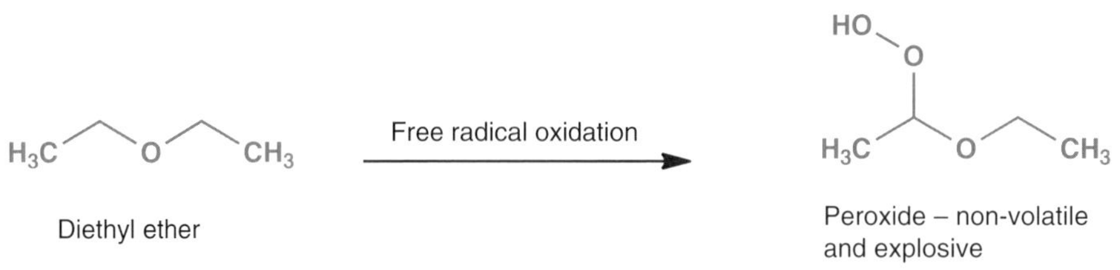

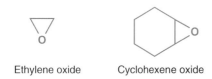

Diethyl ether

Free radical oxidation

Peroxide – non-volatile
and explosive

FIGURE 5.16 Oxiranes or epoxides.

Ethylene oxide   Cyclohexene oxide

The reactivity of epoxides also causes problems within the body's xenobiotic metabolizing processes. The body finds aromatic hydrocarbons (present, for example in cigarette smoke) difficult to metabolize and processes them by forming epoxides, which are extremely reactive. These epoxides can be considered to be **proximate carcinogens**. They remove a hydrogen radical from cellular constituents, becoming converted into a non-toxic hydroxyl compound in the process. However, the substrate from which the hydrogen was abstracted is now a very reactive species, and can cause further unwanted reactions within the cells. Nucleic acids are particularly susceptible to reaction with epoxides in this way, making epoxides potentially carcinogenic. This is described in more detail in Chapter 7 ('Metabolism and toxicity' section in 'Aromatic chemistry in the body').

 Carcinogenesis is described in Chapter 2 'Molecular cell biology' of the *Therapeutics and Human Physiology* book in this series.

KEY POINT

Whereas ethers are generally unreactive, small ring cyclic ethers, like ethylene oxide, are very reactive.

More information can be found on anti-smoking and public health in Chapter 7 'Behavioural and social sciences' of the *Pharmaceutical Practice* book in this series.

BOX 5.9

Cold gaseous sterilization

The usual method of sterilization (killing bacteria) employs heat and high pressure. However, some materials cannot be sterilized by this means because they are heat-sensitive. In these cases cold gaseous sterilization, employing ethylene oxide, can be used.

Ethylene oxide's reactivity ensures that the bacteria are killed, but there is a potential disadvantage in that ethylene oxide is so reactive that it may react with the material itself or its container. Also, ethylene oxide will not penetrate far below the surface of a material because of its reactivity. This means it can only be used for surface sterilization. Its reactivity also makes ethylene oxide highly toxic and so any remaining traces must be removed before the sterilized material can be used.

5.5 Haloalkanes and other organic halogen compounds

If you look for organic halogen compounds (organic compounds containing halogens) in an organic chemistry textbook, you will find almost a whole chapter devoted to two reaction types – nucleophilic substitution and elimination. These are reactions of haloalkanes and they have some relevance in pharmaceutical chemistry, but, in fact, there are other aspects of organic halogen compounds that are far more interesting from a pharmaceutical (see Figure 5.17) and biological point of view.

Nomenclature

This chapter is about functional groups, and when we talk about haloalkanes, we mean the haloalkane functional group (also known as an alkyl halide). The distinction between a haloalkane functional group and a haloalkane molecule is not always made in organic chemistry textbooks; if chemists want to study substitution reactions at R–Cl then R might as well be an alkyl group (made up of sp³ carbon and hydrogen only). However, complex drug molecules such as chlorambucil (see Box 5.10), also contain haloalkane functional groups. R is certainly not a simple alkyl group, but the haloalkane functional group behaves in a predictable way; indeed the reac-

tions of chlorambucil are not very different from those of 1-chloropropane.

Like alcohols, haloalkane compounds can be primary, secondary or tertiary, and they can also be mono- or polyhalogenated. The difference between mono- and polyhalogenations is perhaps less profound than the difference between mono- and polyhydroxylation, but the difference between primary, secondary and tertiary organic halogen compounds is important.

Physical properties

Halogens are very electronegative elements and so haloalkanes and other organic halogen compounds (sometimes called organohalogen compounds or organohalides) tend to self-associate by dipole-dipole attraction. Like ethers, these molecules are poorly soluble in water, because they cannot readily form hydrogen bonds with water molecules. However, in a hydrophobic environment, they *can* form hydrogen bonds (see Figure 5.18). These hydrogen bonds can be important in drug-receptor interactions, leading to prolonged retention of the drug at the receptor site. Consequently, halogen atoms are often introduced into the structure of potential drug molecules in an attempt to improve their biological activity.

133

FIGURE 5.17   Some chlorine-containing drugs.

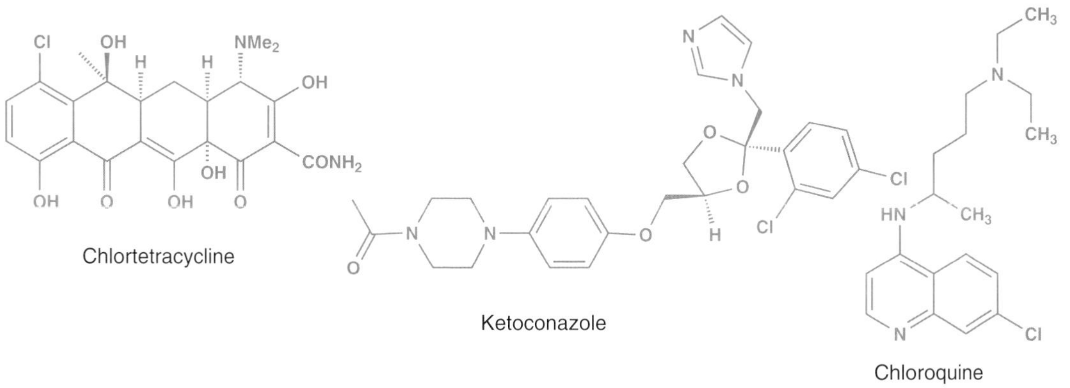

Chlortetracycline

Ketoconazole

Chloroquine

FIGURE 5.18 Hydrogen bonds in an alkyl halide.

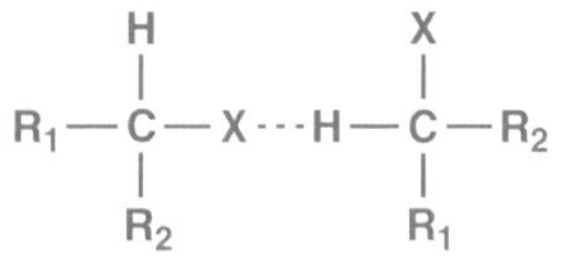

licity, they easily cross lipid membranes and tissue boundaries and can accumulate in fatty tissues. This can give rise to toxicological problems, as discussed in the next section.

Chemical properties

As was stated previously, a key reaction of haloalkanes is nucleophilic substitution, as shown in Figure 5.19.

For an organic chemist, the most interesting part of the molecule in Figure 5.19 is usually R. The nucleophile just modifies the structure R and may well be removed in the next reaction step. In the biological context, however, the nucleophile is an important biological molecule, such as DNA or a protein; halogen compounds are often referred to as alkylating agents, because they have the effect of alkylating (adding an alkyl group to) such biomolecules as described in Box 5.10.

SELF CHECK 5.15

The boiling point of methane (CH_4) is -164 °C. The boiling point of methanol is 65 °C. Estimate the boiling point of chloromethane (CH_3Cl).

These physical properties make many halogen compounds useful as solvents for a wide range of organic molecules. They have the added advantage over ethers in that they are largely non-flammable and, therefore, safer to use. However, because of their high lipophi-

FIGURE 5.19 Nucleophilic (S_N2) substitution of a halogen: general scheme.

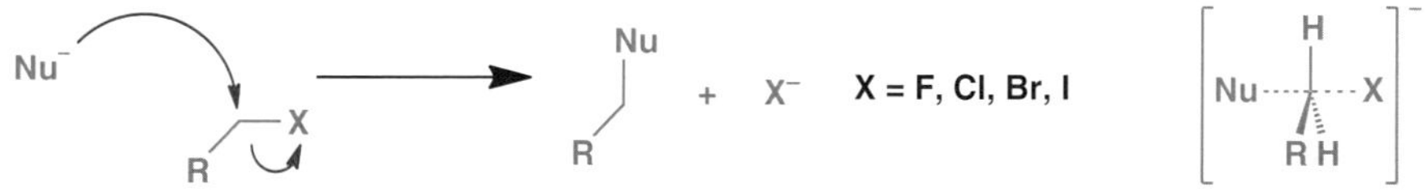

5-membered transition state

BOX 5.10

Alkylating agents in cancer and cancer chemotherapy

Alkylating agents were first used as highly toxic chemical weapons during World War One. Their toxicity stems from the way they can alkylate the guanine bases of DNA, leading to potential gene mutations. They are therefore carcinogenic. However, more selective molecules have now been developed, to the extent that alkylating agents can now be used to treat certain cancers.

Alkylating agents react preferentially with cancer cells, which divide more rapidly than healthy cells. However, they remain toxic to normal cells and so cause severe side effects, particularly in those tissues where cells are rapidly dividing, such as bone marrow and reproductive organs.

Alkylating agents used as drugs include chlorambucil and cyclophosphamide. Note that they are both primary organohalogen compounds.

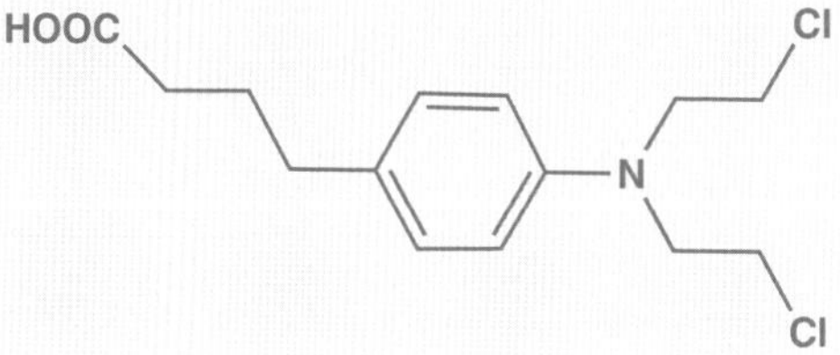

Chlorambucil

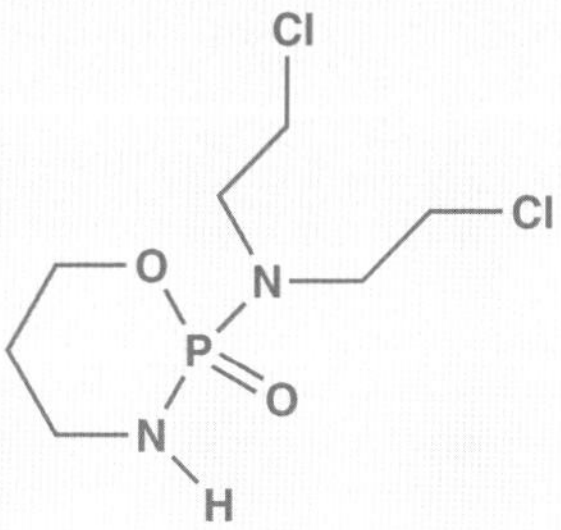

Cyclophosphamide

FIGURE 5.20 S$_N$2 substitution of a chloride ion on chlorambucil by glutathione.

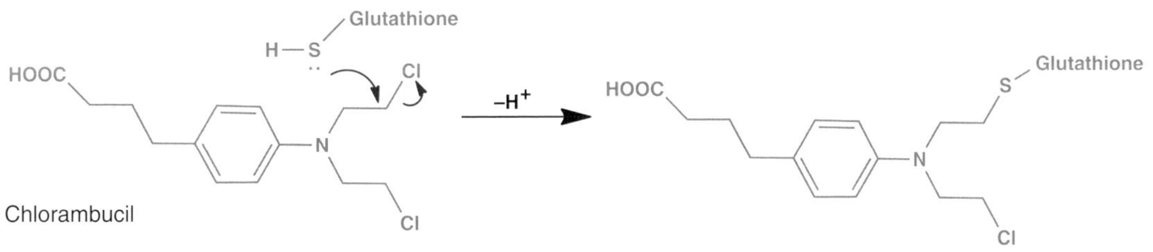

Earlier in this chapter, we talked briefly about S$_N$1 reactions in which a carbocation is an intermediate. The alkylation of biomolecules by organohalogen compounds more usually has the mechanism called S$_N$2. No carbocation is formed. Instead the bond to the leaving group breaks at the same time as the bond to the nucleophile forms. Provided there is a good leaving group (such as a halide ion) and a good nucleophile (the –SH of glutathione is a very good nucleophile, see Figure 5.20), extreme conditions of temperature and pH are not required. For a full description of an S$_N$2 mechanism, you should consult an organic chemistry text book, such as Clayden et al. For our purposes, it is important to note that the S$_N$2 mechanism proceeds best if the haloalkane is primary; the five-membered transition state is crowded and primary haloalkanes take up less space than their secondary and tertiary counterparts.

Because of the potential toxicity of these molecules, it is useful that the human body has a very effective mechanism for inactivating them. The detoxification process involves reaction (by S$_N$2 nucleophilic substitution) with glutathione, a tripeptide that possesses a SH (thiol) group. The SH group is an excellent nucleophile, so the halogen compounds react with glutathione in preference to other cellular constituents (see Figure 5.20).

5.6 Aromatic halogen compounds

Aromatic halogen compounds have similar physical properties to haloalkanes for similar reasons. For example chlorobenzene and 1-chlorohexane have similar boiling points (131 °C and 135 °C) above those of benzene and hexane (81 °C and 69 °C). The chemical properties are quite different however.

Halogen atoms on aromatic rings cannot take part in nucleophilic substitution. This means that they are stable in the body and are much more common in drug molecules than haloalkanes. A halogen in an aromatic ring introduces subtle changes in polarity, which can be advantageous. Examples of halogenated aromatic drugs are the antibacterial agent, chlortetracycline, and the antifungal, ketoconazole.

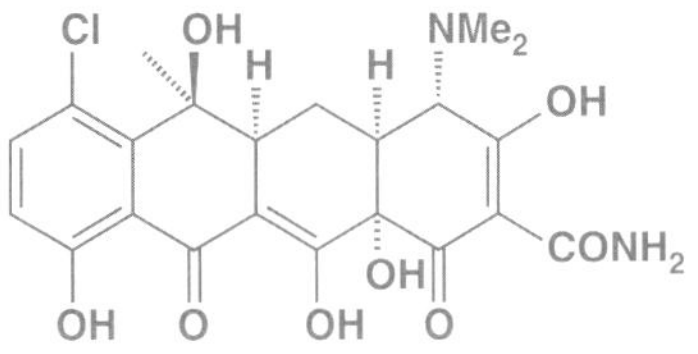

Chlortetracycline

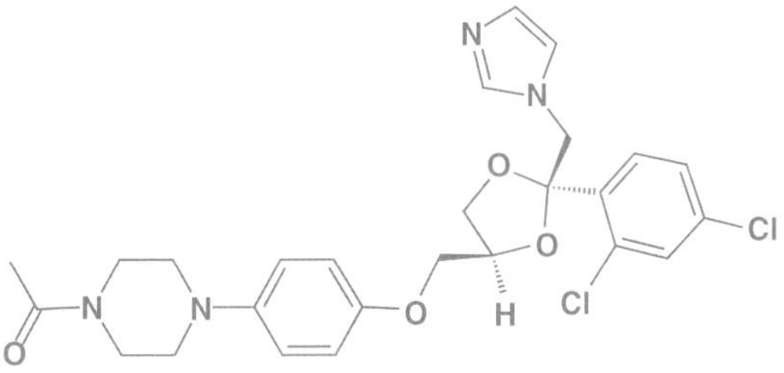

Ketoconazole

The lack of reactivity means that aromatic halogen compounds cannot be detoxified by glutathione. (This is explained in more detail in Chapter 7.) The lack of reactivity of some halogenated molecules (for example, DDT) has led to environmental problems, as explained in Box 5.11. (See Figure 5.21 for some halogenated environmental hazards).

5.7 Polyhalogen compounds

Molecules possessing multiple halogen atoms have found a variety of uses – for example, as solvents, general anaesthetics, refrigerants, and aerosol propellants (which are important components of some drug delivery devices; for example, in the inhalers used for the treatment of asthma). The polyhalogen compounds that have been used for this purpose are known as CFCs (chlorofluorocarbons) and two of these are shown in Figure 5.21. Nowadays, however, drug delivery devices are CFC-free. Why is this?

SELF CHECK 5.16

Two of the three molecules shown in Figure 5.21 have chiral centres. Which are they?

CFCs are good solvents, and are volatile, non-flammable and unreactive – properties that made them attractive in aerosols. Their lack of reactivity, however, caused an environmental problem when they were used widely in this context. While most organic chemicals are broken down in the lower atmosphere, CFCs accumulated in the upper atmosphere. This accumulation became problematic because the C-Cl bonds are broken under the influence of ultraviolet radiation from space, and the chlorine atoms (free radicals) that are formed damage the ozone layer, which protects the Earth from harmful ultraviolet radiation. Pharmaceutical, and other, aerosols, now utilize other propellants and are often labelled 'CFC-free'.

KEY POINT

Organic halogen compounds possess properties that can be potentially useful (solvents, general anaesthetics), but they also have a potential for toxicity (alkylating agents, environmental problems).

FIGURE 5.21 **Halogenated environmental toxins.**

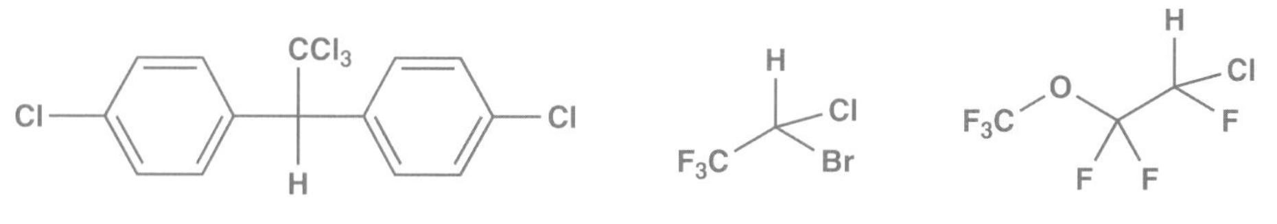

DDT   Halothane   Enflurane

BOX 5.11

DDT

DDT is a highly effective insecticide. It was discovered in 1939 and was regarded as a significant scientific breakthrough. At the time, the stability of DDT was seen as one of its greatest benefits – it would remain active for longer. This, however, proved to be its undoing: DDT began to accumulate in the food chain wherever it was heavily used. DDT is highly lipid soluble and so accumulates in the fatty tissue of species at the top of the food chain (including humans), with potentially toxic effects. Subsequently it received a world-wide ban.

5.8 Amines

A number of organic functional groups contain nitrogen (e.g. nitro, amide, nitrile), but one particularly important nitrogen-containing group is the amine group.

Nomenclature

Amines can be classified as primary (1°), secondary (2°), or tertiary (3°) in a manner similar, but not identical, to alcohols. A primary amine is derived from ammonia (NH_3) by replacement of one hydrogen atom with an alkyl group. Secondary amines have two hydrogens replaced by alkyl groups and tertiary have all three hydrogens replaced by alkyl groups, as illustrated in Figure 5.22.

FIGURE 5.22   Primary, secondary and tertiary amines.

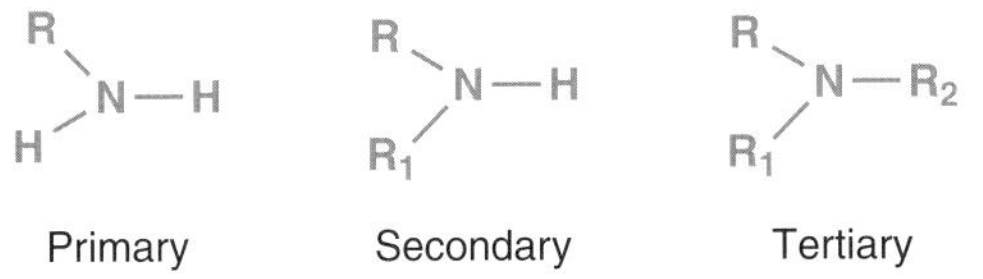

In addition to this classification, amines may be aliphatic (whereby the amine group is attached to an alkyl chain), aromatic (where the amine group is attached directly to an aromatic ring) or heterocyclic (where the nitrogen atom of the amine group is contained in a ring system).

SELF CHECK 5.17

Find the structures of the following amines and classify them as (a) primary, secondary or tertiary and (b) aliphatic, aromatic or heterocyclic: triethylamine, pyrrolidine, aniline, adrenaline.

Hint: draw the structures and check the definitions given in the preceding section. Another hint: Wikipedia is a very good source of drug structures. A third hint: The American English for adrenaline is epinephrine – as a pharmacist you really need to know both words.

Physical properties

Because of the presence of the electronegative nitrogen atom, amines are polar molecules. Primary and secondary amines can self-associate by hydrogen bonding, but tertiary amines cannot because they lack a hydrogen attached to nitrogen. Nitrogen is less electronegative than oxygen so these hydrogen bonds are weaker than those in alcohols. Consequently, the boiling points of amines are lower than those of their corresponding alcohols: whereas methanol boils at 65 °C, methylamine is a gas (with a disgusting smell) with a boiling point of −6 °C.

Because of their ability to hydrogen bond to water molecules, amines are relatively water soluble.

Chemical properties

Amines are basic: they can accept a proton and become positively charged. The lone pair of electrons on the nitrogen acts as a nucleophile and reacts with a positively charged hydrogen ion. Reaction with an acid (a proton donor) will lead to the formation of a salt.

Aliphatic amines are stronger bases than aromatic amines. We usually refer to the pK_a of the conjugate acid to illustrate this point. The conjugate acid of R-NH$_2$ is R-NH$_3^+$ (see below)

For aliphatic amines the dissociation shown above is 50% complete at pH 10, so the pK_a of the conjugate acid of an aliphatic amine is about 10.

It is very important to appreciate that, because amines are bases, they become more protonated at low pH (see Box 5.12).

Eflornithine is used to treat African trypanosomiasis (a protozoal disease related to malaria). The conjugate acid of the side-chain amine has a pK$_a$ of about 10. Estimate the proportion of eflornithine that will be protonated in the blood stream at pH 7.4.

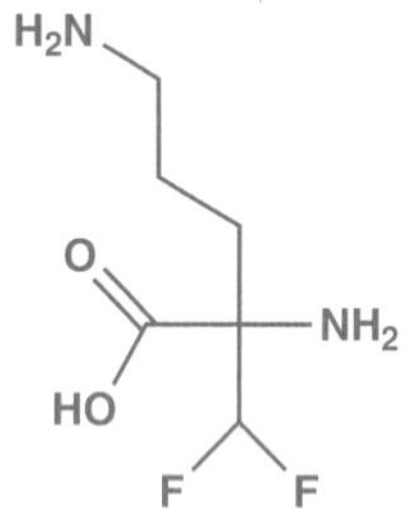

Aromatic amines are less basic than aliphatic amines and their conjugate acids have pK$_a$s of about 5. This means that they are generally protonated in the stomach at pH 2 (the equilibrium reaction, on p 137, lies to the left, favouring the conjugate acid), but exist in the unprotonated (free base) form in the intestine at pH 8 (the equilibrium reaction, on p 137, lies to the right). Why do we see this difference between aliphatic and aromatic amines? The reason is that the lone pair of electrons on the amine nitrogen is delocalized into the aromatic ring and is less readily available to bind to a proton (see Figure 5.23); consequently, it acts as a weaker base. In aliphatic amines, the lone pair is not delocalized and so a proton can be accepted more readily.

Amines in drugs

Many drug molecules possess an amine group, which can be reacted with an acid to form a salt; the salt is generally more water-soluble than the original drug molecule. This can be useful; for example, you cannot

pK$_a$s of bases

A K$_a$ represents an **acid dissociation constant** so a pK$_a$ tells you the pH at which an *acid* is half-dissociated. Therefore, strictly speaking, the pK$_a$ of an amine refers to the following reaction, where the amine is acting as an acid:

$$R-NH_2 \rightleftharpoons R-\overset{-}{NH} + H^+$$

When R = Ph, the pK$_a$ for this reaction is over 30! Because pK$_a$s of over 30 are not relevant in aqueous solution, biologists and pharmacists sometimes get a bit lazy and refer to the pK$_a$ of an amine, when they are really referring to the pK$_a$ of its conjugate acid.

You may find references to the pK$_b$ of an amine, especially in older books. The pK$_b$ of an amine refers to the dissociation of a base, and has a value that is approximately equal to 14 − pK$_a$ of the conjugate acid. The use of the pK$_b$ has the advantage that you do not have to keep saying 'of the conjugate acid'; the disadvantage is that strong bases have lower pK$_b$s and weak bases have higher pK$_b$s, which can be confusing; the use of pK$_b$ is much less common than it once was.

inject a solid, so it is helpful to be able to **formulate** an insoluble amine as a soluble salt.

The amine group in a drug molecule is often a key binding point between a drug and a receptor site. Indeed, a number of the functional groups in this chapter have the capacity, if present in a drug molecule, to be involved in interactions between the drug and a receptor. The hydroxyl group, be it an alcohol or a phenol, may be involved in hydrogen bonding to the receptor. The amine group, which is likely to be ionized at physiological pH, may be involved in an ionic

FIGURE 5.23 The delocalization of the electron pair of an aromatic amine.

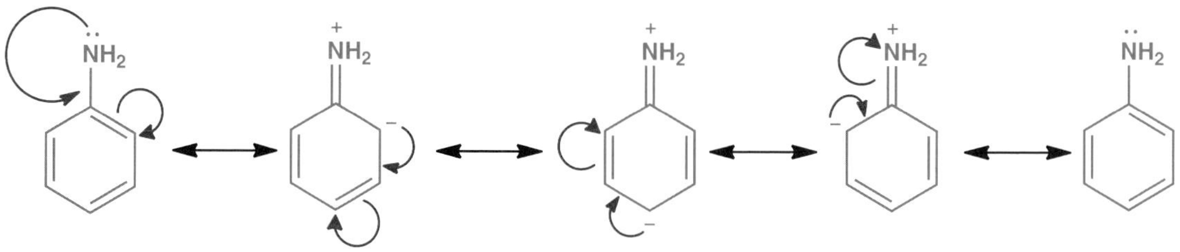

FIGURE 5.24 Oxidative deamination of an amino acid.

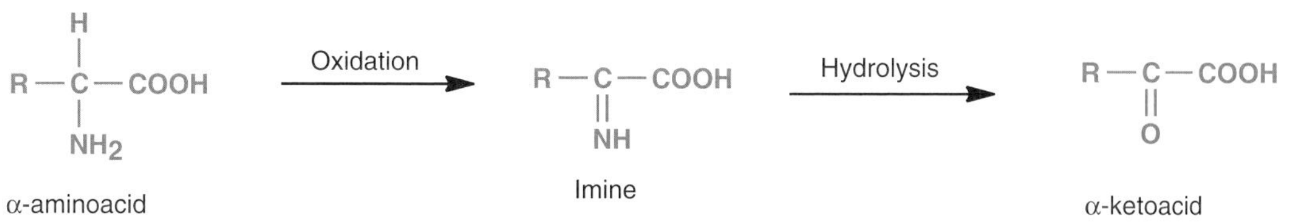

R —C—COOH with H and NH2 (α-aminoacid) → Oxidation → R —C—COOH with NH double bond (Imine) → Hydrolysis → R —C—COOH with O double bond (α-ketoacid)

interaction with an anionic site on the receptor. These are likely to be primary sites of interaction. An ether group may, if present, provide an additional electrostatic interaction.

 More information on drug-receptor interactions can be found in Chapter 4 'Introduction to drug action' of the *Therapeutics and Human Physiology* book of this series.

KEY POINT

The ability of amines to act as bases is an important property, particularly when the group is present in a drug molecule.

Oxidation of amines

In the laboratory, the oxidation of amines is not a particularly important reaction. Biologically speaking, however, it is vital. The majority of energy-producing reactions in the cell involve molecules containing carbon, hydrogen and oxygen only (carbohydrates and fatty acids). Most

of us take in more protein than we need and excess amino acids are metabolized by removal of nitrogen so that they can join the citric acid cycle. The removal of nitrogen happens by a process known as oxidative deamination and is often carried out by monoamine oxidase enzymes, as illustrated in Figure 5.24.

 More detail on the metabolism of amino acids can be found in Chapter 3 'The biochemistry of cells' in the *Therapeutics and Human Physiology* book of this series.

Alkylation of amines

Amines are able to undergo numerous reactions in which they act as nucleophiles, and these reactions are important in the synthesis of drugs. Nucleophilic attack of amines on suitable carbonyl compounds leads to the formation of amides and is described in the next chapter. Amines can be alkylated by nucleophilic attack on haloalkanes. (Can you draw the mechanism?). Further reading about the reactions of amines can be found in Clayden et al.

5.9 Quaternary ammonium compounds

Tertiary amines are formed by progressive replacement of the hydrogens on nitrogen by alkyl groups. If a tertiary amine reacts with a halogenated hydrocarbon, a quaternary ammonium compound, such as that shown in Figure 5.25, is formed.

Quaternary ammonium compounds exhibit different properties from amines. Amines are basic, but quaternary ammonium compounds are neutral. Amines are relatively volatile, but quaternary ammonium compounds are solids. Quaternary ammo-

nium compounds will dissolve in water, but not in organic solvents, unlike amines which dissolve in both. The reason for these changes is that quaternary

FIGURE 5.25 A quaternary ammonium salt.

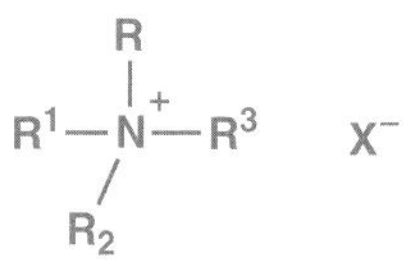

$$R^1—\overset{\overset{\displaystyle R}{|}}{\underset{\underset{\displaystyle R_2}{/}}{N^+}}—R^3 \quad X^-$$

ammonium compounds are ionic and so behave like ionic molecules, unlike the amines which are completely covalent molecules.

If one or more of the alkyl chains in quaternary ammonium compounds is a long alkyl chain (C_{12}–C_{20}), then these molecules will concentrate at the oil-water interface and can act as detergents. (They are known as cationic detergents, because of the positive charge on the nitrogen atom.) This ability to concentrate at an oil/water interface is also responsible for the antibacterial activity of some of these molecules. They penetrate the cell membrane of micro-organisms, particularly those classified as Gram negative organisms. Because of their inability to cross lipid membranes, they are only used as topical antibacterial agents (e.g. antibacterial swabs, throat lozenges).

 More information on molecules which concentrate at the oil-water interface can be found in Chapter 10 'Disperse systems' of the *Pharmaceutics* book of this series.

SELF CHECK 5.19

Find an example of a quaternary ammonium compound that is used for the following purposes: (a) neurotransmitter, (b) topical antibacterial agent, (c) neuromuscular blocking agent.

CHAPTER SUMMARY

➤ The nature of the intermolecular forces of attraction is the major factor in determining the boiling points and water solubilities of molecules.

➤ Hydrogen bonding is a particularly important intermolecular force.

➤ Dehydration and oxidation are important reactions of alcohols, particularly from a biological perspective.

➤ Although both possess an OH group, alcohols and phenols differ considerably in their physical and chemical properties.

➤ Small ring cyclic ethers are very reactive, unlike other types of ether.

➤ Organic halogen compounds have a wide range of potentially useful properties, but are also generally quite toxic.

➤ The basicity of amines is a very important property, both biologically and pharmaceutically.

FURTHER READING

Clayden, J., Greeves, N., and Warren, S., *Organic Chemistry*, 2nd edn. Oxford University Press, 2012.

Holum, J.R., *Organic and Biological Chemistry*, John Wiley and Sons. 1995.

McMurry, J., *Fundamentals of Organic Chemistry*, 7th edn, Cengage Learning, 2011.

The carbonyl group and its chemistry

6

MATTHEW INGRAM

This chapter considers the nature of the carbonyl group and the fantastic chemistry it can undergo. We can find this chemistry in lots of situations: in the body, in the environment, in manufacturing and, of course in pharmaceutical applications. Before reading this chapter, it is important that you are familiar with key concepts covered earlier in this book; these include nomenclature, acids and bases, electrophiles and nucleophiles. If you are unsure, we would recommend re-visiting these before attempting to understand the materials in this chapter.

Learning objectives

Having read this chapter you are expected to be able to:

➤ distinguish between the different types of carbonyl compounds, draw their chemical structures and name simple derivatives

➤ predict simple reactions and their likely products

➤ understand how nature uses carbonyl chemistry

➤ understand how the medicinal chemist or pharmacist can manipulate different types of carbonyl compounds for a desired clinical outcome.

6.1 Carbonyl structure and nomenclature

At the heart of the chemistry of the carbonyl group is the carbonyl bond, a double bond comprising one σ and one π bond, which joins the carbon and the oxygen. The difference in electronegativity between the carbon and oxygen (with oxygen being strongly electronegative) means a dipole moment exists between the two atoms.

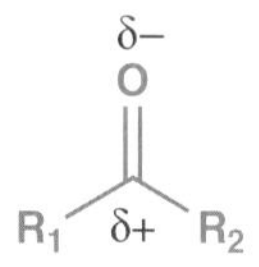

 More details on this type of bonding can be found in Chapter 2 'Organic structure and bonding' of this book.

The carbonyl group is central to pharmaceutical chemistry and is present in many drug molecules and **excipients** that you may already be familiar with.

SELF CHECK 6.1

Find the carbonyl group(s) in each of the molecules in Figure 6.1.

SELF CHECK 6.2

Briefly explain the use of each of these carbonyl compounds in pharmacy.

As you can see from Figure 6.1, the carbonyl group is present in many different types of compound. If we draw the carbonyl in the form:

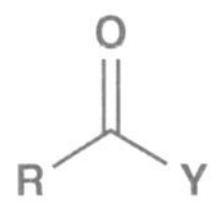

$$\underset{R \quad\quad Y}{\overset{O}{\|}}$$

then Y can be H, or C, or an electronegative element. In nature and in drug molecules, the position Y is very often occupied by O (as in aspirin) or N (as in penicillin)

(see also Box 6.1). The family of carbonyl compounds is as shown in Table 6.1.

We need to be able to name these families of molecules using the IUPAC system. The names are made up from the prefix, which is described in Chapter 2, and the ending, which depends on the type of carbonyl compound. The name endings are given in Table 6.1.

BOX 6.1

Phosphate esters

Table 6.1 shows that if we take an ester, such as ethyl acetate (ethyl ethanoate), we can replace one of the oxygens with sulfur or even selenium to give a thioester or a selenium ester. So we might expect to be able to replace the nitrogen of an amide by phosphorus (which is immediately below nitrogen in the periodic table). In fact, phosphate esters are based on the P=O group, not on the C=O group, because phosphorus has similar electronegativity to carbon and acts as an electrophile rather than as a leaving group. Phosphate esters are incredibly important in biochemistry and are considered in Chapter 9.

FIGURE 6.1 Common molecules in pharmaceutical chemistry that all contain the carbonyl functionality.

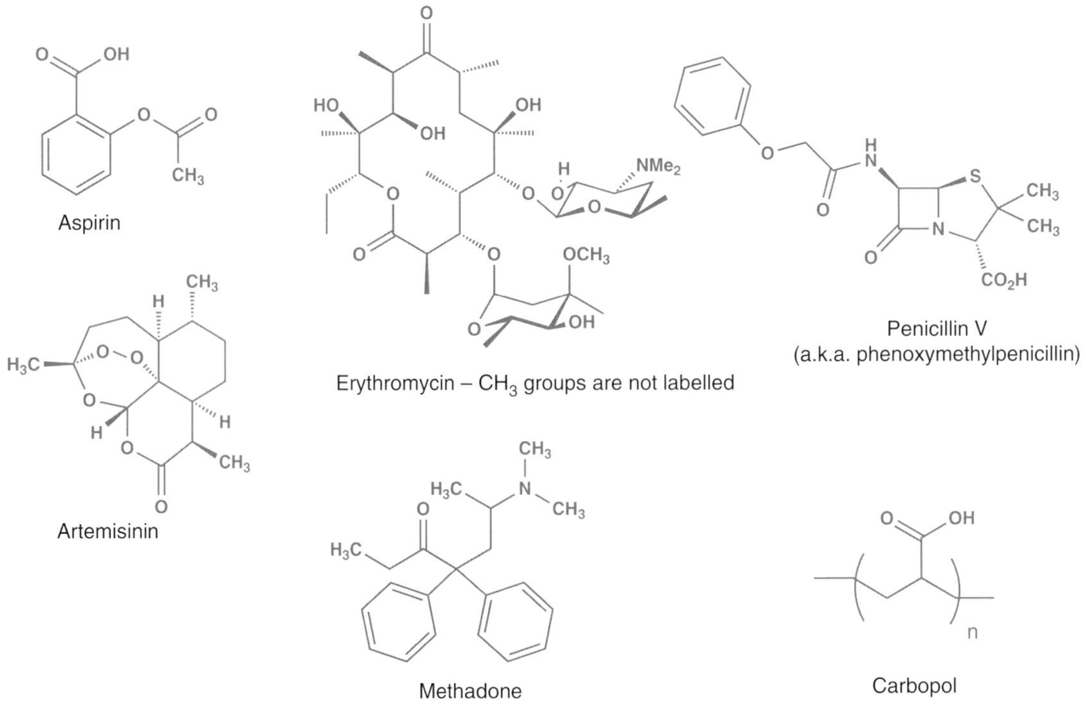

Aspirin

Erythromycin – CH₃ groups are not labelled

Penicillin V
(a.k.a. phenoxymethylpenicillin)

Artemisinin

Methadone

Carbopol

TABLE 6.1 Different types of carbonyl compound

Name	Name ending	Structure	Typical example with systematic name. Trivial names (where appropriate) are given in brackets
Aldehyde	-al		Ethanal (acetaldehyde)
Ketone	-one		Propan-2-one (acetone)
Carboxylic acid	-oic acid		Propanoic acid
Acyl halide	-onyl chloride, bromide or iodide		Propanoyl chloride
Acid anhydride	-oic anhydride		Ethanoic anhydride (acetic anhydride)
Ester	-oate		Ethyl propanoate
Thioester	thioate		S-Ethyl ethanethioate (Ethyl thioacetate)
Amide	-amide		Propanamide

SELF CHECK 6.3

Name these compounds using the IUPAC system. Tip: Revise the nomenclature of alkenes in Chapter 4.

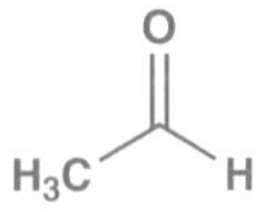

(A)

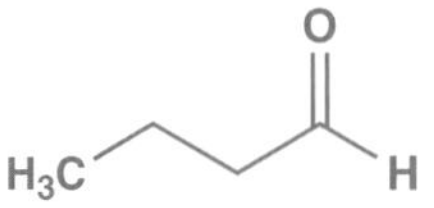

(B)

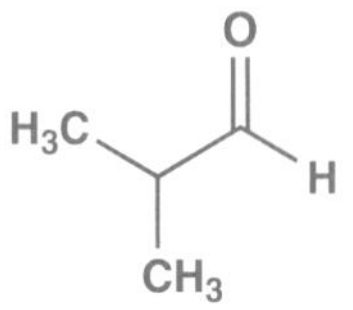

(C)

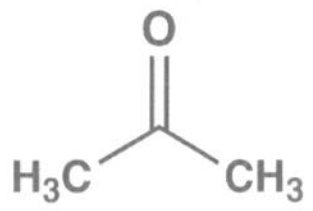

(D)

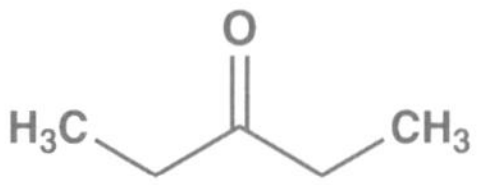

(E)

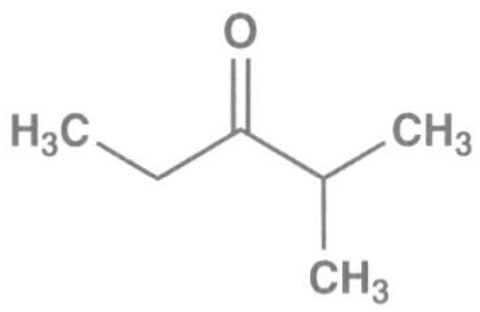

(F)

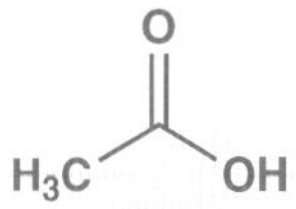

(G)

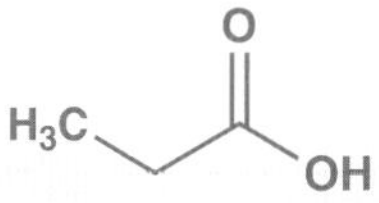

(H)

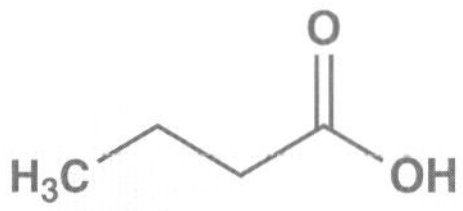

(I)

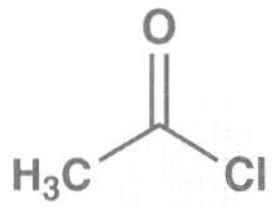

(J)

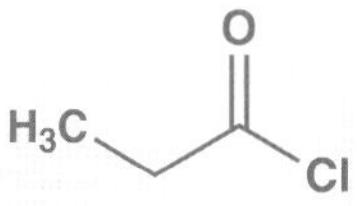

(K)

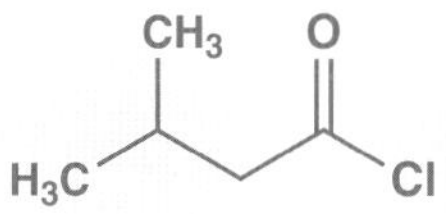

(L)

In addition to the systematic names, there are common names that are still used and you will come across these during your studies and in your future careers. Figure 6.2 gives the common names for some important molecules that contain carbonyl groups.

The common names existed before the creation of the IUPAC system and are still widely used. For example, acetic acid is still the preferred name in the laboratory, and is even preferred by IUPAC over the systematic name. Sportsmen and women, as well as scientists and health care professionals, refer to (S)-2-hydroxypropanoic acid as L-lactic acid. Pharmacists need to be aware of both systems in order to bridge the gap between chemists, health care professionals and the general public.

SELF CHECK 6.4

Can you convert these names into structures?

1. Acetaldehyde (ethanal)
2. Pentanal
3. Octan-2-one
4. 3-Methylpentanoic acid
5. 2,2-Dimethylbutanedioic acid
6. Propanoic anhydride
7. Hexanoic acid
8. Butanal
9. Benzophenone (diphenyl ketone)
10. Acetyl chloride (ethanoyl chloride)

 Common (trivial) names were discussed in Chapter 1 'The importance of pharmaceutical chemistry'.

FIGURE 6.2 Some important carbonyl compounds and their trivial names.

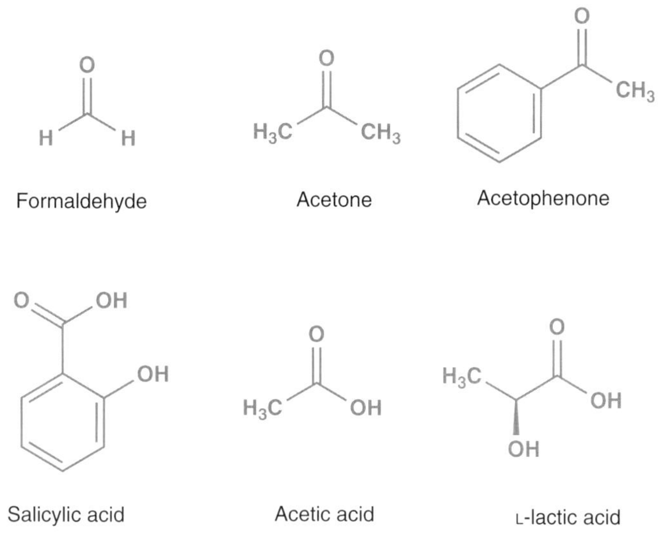

Formaldehyde

Acetone

Acetophenone

Salicylic acid

Acetic acid

L-lactic acid

6.2 **The power of the carbonyl group**

The central importance of carbon is discussed in Chapter 2. A moment's thought (or breathing) will remind you of the importance of oxygen. A functional group containing both carbon and oxygen is, therefore, likely to be important. Spend a moment trying to think of a drug *without* a carbonyl group, and a biochemical intermediate *without* a carbonyl group. This exercise should convince you that the carbonyl group holds a special place in pharmaceutical chemistry. It is, indeed, powerful!

Carbonyl groups and physical properties

Oxygen is **electronegative** and draws electron density away from the carbon, leaving the carbon with a δ+ charge and the oxygen with a δ- charge. This is described as a **chemical dipole moment** (see Figure 6.3).

In aldehydes, ketones, acyl chlorides, acid anhydrides and esters, dipole-dipole attraction is the intermolecular force found between adjacent molecules and, thus, influences the boiling/melting points of these compounds. Because dipole-dipole attractions

FIGURE 6.3 The chemical dipole moment of a carbonyl group and its reactivity.

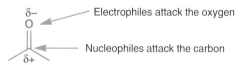

Electrophiles attack the oxygen

Nucleophiles attack the carbon

are relatively weak, these classes of carbonyl compounds are predominantly liquids with relatively low boiling points (see Table 6.2). With carboxylic acids and amides, however, the key intermolecular force is hydrogen bonding (see Figure 6.4). Hydrogen bonding is a much stronger intermolecular force than dipole-dipole attraction. Consequently, carboxylic acids and amides have higher boiling points – indeed many are solids rather than liquids (see Table 6.2).

 More detail can be found on intermolecular interactions in Chapter 2 'Organic structure and bonding'.

The ability to form hydrogen bonds also influences water solubility. Aldehydes, ketones and esters cannot form hydrogen bonds with each other, but they can act as hydrogen bond acceptors with water. Simple aldehydes and ketones, therefore, are readily soluble in water, but the solubility rapidly decreases with an increasing hydrocarbon component (see Table 6.3).

TABLE 6.2 Boiling points of some C3 organic compounds

Compound	Boiling point (°C)
Propene	−47.4
Acetone	56.6
Ethyl acetate	77
Propanoic acid	141
Dimethylformamide (DMF)	153

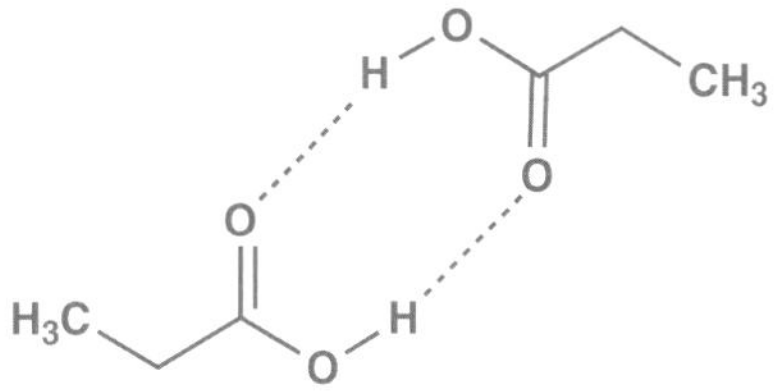

FIGURE 6.4 Hydrogen bonding in propanoic acid. Dimers are formed, raising the boiling point, relative to propanone.

TABLE 6.3 Water solubility of carbonyl compounds

Compound	Water solubility (%)
Acetone	Completely miscible
Butan-2-one	25.6
Pentan-2-one	5.5
Hexan-2-one	1.6
Acetic acid	Completely miscible
Hexanoic acid	1.0
Dimethylformamide	Completely miscible

The increased hydrocarbon component already present in esters (RCOOR′) means that their water solubility is compromised from the start (because R′ is usually a hydrocarbon).

Carboxylic acids can interact with water as both a hydrogen bond acceptor and a donor, and so low molecular weight carboxylic acids are very soluble in water. Acids up to four carbons are completely miscible with water. This solubility, again, rapidly decreases with an increasing hydrocarbon component, so that hexanoic acid has only 1% water solubility.

Amides exhibit similar solubility profiles because they, too, can form effective hydrogen bonds with water. We do not have to worry about the water solubility of acid anhydrides and acyl chlorides because they react with water (see later in this chapter).

Why do carbonyls undergo useful reactions?

Aspirin is a painkiller that has been used for well over 100 years, and it actually contains two carbonyl groups. The reaction that forms aspirin is reviewed several times in this book, particularly in Chapter 11 when we consider the importance of impurities in pharmaceutical analysis. However, for now let us celebrate the coming together of two carbonyl containing compounds.

In Figure 6.5, salicylic acid, which already has one carbonyl group, acquires another. The hydroxyl function of salicylic acid reacts with acetic anhydride (ethanoic anhydride) and is acetylated, to give an acetate ester. The transfer of acyl (CH$_3$C=O) groups from one molecule to another is also important in biological systems. For example, food enters the citric acid cycle

FIGURE 6.5 The synthesis of aspirin from acetic anhydride and salicylic acid.

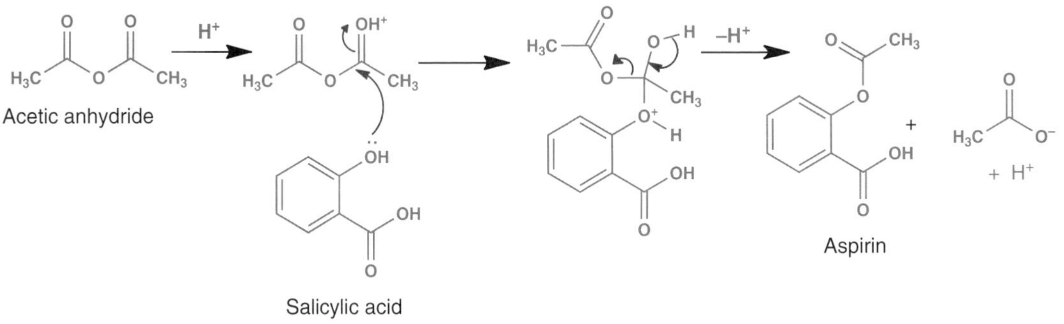

in the form of acetyl coenzyme A; the acetyl group is transferred to oxaloacetate to give citrate. However, anhydrides are very reactive and are incompatible with biological systems, so nature uses alternative carbonyl compounds, typically thioesters, for acyl transfer in cells.

 Further details on the citric acid cycle can be found in Chapter 3 'The biochemistry of cells' in the *Therapeutics and Human Physiology* book in this series.

If we understand why carbonyl compounds behave as they do, we should be able to predict how they will react with other molecules and how they might react in a biological situation.

It's all in the electrons: the power of electronegativity

From a thermodynamic point of view, the carbonyl group is very stable. The presence of carbonyls in drugs and in nature therefore comes as no surprise. This thermodynamic stability does not, however, prevent carbonyls from taking part in reactions; this is something they do rather eagerly. However, the endpoint of a reaction involving a carbonyl is very often another carbonyl. As we read further through the chapter we will come to understand why this is.

If we examine Figure 6.3 again we can see how the dipole moment influences the reactivity of carbonyl compounds. Electrophiles (such as H⁺) seek out areas of high electron density (the oxygen), whereas nucleophiles (such as amines) are attracted to areas of low electron density (the carbon).

Carbonyls are subject to nucleophilic attack

In practice, the chemistry of carbonyl groups is dominated by nucleophilic attack on carbon, as shown in Figure 6.6. Nucleophiles may either be negatively charged or neutral, and Table 6.4 shows some examples.

The action of the nucleophile, on the carbonyl group, gives us an intermediate in which the π bond breaks and the carbon is single-bonded to four species. The four bonds arrange themselves to be as far apart as possible and the intermediate is tetrahedral.

TABLE 6.4 Some nucleophiles capable of reacting with carbonyl groups

Examples of negative nucleophiles	Examples of neutral nucleophiles
HO^-	H_2O
RO^-	H_3N
R_3C^-	ROH
NC^-	RNH_2

FIGURE 6.6 Nucleophilic attack on the carbonyl group. Redrawn with permission from Burrows et al. (2009) Chemistry³, OUP.

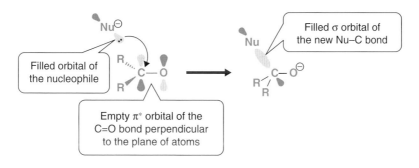

For more about tetrahedral carbon atoms see Chapter 2 'Organic structure and bonding'.

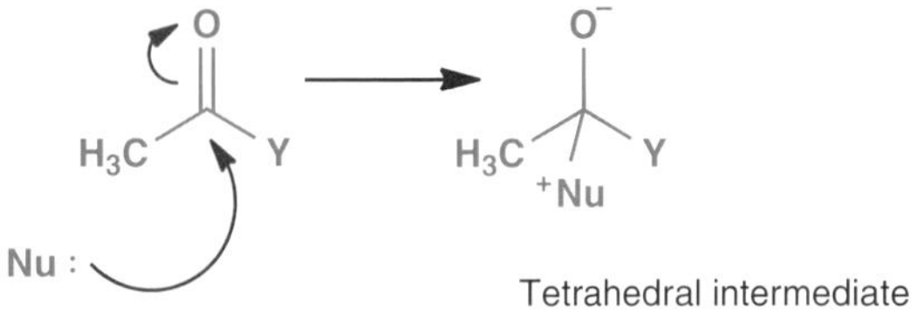

The fate of the tetrahedral intermediate is determined by the best **leaving group**. Let us consider what happens when:

- Nu is a good leaving group (better than Y)
- Y is a good leaving group (better than Nu)
- neither Nu nor Y is a good leaving group.

For a detailed description of leaving groups see Section 6.3 on 'Reactions of carbonyl compounds – nucleophilic attack on carbon' in this chapter.

Nu is a good leaving group

If Nu is a good leaving group, it simply leaves at the end of the reaction. In other words, the reaction simply reverses (see Scheme 6.1).

SCHEME 6.1

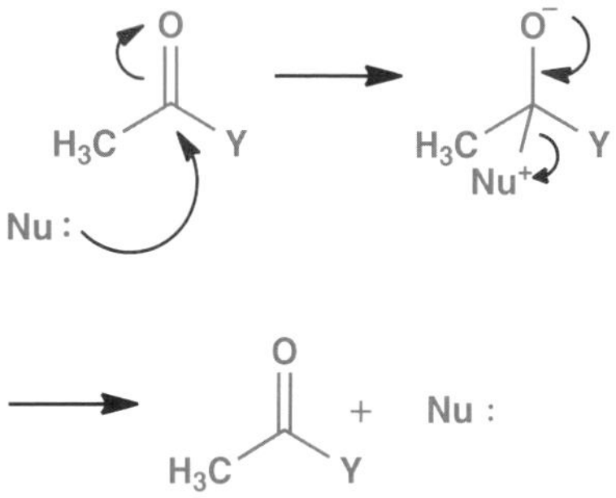

Cl^- is an example of a good leaving group and you cannot make acyl chlorides by treating another carbonyl compound with chloride ion. Similarly, acetate is a good leaving group (see the synthesis of aspirin, Figure 6.5) and you cannot make acid anhydrides by attacking another carbonyl with an acetate ion.

An everyday example is shown in Figure 6.7. Chloride ion, a good leaving group, does not react with acetic acid. So, for example, the chloride ions in salt do not react with acetic acid in vinegar, because Cl^- is a good leaving group. If they did, an acyl chloride would be generated – not something you would want with your chips! If you eat chips with salt and vinegar too often, you might get fat from the fried food, or you might get hypertension from the salt, or you might get tooth decay from the acid, but otherwise it is perfectly safe.

Y is a good leaving group

If Y is a good leaving group, Y leaves at the end of the reaction (see Scheme 6.2). This behaviour leads us to some very important reactions of carbonyl compounds:

- formation of esters from carboxylic acids
- hydrolysis of esters to carboxylic acids
- formation of esters from acyl chlorides (acid chlorides) or acid anhydrides (acyl anhydrides)
- formation of amides from acyl chlorides or acid anhydrides
- hydrolysis of amides.

These reactions will be discussed in Section 6.3 'Reactions of carbonyl compounds – nucleophilic attack on carbon'.

FIGURE 6.7 Lack of reaction of Cl^- with acetic acid: there is no reaction between the salt and vinegar on your chips.
Source: Viktor Fischer.

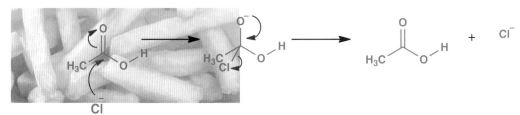

SCHEME 6.2

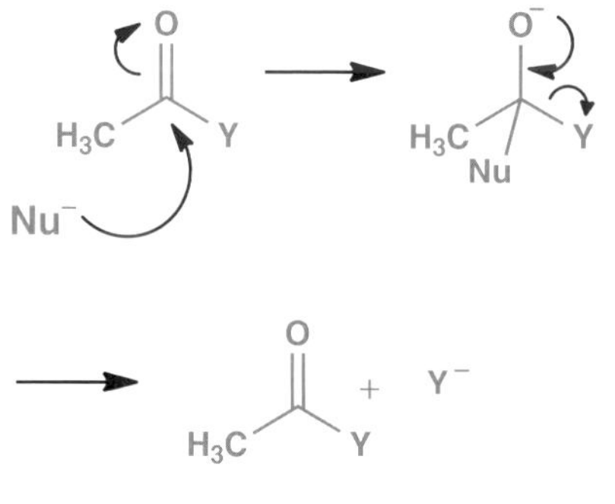

SCHEME 6.3

Neither Y nor Nu is a good leaving group

Aldehydes and ketones do not have good leaving groups. When a nucleophile attacks an aldehyde or ketone, the carbonyl π-bond breaks and does not reform (see Scheme 6.3). There are three major classes of reaction of this type:

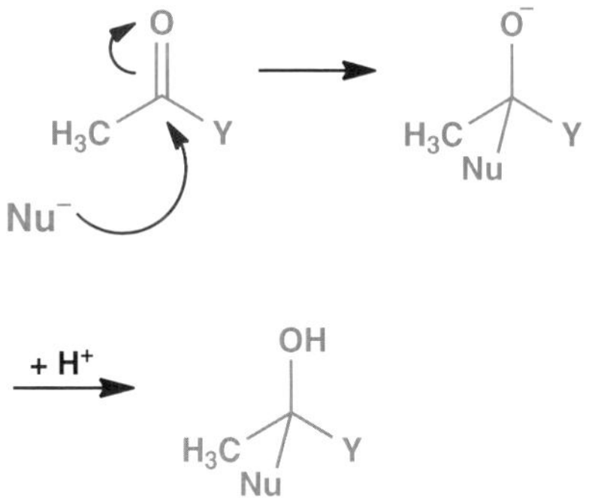

- addition of water or alcohols to give hemiacetals and acetals
- addition of the equivalent of H^-, a reduction reaction
- addition of a R_3C^- equivalent; formation of a C–C bond.

  These are discussed in Section 6.3 too.

6.3 Reactions of carbonyl compounds – nucleophilic attack on carbon

The most important reactions of carbonyl compounds involve attack of the carbon with the δ+ charge by a nucleophile. This section describes reactions of this type which have significant pharmaceutical and biological relevance.

Making carbonyls reactive

Many carbonyl compounds are stable. Acetone can be used as a solvent because it dissolves reactants with-out itself reacting. The 'best before' date on a bottle of vinegar (dilute acetic acid) is months after you buy it. Many of the proteins in your body (which are poly-amides) are metabolized only very slowly. To make carbonyl compounds react, we often need to add an acid or base catalyst to boost the reactivity (see Figure 6.8).

By adding acid we can build up the positive charge at the carbon of the carbonyl group. This type of catalysis has been used in carbonyl chemistry extensively.

FIGURE 6.8 Activation of a carbonyl by an acid catalyst.

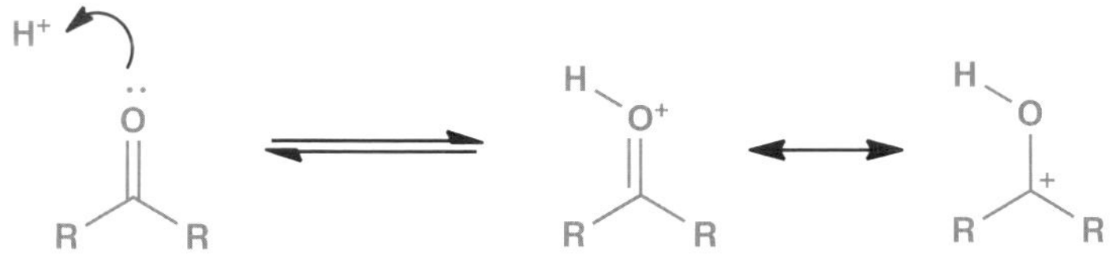

You may have come across this simple reaction, the Fischer esterification, in your pre-university courses.

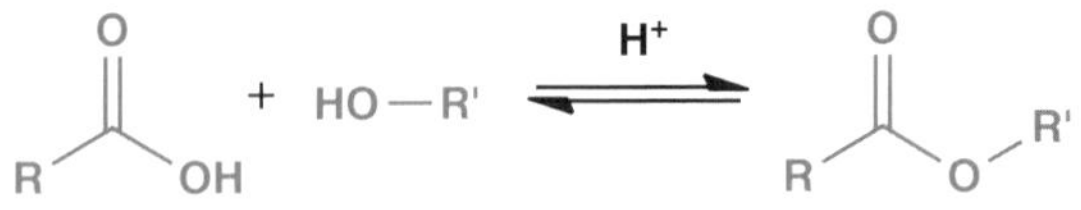

Acid + alcohol gives you an ester+water.

The Fischer synthesis of esters – an example of acid catalysis

In the first step of the reaction, acid catalysis makes the carbonyl more reactive. This activation of the carbonyl group then allows nucleophilic attack by the alcohol, forming a tetrahedral intermediate. Protonation of the oxygen from the acid (pink), and deprotonation of the oxygen from the alcohol, (green) makes the hydroxyl group a good leaving group (see Scheme 6.4).

All of these steps are reversible. All other things being equal, this reaction will give a mixture of ester and carboxylic acid. This reaction is often used to illustrate Le Chatelier's principle:

If a chemical system at equilibrium experiences a change in concentration, temperature, volume, or partial pressure, then the equilibrium shifts to counteract the imposed change and a new equilibrium is established.

The position of equilibrium may be driven to the right (that is, towards completion of the reaction), by adding a large excess of alcohol (normally by using the alcohol as a solvent), or by removing water from the reaction mixture.

The problem with the Fischer esterification is that it does not normally go to completion. Because all the reactions are in equilibrium, the final ester needs to be purified from a complex mixture of starting materials and products. For small and medium scale synthesis, we generally use more sophisticated reagents and organic solvents, which are discussed below.

Equilibrium reactions of carbonyl compounds are, however, very important in pharmacy. The body makes extensive use of equilibrium reactions, at low temperature and in aqueous solution.

Base-catalysed hydrolysis of esters

We have already encountered esters as pro-drugs, in Chapter 1. Because esters are generally more hydrophobic than the corresponding alcohols, they can be used to enable drugs to cross membranes and reach their sites of action. They then need to be hydrolysed to obtain the active drug. The bloodstream is sufficiently alkaline to hydrolyse many esters (pH 7.4 does support some base-catalysed reactions!). In addition, the bloodstream contains many esterases, which catalyse ester hydrolysis. Figure 6.9A shows the hydrolysis of aspirin in basic conditions.

The hydroxide ion attacks the carbonyl group to give a tetrahedral intermediate, and when the carbonyl is reformed, salicylate leaves. Now, in this case, loss of salicylate is preferred compared with the alternative loss of hydroxide because salicylate is a better leaving group.

The hydrolysis of erythromycin A acetate proceeds by the same mechanism (see Figure 6.9B), but now the intermediate loses the erythromycin ion, which is not really a better leaving group than a hydroxide.

For more details on esters as pro-drugs, see Chapter 1 'The importance of pharmaceutical chemistry'.

SCHEME 6.4

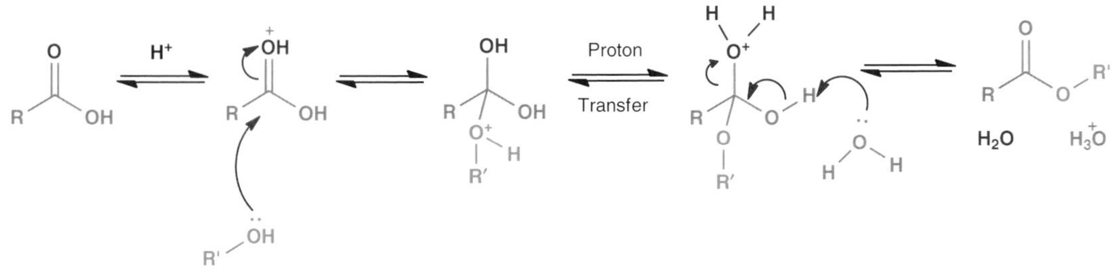

FIGURE 6.9 Hydrolysis of (A) aspirin and (B) erythromycin A acetate in base (Note that the terminal methyl groups in erythromycin are not drawn in).

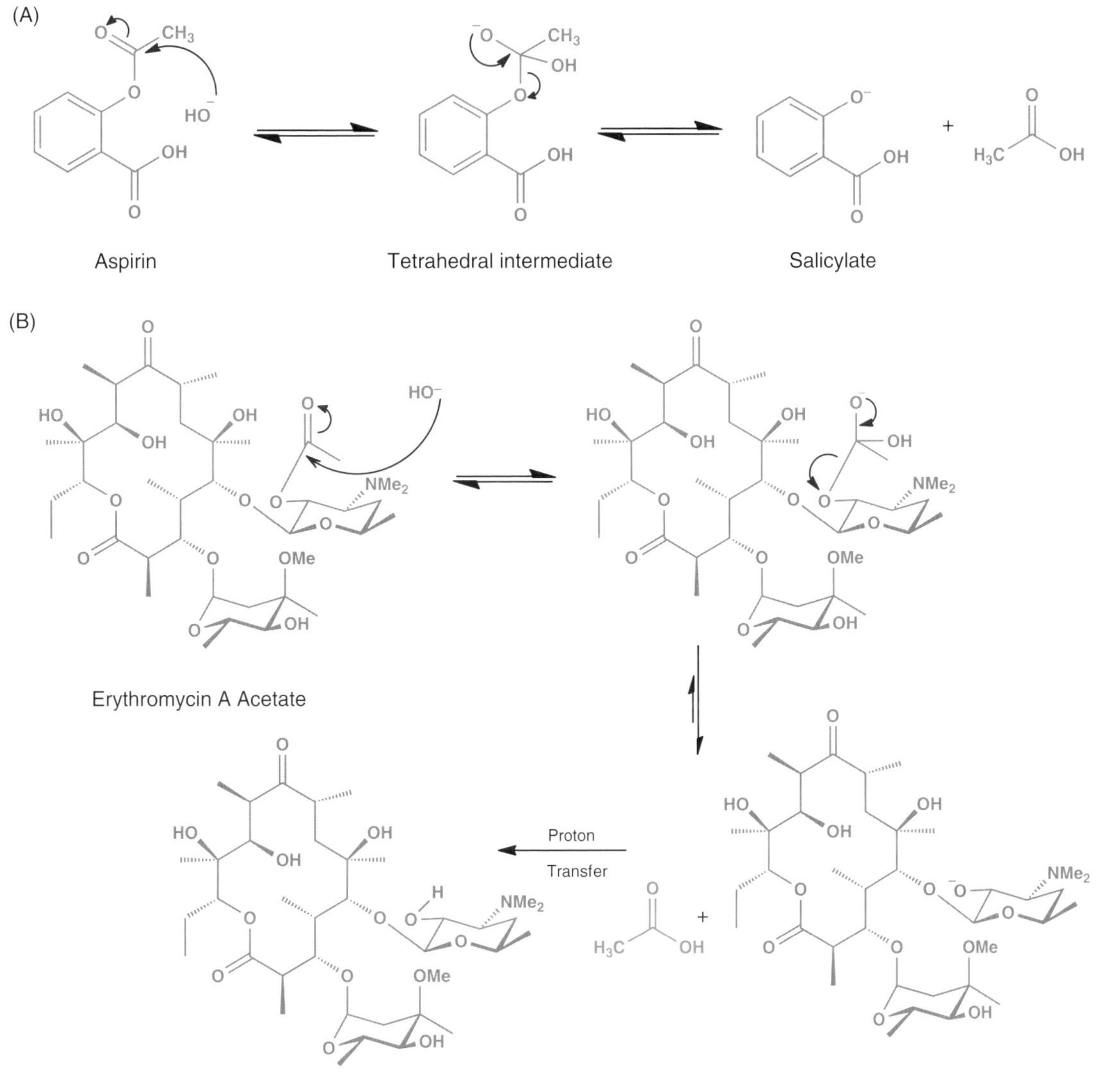

(A)

Aspirin   Tetrahedral intermediate   Salicylate

(B)

Erythromycin A Acetate

Proton Transfer

151

SELF CHECK 6.5

Why is salicylate a better leaving group than hydroxide? (Note: you should draw the structures and mechanisms – this becomes much easier with practice.)

SELF CHECK 6.6

The hydrolysis of erythromycin A acetate is, however, favoured. Look at the final products and see if you can work out why. (Hint: what will the basic solution do to the molecule coloured green?)

Interconversion of carbonyl compounds

The Fischer esterification is an example of the conversion of one type of carbonyl compound into another – namely a carboxylic acid into an ester. Protonation (acid catalysis) means that the leaving group is HY, rather than Y⁻. The hydrolysis of an ester in basic conditions is the reverse interconversion: an ester is converted to a carboxylic acid. Both of these reactions, and the other reactions described in this section, are *substitution* reactions during which Nu substitutes for Y⁻ (see Scheme 6.5).

SCHEME 6.5

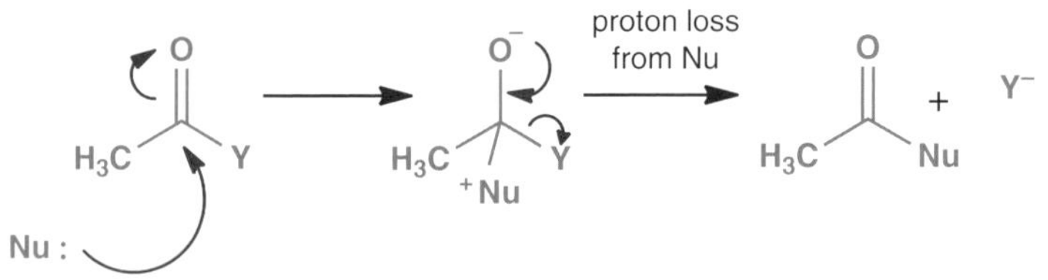

SCHEME 6.6

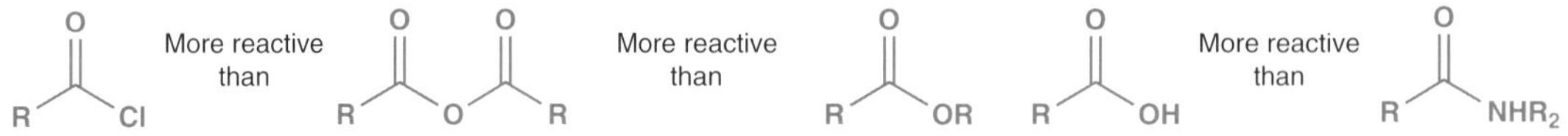

The interconversion of carbonyl compounds is extraordinarily important, both in pharmaceutical chemistry and in the body. You have already seen numerous examples of the interconversion of carboxylic acids and esters. If you have studied the companion volume on *Therapeutics and Human Physiology*, you will have encountered the interconversion of carboxylic acids, esters and amides in such essential processes as the synthesis of proteins. There is a clear order of reactivity among the various carbonyl compounds, as illustrated by their reactivity towards water (see Scheme 6.6).

Acetyl chloride (ethanoyl chloride) reacts violently with water to form acetic acid and HCl. Acetic anhydride reacts rapidly with water at neutral pH. Esters, such as ethyl acetate, do not react (ethyl acetate mixes poorly with water and forms a layer on top of the water layer). Acetic acid just dissolves quietly in water, and with a few trace flavourings, you put it on chips. Finally, the amide bond is a very strong bond, which is why your hair does not fall out in the rain.

SELF CHECK 6.7

What is the main constituent of hair? Why is the strength of the amide bond important to the stability of hair?

What makes a good leaving group?

The order of reactivity of carbonyls depends upon the stability of the leaving group. Cl$^-$ is an excellent leaving group, RCOO$^-$ is pretty good, HO$^-$ and RO$^-$ are poor, and NR$_2^-$ is terrible. Good leaving groups are groups that do not require protonation to stabilize them. Put another way, their conjugate acids have low pK$_a$ values. So the pK$_a$ of HCl is −7 and Cl$^-$ is a very good leaving group.

The pK$_a$ of RCOOH is about 4.5–5, making acetate a good leaving group. By contrast, H$_2$O and ROH have pK$_a$ values around 15–16; HO$^-$ and RO$^-$ are poor leaving groups. Finally, the pK$_a$ value for RNHR′ is typically 30–40 and RN–R′ is not a leaving group at all. Acid anhydrides and acyl chlorides are too reactive to be found in nature, but they are very useful synthetic intermediates.

For more information on pK$_a$s, see Chapter 6 'Acids and Bases' of the *Pharmaceutics* book in this series.

SELF CHECK 6.8

Thioesters and thioacids, such as thioacetic acid, are seldom used in synthetic chemistry, but nature makes enormous use of them. Where would you expect thioesters to come in the order of reactivity described above?

Synthesis of esters from acyl chlorides

An example of the usefulness of acyl chlorides is in the preparation of diacetylmorphine, also known as diamorphine, or heroin. Esterification is often used to make a drug more lipid soluble, enabling it to cross membranes more readily, and so helping it to reach its cellular target. Diamorphine is used for palliative care in end-stage terminal disease. When the diacetate is given as an injection, the increased lipid solubility increases the potency and means less drug needs to be used to achieve the analgesic effect.

In Figure 6.10, we have introduced the acetyl groups using two equivalents of acetyl chloride (ethanoyl chloride). The hydroxyl groups of morphine act as nucleophiles and the weakly basic pyridine acts as both catalyst and solvent.

FIGURE 6.10   Synthesis of diamorphine from morphine.

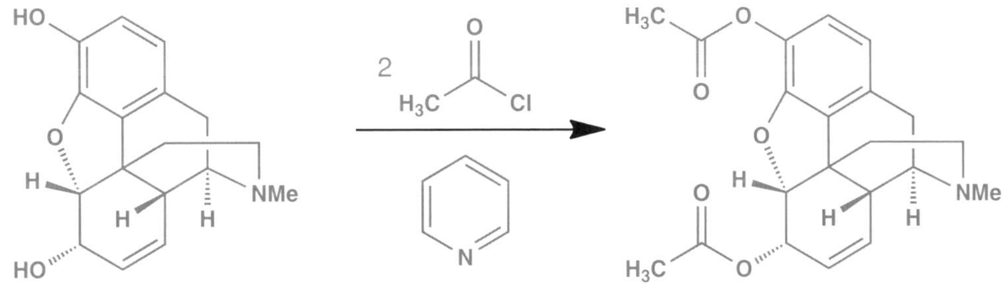

SELF CHECK 6.9

Draw a simple mechanism for the formation of diamorphine from morphine. Note that pyridine is able to drive the reaction by reacting with the HCl produced as a by-product.

The complete mechanism, shown in Figure 6.11, shows how pyridine acts as a catalyst. It is a very advanced mechanism for introductory pharmacy students, and in an examination you would be very likely to get full marks if you drew the morphine hydroxyl group attacking acetyl chloride directly. Pyridine acts

FIGURE 6.11   Complete mechanism of synthesis of diamorphine from morphine.

Diamorphine

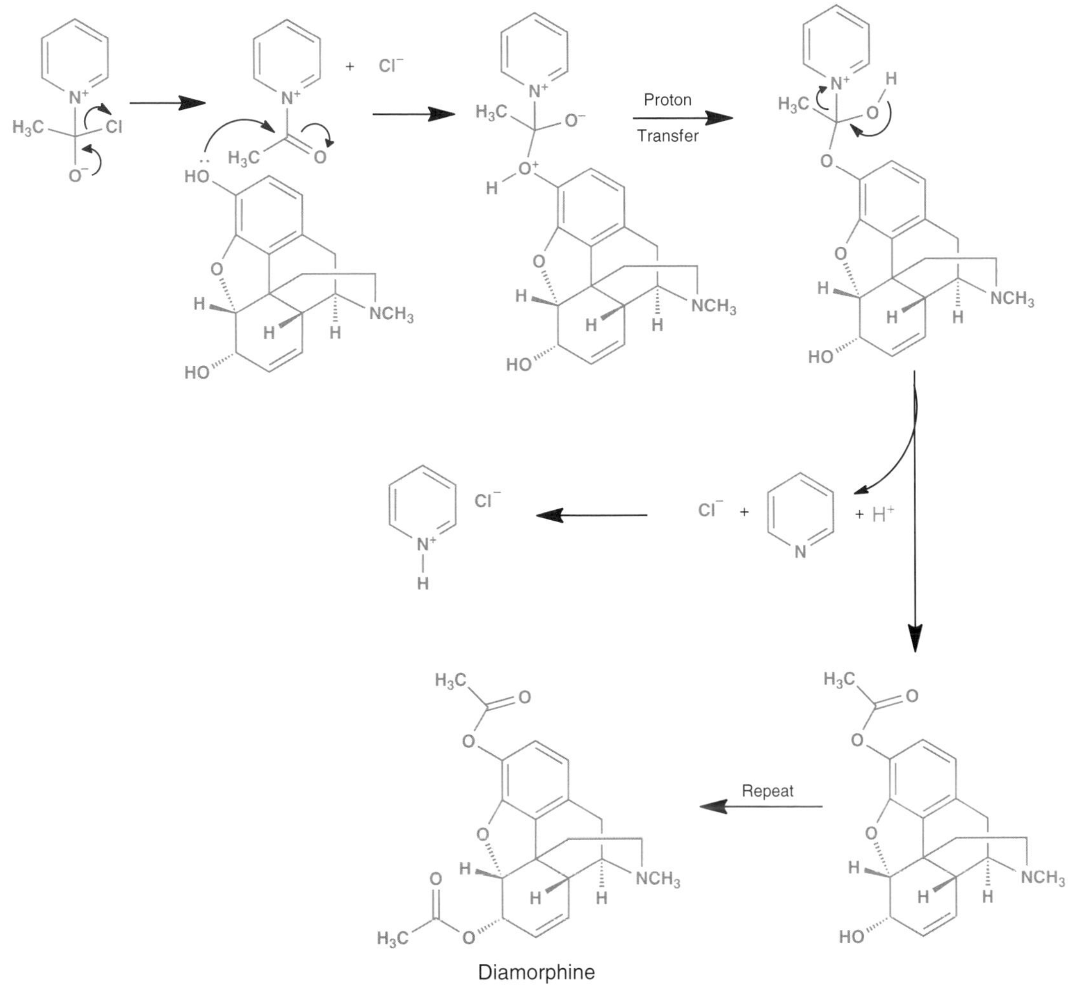

as a catalyst because it is both a good nucleophile and a good leaving group.

KEY POINT

You can draw chemical structures in various different orientations, without them being wrong. (The acetyl groups in Figure 6.11 are drawn in different orientations from Figure 6.10 for convenience.)

Synthesis of amides from acyl chlorides

Acyl chlorides can also be used to make amides. If you look in the BNF, you will see that there are many penicillins used clinically, and for each successful penicillin, there are many more that were synthesized, but proved to be unsuccessful in a clinical context. Unsuccessful drugs may be inactive; or active but toxic; or active, non-toxic but no better than existing drugs; or active, non-toxic, with some unique properties but simply too expensive to bring to market.

The various penicillins available are manufactured from the fungal product 6-aminopenicillanic acid. This compound has negligible antibacterial activity, but when it is treated with a suitable acyl chloride, a penicillin is produced, as shown in Figure 6.12.

SELF CHECK 6.10

Find the new amide bonds in methicillin and oxacillin.

Methicillin and oxacillin are just two examples of the plethora of penicillins that can be made from acyl chlorides. Both have activity against the most common penicillin-resistant bacteria, and are used to treat serious infections.

SELF CHECK 6.11

There is an alternative method of making penicillins from 6-aminopenicillanic acid. Use a specialist database, such as PubMed (www.ncbi.nlm.nih.gov) to find it.

FIGURE 6.12 Synthesis of penicillins from 6-aminopenicillanic acid and acyl chlorides.

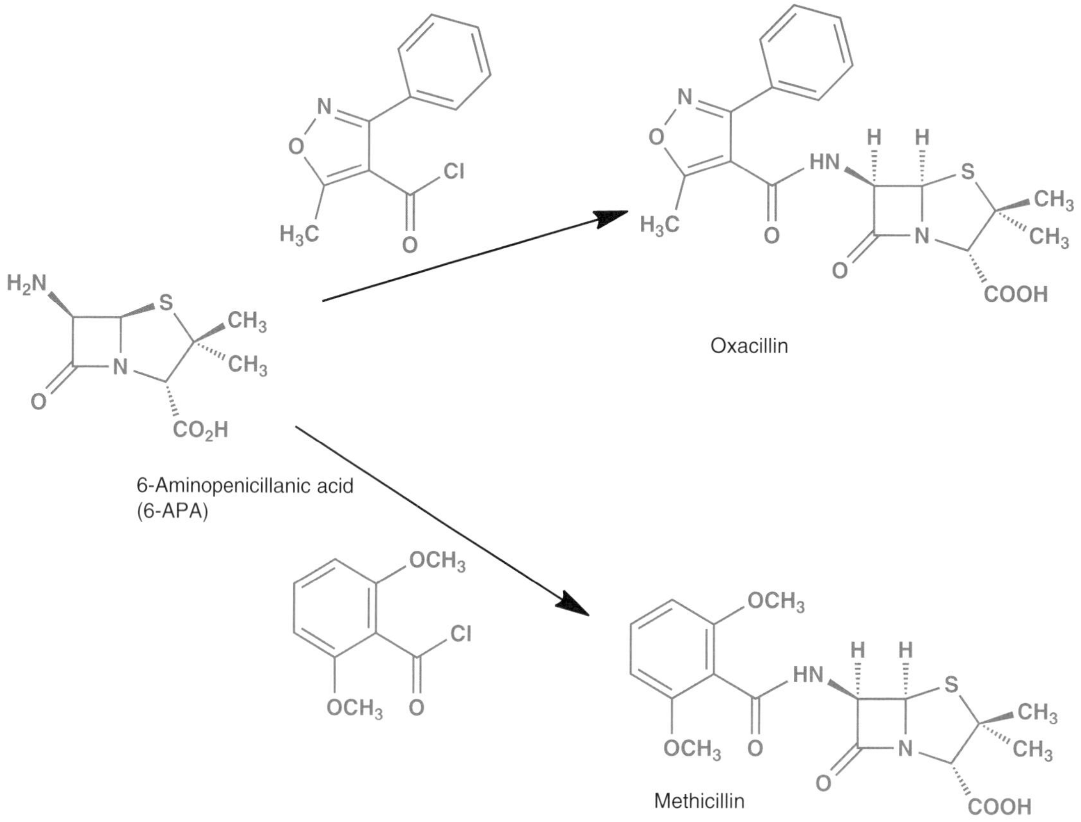

Synthesis of esters and amides from acid anhydrides

Acid anhydrides are less reactive than acyl chlorides, but are normally reactive enough for the synthesis of esters and amides. The most readily available acid anhydride is acetic anhydride (ethanoic anhydride), and this reagent can be used to synthesize diamorphine from morphine, or erythromycin A acetate from erythromycin A. Probably its most familiar use is in the synthesis of aspirin from salicylic acid, as shown in Figure 6.5.

When it is used to prepare esters, acetic anhydride is used in strictly non-aqueous conditions because water competes effectively with salicylic acid or morphine for acetic anhydride. Acetic anhydride can, however, be used in aqueous solution at pH 9–11 to prepare amides. For example, in studies of protein structure and abundance, it can be very useful to modify one type of amino acid residue, as explained in Integration Box 6.1. Lysine is the only amino acid with a primary amine side chain, and it is the only residue modified by acetic anhydride in aqueous solution (see Figure 6.13).

 Further details on amino acids can be found in Chapter 9 'The chemistry of biologically important molecules'.

Derivatizing proteins for structural determination

The digestive enzyme trypsin cleaves proteins adjacent to the most basic amino acids, arginine and lysine. The resulting peptides can be analysed by mass spectrometry. Some proteins, however, are very rich in basic residues and trypsin cleavage is inefficient where there are clusters of arginines and lysines. Treatment with acetic anhydride converts the basic amine side chain of lysine to a neutral amide, and trypsin does not cleave adjacent to the modified lysine residue. This results in more efficient reactions at arginine only.

Lysine is an essential amino acid. Mammals cannot manufacture it and require it in food. The protein in grains, such as wheat, are low in lysine, so lysine deficiency can occur in vegetarians, and there is some evidence that lysine deficiency can cause clinical anxiety. Milk, cheese, peas and beans are all rich in lysine, however, so a varied vegetarian diet should not lead to lysine deficiency.

How do you make acyl chlorides and anhydrides?

We have discussed making useful ester and amide drugs from acyl chlorides, but this raises the question:

155

FIGURE 6.13 The conversion of a lysine side chain to an amide using acetic anhydride.

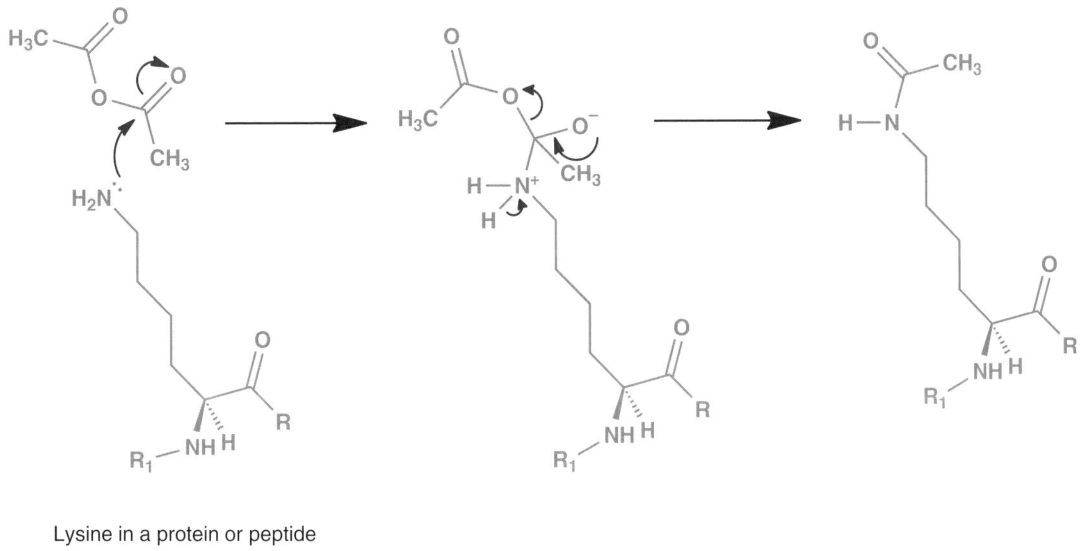

Lysine in a protein or peptide

how do we make the reactive carbonyls – acyl chlorides and anhydrides?

Acyl chlorides are not made by simple substitution at other carbonyls because the chloride ion does not react with the carboxylic acid to give an acyl chloride. To make an acyl chloride requires treatment of a carboxylic acid with thionyl chloride ($SOCl_2$) or a similar reagent. Figure 6.14 shows the overall reaction (a) and the mechanism (b).

Again, Le Chatelier's principle is at work. Thionyl chloride is a liquid; it is added to the carboxylic acid, usually in a 'dry' solvent – that is a solvent from which all traces of water have been removed. SO_2 and HCl are gases that diffuse away from the reaction mixture, so the reaction cannot reverse. The acyl chloride forms quite rapidly at room temperature. Because acyl chlorides are very reactive (and smelly), they are very often used without isolation or purification. For example, numerous esters of the antibiotic erythromycin have been made in this way (see Figure 6.15).

 More information on the esters of erythromycin can be found in Chapter 1 'The importance of pharmaceutical chemistry'.

SELF CHECK 6.12

There are five hydroxyl groups in erythromycin, three of them secondary, two tertiary (see Figure 6.15). You have learnt in Chapter 5 that secondary hydroxyl groups are more reactive than tertiary hydroxyl groups. Now look back at the role of pyridine in the esterification of morphine and see if you can work out why the hydroxyl coloured in red, in erythromycin is the only one to be esterified at room temperature if no catalyst is added.

Synthesis of acid anhydrides

In the laboratory it is generally much easier to make acyl chlorides than acid anhydrides. If you need the anhydride (because the acyl chloride is unstable or too reactive to work with) you treat the corresponding carboxylic acid with a drying agent, such as phosphorus pentachloride, and distil the anhydride, leaving the drying agent in the reaction flask, Figure 6.16. This may need to be repeated to ensure complete reaction.

SELF CHECK 6.13

Predict the products of the following reactions giving both their names and their structures.

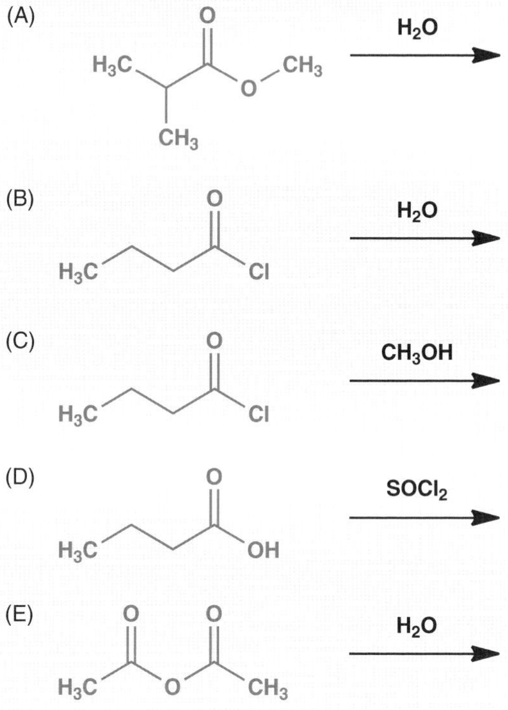

(A)

(B)

(C)

(D)

(E)

FIGURE 6.14 The reaction of thionyl chloride with a carboxylic acid to produce an acyl chloride (A) overall reaction (B) mechanism.

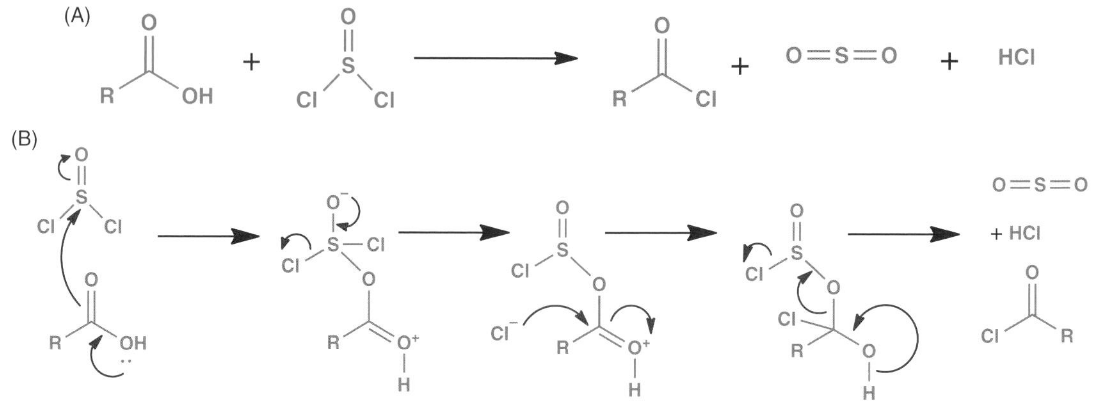

(A)

(B)

FIGURE 6.15 The synthesis of esters of erythromycin A.

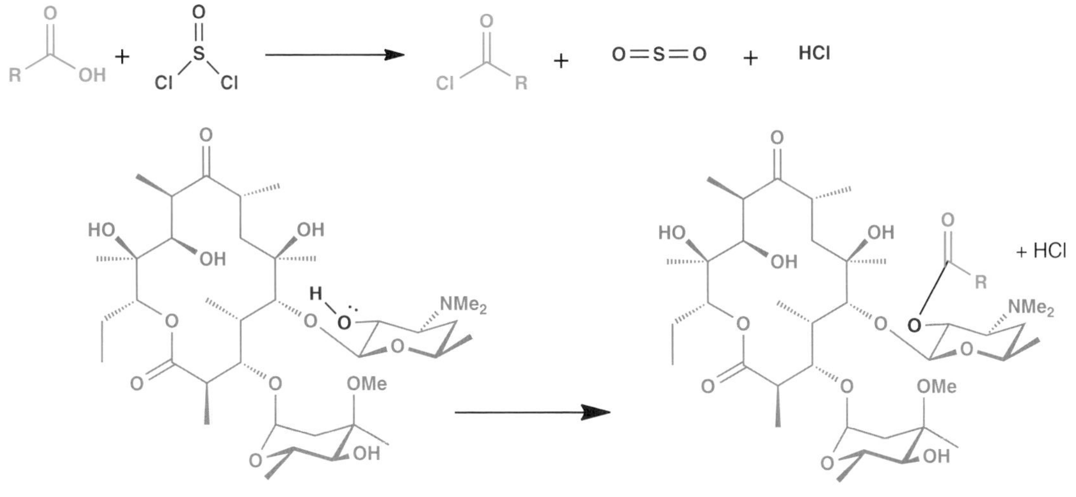

Erythromycin A   Erythromycin A Ester

FIGURE 6.16 Synthesis of trifluoroacetic anhydride by dehydration of trifluoroacetic acid.

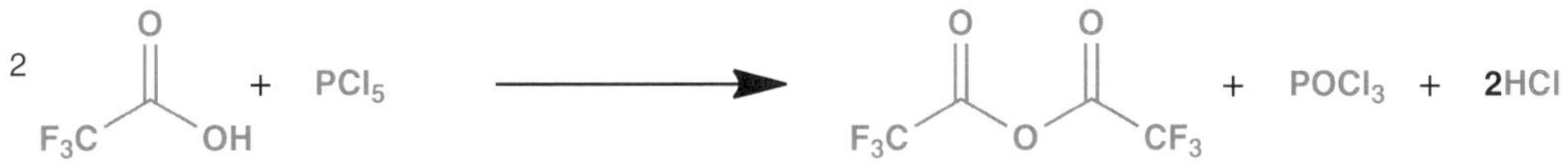

Aldehydes and ketones: What can you do without a good leaving group?

Aldehydes and ketones do not have good leaving groups. Neither hydrogen nor carbon readily accommodates a negative charge, so these compounds cannot undergo the substitution reactions we have seen previously. We have lost the ability to exchange groups and so our only option is to add groups, leading to nucleophilic addition (see Figure 6.17).

Here we are concerned with three situations (see Figure 6.18 for examples):

- when Nu is an oxygen nucleophile, leading to hemiacetals

- when Nu^- is a hydrogen nucleophile, leading to reduction reactions
- when Nu^- is a carbon nucleophile, enabling us to form carbon-carbon bonds.

Formation of acetals and hemiacetals

If you add an aldehyde to an alcohol, an equilibrium is set up between the aldehyde and the addition product, known as a hemiacetal (see Figure 6.19). A hemiacetal is characterized by a carbon bonded to OH, OR^2, R and either H or R^1.

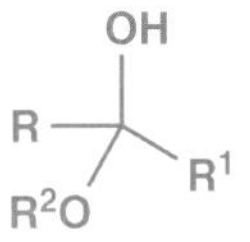

Hemiacetal

FIGURE 6.17 Mechanism of addition to aldehydes (R′ = H) and ketones. The nucleophile may be either a neutral or negatively charged species.

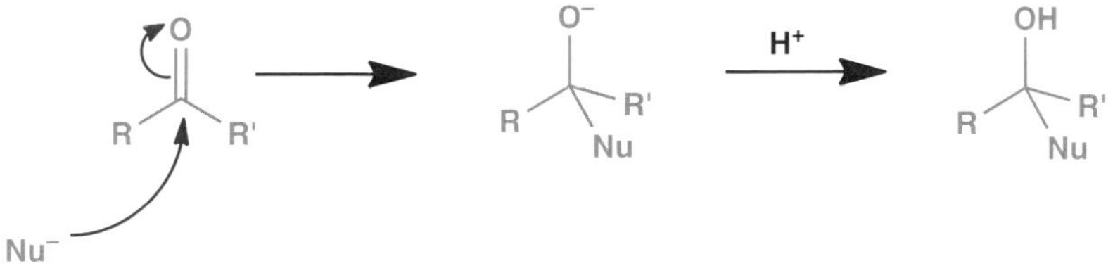

FIGURE 6.18 Common transformations of aldehydes and ketones.

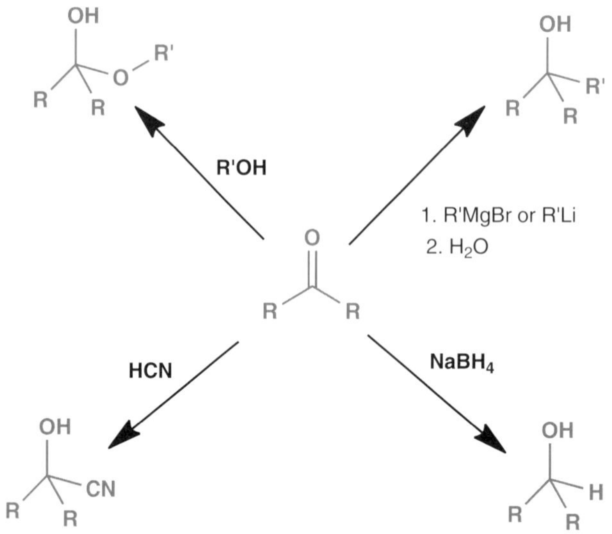

R'OH

1. R'MgBr or R'Li
2. H₂O

HCN

NaBH₄

FIGURE 6.19 Formation of a hemiacetal from acetaldehyde (ethanal) and ethanol.

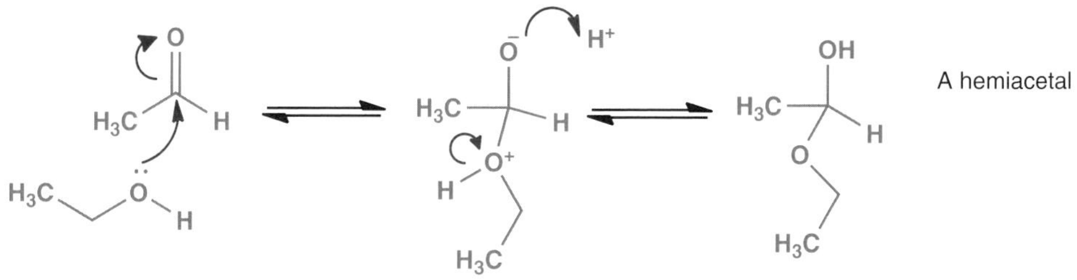

A hemiacetal

The hemiacetal cannot normally be isolated, but we know that it is there because certain forms of spectroscopy, such as infra-red and NMR spectroscopy, can detect hemiacetals in solution.

Human beings are absolutely dependent on a group of hemiacetals known as sugars (see Figure 6.20). Sugars are internal hemiacetals – the aldehyde and the alcohol are in the same molecule – and they are more stable than hemiacetals formed from two different molecules.

 More details on sugars and carbohydrates can be found in Chapter 9 'The chemistry of biologically important macromolecules'.

Ketones can also undergo reactions with alcohols in the same way. The products are also known as hemiacetals, although you can use the term hemiketal if you prefer. (IUPAC considers hemiketals to be a sub-class of hemiacetals.)

SELF CHECK 6.14

Can you think of an important sugar that normally exists as a hemiketal? (Hint: your sugar bowl might be a source of inspiration.)

Aldehydes and ketones can react with alcohols under acid catalysis to produce acetals, as shown in Figure 6.21. Acetals have the general structure shown here, and are characterized by a carbon bonded to OR^2, OR^2, R and either H or R^1

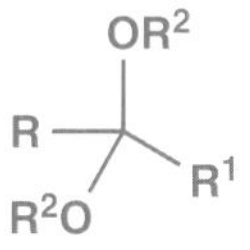

OR^2

R
R^1

R^2O

Acetal

Organic chemists love this mechanism (see Figure 6.21) because it illustrates lots of important points – for

FIGURE 6.20 Some important hemiacetals: ribose is found in RNA and in ATP; deoxyribose is found in DNA; glucose is the entry point of glycolysis, the metabolic pathway that enables us to derive energy from food.

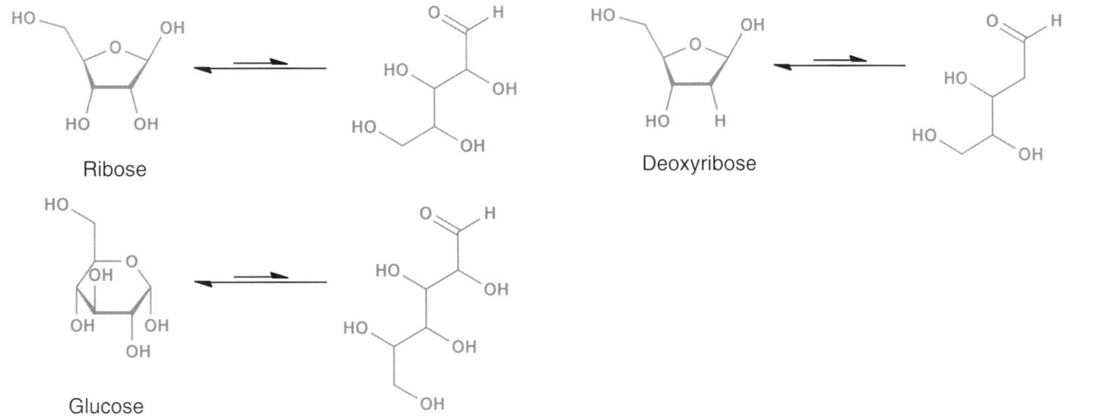

Ribose

Deoxyribose

Glucose

FIGURE 6.21 Formation of an acetal by reaction of an alcohol and a ketone under acid catalysis.

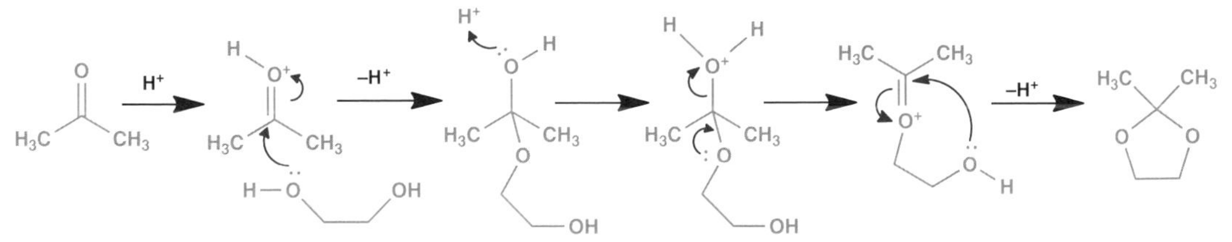

example, that the electrophile is always a protonated carbonyl (an oxonium ion). Pharmacy students are more likely to ask 'What is it for?'

Acetals:

- can be used as **protecting groups** in the synthesis of drugs and other molecules
- are found in carbohydrates in all life forms
- are found in drugs, for example, the aminoglycoside antibiotics.

If you wanted to modify the carboxylic acid functional group of the painkiller ketoprofen, but preserve the ketone, you might need to protect the ketone, as shown in Figure 6.22(a). Figure 6.22 also shows some important acetals found in nature. Figure 6.22(b) shows a polymer of glucose. This type of polymer is found in food stores, starch in plants and glycogen in animals. Figure 6.22(c) shows streptomycin, an aminoglycoside antibiotic. It has two acetal functions, connecting the sugar rings.

SELF CHECK 6.15

Find the acetal functions in the glucose polymer and in streptomycin and label them.

SELF CHECK 6.16

There is an unmodified carbonyl in streptomycin. What is it?

 More information on carbohydrates, including glucose polymers, can be found in Chapter 9 'The chemistry of biologically important macromolecules'.

Reduction of aldehydes and ketones

In order to reduce an aldehyde or a ketone to the corresponding alcohol, we have to add the equivalent of H^-. Hydride (H^-) is a hopeless nucleophile; it acts only as a base and normally abstracts a proton from a carbonyl compound to give H_2. To make H^- into a good

FIGURE 6.22 Acetals in pharmacy: (A) use of an acetal protecting group. Note the reagent LiAlH$_4$, which we will study in Section 6.3. This reagent would react with the ketone, which needs to be protected; (B) a polymer of glucose units; (C) streptomycin.

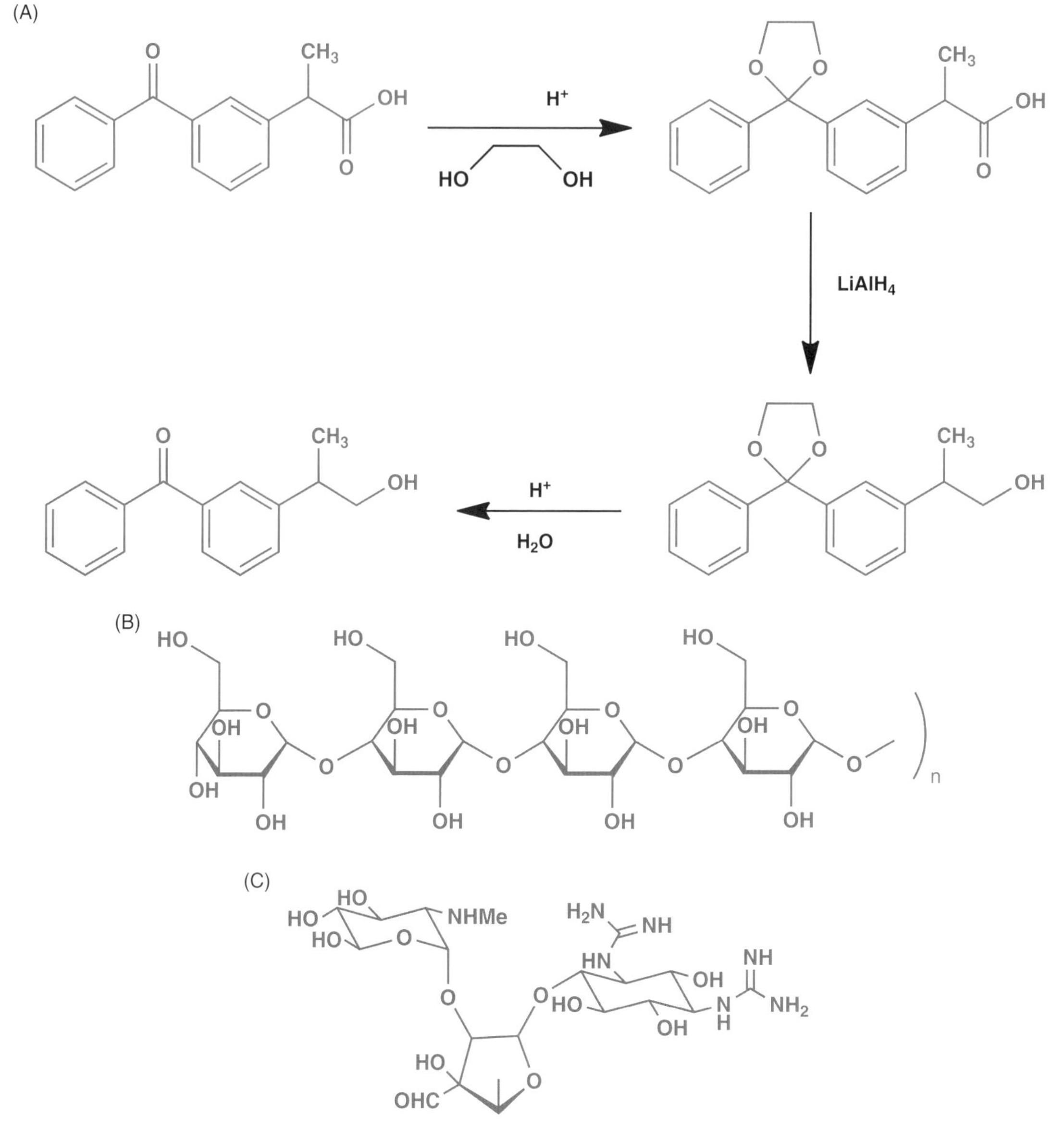

(A)

(B)

(C)

nucleophile, we make it bigger, and spread the negative charge over several hydrogens.

 The theory that underpins this is very interesting, and is covered briefly on the Online Resource Centre (www.oxfordtextbooks.co.uk/orc/ifp).

Typically, the hydride equivalents that we use are sodium borohydride (NaBH$_4$) or lithium aluminium hydride (LiAlH$_4$). Sodium borohydride is a mild reducing agent; it is easy to use and is compatible with water or alcohols in the solvent. Lithium aluminium hydride will reduce even the most stubborn ketones (indeed it reduces esters to alcohols), but must be used

in very dry (water-free) solvents. Figure 6.23 shows how sodium borohydride is used to reduce a ketone to an alcohol, a key step in the synthesis of duloxetine. Duloxetine is an antidepressant, used to treat major depressive disorder.

SELF CHECK 6.17

See if you can draw the mechanism for the reduction of an ester to an alcohol by lithium aluminium hydride. (Hint: you have seen two mechanisms – interconversion and addition – for the reaction of carbonyls with nucleophiles and you will need both of them.)

FIGURE 6.23 Reduction of a ketone to an alcohol using sodium borohydride. This is an important step in the synthesis of duloxetine.

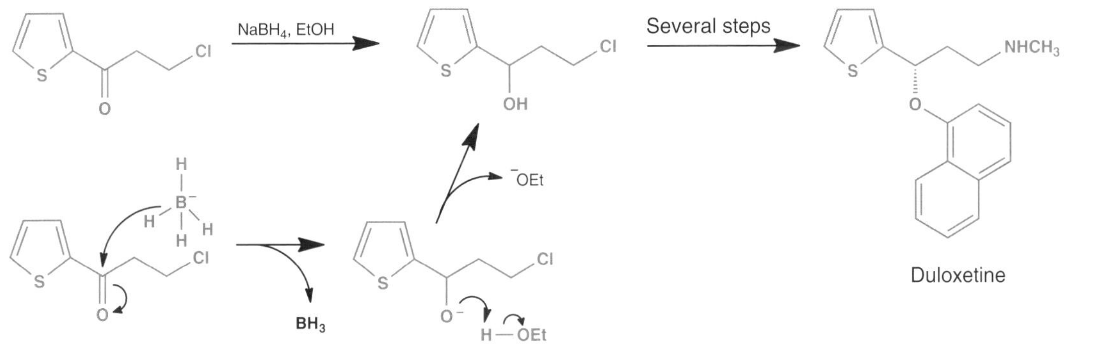

Duloxetine

SELF CHECK 6.18

It is very dangerous to use a carbon dioxide fire extinguisher if lithium aluminium hydride is present. Why?

One very good way of finding new drugs is to make small modifications to old drugs (that is, to make analogues of the old drugs). With very complex molecules, the number of analogues that can be produced is often severely restricted by the chemistry.

SELF CHECK 6.19

Lots of analogues of erythromycin (most of them unsuccessful) have been made over the years and 9-S-dihydroerythromycin is one of them. What reagent would you use to make this compound from erythromycin?

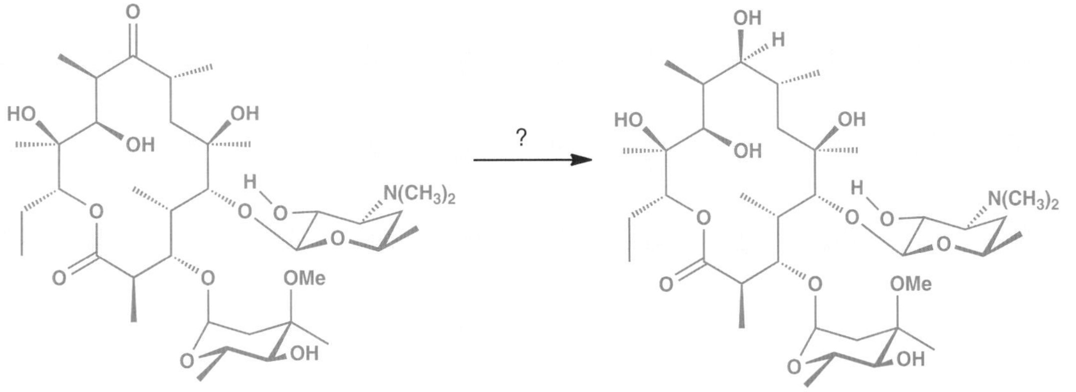

You might have expected the reaction featured in Self check 6.19 to produce a mixture of 9S- and 9R-dihydroerythromycin. Erythromycin, however, has 18 chiral centres, so the mixture would be of diastereomers, not of enantiomers. The other chiral centres have a profound influence on the course of the reaction, and in practice the 9R-compound cannot be detected.

 Enantiomers and diastereomers are covered in more detail in Chapter 3 'Stereochemistry and drug action'.

Erythromycin A

?

9S-Dihydroerythromycin A

Formation of carbon-carbon bonds

Pre-university courses often include the action of cyanide on ketones and aldehydes to give cyanohydrins. This reaction does form a carbon-carbon bond but the reaction is rather limited; it is frustrating that only a single carbon can be added. More importantly, however, huge amounts of safety legislation now surround the use of cyanide, and this reaction is only permitted if there is an exceptionally good reason for it. Fortunately, there are better ways of making carbon-carbon bonds.

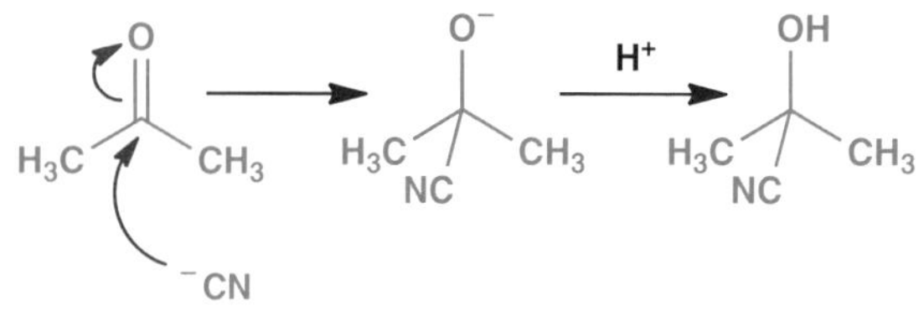

To make a carbon-carbon bond, one carbon (usually an alkyl or aryl group) needs to support a negative charge. Carbocations (carbons supporting positive charges) are explored in Chapters 4 and 5, and should be familiar, but we now require the equivalent of C^-. To make carbon carry a negative charge, we add it to a less electronegative element, typically a metal. Lithium, zinc and magnesium are common choices. The mechanisms of action are also closely similar, so we will consider the reactions of organomagnesium compounds, known as Grignard reagents, as a representative example.

A Grignard reagent (Grignard was a French chemist who published this reaction in 1900) is made by the treatment of an alkyl halide with magnesium in an inert solvent. The Grignard reagent is then equivalent to C^-, because magnesium is less electronegative than carbon. A typical Grignard reagent, derived from bromoethane, can be drawn in full like this:

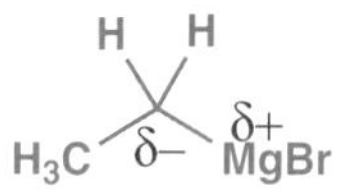

Here is the type of reaction it can undergo (Scheme 6.7).

Now this really is a fantastic reaction. If you can make carbon-carbon bonds, you can make almost anything.

SELF CHECK 6.20

Predict the outcome of the following reactions

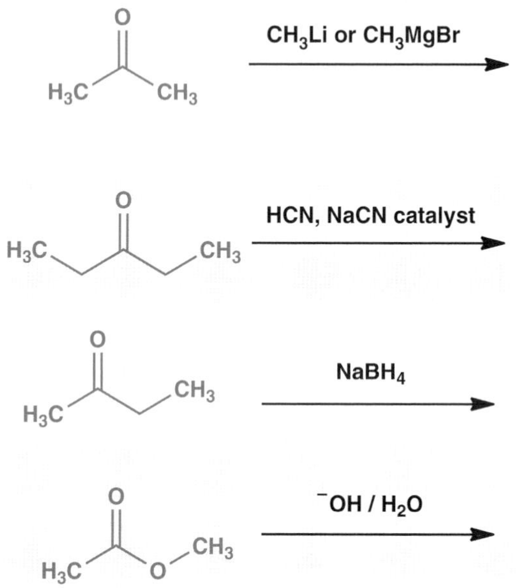

KEY POINT

Carboxylic acids and derivatives are able to undergo substitution reactions. Aldehydes and ketones possess functional groups that cannot act as leaving groups so any product must be an addition product.

SCHEME 6.7

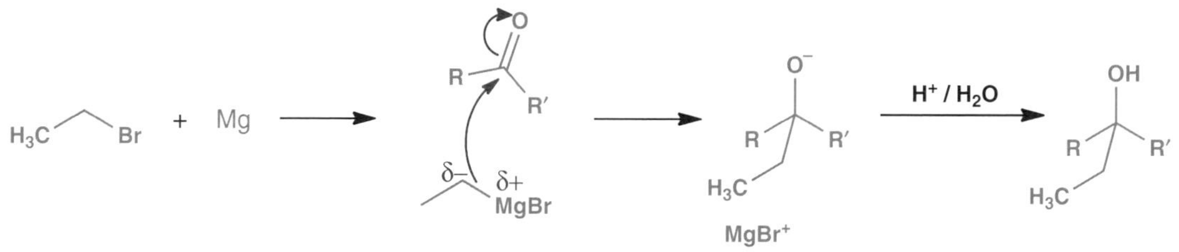

6.4 α-substitution reactions

A third major type of carbonyl reaction is the α-substitution reaction, which occurs at the position next to (or 'α' to) the carbonyl group. H^+ is removed from this position and is substituted by another atom or group. These reactions not only have a place in medicinal and organic chemistry but they are prevalent in biochemical processes also. Let us consider the overall factors that can bring about these reactions.

Keto-enol tautomerism

In the presence of acid, a carbonyl with an α-hydrogen is able to undergo an equilibrium reaction, known as keto-enol tautomerism. The enol tautomer is only present in very low concentrations, and its formation occurs very slowly under neutral conditions but, as we have learnt from other areas of reactivity, the presence of either an acid or a base can be a game changer in terms of chemical reactivity (see Scheme 6.8).

 For more about keto-enol tautomerism, see Chapter 4 'Properties of aliphatic hydrocarbons'.

In the enol form of a carbonyl, the α-carbon takes part in a π-bond, and is therefore able to act as a nucleophile, a C⁻ equivalent. Something similar occurs in base, with an enolate rather than an enol being formed (see Scheme 6.9).

Sodium ethoxide (made by dissolving sodium in ethanol) is a suitable base for catalysing keto-enol tautomerism.

The ability to form enols or enolates makes the α-hydrogen relatively acidic; it has a pK_a of about 19. ('Relatively' in this context means relative to an alkene with a pK_a of 44; it is still nothing like as acidic as acetic acid, pK_a 4.7.) There are plenty of strong bases capable of removing a proton from the α-carbon of a carbonyl compound, enabling it to act as a nucleophile.

 pK_a is discussed in more detail in Chapter 4 'Properties of aliphatic hydrocarbons'.

Reactions in which the carbonyl is the nucleophile

When a carbonyl forms an enol or an enolate, it acquires the ability to act as a nucleophile; in order to form a double bond, the α-carbon loses a proton, and the π-electrons can now react with an electrophile. We have met several electrophiles so far in this book, and many of them can react with enols or enolates. We will now consider some of the most useful of these reactions.

Halogenation

Halogenation is one of the most useful α-substitution reactions of carbonyls, because the α-halocarbonyls

SCHEME 6.8

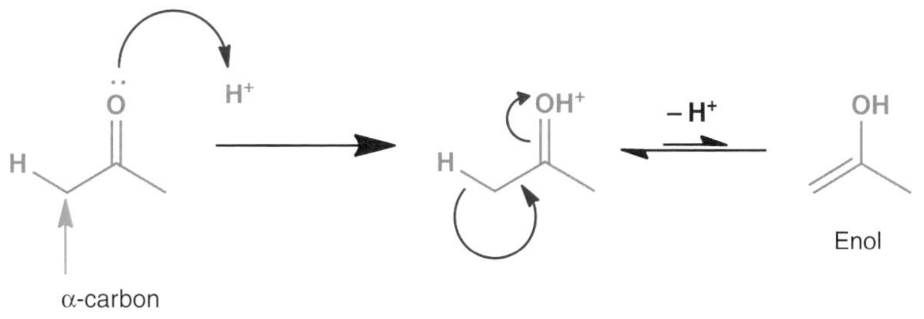

SCHEME 6.9

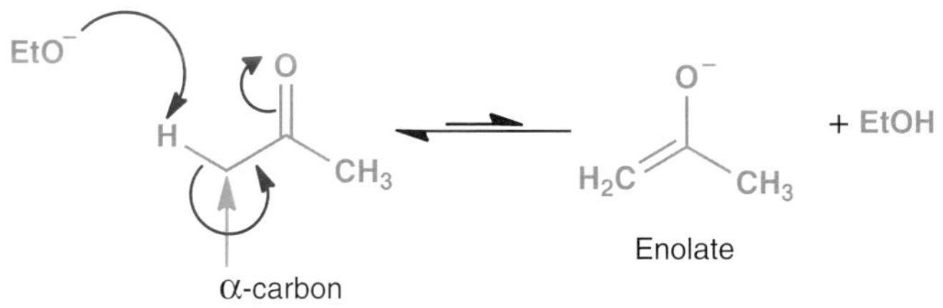

formed in these reactions are excellent synthetic intermediates. Let us start by considering the bromination of acetone (propanone), which is shown in Figure 6.24.

The reaction is initially very slow in neutral conditions, but eventually speeds up because HBr is generated in the reaction and can act as a catalyst. It is more usual to add some acid (such as acetic acid) to the reaction mixture at the outset.

The electronegative bromine in bromopropanone makes enol formation more difficult, so if one equivalent of bromine is used to brominate acetone, the product is almost exclusively bromopropanone. As a consequence, this reaction is considered to be very 'clean'.

SELF CHECK 6.21

If acetone is treated with two equivalents of bromine in acid conditions, what would you expect the product to be?

Benazepril is an ACE (angiotensin converting enzyme) inhibitor used to treat high blood pressure. Its synthesis starts with 1-tetralone, which looks rather unpromising (see Figure 6.25). However, monobromination allows the position α to the carbonyl to be built upon. After several synthetic steps, a large group (almost half the molecule) is built onto the α-carbon, resulting in a very profitable drug.

SELF CHECK 6.22

Draw a mechanism for the acid-catalysed bromination of 1-tetralone.

The halogenation of carbonyls in base is rather different as shown in Scheme 6.10.

The first bromination proceeds in a similar way to the acid-catalysed reaction (though note that the OH⁻ is consumed; it is not strictly a catalyst). The hydrogens coloured green are now, however, relatively acidic. The enolate generated from bromopropanone

FIGURE 6.24 The monobromination of acetone (propanone).

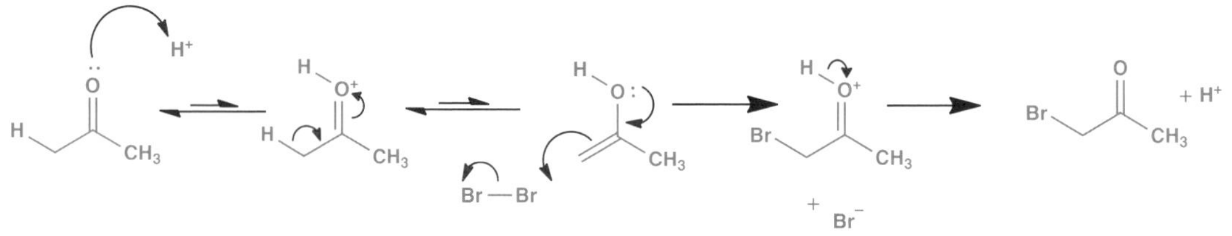

FIGURE 6.25 The synthesis of benazepril. The part of the molecule that replaces the bromine is coloured in purple. If the rest of this synthesis (such as the ring expansion or the separation of enantiomers) interests you, you can find it in *The Art of Drug Synthesis,* Ed. Johnson and Li, Wiley, 2007, p.150.

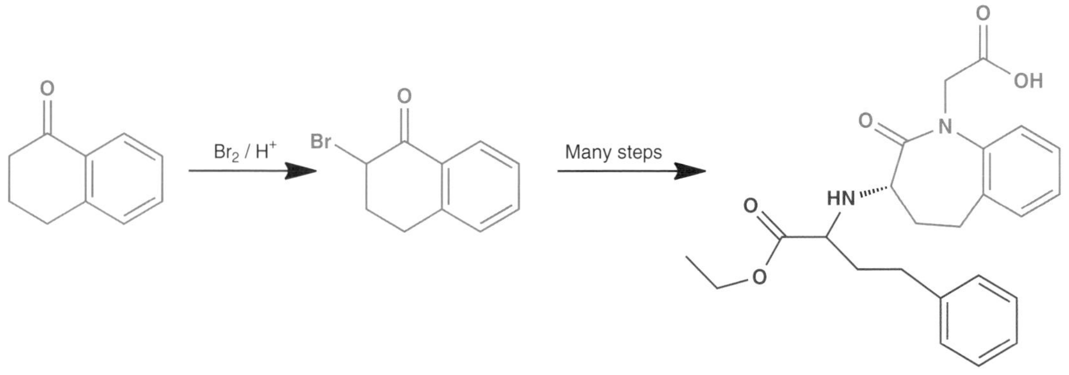

Benazepril

SCHEME 6.10

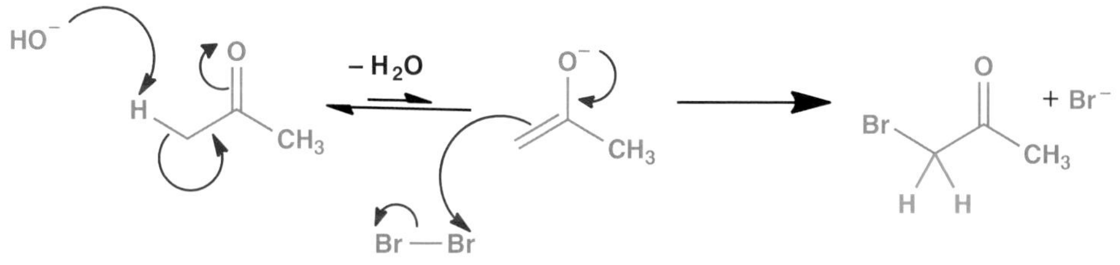

SCHEME 6.11

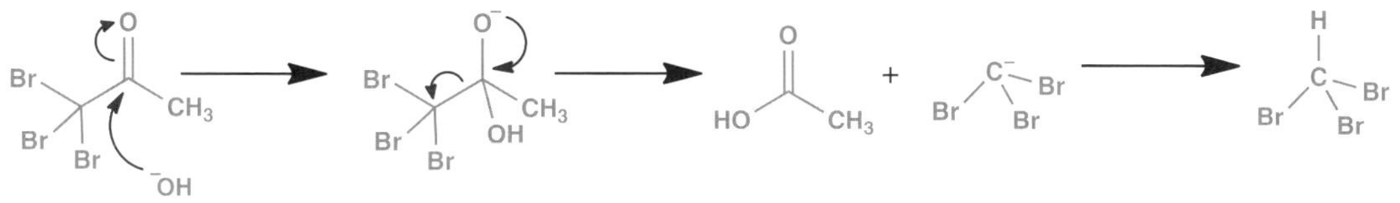

is stabilized by the electron-withdrawing bromine. The second bromination is faster than the first, and the third is faster again.

And now, $\underset{Br}{\overset{C^-}{\underset{Br}{|}}}Br$ is actually a better leaving group than OH⁻, so the reaction shown in Scheme 6.11 occurs.

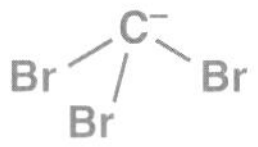

SELF CHECK 6.23

What happens if you treat 1-tetralone with iodine and sodium hydroxide? Draw the mechanism.

When iodine is used in place of bromine, we get the so-called 'iodoform' reaction. Iodoform, CHI_3, is pale yellow and insoluble in water. A methyl ketone gives iodoform, on treatment with sodium hydroxide and iodine; other ketones do not. So if you want to test for a methyl ketone you can simply conduct an iodoform test rather than investing half a million pounds in an NMR spectrometer.

The aldol reaction

Enols and enolates are nucleophiles, but their parent carbonyl compounds are electrophiles. If you add just a tiny amount of base to an aldehyde or ketone, a small amount of enolate forms and this nucleophile reacts with a carbonyl molecule. The reaction of acetaldehyde with base is shown in Figure 6.26. This reaction is known as an aldol reaction because the historical name for the product (3-hydroxybutanal) was aldol.

Aldehydes, ketones and carboxylic acid derivatives can all undergo aldol reactions provided they have at least one α-hydrogen. The important step is the formation of the enolate ion, which can then act as a nucleophile and form a new carbon-carbon bond.

FIGURE 6.26 The aldol reaction of acetaldehyde. Note that a new carbon-carbon bond is formed (shown in black).

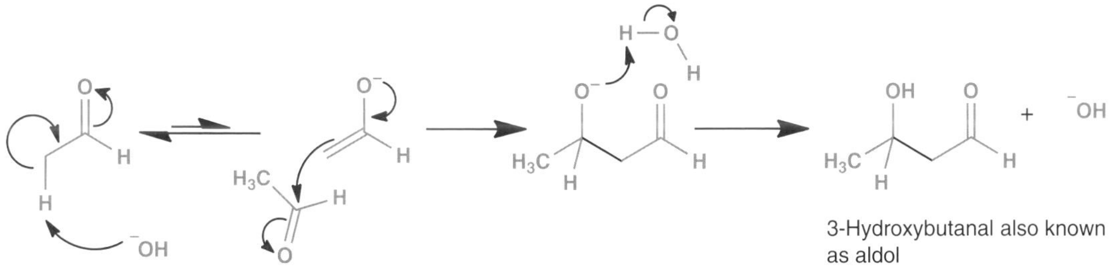

3-Hydroxybutanal also known as aldol

FIGURE 6.27 Loss of water from an aldol product to give an unsaturated aldehyde.

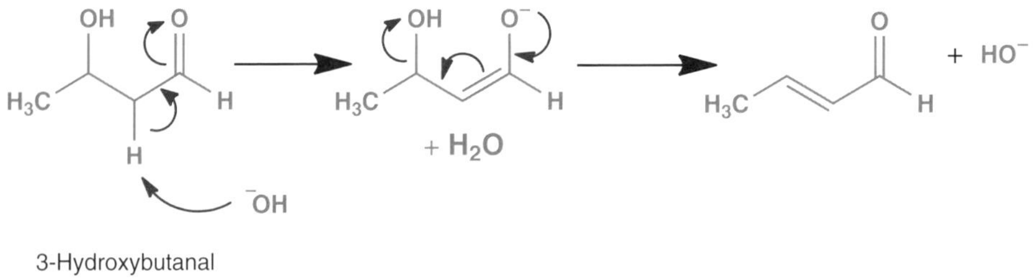

3-Hydroxybutanal

SELF CHECK 6.24

Draw a mechanism for the aldol reaction of acetone (propanone).

Aldol products can rather easily lose water as shown in Figure 6.27, especially if too much base is added. This is why the aldol reaction is sometimes termed the aldol condensation. (Loss of water or another small molecule = condensation).

The Claisen condensation

The Claisen condensation is a close relative of the aldol reaction; however, esters – rather than aldehydes or ketones – are the reactants. The overall reaction is shown in Scheme 6.12.

SCHEME 6.12

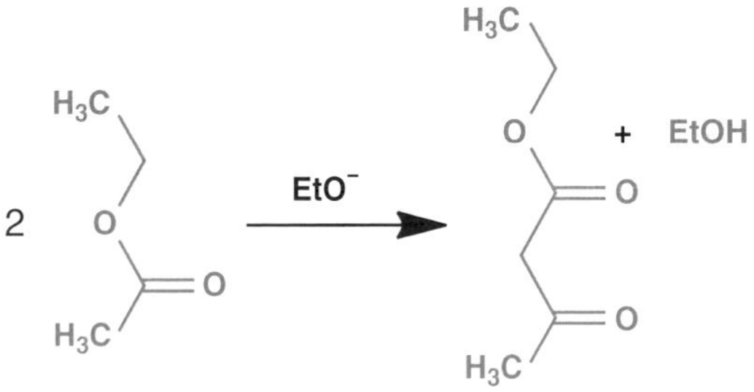

Let us explore it step-by-step, starting with the enolate ion (see Scheme 6.13).

SCHEME 6.13

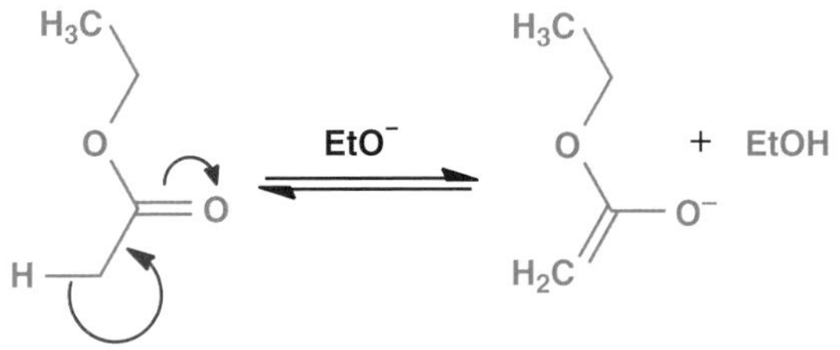

Now we have formed our nucleophile (the enolate), we can seek out an electrophile, which is, of course, the carbonyl of another ester molecule (see Scheme 6.14).

SCHEME 6.14

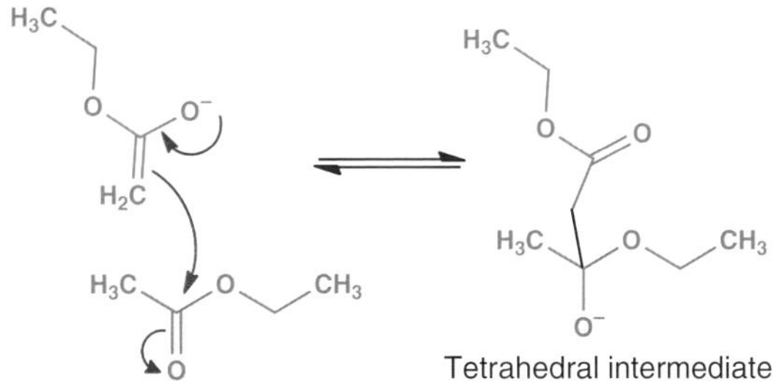

Tetrahedral intermediate

SCHEME 6.15

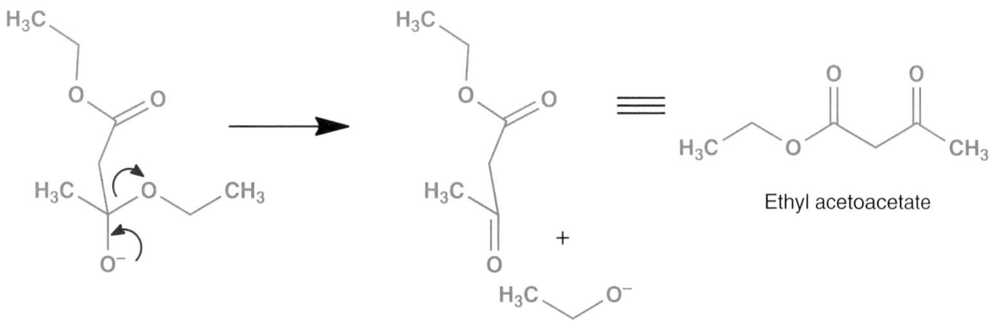

A tetrahedral intermediate is formed, with the new bond shown in black. Actually it is quite easy to forget the new bond when you draw the intermediate. Try counting the carbons when you draw intermediates, or use colours as above. If all else fails, draw out the complete structures with all the carbons and hydrogens; this is not wrong, remember, just a bit laborious.

Finally, the tetrahedral intermediate loses EtO⁻ (see Scheme 6.15).

Reading most textbooks, it is easy to imagine that other people (teachers and fellow students) draw their structures perfectly, with nice bond angles, every time. They do not. At the end of a complex mechanism, you may well have a dreadful-looking structure like the one in the middle of the scheme. If so, just redraw it, as shown.

The entire Claisen reaction scheme is summarized in Figure 6.28.

Aldol and Claisen reactions: summary

Aldol	Claisen
Aldehyde or ketone	Ester
↓	↓
Enolate ion	Enolate ion
↓	↓
Tetrahedral product	Tetrahedral intermediate
	↓
	Product (ketoester)

FIGURE 6.28 The Claisen condensation in full.

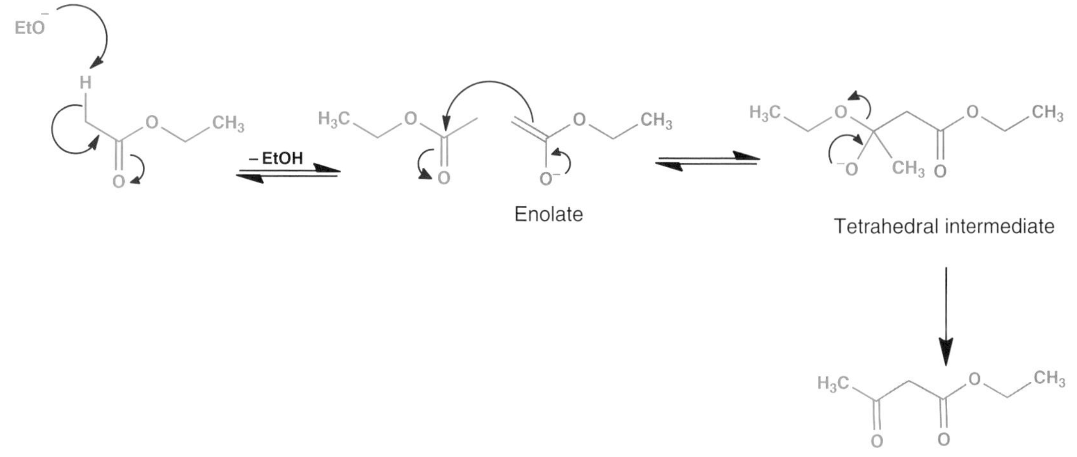

FIGURE 6.29 The synthesis of elzasonan.

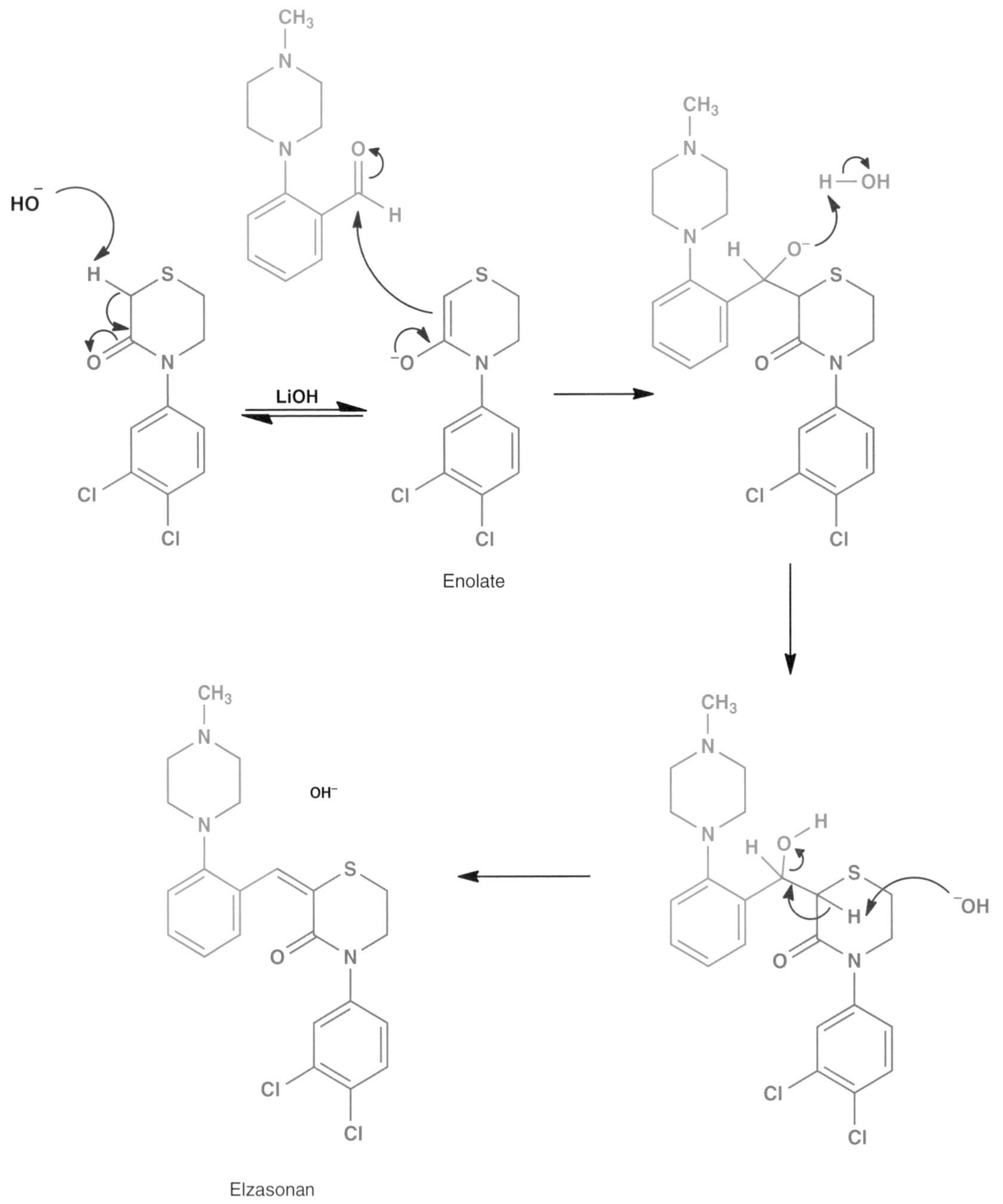

Elzasonan

The aldol and Claisen reactions in focus

Aldol and Claisen reactions are used in the synthesis of drugs, but they are normally limited to cases where the mechanism and hence the reaction products are unambiguous. The problem is that a molecule like butanone has five α-hydrogens. An initial deprotonation may take place at carbon-1 or carbon-3 – but even after this reaction four α-hydrogens remain. A complex mixture of products is likely to be formed.

SELF CHECK 6.25

Consider the effect of adding base to a mixture of acetone and butanone. How many aldol products can you draw?

Elzasonan was a promising antidepressant that was withdrawn after phase II clinical trials. The final stages in the synthesis involved an aldol condensation reaction of a benzaldehyde derivative with a thiamorpholine (see Figure 6.29).

This reaction works well because only one of the two carbonyls can form an enolate, and it can only form *one* enolate (there are hydrogens on only one side of the C=O function). In the laboratory, the aldol and Claisen reactions are quite limited because mixtures of products are often formed. Nevertheless, a great many drugs are made by Claisen condensations, and all are natural products known as polyketides.

Polyketide synthesis – Claisen-type reactions in biology

Polyketides are made by plants, fungi and bacteria, and consist of acetate or propionate units (or sometimes other simple carboxylic acid derivatives), which are polymerized using Claisen-type reactions. Many of these compounds have biological activity. The polyketides include the drugs mentioned previously – erythromycin, nystatin and tetracycline – but they also include environmental toxins such as the aflatoxins.

The biosyntheses (that is the syntheses by biological systems) of polyketides are dependent on huge multi-enzyme complexes. The multi-enzyme complex that synthesizes the lactone ring of erythromycin is made up of over 20 subunits; these 20 proteins act together to catalyse Claisen-like reactions. Strip away the sugars and the hydroxyl group at carbon-12 from erythromycin A and you get erythronolide B, the portion of the antibiotic made on the multi-enzyme complex called a polyketide synthase. Erythronolide B is made by joining seven propanoate (3-carbon) units together. Figure 6.30 shows how the first two units are joined together.

Propanoate (the starter unit in pink) is bound to the enzyme using a thioester bond. As we have seen previously, thioesters are electrophiles, whose reactivity is intermediate between that of esters and acid anhydrides. Using thioesters is biology's way of making an active species. The pink starter unit is poised to react with a second molecule of propanoate (in dark green), which will act as the nucleophile. The soil bacterium that makes erythromycin does not add an acid or base catalyst. Instead, it activates propanoate as a nucleophile by replacing a proton with a carboxylic acid functional group to give methyl malonate.

With a carbonyl on either side, the central carbon is now a very reactive nucleophile; it reacts with the starter unit with loss of carbon dioxide to give an enzyme-bound thioester, made up from two propanoate units. A chiral centre is produced in this reaction, and it has *R*-stereochemistry.

Finally, there is a reduction, involving nature's reducing agent, **NADPH**, to give the alcohol with *R*-stereochemistry.

A further five propanoate units are added to complete the erythronolide molecule. Each addition is both regiospecific (the reactants go to the right place) and stereospecific (only one stereoisomer is formed). Nature has spent millions of years evolving enzymes that carry out very clean Claisen-like reactions.

 Polyketide synthesis is very similar to fatty acid synthesis, which is described in Chapter 3 'The biochemistry of cells' in the *Therapeutics and Human Physiology* book in this series. Fatty acid synthesis involves polymerization of acetate units, and, indeed, many polyketides are made up of acetate, rather than propanoate units. Polyketides also have much greater structural variety.

SELF CHECK 6.26

Draw the mechanism for the reaction of the next propanoate unit (the blue one) to the growing erythronolide molecule. Notice that there is no reduction step.

SELF CHECK 6.27

Can you see how the completed chain cyclizes to give enzyme-free erythronolide B?

FIGURE 6.30 (A) Erythromycin A, (B) Erythronolide B stick structure and (C) Erythronolide B with methyl groups marked – you can see how seven three-carbon units make up this structure; (D) the first stage in the biosynthesis of erythronolide B – a Claisen-type reaction.

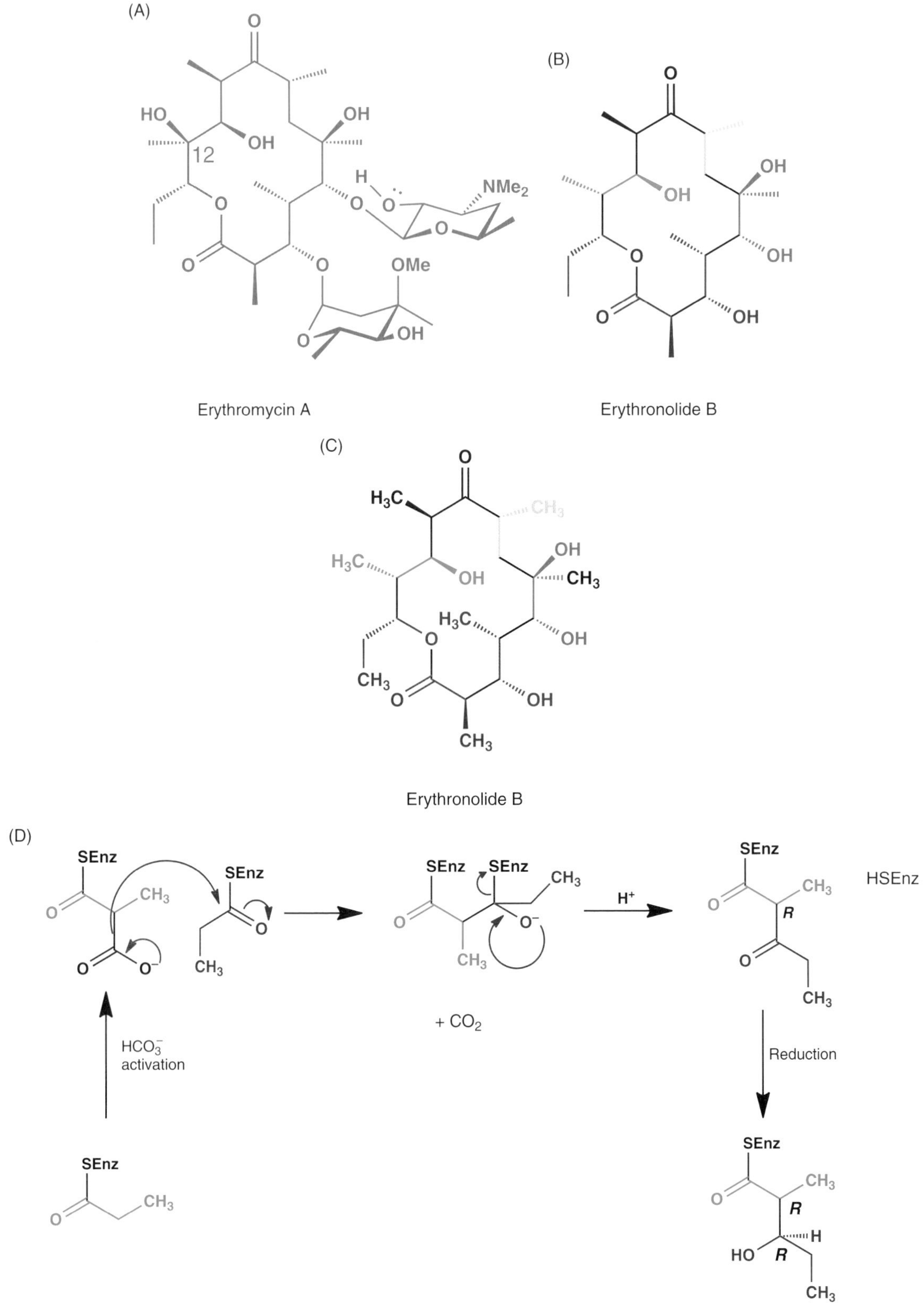

(A)

Erythromycin A

(B)

Erythronolide B

(C)

Erythronolide B

(D)

6.5 Carbonyls in the body

So far we have looked at the potential of the carbonyl group to be transformed. We have seen how the powerful acyl chlorides and acid anhydrides are the most reactive amongst this useful family. Acyl chlorides and acid anhydrides, however, have no place in the body. They react violently with water and with many other nucleophiles. So which carbonyls *are* of value in biological systems?

Natural carbonyl compounds: the importance of thioesters

Thioesters are very versatile compounds. During the biosynthesis of polyketides they serve as both nucleophiles and electrophiles in Claisen-like reactions; as such, they behave as nature's equivalent of esters. Their intermediate reactivity also enables them to act as nature's equivalent of acid anhydrides.

Coenzyme A (see Figure 6.31) is readily acylated (often with an acetyl group) to give a thioester, which can be used in acyl transfer reactions.

An example of how nature uses these powerful thioesters is in the synthesis of N-acetylglucosamine. This sugar is used in many biochemical processes, and is a component of bacterial cell walls. Here it is synthesized by an aminolysis reaction between glucosamine and acetyl CoA (see Figure 6.32).

> **KEY POINT**
>
> Thioesters are less reactive than acid anhydrides, so they can be used in biological systems, but they are more reactive than esters, and much more reactive than amides.

FIGURE 6.31 Formation of acyl CoA from coenzyme A. When R=CH$_3$, the molecule is acetyl CoA.

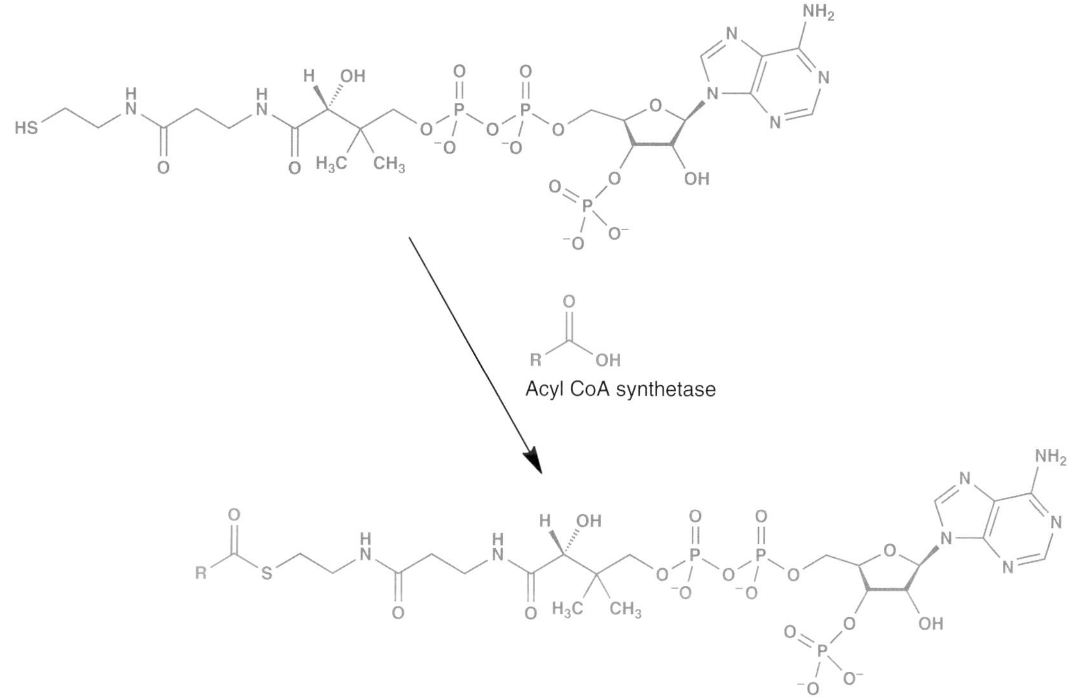

Acyl CoA synthetase

FIGURE 6.32 The synthesis of *N*-acetylglucosamine from glucosamine and acetyl CoA. Note the abbreviation for coenzyme A – you would not normally draw out the whole molecule in a mechanism!

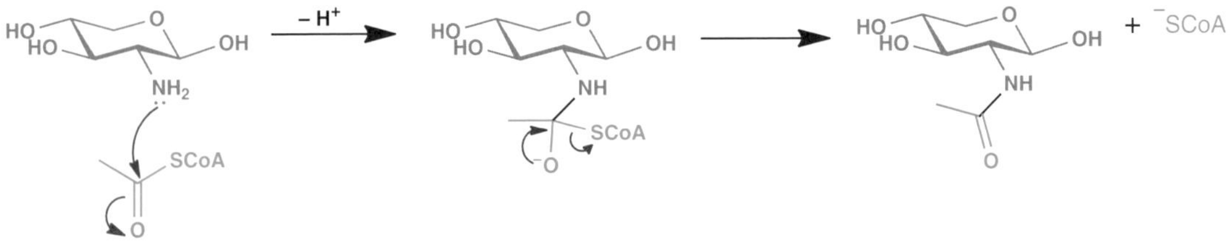

6.6 Carbonyls in drugs – opportunities and problems

We have seen throughout this book, and especially in this chapter, that many drugs contain carbonyls. We have also seen that carbonyls can be reactive. This means that their presence in drugs is often vital but can lead to problems. We have already discussed many of the opportunities and problems of carbonyl chemistry in pharmacy. Here are two specific examples.

Drug storage – aspirin

Acetyl salicylic acid is stable enough to be compressed into a tablet and used as a pharmaceutical. But would you advise storing aspirin tablets in a bathroom cabinet for a long time? We have seen hydrolysis of esters in base in Figure 6.9, but Figure 6.33 shows how esters

can also be hydrolysed in acid, in a reversal of the Fischer ester synthesis. So aspirin, which is an acidic molecule, can react with moisture to be converted back to salicyclic acid, so it would not be a good idea to leave a bottle of aspirin tablets in the bathroom cabinet!

It used to be common to smell vinegar on old aspirin tablets. Now, however, aspirin is often coated to enhance its stability. Of course, aspirin works in the body as a pro-drug. In the intestine, it is converted to salicylic acid by base-catalysed hydrolysis.

Proteins and other amides

A visit to your hairdresser should convince you that it is difficult to break amide bonds. Hair is made of

FIGURE 6.33 The acid catalysed hydrolysis of aspirin.

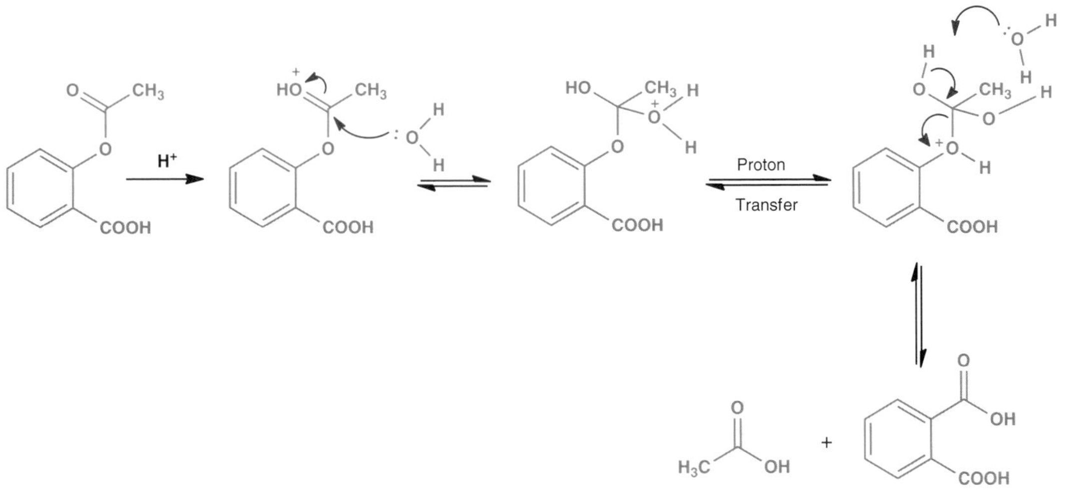

protein, and protein is made of amino acids linked by amide bonds (known as peptide bonds when they are found in proteins). Hair can be straightened, curled, bleached and dyed, but the amide bonds remain intact. Hairdressers have to study biology and are often very knowledgeable about protein chemistry.

Similarly if you boil an egg, the protein coagulates and solidifies, but the peptide bonds do not break. This raises questions about nutrition! You can boil, scramble, fry, or bake eggs and the protein remains intact, so how does egg work as a nutrient?

The answer is that our digestive tracts are full of enzymes that catalyse the hydrolysis of peptide bonds. Pepsin in the stomach cleaves peptide bonds adjacent to aromatic residues; trypsin in the intestine cleaves peptide bonds adjacent to basic residues; carboxypeptidase cleaves peptide bonds at one end of a protein, and so on. Because of these enzymes, proteins are processed very efficiently. Each individual enzyme is, however, quite specific, which means that some amide bonds are able to evade all these enzymes.

Consider methicillin, for example.

This molecule contains two amide bonds. The β-lactam bond is rather reactive for an amide. This is because amide carbons, like all sp² carbons, prefer bond angles of 120°, but the four-membered ring constrains the β-lactam bond to a 90° bond angle. Despite this reactivity, both the green and the pink amide bonds escape attack by digestive enzymes and so methicillin can be delivered orally.

It is a different story with insulin, a medicine required daily by patients with Type 1 diabetes. Diabetics have to inject themselves regularly with insulin. Despite modern 'pens', it would be much more convenient to take the drug orally. Unfortunately, the enzymes in the digestive tract recognize insulin as food (see Figure 6.34).

The hydrolysis of a peptide bond proceeds nucleophilic attack on the carbonyl as described in Section 6.3. For the tetrahedral intermediate to lose an amine rather than water requires a complex enzymic mechanism (see Figure 6.35). In the laboratory, amide bonds can be cleaved by heating at boiling point in 6 M acid.

ACE inhibitors

A very important class of drugs in use today are the ACE (angiotensin converting enzyme) inhibitors. ACE catalyses the hydrolysis of a specific peptide bond. By doing this, it converts angiotensin I to angiotensin II. Angiotensin II is a potent vasoconstrictor, which

173

FIGURE 6.34 Insulin – a small protein with many peptide (amide) bonds. (A) A structure highlighting the amide bonds; (B) ribbon structure showing the secondary structure of insulin. Source: life-enhancement.com.

(A)   (B)

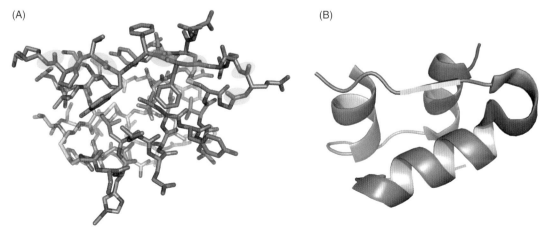

A patient who suffers from diabetes comes into the pharmacy with a prescription for 'Ketostix'. He asks what they are for and how do they work.

REFLECTION QUESTIONS

1. What is the reason for diabetes patients using 'Ketostix'?

2. How do 'Ketostix' work?

Answers

1 Ketones build up in the body when the body metabolizes fats instead of carbohydrates to provide energy. This occurs when the body has insufficient insulin to allow the transport of glucose from the bloodstream into the cells. In this circumstance ketones build up in the blood and are excreted in the urine and this is a common complication of poorly controlled diabetes. This situation, if left untreated, can lead to ketoacidosis, a condition which may be life-threatening. It is appropriate, therefore, to test the urine for ketones at regular intervals.

2 'Ketostix' are thin strips of plastic, impregnated with a nitroprusside reagent, which are dipped into the patient's urine. Acetoacetic acid (the ketones – check the structure) reacts with sodium nitroferricyanide (nitroprusside) and glycine to produce a purple complex. The depth of the colour of the complex indicates the level of ketones in the urine.

causes high blood pressure and related conditions when present in excess.

ACE inhibitors are currently used in the treatment of hypertension. They work by stopping the production of angiotensin II from angiotensin I by the zinc-containing ACE (see Figure 6.36). The first drug in this family was captopril. It was proposed that it binds at the active site of ACE, with the S^- of the drug interacting with Zn^{2+} in the enzyme and oxygens in both carbonyl groups also involved in binding. These oxygens (whether charged or not) are able to act as hydrogen bond acceptors.

 Hydrogen bonds are discussed in more detail in Chapter 2 'Organic Structure and Bonding'.

Captopril is a relatively simple molecule (count the chiral centres and calculate the molecular weight) and many analogues of this compound have been synthesized. A particular motivation has been to improve the **bioavailability**. Captopril needs to be taken two or three times per day yet, for long term therapy, it is far preferable to have a drug that is taken just once per day. In addition, captopril has a nasty taste, due to the sulfur in the molecule.

Further studies led to the development of enalaprilat. However, as is often the case with drug development, there was one step forward and one step back. Enalaprilat was not orally active and was only suitable for use as an injectable. The scientists at Merck were not deterred by this, however; they esterified enalaprilat

FIGURE 6.35 The action of digestive enzymes in cleaving peptide bonds.

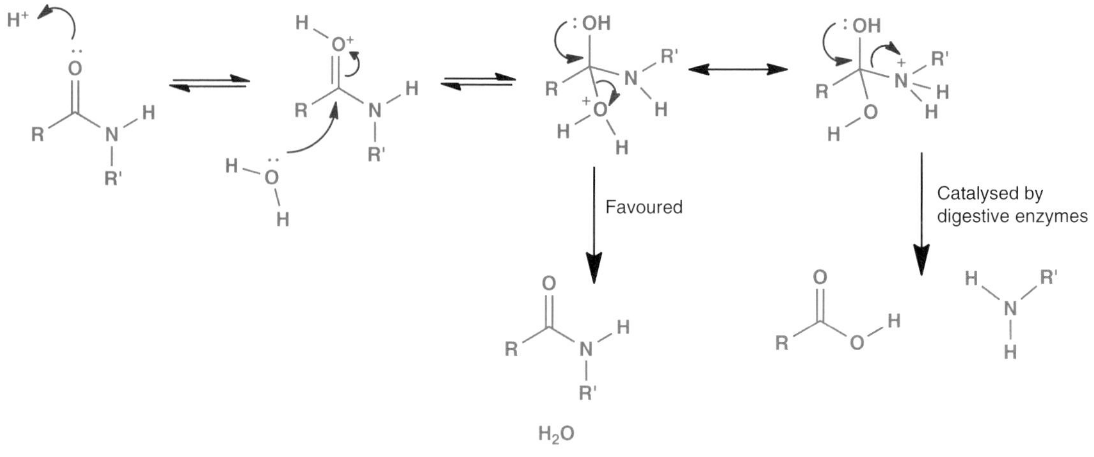

FIGURE 6.36 (A) Conversion of angiotensin I to angiotensin II by angiotensin converting enzyme; (B) Ionized form of captopril.

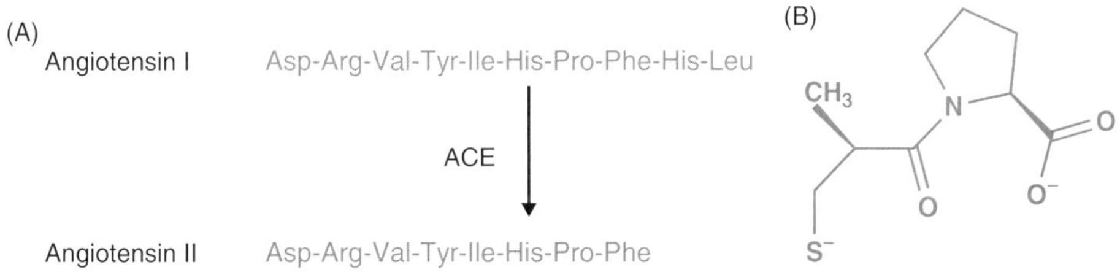

(A)
    Angiotensin I   Asp-Arg-Val-Tyr-Ile-His-Pro-Phe-His-Leu

                               ACE ↓

    Angiotensin II   Asp-Arg-Val-Tyr-Ile-His-Pro-Phe

(B)

FIGURE 6.37 Conversion of the pro-drug enalapril to the active drug enalaprilat.

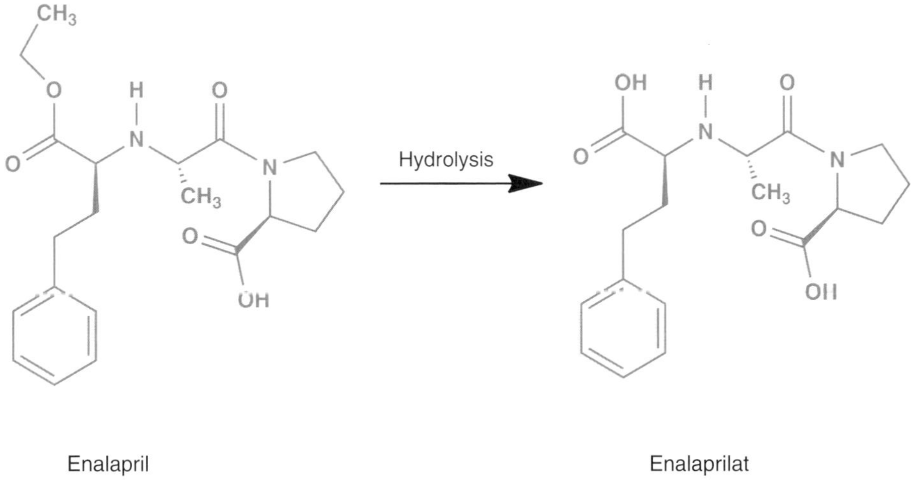

Enalapril   Enalaprilat

175

to give enalapril. Enalapril is an inactive pro-drug; for the ACE inhibitor activity to take effect the ethyl ester is hydrolysed back to enalaprilat after the pro-drug has passed through the gut wall and entered the blood stream. Truly fantastic carbonyl chemistry: a carbonyl compound has been converted to another to make a drug that is suitable for oral delivery then metabolism has converted it back to the active form to save lives (see Figure 6.37).

SELF CHECK 6.28

Draw a mechanism for the conversion of enalapril to enalaprilat. Hint: when drawing mechanisms in assessments, it is usually acceptable to refer to most of the molecule as 'R' and to focus on the important functional groups!

CHAPTER SUMMARY

Much of carbonyl chemistry can be understood with just three mechanisms: substitution reactions of carboxylic acid derivatives, addition reactions to ketones and aldehydes, nucleophilic reactions of enols and enolates.

These mechanisms can help us to understand how drug molecules are made in the laboratory and in nature, how drugs work, how they are metabolized and how they are degraded on storage.

In the 1970s, Stuart Warren wrote a ground-breaking book entitled 'The Chemistry of the Carbonyl Group', which you may be able to find in your University Library. The fact that a whole textbook could be written about it illustrates just how fantastic carbonyl chemistry is. There is further reading, in particular a discussion of the infrared spectra of carbonyl compounds, on the website associated with this book (http://www.oxfordtextbooks.co.uk/orc/ifp).

Further information on aspirin can be found on the following website: http://www.aspirin.com

Introduction to aromatic chemistry

MIKE SOUTHERN

What do aspirin, paracetamol, salbutamol and chloramphenicol have in common? You might know that they are all synthetic drugs, but additionally they are members of a class of organic molecules known as aromatic compounds. The chemistry of aromatic compounds started with the discovery of benzene (the basic unit of aromatic chemistry) by Michael Faraday in 1825. The precise structure of benzene was elusive and it was not until 1865 that the correct structure was first described by Kekulé. The importance of aromatic chemistry cannot be overstated; it was the driving force behind the dyestuff industry in the mid-to late 1800s, and that was the start of the chemical industry that we know today. In the intervening years, we will see that aromatic compounds have been adopted by the pharmaceutical industry and feature heavily in lists of the best selling drugs.

Learning objectives

Having read this chapter you are expected to be able to:

➤ appreciate the importance of aromatic compounds to the drug industry

➤ describe the structure of benzene and other aromatic hydrocarbons

➤ predict how benzene and its derivatives react with electrophiles in electrophilic aromatic substitution reactions

➤ describe key points in the synthesis of paracetamol, aspirin and salbutamol from phenol

➤ show awareness of the influence that some aromatic compounds have in the body and the role they play either as endogenous compounds or as drugs

➤ provide an overview of how the body metabolizes aromatic compounds, how this can deactivate drugs and sometimes form compounds that are hazardous to health.

7.1 What is aromatic chemistry?

Aromatic compounds were initially identified by their characteristic odour. Many of these early compounds were derivatives of benzene and we will focus on the chemistry of benzene and its derivatives in this chapter. However, there is much more to aromatic chemistry than smell, and the term 'aromatic' has now been extended to cover a wide range of molecules, many of which have little detectable odour.

Benzene is a volatile liquid that has a boiling point of 80 °C, a melting point of 5.5 °C, a density of 0.874 g/mL and a molecular formula of C_6H_6. The structure of benzene was the topic of considerable debate until Kekulé described the currently accepted structure in 1865. Some of the early proposed structures, including Kekulé's, are shown in Figure 7.1. The structure is an average of the two forms shown and this accounts for the structural observations described.

Benzene is a planar molecule in which all the carbon-carbon bond lengths are equal at 1.40 **angstroms** (Å) (a value between the C–C single bond length of 1.54 Å and the C–C double bond length of 1.34 Å). The σ-framework of benzene consists of a planar six-membered ring in which each carbon is sp^2 hybridized. This leaves a single electron in a p-orbital on each carbon

atom of the six-membered ring; these orbitals and electrons combine to form the characteristic π-cloud shown in Figure 7.2. The π-system of benzene (and other aromatic molecules) dominates their chemistry and reactivity. The electrons of the π-system are delocalized; they can take part in resonance or conjugation.

 Aspects of delocalization are discussed in Chapters 2 'Organic structure and bonding', 5 'Alcohols, phenols, ethers, organic halogen compounds, and amines' and 6 'The carbonyl group and its chemistry' of this book.

A common representation of benzene is shown in Figure 7.3. The circle within the six-membered ring represents the π-system. While this representation is structurally satisfactory, because it illustrates that the six π-electrons are delocalized across the six carbon

FIGURE 7.1 Proposed structures of benzene.

Dewar Benzene

Prismane

Early representations of benzene that have subsequently been synthesized –but behave nothing like benzene itself

Kekulé's rapidly interconverting structures of benzene

FIGURE 7.2 The σ- and π-bonding of benzene.

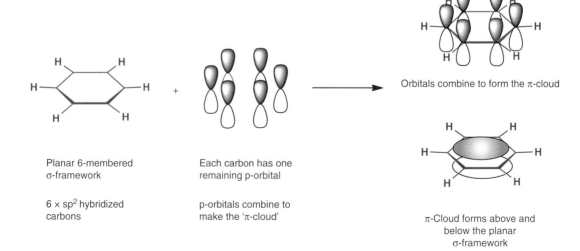

Planar 6-membered σ-framework

$6 \times sp^2$ hybridized carbons

Each carbon has one remaining p-orbital

p-orbitals combine to make the 'π-cloud'

Orbitals combine to form the π-cloud

π-Cloud forms above and below the planar σ-framework

atoms of the ring, it is not amenable to mechanistic descriptions of aromatic chemistry. In order to predict reaction pathways and plan sensible syntheses, we must understand the mechanisms of aromatic chemistry – and to do this we must employ the single and double bonded model shown in Figure 7.3.

Although the term 'aromatic chemistry' originally stemmed from the characteristic odour of compounds, as our understanding has increased this definition has changed and more stringent criteria are now employed. To be considered aromatic a compound must:

- be planar (or close to planar)
- have 4n+2 π-electrons; this criterion is known as Hückel's rule
- be cyclic and fully **conjugated**.

In Hückel's rule 'n' is a positive integer, meaning that aromatic compounds can have 2 π-electrons (n=0), 6 π-electrons (n=1 e.g. benzene); 10 π-electrons (n=2, e.g. naphthalene); 14 π-electrons (n = 3, e.g. anthracene); 18 π-electrons (n=4 e.g. [18]-annulene) and so on (see Figure 7.4).

Nuclear Magnetic Resonance (NMR) experiments have given us a new criterion for aromaticity: aromatic compounds are able to form a ring current in a magnetic field. This means that their protons produce signals in a particular region of the spectrum, about δ 7 ppm. Note that aromatic compounds do not need to be completely carbon based; pyridine is an example of a heteroaromatic compound (nitrogen is considered to be a heteroatom).

 Further details about NMR spectroscopy are given in Chapter 11 'Introduction to pharmaceutical analysis' of this book.

When the criteria for aromaticity are fulfilled (i.e. when the compound is planar, cyclic and obeys Hückel's rule), the molecule achieves a degree of extra stability. This is known as 'aromatic stabilization energy' and is 151 kJ mol^{-1} or 36 kcal mol^{-1} in the case of benzene. Aromatic stabilization energy has a dramatic effect on the reactivity of aromatic systems and is discussed later (see Section 7.3).

FIGURE 7.3 Representations of the benzene molecule.

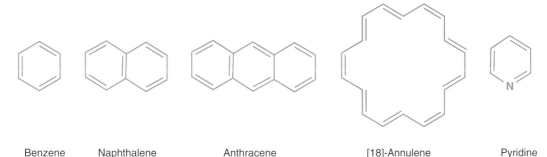

A common and realistic representation of benzene.

The preferred representation for any mechanistic consideration.

The double-heater arrow represents the interconversion or resonance forms.

FIGURE 7.4 Some simple aromatic compounds.

Benzene   Naphthalene   Anthracene   [18]-Annulene   Pyridine

7.2 Why is aromatic chemistry important?

Aromatic chemistry has a long and distinguished history and was the driving force behind the rise of the chemical industry in the 1850s. Early industrial chemistry focussed on dye production: dyes require extended **chromophores**, and chromophores are often aromatic. The first synthetic dye, mauveine, was prepared accidentally by William Perkin while he was trying (unsuccessfully) to synthesize the antimalarial compound quinine. Mauveine's spectacular purple colour is illustrated in Figure 7.5. Although mauveine was initially only produced in small quantities, Perkin was able to refine the process and improve the yield. He was also astute enough to patent the invention, and, in 1857, to set up a factory to manufacture the dye. Industrial chemistry began here, and despite many changes, continues to be important today.

 More information on chromophores can be found in Chapter 1 'The importance of pharmaceutical chemistry' of this book.

Prontosil and the sulfa drugs

Mauveine was the by-product of an attempt to produce a drug, but some of the earliest successful drugs

FIGURE 7.5 An early sample of mauveine and a shawl dyed with an original sample.

were actually dyes! Prontosil, shown in Figure 7.6A, is a red azo-dye (azo refers to the characteristic nitrogen-nitrogen double bond in these dyes). Scientists at the Bayer company in Germany had shown that prontosil could kill streptococcal bacteria in mice, and, in a leap of faith, Gerhard Domagk (Bayer's Director of Pathology and Bacteriology) saved his daughter from serious complications associated with a bacterial infection by treating her with it. Domagk was awarded the Nobel Prize in Physiology or Medicine in 1939 'for the discovery of the antibacterial effects of prontosil', but, because of orders from the Nazi regime, he was unable to accept it until 1947. The compound went on to become the first commercially available antibiotic.

The active agent in prontosil is actually a metabolite, sulfanilamide and this discovery led to the development of the so-called 'sulfa' antibacterial drugs. The sulfa drugs have been almost completely superseded by the pencillins, which have better activity and fewer side-effects, although sulfamethoxazole is still used in the treatment of certain types of pneumonia.

Many familiar drugs contain aromatic structures. These include OTC (over-the-counter) analgesics, aspirin, paracetamol and ibuprofen, which ease pain by interfering with the pain response within the body. Their structures are shown in Figure 7.7; notice how they all feature an aromatic ring.

 More detailed information on aspirin, COX, and prostaglandins can be found in Chapter 7 'Communication systems in the body' of the *Therapeutics and Human Physiology* book of this series.

Aspirin

The story of aspirin is an interesting one. Willow bark has been known to treat mild pain and fever for hundreds of years and Hippocrates of Kos advocated its use for fever, pain and childbirth around 2400 years ago. In 1763, Edward Stone published the first recorded clinical trial via The Royal Society of London in which

FIGURE 7.6 (A) The production of sulfanilamide in the body from the azo-dye prontosil; (B)The structure of sulfamethoxazole, a sulfa drug still in use today.

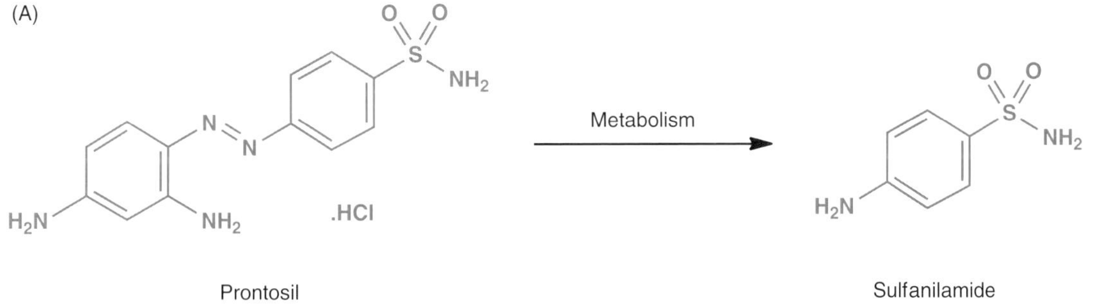

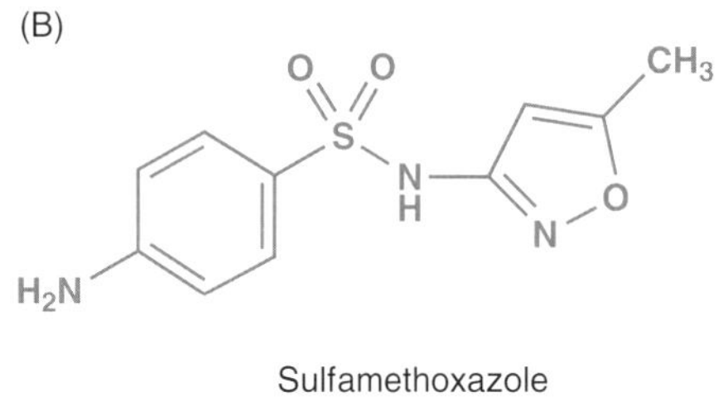

(A)

Prontosil

Metabolism

Sulfanilamide

(B)

Sulfamethoxazole

FIGURE 7.7 The familiar over-the-counter analgesics aspirin, paracetamol and ibuprofen.

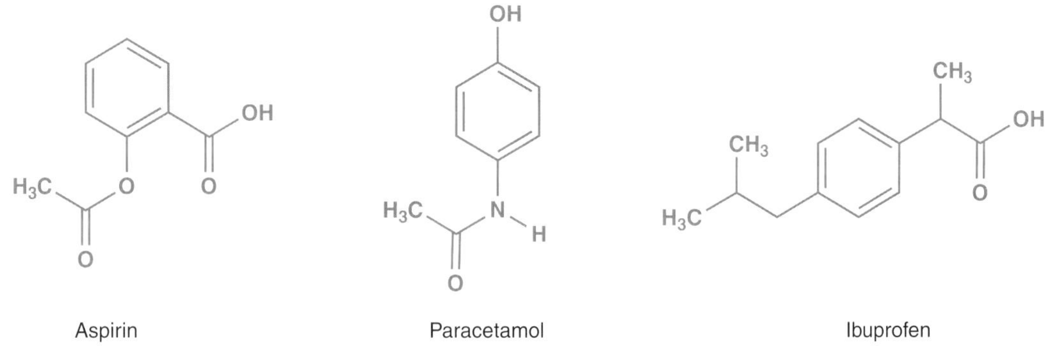

Aspirin

Paracetamol

Ibuprofen

he administered dried willow bark extracts in tea, water or beer to fifty patients who had fever and then observed the effects. In 1828, Johan Andreas Buchner in Munich purified willow bark extracts and obtained a yellow powder that he called salicin (see Figure 7.8). We now know that salicin is converted to salicylic acid in the body; it is salicylic acid that causes the beneficial effects associated with willow bark extract.

Generally, trees are not a sustainable source of drugs; they take many years to grow. Furthermore, the concentration of drug in a material like willow bark is hard to predict. (This is often a problem with plant preparations.) It was straightforward to replace willow

bark extract with pure salicylic acid but, unfortunately, the pure acid causes severe irritation to the mouth, throat and stomach. The side-effects of salicylic acid were vastly reduced when the hydroxyl group was acetylated to form aspirin. This process was mastered in 1897 by Felix Hoffmann at Bayer who was able to produce aspirin with suitable purity for therapeutic use. Incidentally, the acetylation of morphine to give diamorphine (heroin) was also first performed around the same time.

 For more information on these drugs, see Chapter 10 'Origins of drug molecules'.

FIGURE 7.8 The path from willow bark to aspirin.

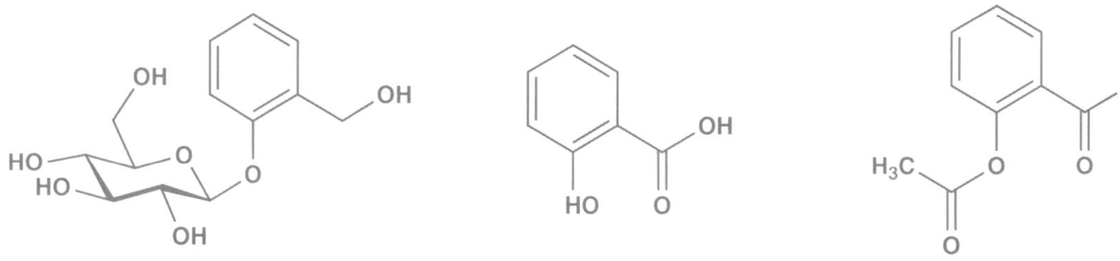

Salicin   Salicylic acid   Aspirin

Prozac

Aromatic compounds have even altered the way certain diseases are perceived by society, as illustrated by the drugs shown in Figure 7.9. Prior to the development of Prozac and other selective serotonin reuptake inhibitor (SSRI) antidepressants, depression was rarely discussed in public. Sufferers and their families were often reluctant to admit they had a problem. The SSRIs made depression a much more easily treatable disease, benefitting both sufferers and the economy. This in turn led to much more widespread discussion of the disease.

Viagra

Similarly, Viagra changed the way society views erectile dysfunction. Before the release of the drug, the topic was rarely discussed. Erectile dysfunction is now treatable in many cases and is much more easily

discussed and can even form the basis of jokes. Viagra was initially developed by Pfizer as a treatment for angina. It was unsuccessful in this context but, during clinical trials in humans, a number of men were honest enough to report the side-effects that led to its release as a treatment for erectile dysfunction.

Prozac and Viagra are particularly well-known aromatic drugs, but, in fact, all the top five drugs (in terms of worldwide sales) in 2009 contained substituted benzene rings. The drugs in question were: Lipitor, Plavix, Nexium, Seretide and Seroquel with total sales in the region of $45 billion (see Figure 7.10).

SELF CHECK 7.1

What do you think the top five selling drugs are used to treat? Are you surprised?

The appearance of aromatic rings in the top five is not coincidence: aromatic structures feature strongly in

FIGURE 7.9 Prozac and Viagra: aromatic molecules that have reduced the stigma of depression and erectile dysfunction.

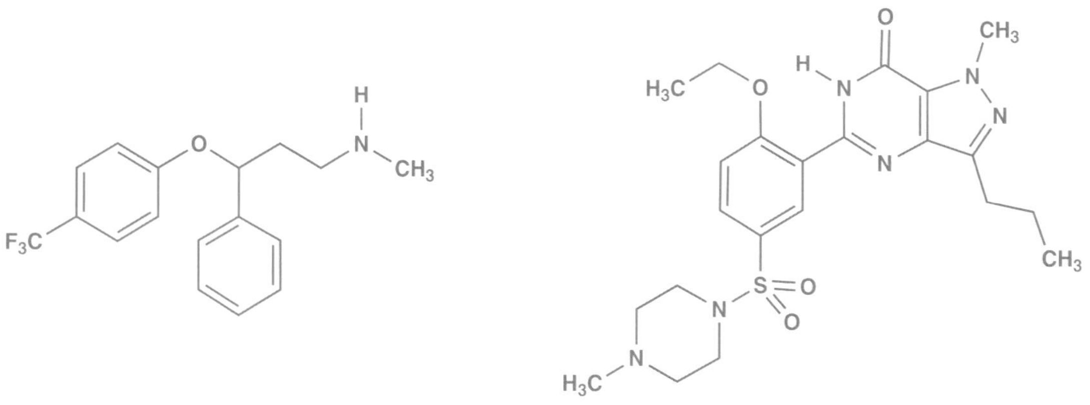

Prozac   Viagra

FIGURE 7.10 The top five drugs in terms of 2009 sales worldwide.

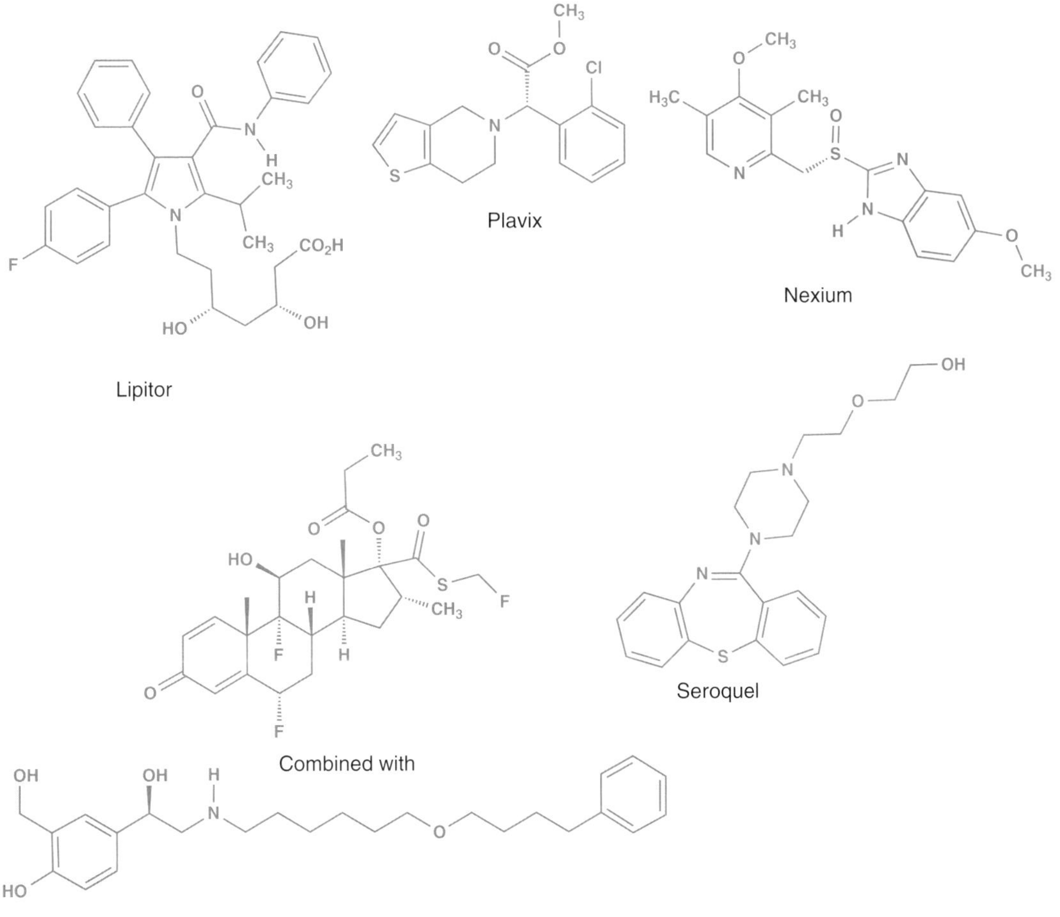

Lipitor

Plavix

Nexium

Combined with

Seroquel

Seretide (a mixture of fluticasone propionate and salmeterol)

a large percentage of successful drugs, indicating the importance of aromatic chemistry to global health.

Ecstasy and cannabis

Aromatic compounds also feature in illegal substances such as MDMA (ecstasy) and Δ^9-tetrahydrocannabinol (Δ^9-THC, one of the active ingredients of cannabis), shown in Figure 7.11. There is long-standing anecdotal evidence that a number of illegal compounds have therapeutic benefits and these anecdotes have led to proper scientific trials being conducted.

Sativex is a cannabis extract (containing THC and cannabidiol) that is now available (in the UK and Spain) as an oromucosal spray to treat the spasticity associated with multiple sclerosis. In Canada, it is also used for the treatment of neuropathic pain. Small

scale clinical trials have also indicated that MDMA may be of benefit in the treatment of post-traumatic stress disorder.

Aromatic compounds are not only found in legal and illegal drugs; see Box 7.1 to discover how they also give a spicy kick to the food we eat.

It is clear that many biologically active compounds are aromatic. If we want to be able to maximize the impact of the beneficial effects of this class of compound we must be able to construct such molecules in a predictable and controlled manner. This will allow us to discover new chemical entities with beneficial therapeutic effects and to optimize the therapeutic aspects and minimize the side-effects of current or potential new drugs. In order to do this successfully, we need to understand the chemistry of benzene and related aromatic compounds.

FIGURE 7.11 Illegal drugs: (A) The cannabis plant and one of its active ingredients, Δ⁹-tetrahydrocannabinol (Δ⁹-THC); (B) the synthetic drug MDMA.
Licensed under the Creative Commons Attribution-Share Alike 3.0 Unported license. Copyright cânepă.

(A)

Δ⁹-Tetrahydrocannabinol

(B)

MDMA 'ecstasy'

BOX 7.1

Feeling the heat of chilli peppers

There is no doubting the popularity of chilli peppers as an addition to food. They have found fans across the globe since their export from South America. Chilli peppers are thought to have been cultivated in South America for a few thousand years and it is likely that Christopher Columbus first introduced chillis to Europeans after encountering them in the Caribbean.

The main compound responsible for the 'heat' of chillis is the benzene derivative capsaicin (see Figure 7.12). The heat of chillis is measured on the Scoville scale, which gives a measure of how many times a standard extract of a particular chilli has to be diluted with dilute syrup

until its heat becomes undetectable by taste. Clearly this is a little subjective, so high performance liquid chromatography (HPLC) is now employed to provide a more accurate measurement. The common Jalapeño generally has a heat of around 3,500 Scoville Heat Units (SHU), the Scotch Bonnet ranges from 100,000 to 350,000 SHU and the 2011 Guinness Book of Records states that the world's hottest chilli, the Trinidad Scorpion 'Butch T', was measured at an eye-watering 1,463,700 SHU.

 HPLC is discussed in more detail in Chapter 11 'Introduction to pharmaceutical analysis' of this book.

FIGURE 7.12 Red chillis and the aromatic compound capsaicin – the main compound that gives chillis their 'heat'.
Licensed under the Creative Commons Attribution-Share Alike 3.0 Unported license. Copyright Norbert Nagel.

Capsaicin

7.3 The chemistry of benzene

The most important chemistry of aromatic compounds is electrophilic aromatic substitution (EAS or S_EAr – <u>S</u>ubstitution <u>E</u>lectrophilic <u>A</u>romatic) in which the aromatic compound reacts with an electrophile. Nucleophilic aromatic substitution reactions (NAS or S_NAr) in which the aromatic species react with a nucleophile are also important in synthetic chemistry. Most of these reactions are beyond the scope of this book, but we will briefly consider one example. Further information about NAS can, however, be found in the Online Resource Centre.

Note that, in the following reactions, we will show hydrogen atoms on the aromatic rings where it helps to explain the mechanistic details being discussed.

Electrophilic aromatic substitution

Firstly, let us consider the general reactivity of benzene. We can represent it as a six membered ring containing three double bonds – but the reactivity of benzene is very different from that of simple alkenes. This can be illustrated by the reactivity of benzene with bromine compared with that of a simple alkene. Bromine reacts with alkenes extremely rapidly with the red/brown colour of bromine disappearing almost instantaneously (until all the alkene is consumed). The resulting product is a dibromide formed by an **addition reaction**. By contrast, if benzene is treated with bromine no reaction takes place and the colour of bromine persists. Benzene can be forced to react with bromine by employing a catalyst that makes the bromine more reactive (more electrophilic in this case). The product, however, is not a dibromide resulting from an addition reaction, but bromobenzene, and the reaction is a **substitution reaction**.

The difference in reactivity between alkenes and aromatic compounds arises from the aromatic stabilization energy (discussed in Section 7.1). Remember that this is a result of the six carbon atoms of the benzene ring sharing their six p-electrons in a cyclic delo-

calized system or π-cloud. For benzene to react, this delocalization has to be interrupted, which requires a relatively large amount of energy in order to compensate for the loss of aromaticity and its associated aromatic stabilization energy.

Benzene reacts with bromine only when the bromine is activated by a catalyst, and when the reaction does occur it yields an aromatic compound. The intermediate cationic species does not react with a bromide ion (as is the case in the bromination of alkenes), but instead it loses a proton so that the molecule can re-establish its aromatic character and regain the associated stabilization energy. In effect, a hydrogen atom is substituted by a bromine atom.

In short, bromine reacts with an alkene by an addition reaction, but activated bromine reacts with benzene by a substitution reaction, as shown in Figure 7.13.

 The bromination of alkenes and also addition and substitution reactions are discussed in more detail in Chapter 4 'Properties of aliphatic hydrocarbons' of this book.

Mechanism of electrophilic aromatic substitution

In the bromination of benzene illustrated in Figure 7.14, bromine first combines with aluminium trichloride to form the active electrophile. The π-system of benzene then acts as a nucleophile and attacks the bromine complex to form the high energy, positively charged, non-aromatic sigma-complex (σ-complex), also known as the Wheland intermediate. Loss of a proton neutralizes the intermediate to restore aromaticity, forming HBr as a by-product and regenerating the catalyst. The addition of an electrophile, formation of the σ-complex and restoration of aromaticity are common to all EAS reactions; it is only the nature of the electrophile that changes.

FIGURE 7.13 (A) The addition of bromine to prop-2-ene; (B) bromine does not react with benzene; (C) electrophilic aromatic substitution reaction of bromine and benzene.

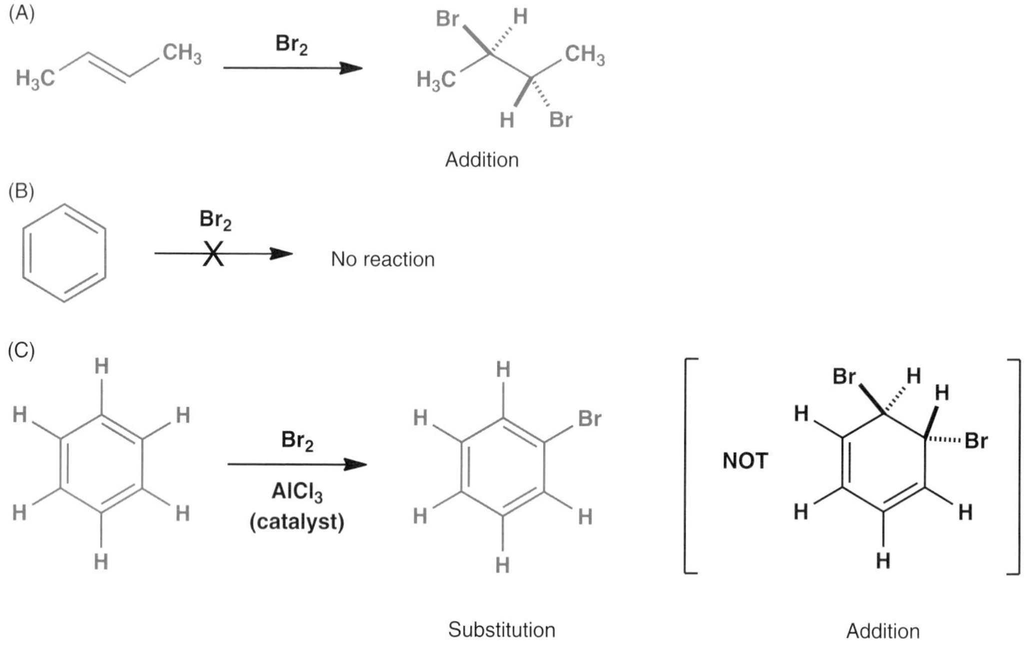

(A)

Addition

(B)

No reaction

(C)

Substitution   NOT   Addition

It is extremely important to understand the mechanism of EAS, because if we understand one example then we understand them all. Let us therefore look at the mechanism in more detail, focussing on the aromatic ring. We will use a generic electrophile, E⁺, and examine the fate of the double bonds of the benzene ring.

Look at Figure 7.15, which illustrates the reaction of benzene with a generic electrophile. As benzene attacks the electrophile one carbon of the double bond (C2)

becomes positively charged whereas the other (C1) forms a bond to the electrophile. For C1, little has changed; it has four bonds in the starting material and four in the intermediate (although one bond in the π-system has been replaced by the new σ-bond). Consequently, C1 still has a full octet of electrons. C2, however is now electron deficient. The electron it donated to the π-bond is now used to form the new σ-bond from C1 to E. C2 has only six valence electrons (three bonding pairs) and therefore has a positive charge.

FIGURE 7.14 Electrophilic aromatic substitution – the bromination of benzene.

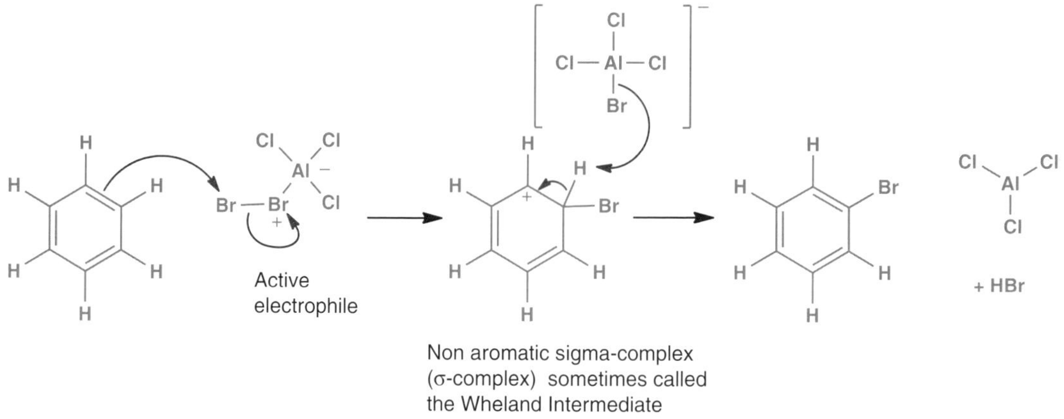

Active electrophile

Non aromatic sigma-complex (σ-complex) sometimes called the Wheland Intermediate

+ HBr

FIGURE 7.15 Electrophilic aromatic substitution – reaction of benzene with a generic electrophile.

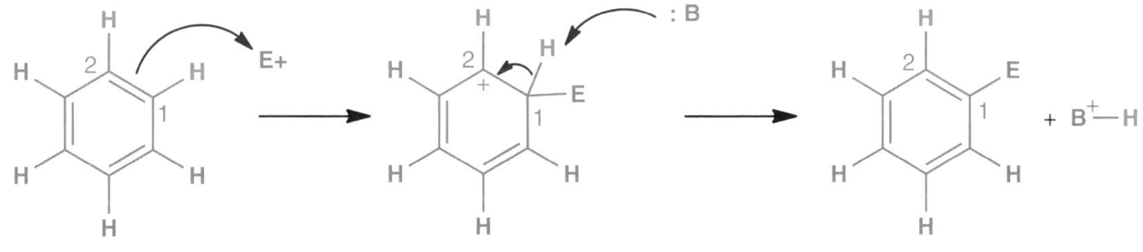

σ-complex 1

There is a large driving force for re-aromatization, because of the aromatic stabilization energy, and any basic species in the reaction mixture, even a weak base, can remove the proton so that aromaticity is restored (see Figure 7.15).

Another aspect of the reaction that needs consideration is the delocalization of electrons in the σ-complex, as illustrated in Figure 7.16. Remember that the electrons in the π-system of benzene are delocalized; we used curly arrows to represent the movement of electrons around the ring to interconvert the two resonance forms. We can also use curly arrows to represent the delocalization of double bonds within the σ-complex, which, in this case, moves the positive charge around the ring.

The σ-complex is not aromatic: it consists of a cyclic, planar carbon framework but it no longer obeys Hückel's 4n+2 π-electrons rule. However, the delocalization of the remaining electrons stabilizes the high energy intermediate by spreading the positive charge across a number of atoms.

Consider σ-complex 1. There is a double bond (C3–C4) adjacent to a positively charged carbon, C2. Using a curly arrow, we can show that the two electrons of the C3–C4 double bond move towards the positively charged carbon to form a new carbon-carbon double bond (C2–C3) leaving a positive charge on C4. We have now formed σ-complex 2. If you have trouble convincing yourself of this, count the valence electrons on each of the carbons discussed (C2–C4). Now σ-complex 2 once again has a double bond (C5–C6) adjacent to a positively charged carbon (C4). Using another curly arrow, we can represent further delocalization of the electrons and move the double bond from C5–C6 to C4–C5, leaving carbon C6 with a positive charge. A proton can then be lost from σ-complex 3 to restore the aromaticity.

In total, the three resonance forms shown in Figure 7.16 can be drawn to illustrate the delocalization of the electrons.

For the remainder of this section, only one resonance form will normally be shown, but you should remember that another two can be drawn.

SELF CHECK 7.2

We have seen how a proton can be lost from σ-complex 1 and σ-complex 3 to restore aromaticity. Can you see how a proton can be lost from σ-complex 2 to restore aromaticity?

FIGURE 7.16 Interconversion of σ-complexes during electrophilic aromatic substitution.

σ-complex 1   σ-complex 2   σ-complex 3

187

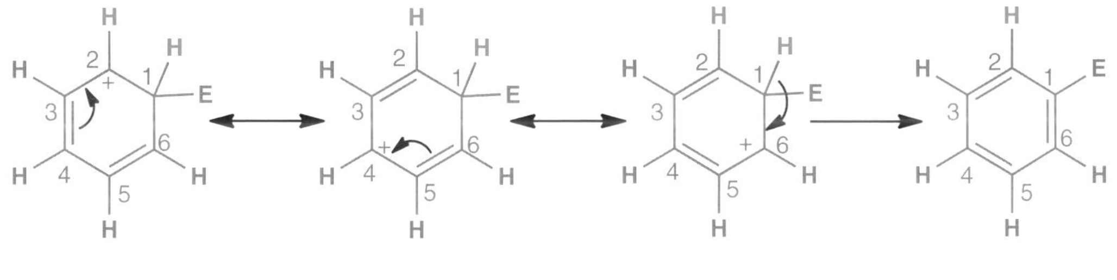

If you are having trouble understanding the advantages of delocalization then consider a potato that has finished baking and needs to be removed from the oven, in the absence of oven gloves or similar equipment. The hot potato represents the high energy cationic species and your hands represent positions of delocalization. If you attempt to remove the potato with one hand then it is highly likely that a severe burn will ensue and the potato will end up on the floor. However, if you make use of two hands to juggle the potato and delocalize the heat/energy, then you have a much greater chance of getting the potato onto a plate safely.

SELF CHECK 7.3

Use the alternate initial resonance form of benzene shown in Figure 7.15 to verify that the overall mechanism and associated resonance forms (see Figure 7.16) can still be employed to form a substituted benzene.

Specific reactions which employ EAS

Now, it is time to apply EAS to real reactions. Note that, as far as the benzene ring is concerned, the mechanism is always the same. Bromination has already been discussed and we will now consider four more important EAS reactions.

Nitration

Arguably the most important reaction in aromatic chemistry is nitration, because of the versatility of the nitro group (considered in more detail later). Mixing

nitric and sulfuric acid together forms the active electrophile $[NO_2]^+$ and this reacts with benzene by the EAS reaction mechanism to form nitrobenzene (see Figure 7.17).

Sulfonation

A similar reaction employing benzene and concentrated sulfuric acid or oleum (SO_3 dissolved in sulfuric acid) gives benzene sulfonic acid as the product, as illustrated in Figure 7.18. In this case, protonated SO_3, $[SO_3H]^+$, is the active electrophile. Sulfonic acids are very strong acids (similar in strength to sulfuric acid) and are useful for organic synthesis because they are very soluble in organic solvents (unlike inorganic acids). One example of their use is that they serve as precursors to the sulfa antibiotics.

SELF CHECK 7.4

Draw all the resonance forms of the intermediate shown in Figure 7.18.

Friedel-Crafts reactions – the formation of carbon-carbon bonds

The formation of carbon-carbon bonds is of paramount importance when constructing organic molecules. Towards the end of the 1800s, Charles Friedel and James Crafts developed useful methods to couple acyl and alkyl chlorides with benzene and other aromatic species. The fact that these methods are still widely employed today illustrates their versatility, utility and importance. In both cases aluminium trichloride (or a related compound such as tin tetrachloride) is employed to generate the active electrophile.

FIGURE 7.17 **Nitration of benzene.**

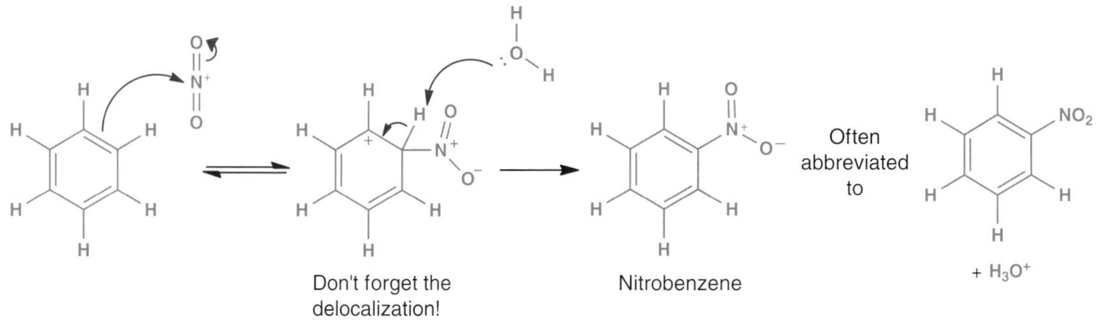

Don't forget the delocalization!

Nitrobenzene

Often abbreviated to

$+ H_3O^+$

FIGURE 7.18 Sulfonation of benzene.

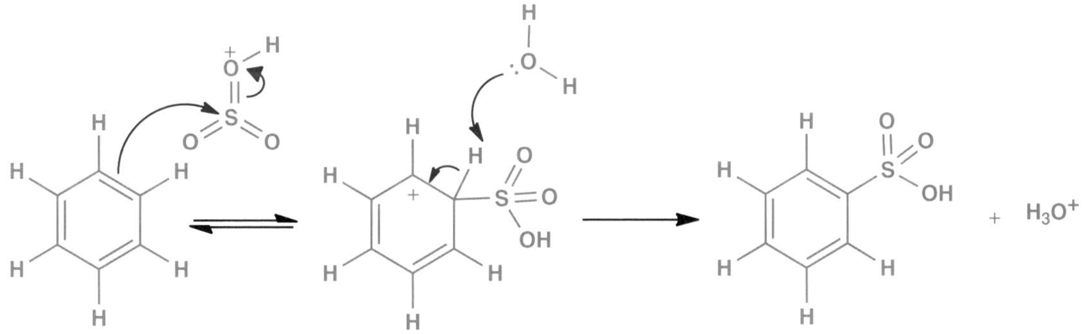

Don't forget the
delocalization!

Benzene sulfonic acid

189

The Friedel-Crafts **acylation** is shown in Figure 7.19. The active electrophile is the oxonium ion (the odd looking species with the positive charge on oxygen), which is formed by combining an acyl chloride (sometimes called an acid chloride) with aluminium trichloride. The EAS reaction proceeds in the usual manner to form an aromatic ketone. The reaction is very general and many R groups can be employed.

The Friedel-Crafts alkylation also employs aluminium trichloride to form a cationic species which is the active electrophile as shown in Figure 7.20.

The Friedel-Crafts alkylation is a less useful reaction than the Friedel-Crafts acylation because it can give a mixture of products. There are two possibilities for these extra products and either, or both, may occur. Firstly, the introduction of an alkyl group activates the benzene ring towards further alkylation, so that products with extra alkyl groups may be formed,

as discussed in more detail below. Another possibility is that the active electrophile (a carbocation) may rearrange before it can react with benzene. The stability of carbon based cations follows a trend: tertiary carbocations are more stable than secondary carbocations which are considerably more stable than primary carbocations, as depicted in Figure 7.21.

Stabilities of carbocations are discussed in more detail in Chapters 2 'Organic structure and bonding' and 4 'Properties of aliphatic hydrocarbons' of this book.

In practice, if the electrophile required for a Friedel-Crafts **alkylation** is a primary carbocation, it is likely to undergo rearrangement to the more stable secondary cation before reacting with benzene; this results in a mixture of products. For example, if 1-chloropropane were employed in a Friedel-Crafts alkylation

FIGURE 7.19 Friedel-Crafts acylation of benzene.

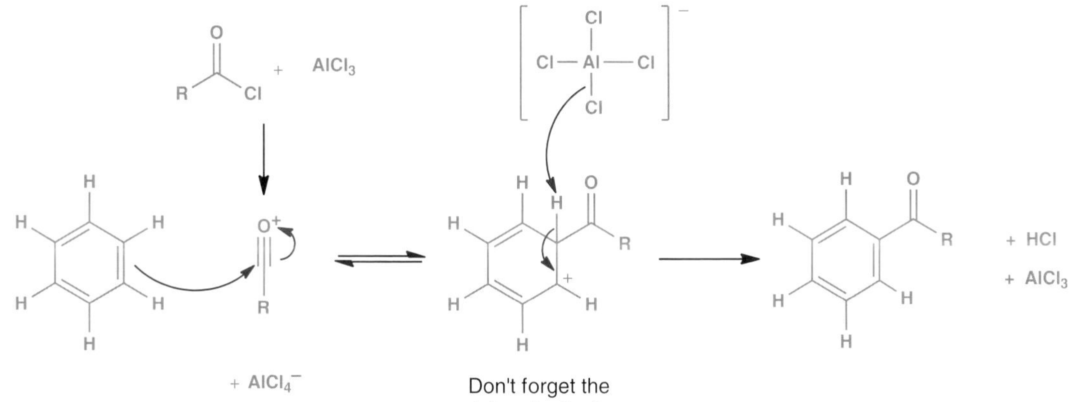

+ AlCl₃

+ HCl

+ AlCl₃

+ AlCl₄⁻

Don't forget the
delocalization!

FIGURE 7.20 Friedel-Crafts alkylation of benzene.

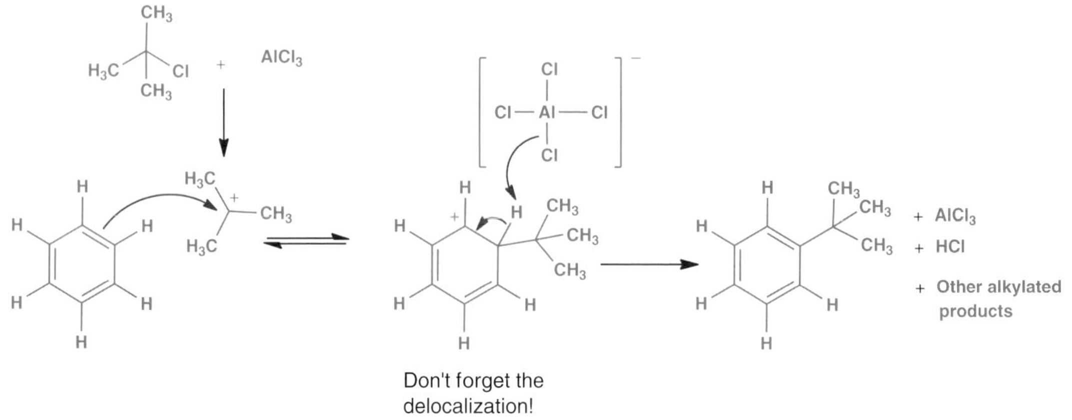

Don't forget the
delocalization!

FIGURE 7.21 Stability of carbocations.

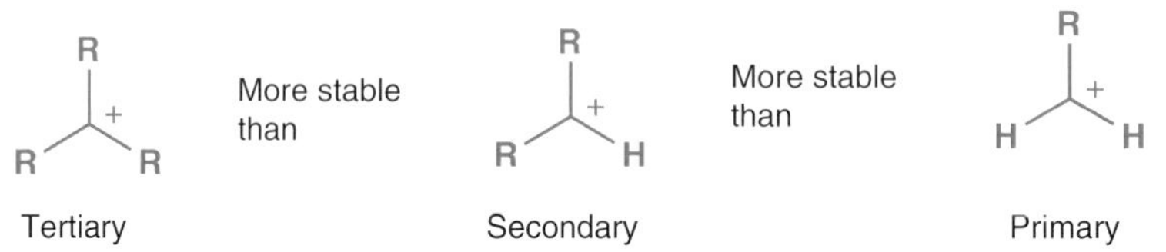

Tertiary   Secondary   Primary

FIGURE 7.22 Friedel-Crafts alkylation of 1-chloropropane with benzene.

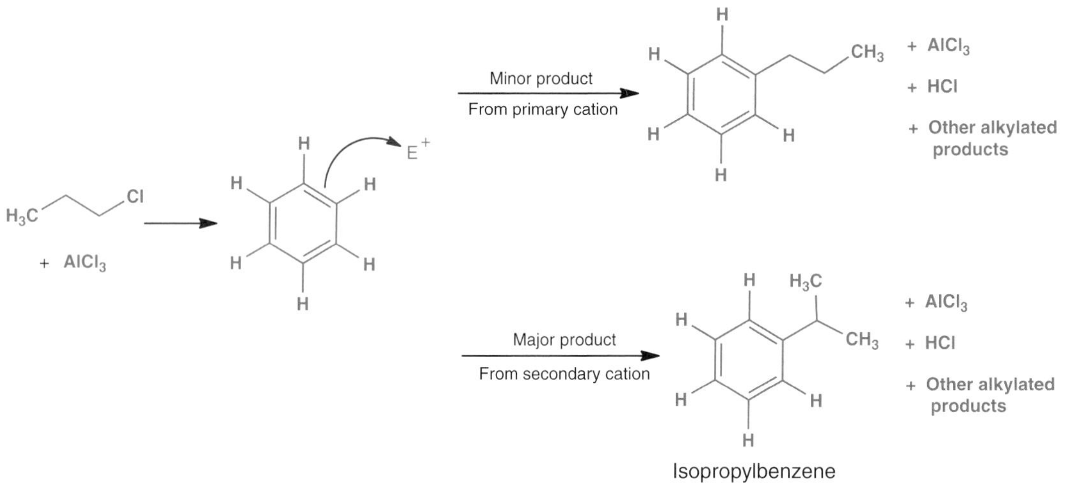

Isopropylbenzene

then the major product from the reaction would be isopropylbenzene (2-phenylpropane), which would result from the rearrangement of the primary carbocation as shown in Figure 7.22.

Fortunately there is a simple solution to this problem, although an extra reaction is required. The Friedel-Crafts acylation can be employed, and the ketone produced then reduced to the alkane. This sort of reduction can be carried out using various methods: for example, the Wolff–Kishner reduction uses hydrazine and potassium hydroxide while the Clemmensen reduction employs zinc, mercury (often called zinc amalgam) and concentrated hydrochloric acid (see Figure 7.23).

Electrophilic aromatic substitution of substituted benzenes – directing effects

The EAS reactions that we have seen so far are not only used on benzene but are also useful for further functionalizing substituted benzenes i.e. benzene that

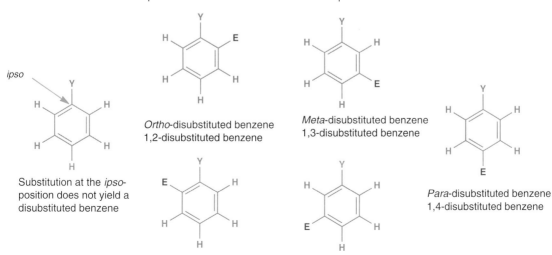

already has at least one substituent on the ring. The substituent on the ring has an effect on the outcome of the EAS reaction, known as the **directing effect**. There are three distinct classes of directing effect and in order to understand them we must first consider the nomenclature used to describe substituted benzenes, as illustrated in Figure 7.24.

Ortho, *meta* and *para* are relatively old Greek-derived terms. They are so far removed from their original meanings that English words such as ortho-dox, metamorphosis and parachute are unlikely to help you to remember them! The more modern way to describe aromatic substitution is numerical, and descriptors such as 1,2 (*ortho*), 1,3 (*meta*), and 1,4 (*para*) to denote the relationship between the two substituents on a benzene ring are perhaps more coherent. However, the language of aromatic chemistry is dominated by *ortho*, *meta* and *para* so we will use them where appropriate. In more complex trisubstituted systems it is usually easier to use the numerical terminology (see Box 7.2).

Note that there is a plane of symmetry down the middle of monosubstituted benzene molecules such as toluene (methylbenzene) or chlorobenzene mean-

BOX 7.2

ortho, meta, para substitution

As explained in Chapter 1, when old-fashioned nomenclature is retained, there is usually a good reason. The *ortho*, *meta*, *para* nomenclature tells you where substituents are relative to one another, which is useful for talking about mechanisms. To use numbers, you have to decide where number 1 is, which is useful for describing a final molecule. It is a bit like left and right vs north and south. A map shows you that Manchester is almost due east of Liverpool, but a SatNav gives directions in terms of right and left.

ing that there are two equivalent *ortho*-positions, two equivalent *meta*-positions but only one *para*-position (see Figure 7.24).

The directing effects of electron-donating groups

Consider the second substitution in a monosubstituted aromatic ring. The first important question is whether any particular isomer is favoured during the reaction,

FIGURE 7.24 Substitution patterns in disubstituted aromatic compounds.

Substitution at the *ipso*-position does not yield a disubstituted benzene

Ortho-disubstituted benzene
1,2-disubstituted benzene

Meta-disubstituted benzene
1,3-disubstituted benzene

Para-disubstituted benzene
1,4-disubstituted benzene

FIGURE 7.25 The resonance patterns of Y-benzene reacting with a generic electrophile at the *ortho-, meta-* and *para-*positions.

Ortho

Meta

Para

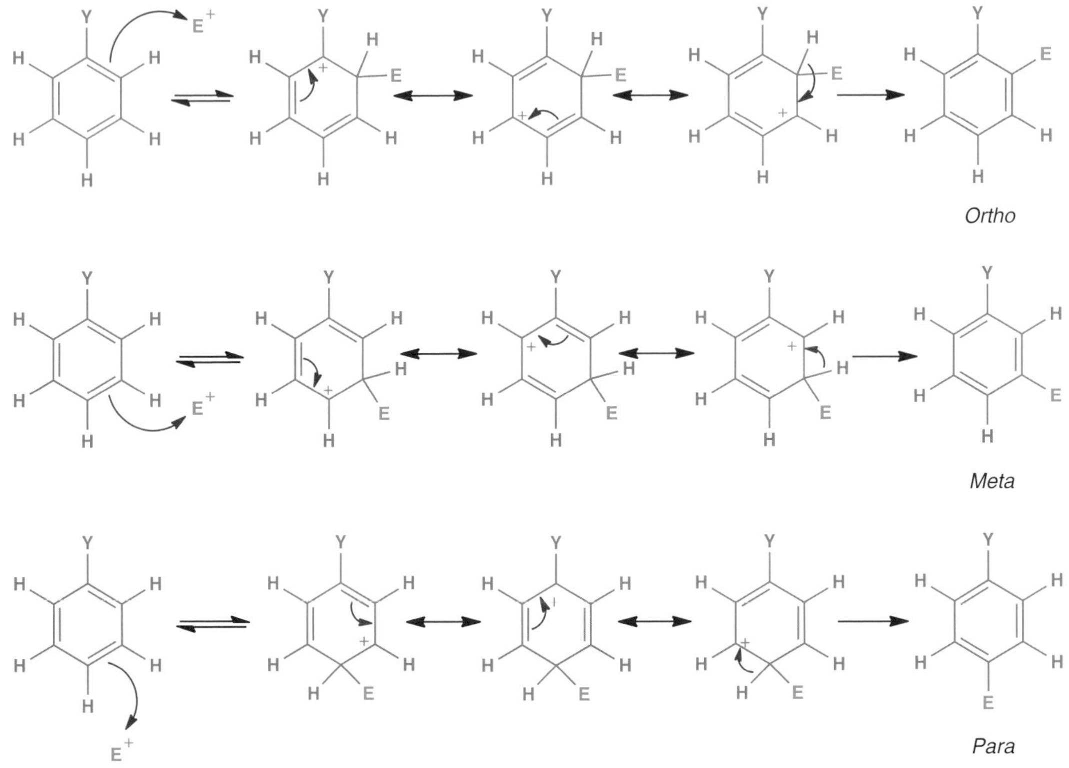

or whether an even distribution of different isomers is produced. To address this point we need to consider the intermediates in the reaction. Figure 7.25 shows the reaction of a monosubstituted Y-benzene with a generic electrophile E$^+$ at the *ortho- meta-* and *para-*positions. The three possible resonance forms associated with the reactions at each position are shown and consideration of these will help us determine the directing effects of different groups.

Notice that the patterns of delocalization are the same when substitution is *ortho-* and *para-* to Y: the same three carbons (with respect to Y) carry the positive charge. In each case, one of the resonance forms has the cation on the carbon bearing the group Y (the *ipso* carbon). Reaction at the *meta-*position, conversely, generates a delocalization pattern in which the other three carbons carry the positive charge. You might expect therefore that Y could have an effect on the product of the reaction. We could postulate that if Y has characteristics that stabilize an adjacent positive charge, then the *ortho-* and *para-*isomers will be favoured. However, if Y destabilizes an adja-

cent positive charge, then the *meta-*isomer is likely to be favoured because reaction at this position gives a delocalization pattern without the positive charge on the carbon bearing Y.

SELF CHECK 7.5

Close the page and draw the intermediates formed by bromobenzene reacting with a generic electrophile E$^+$.

We have already seen a group that stabilizes an adjacent positive charge – the alkyl group. Recall the order of carbocation stability: the more alkyl groups attached to a carbocation the more stable the cation. This is because electron density in one of the C–H σ-bonds of an attached alkyl group (methyl groups in the example shown), can interact with the positive charge (empty p-orbital) and help stabilize it, a phenomenon called σ-conjugation (**sigma conjugation**). The situation with EAS is remarkably similar and one of the C–H bonds of the methyl group of toluene can align itself with the π-system and help stabilize the positive charge (see Figure 7.26).

FIGURE 7.26 The stabilization of carbocations by σ-conjugation.

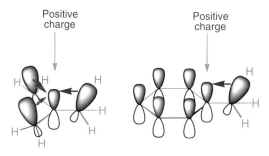

 σ-conjugation is discussed in more detail in Chapter 4 'Properties of aliphatic hydrocarbons'.

So, we can hypothesize that electrophilic aromatic substitution of toluene will form predominantly the *ortho-* and *para-products*. But does this hypothesis bear any resemblance to reality? Yes, it does! EAS reactions of toluene (methylbenzene) have a typical distribution pattern of approximately 60% *ortho*, 5% *meta* and 35% *para*, as depicted in Figure 7.27. Bear in mind that there are two *ortho*-positions, two *meta*-positions but only one *para*-position so, statistically, twice as much incorporation at the *ortho*- and *meta*-positions vs the *para*-position would be expected.

Alkyl groups are described as mildly ***ortho/para*-directing** and as **activating groups**. This explains the formation of other alkylated products in the Friedel-Crafts alkylation reactions.

SELF CHECK 7.6

Can you explain why the addition of an alkyl group to a benzene ring will facilitate the formation of 'other alkylated products'? What do you expect they might be?

The stabilization of the cation by the electron density of the C–H bond (σ-conjugation) is not fantastic, but it is certainly better than no stabilization at all. However, there are other (better) ways to provide stabilization – for example, by conjugation with the lone pair of an oxygen atom. The orbital containing the lone pair can align itself parallel with the π-system, and interact with it (see Figure 7.28). This means that an oxygen based functionality such as a hydroxyl group or an ether can effectively donate electrons and stabilize an adjacent positive charge by delocalization.

How can an element as electronegative as oxygen be electron donating? This can seem confusing but it is important to understand that there are two separate effects at work. The electronegative oxygen exerts an electron withdrawing **inductive effect** on the σ-framework of the molecule. However, the oxygen lone pair is able to interact with and donate electron density into the aromatic π-system. Since the EAS reaction occurs through the π-system the inductive effect is much less significant and consequently alcohol and ether groups are, overall, strongly electron donating in aromatic systems, as depicted in Figure 7.28.

The delocalization patterns of methoxybenzene (anisole) reacting with the generic electrophile E⁺ are shown in Figure 7.29. Note the extra resonance forms arising from participation of the oxygen when the electrophile reacts at either the *ortho*- or *para*-position. Remember there is an energetic advantage to having extra resonance forms in the non-aromatic intermediate (the hot potato analogy) and this has implications for aromatic compounds containing such groups. These functionalities are strongly ***ortho/para*-directing**, and are also **activating**: the aromatic compounds containing them are significantly more reactive in EAS reactions than benzene.

FIGURE 7.27 Representative distribution pattern for the reaction of toluene with an electrophile.

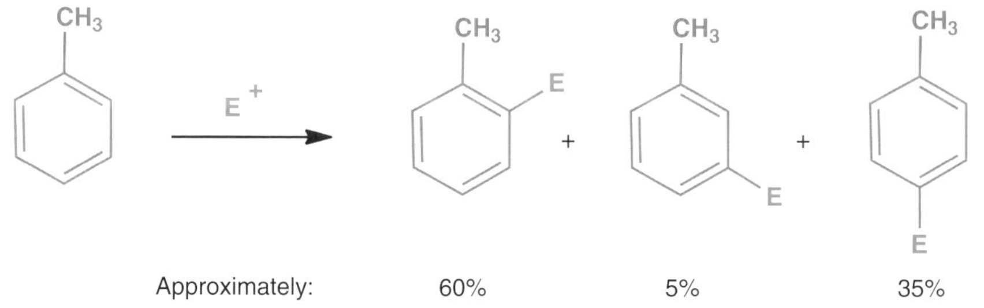

Approximately: 60%   5%   35%

FIGURE 7.28 The resonance and inductive effects of the oxygen of a phenol derivative.

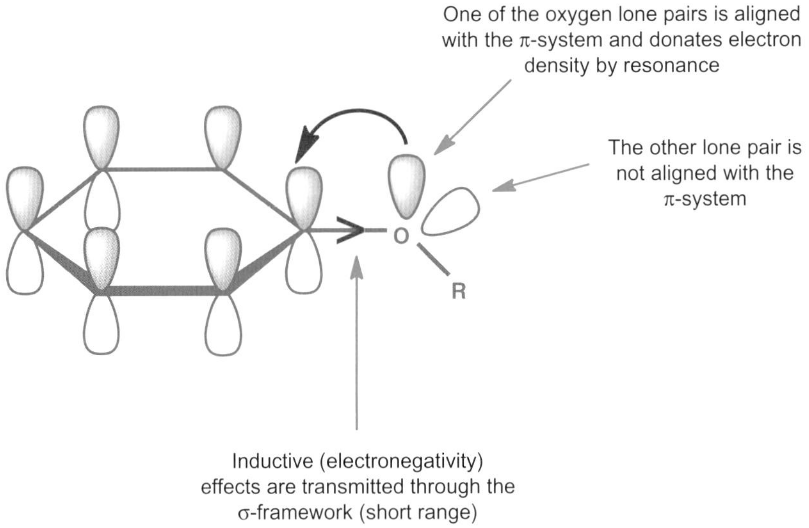

One of the oxygen lone pairs is aligned with the π-system and donates electron density by resonance

The other lone pair is not aligned with the π-system

Inductive (electronegativity) effects are transmitted through the σ-framework (short range)

FIGURE 7.29 Distribution and stabilization of the carbocations formed when anisole reacts with an electrophile.

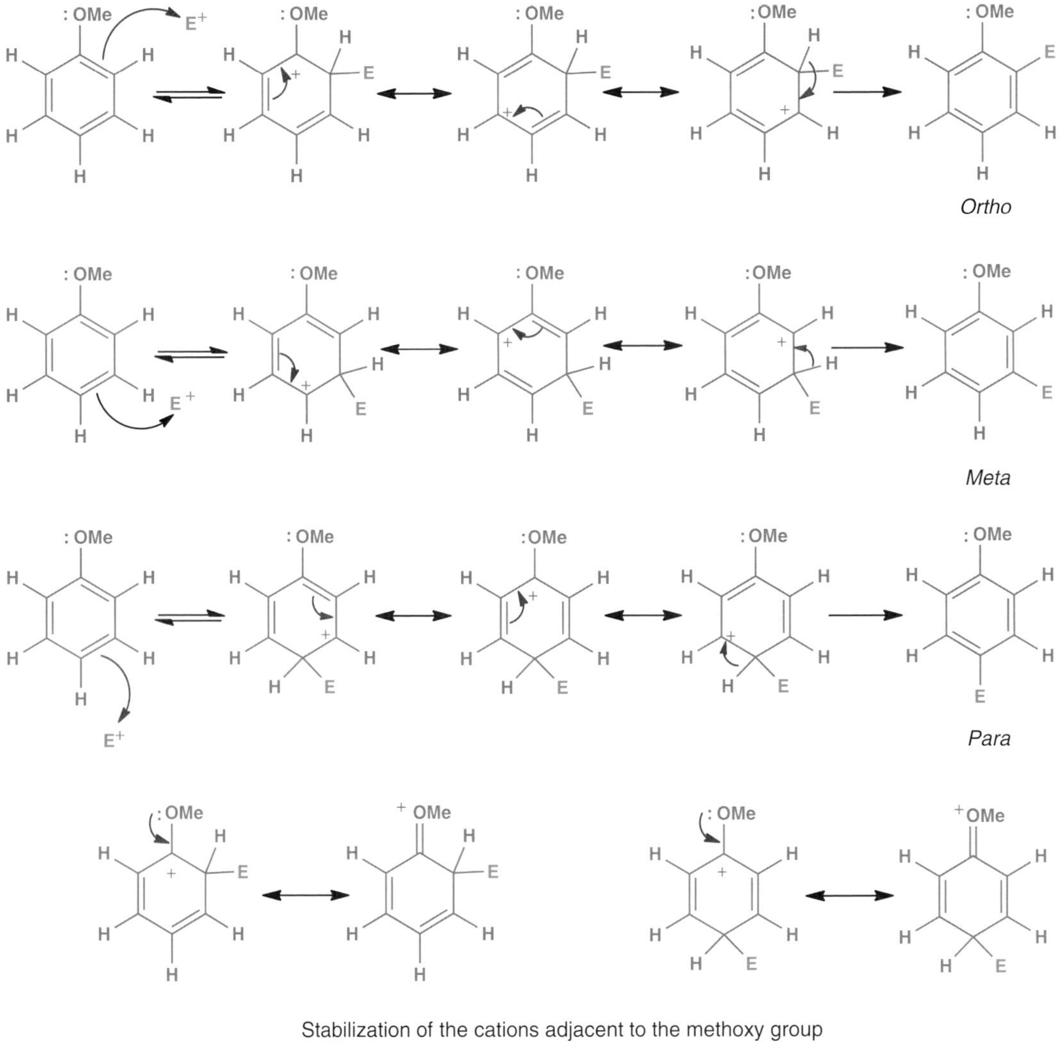

Ortho

Meta

Para

Stabilization of the cations adjacent to the methoxy group

Oxygen can also act as an electron donating group in non-aromatic systems. Examples include the chemistry of acetal formation and hydrolysis.

 Acetals and hemiacetals are discussed in Chapter 6 'The carbonyl group and its chemistry'.

The effect of adding a powerful electron-donating group is considerable. For example, the bromination of phenol (hydroxybenzene) does not require a catalyst (unlike the bromination of benzene or toluene) and it is difficult to control the reaction – the tribromide is readily formed at room temperature. When three equivalents of bromine are employed, the tribromide is the only product; note that only the *ortho*- and *para*-positions (with respect to the hydroxyl group) are brominated, as shown in Figure 7.30.

Replacing the hydroxyl group with the nitrogen-based amine group generates anilines (aminobenzenes) that are even more reactive than their phenol analogues. Again the lone pair is involved in the enhanced reactivity, but the lower electronegativity of the nitrogen makes it better equipped to carry the positive charge. Anilines are often too reactive and making simple disubstituted aromatics (by the addition of a single electrophile) is even more difficult than in the case of phenol. Fortunately, it is possible to reduce the reactivity of these types of highly reactive aromatic compounds by converting the amine to an amide or the alcohol to an ester.

Controlling the reactivity of activated aromatics

If we were unable to use anilines or phenols in the synthesis of potential drugs we would be excluding a vast range of useful and important compounds. Chemists in the past have realized this and have developed methods to circumvent the problem. In the case of aniline, one method is to derivatize the amine to make an amide. If derivatization is carried out using acetic anhydride (ethanoic anhydride), the product is acetanilide (also known as *N*-phenylacetamide).

The electron withdrawing nature of the carbonyl (see Figure 7.31) creates competition for the nitrogen's lone pair, making it less available to interact with, and stabilize the positive charge of the σ-complex. The carbonyl calms the overall reactivity of the aniline such that a single alkylation (mono-alkylation) is possible. Notice that the amide is still activating and no catalyst is required to activate the bromine.

Of course this approach would be useless if it were not possible to convert amide back to amine (which is achieved by treatment with acid). Do not forget that amines are basic so they will be protonated in acidic medium to form the ammonium salt; the acidic conditions will then need to be neutralized if the free amine is required. A similar approach can be used in the case of phenols by making an ester rather than an amide.

 The basicity of amines is discussed in Chapter 5 'Alcohols, phenols, ethers, organic halogen compounds, and amines'.

 Protecting groups are discussed in more detail in Chapter 6 'The carbonyl group and its chemistry'.

You may be wondering why the *para*-isomer is favoured over the *ortho*-isomer when acetanilide is brominated (or indeed substituted with other electrophiles). This is because of steric hindrance: the amide group blocks the reaction at the *ortho*-positions. Do not forget that molecules are not fixed, rigid structures. Rotation around the benzene-nitrogen bond is possible and so

FIGURE 7.30 **The bromination of phenol.**

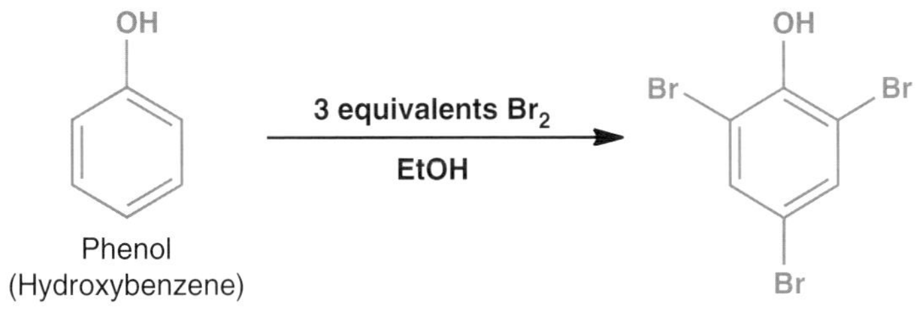

FIGURE 7.31 The use of an electron withdrawing group to control the reactivity of aniline.

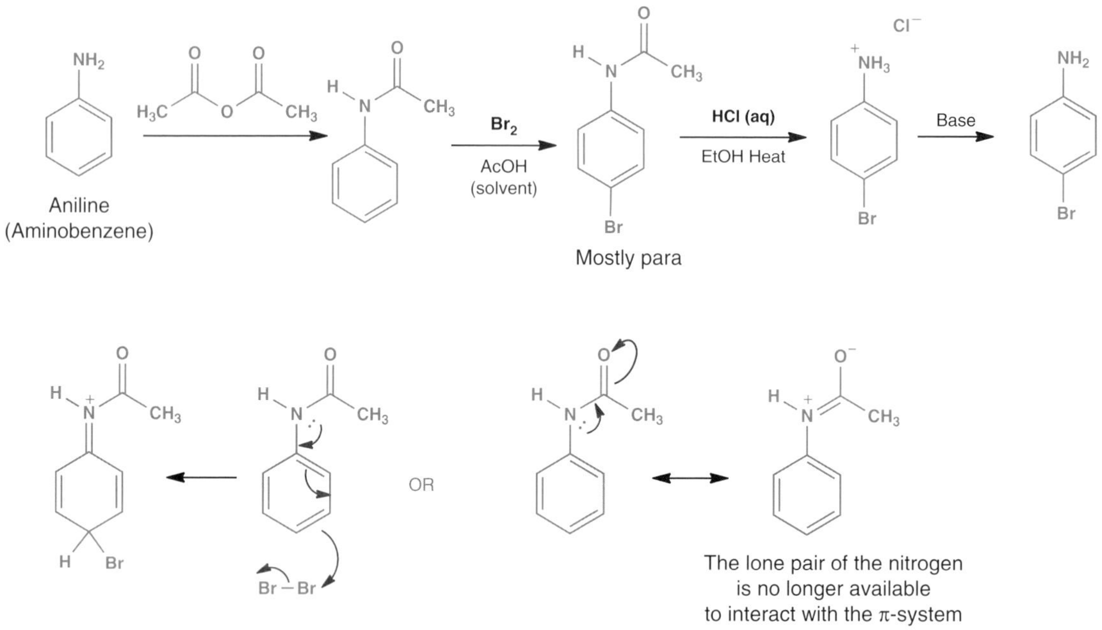

The directing effects of electron withdrawing groups

We have seen that having a cation adjacent to an electron donating group is energetically favourable. Naturally, it is also important to be able to describe the directing effects of an electron withdrawing group, such as the extremely versatile nitro (NO$_2$) group. Since an electron donating group stabilizes a positive charge, it is not really surprising that an electron withdrawing group does the opposite: it acts to make the charge more positive – an unfavourable situation. This means that it is advantageous if the cation does not fall on the carbon bearing the substituent; if necessary use Figure 7.25 to remind yourself of the resonance patterns.

Also if you look at the resonance forms shown in Figure 7.32, you can see that the carbons in the *ortho*

both *ortho*-positions are hindered, making reaction at that position more difficult.

and *para*-positions carry the cationic charge. If a carbon is to act as a nucleophile, as in electrophilic aromatic substitution, then it must be a provider of electrons. This is much more difficult if it is carrying a degree of positive charge. Therefore, when an electron withdrawing group is present on the aromatic ring, electrophilic aromatic substitution occurs at the *meta*-position, which is the least deactivated position.

The presence of the electron withdrawing group means that the formation of the σ-intermediate is more difficult because the electron withdrawing group competes for the electron density required to make the π-system nucleophilic. Therefore, we would expect a benzene substituted with an electron withdrawing substituent to react more slowly with electrophiles than benzene itself. This is observed experimentally. Electron withdrawing groups are both **meta-directing** and **deactivating**, as shown in Figure 7.33.

FIGURE 7.32 The withdrawing effect of the nitro group (by conjugation).

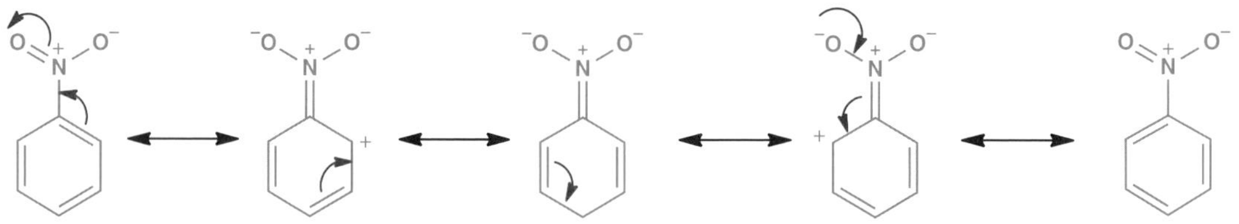

FIGURE 7.33 Representative distribution pattern for the reaction of nitrobenzene with an electrophile.

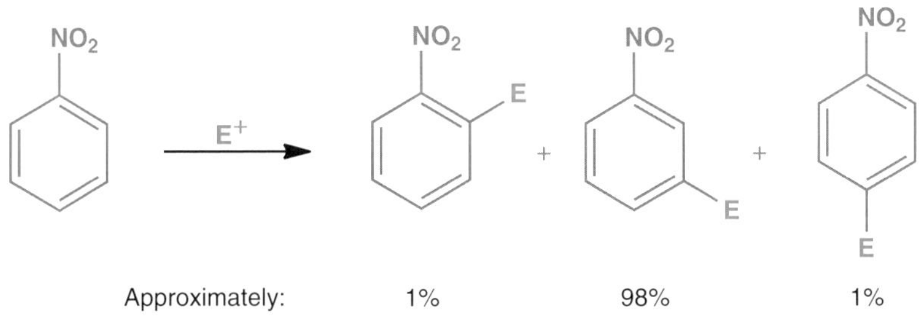

Approximately: 1% 98% 1%

The directing effects of halogens

There is a final class of directing group that needs to be considered, namely the halogens. They have their own unique directing effect: *ortho/para*-directing but deactivating. The inductive effect of the electronegative halogen is deactivating, but the activating effect of delocalization of the lone pair is of similar size. Overall, there is a small deactivating effect, but the *meta*-position does not benefit from activation by conjugation so it is deactivated more than the *ortho*- or *para*-position.

The inductive (deactivating) effect works through the σ-bonds and is therefore much more pronounced at the *ortho*-position than at the *para*-position. This is most important for the most electronegative halogens, especially fluorine. The halogens lower down the periodic table have lower electronegativities but larger atomic radii and this also reduces reactivity at the *ortho*-position. Although halogens are described as being *ortho/para*-directing it is worth remembering that they have a preference for reactivity at the *para*-position!

Table 7.1 shows the distribution and relative rates of nitration of halobenzenes, with respect to benzene. To give you an idea of the effects of different substituents on the rates of EAS reactions, the relative rates of reaction of some substituted benzenes with a generic electrophile in an EAS reaction are shown below:

> **KEY POINT**
>
> **Electron donating groups** are described as ***ortho/para*-directing** and **activating** (i.e. the substituted benzenes react more readily with electrophiles than benzene). Common electron donating groups include alkyl, hydroxyl, ethers, amines, amides, **thiols** and **thioethers**.

> **KEY POINT**
>
> **Electron withdrawing groups** are described as ***meta*-directing** and **deactivating**. Common electron withdrawing groups include nitro, aldehydes, ketones, esters, carboxylic acids, nitriles and sulfonic acids.

> **KEY POINT**
>
> **Halogens** are described as ***ortho/para*-directing** and **deactivating**. They have a preference for reaction at the *para*-position and this can be exploited.

TABLE 7.1 Distribution and relative rates of nitration of halobenzenes.

X	Ortho	Meta	Para	Rate
H	–	–	–	1.00
F	13	0.6	86	0.18
Cl	35	0.9	64	0.06
Br	43	0.9	56	0.06
I	45	1.3	54	0.12

> **SELF CHECK 7.7**
>
> Consider Figure 7.30 – the tribromination of phenol
>
> 1. Draw the mechanism of monobromination and indicate the major products.
>
> 2. Draw a mechanism for the formation of 2,4-dibromo-phenol.
>
> 3. Arrange the following reactions in order of rate: bromination of phenol, bromination of 4-bromophenol, bromination of 2,4-dibromophenol.

R =	NO_2	H	CH_3	OCH_3	$N(CH_3)_2$
Rate	1×10^{-7}	1	4×10^3	1×10^9	1×10^{14}

Note that in cases where conflicting directing effects exist, strong directors such as hydroxyl, amino and nitro will dominate weaker directors such as halogens. This is illustrated in the bromination of phenol in Figure 7.30.

Post-substitution functional group interconversion and metal-based reactions

Having prepared the desired aromatic skeleton by aromatic substitution reactions, it is possible that further reactions are required. Many reactions are possible – for example, the reduction of nitro groups to amines, or ketones to alcohols or alkyl groups. Another option that is particularly versatile and widely used is organometallic chemistry. Aromatic bromides in particular are useful reagents in this regard. Of particular note are Grignard (organomagnesium) reactions and palladium catalysed sp^2-sp^2 couplings.

Grignard reactions

You may not have realized that aromatic bromides can be converted into Grignard reagents (which we discuss in Chapter 6), and can, therefore, be used to make new carbon-carbon bonds. The advantage of these reactions is that they can be used to generate a series of compounds that have the same basic framework but in which functionality at a specific position is varied. This allows medicinal chemists to probe the significance of a specific portion of a potential drug molecule and to fine-tune its biological activity; this is the basis of a **structure activity relationship (SAR)** study.

 More information on Grignard reagents can be found in Chapter 6 'The carbonyl group and its chemistry'.

 Structure activity relationships are also discussed in Chapter 12 'The molecular characteristics of good drugs'.

SELF CHECK 7.8

Use your knowledge of aromatic chemistry (this chapter) and your knowledge of carbonyl chemistry (Chapter 6) to design a synthesis of 1-phenylethanol, starting from benzene and acetaldehyde.

The versatility of the nitro group

We are now in a position to consider the versatility of the nitro group. The nitro group is easily installed by nitration, whereupon it has a *meta*-directing effect for the installation of further electrophiles by EAS. It is easily reduced to the corresponding amine by hydrogenation with hydrogen and a palladium catalyst on charcoal or by a mixture of tin and hydrochloric acid. The *ortho/para*-directing effects of the amino group can be exploited at this point to install another electrophile.

Alternatively, if an aromatic amine is treated with cold sodium nitrite and an acid (often HCl), a diazonium salt results, as illustrated in Figure 7.34. The term diazonium refers to the $-^+N\equiv N$ group. The exceptional leaving group ability (neutral gaseous nitrogen is formed) of the diazonium group dictates its chemistry. These salts are often unstable at or above room temperature; nitrogen gas leaves spontaneously leaving a cation which can react with a very wide variety of nucleophiles, see Figure 7.34. The substitution of N_2^+ by a nucleophile is an example of nucleophilic aromatic substitution (for further examples see the web-based material).

Now we have seen some reactions of aromatic systems, it is time to see these reactions exploited in the synthesis of drugs.

7.4 The synthesis of drugs

Numerous important drugs are aromatic, as was mentioned previously. We can now consider some of their syntheses, starting with paracetamol, a derivative of acetanilide. Acetanilide was itself employed

FIGURE 7.34 Reactions of phenyldiazonium chloride.

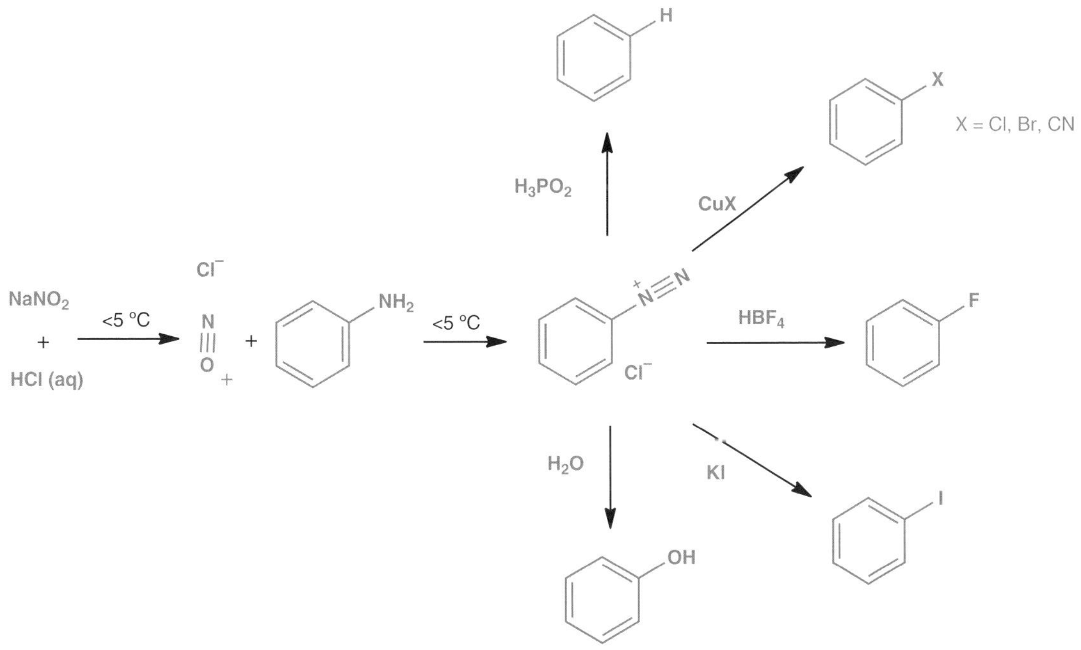

as an analgesic in the past (one of its metabolites is paracetamol, see Section 7.5), but it was withdrawn because of concerns about its toxicity. The preparation of paracetamol provided a medicine that is an effective analgesic with few side-effects at therapeutic doses (see Integration Box 7.1).

Paracetamol biochemistry is discussed in Chapter 8 'Homeostasis' of the ***Therapeutics and Human Physiology*** book in this series.

Paracetamol synthesis

The first step in the synthesis of paracetamol is the nitration of phenol by electrophilic aromatic substitution, as illustrated in Figure 7.35. The high reactivity of phenol means that even the low concentration of $[NO_2]^+$ produced in dilute nitric acid is sufficient for the EAS reaction to proceed within a reasonable timescale. Note that the powerful deactivating effect of the nitro group means that it is possible to obtain workable yields of the mono-nitrated material.

As expected, a mixture of *ortho*- and *para*-isomers is produced, which need to be separated. In this case, the *ortho*-isomer (2-nitrophenol) has a much lower boiling point than the desired 4-nitrophenol, and can be readily removed by distillation (Integration Box 7.2). If you have noticed that the yields of the *ortho*- and *para*-products do not add up to 100% you may be wondering what has happened to the remainder. It is rare that chemical reactions proceed with 100% yield because there are often side-reactions that can occur (in this case over nitration is the main issue) and there can be problems with isolating the pure product. One of the skills required for synthetic chemistry is to be able to minimize the formation of side-products and maximize the isolation of the desired compound.

INTEGRATION BOX 7.1

Paracetamol

Although paracetamol has few side-effects at therapeutic doses it is extremely dangerous in overdose, and one of its metabolites is a major cause of liver damage. Interestingly, it has also been shown that acetanilide is metabolized to paracetamol.

SELF CHECK 7.9

Why is a mixture of *ortho*- and *para*-isomers formed in the nitration of phenol?

FIGURE 7.35 The synthesis of paracetamol.

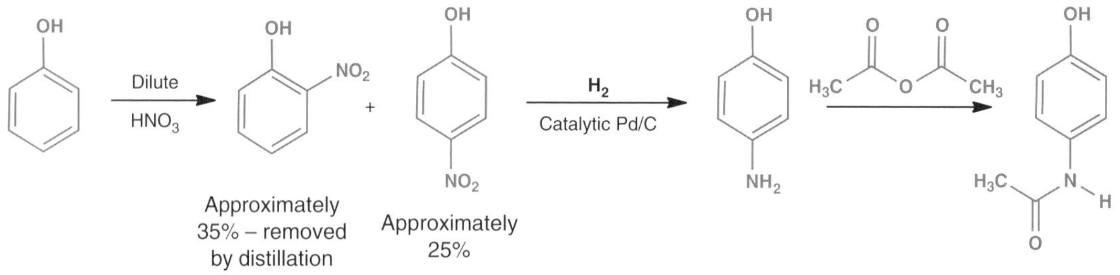

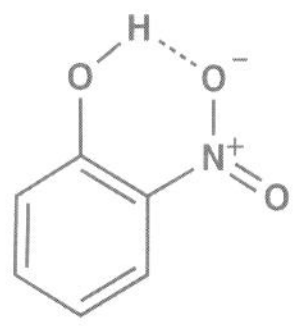

Approximately 35% – removed by distillation

Approximately 25%

Intramolecular Hydrogen bonding

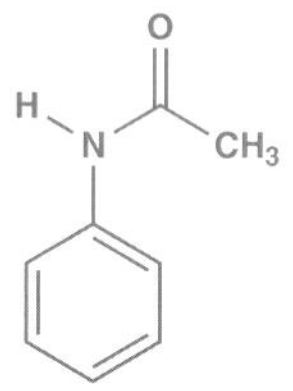

Acetanilide

INTEGRATION BOX 7.2

The separation of the *ortho*- and *para*-isomers

The separation of the *ortho*- and *para*-isomers of nitrophenol is easier than normal because there is an intramolecular hydrogen bond between the hydroxyl and nitro group in the *ortho*-isomer that cannot form in the *para*-isomer. Consequently, the intermolecular attractive interactions are reduced in the *ortho*-isomer such that its melting and boiling points are considerably lower than the *para*-isomer. This difference makes separation by distillation relatively easy.

 Intra and intermolecular hydrogen bonds are discussed in greater detail in Chapter 2 'Organic structure and bonding'.

The next task is to reduce the nitro group (NO_2) to an amino group (NH_2). A number of reagents can be used; tin and hydrochloric acid or catalytic hydrogenation (used in this case) are common. The final step in the synthesis is the acetylation of the amine with acetic anhydride (or acetyl chloride). Note that reaction occurs at the more nucleophilic amino group rather than at the hydroxyl group. The aromatic ring is not affected by this reaction.

Aspirin synthesis

The synthesis of aspirin is illustrated in Figure 7.36. This synthesis also involves the electrophilic

aromatic substitution of phenol; the electrophile is carbon dioxide and the desired product is *ortho*-substituted. Sodium hydroxide is used to deprotonate phenol, forming the expected sodium phenoxide (an extremely active participant in EAS reactions), which reacts with electrophilic carbon dioxide. The clever aspect of this reaction is that the sodium cation associated with the negatively charged oxygen of phenoxide can coordinate to the carbon dioxide and hold it in close proximity to the *ortho*-position. This behaviour ensures a high ratio of the desired *ortho*-product. The reaction is known as a Kolbe Reaction.

Notice the keto-enol tautomerism in the mechanism. In this case the aromatic stabilization energy means that the enol form is completely dominant. (Compare this with the keto-enol ratio of acetone.) The reaction of salicylic acid with acetic anhydride then forms aspirin.

 More information on keto-enol tautomerism can be found in Chapter 6 'The carbonyl group and its chemistry'.

Salbutamol synthesis

Aromatic chemistry is not just useful for analgesics. The asthma treatment salbutamol (see Figure 7.37) also has an aromatic core; its structure is, in fact, a modified form of adrenaline. Adrenaline is a hormone that forms an important component of the '*fight or flight*' response. Adrenaline has been used to treat

FIGURE 7.36 The synthesis of aspirin.

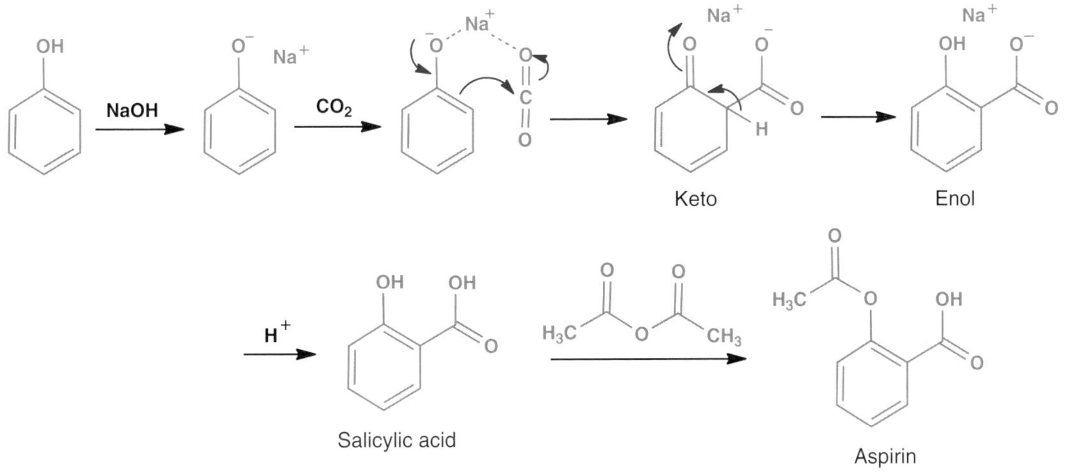

Keto

Enol

Salicylic acid

Aspirin

FIGURE 7.37 Structures of adrenaline and salbutamol.

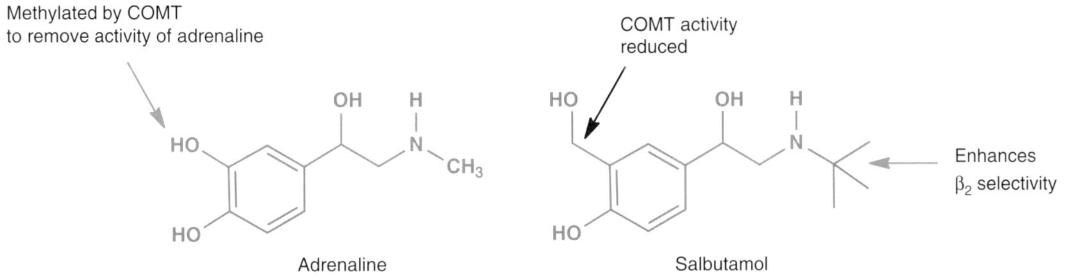

Methylated by COMT
to remove activity of adrenaline

COMT activity
reduced

Enhances
β_2 selectivity

Adrenaline

Salbutamol

asthma because it causes dilation of the bronchial airways.

However, the effects of adrenaline on the body are dramatic and short-lived. It would be more desirable for an anti-asthma drug to exhibit greater selectivity for the β_2-adrenoreceptors to maximize dilation of the **bronchial tree** (while minimizing the detrimental cardiovascular interactions), and also to have an increased duration of action. Two modifications to adrenaline fulfil these objectives:

- increasing the steric bulk on the amine by replacing a methyl group with a *tert*-butyl group enhances selectivity for the β_2-adrenoreceptor

- the installation of a CH_2 between the aromatic ring and the hydroxyl group methylated by catechol-*O*-methyltransferase (COMT) reduces the activity of that enzyme on the drug and so increases its duration of action.

The synthetic route, which is illustrated in Figure 7.38, starts with aspirin. Treatment with $AlCl_3$ initiates a reaction known as the Fries Rearrangement which involves formation of an acylium ion (the active electrophile in the Friedel-Crafts acylation), and an equivalent to a Friedel-Crafts acylation ensues. The *ortho-*, *para*-directing effect of the oxygen of the aluminium alkoxide is dominant, but the *meta*-directing carboxyl group also favours substitution at carbons 3 and 5. So where will the electrophile attach? Both substituents direct the incoming electrophile to positions 3 and 5 but a 1,2,3-trisubstituted benzene is sterically congested so 5-substitution is favoured and the major product by far is the isomer shown in Figure 7.38.

When the acid is treated with methanol in the presence of catalytic acid, the expected ester is formed. Bromination of the ketone with bromine in chloroform gives the mono bromide by reaction via the enol form (aided by traces of acid in the chloroform and

FIGURE 7.38 Synthesis of salbutamol and the structure of salmeterol.

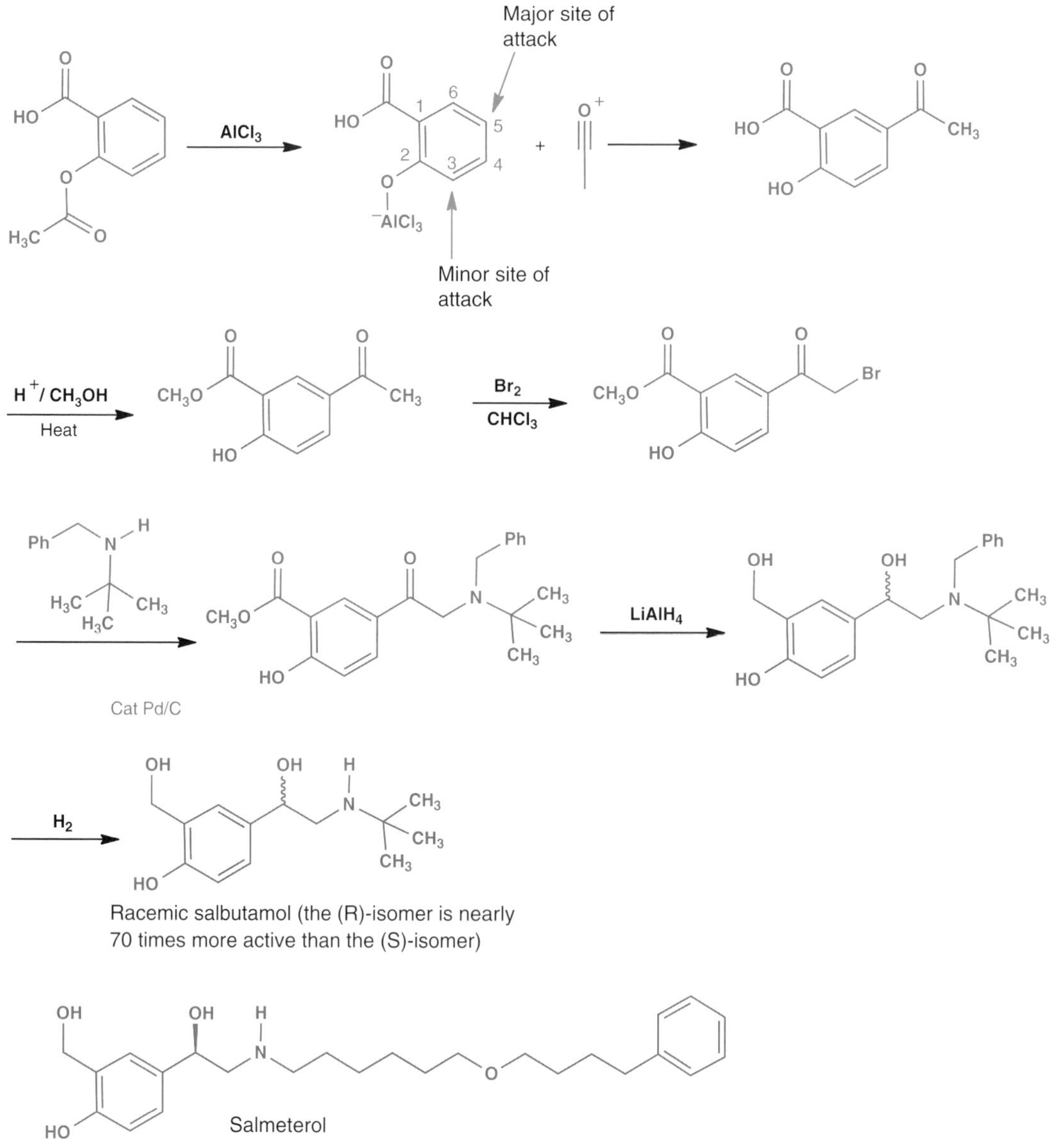

bromine). (Refer to Chapter 6 if this is not clear.) The addition of *N*-benzyl-*N-tert*-butylamine by nucleophilic substitution of bromide installs the amine functionality.

The benzyl group (PhCH$_2$-) may seem unnecessary, as it does not feature in the final product, but it serves to prevent addition of the amine to more than one bromoketone molecule in this step and can be cleanly removed later. (Over alkylation of nitrogen is a common problem in synthesis.) Reduction of the ketone and ester with lithium aluminium hydride, followed

by the removal of the benzyl group with hydrogen and palladium on charcoal catalyst, generates salbutamol in racemic form.

Although salbutamol is an extremely effective asthma treatment (most people will have seen the blue inhalers), its duration of action is shorter than ideal. Further work on the side-chain led to the production of salmeterol which has an extended duration of action (around 16 hours) compared with 4 hours for salbutamol. Notice how the aromatic core has remained intact in the structure of salmeterol.

Source: esolla/iStock

This coloured inhaler is used for salbutamol; other asthma drugs are provided in different coloured inhalers.

 Amine alkylation is discussed in more detail in Chapter 5 'Alcohols, phenols, ethers, organic halogen compounds, and amines' of this book.

 More details on reactions of ketones can in found in Chapter 6 'The carbonyl group and its chemistry'.

7.5 Aromatic chemistry in the body

The body requires many aromatic compounds. Some of these are synthesized from preformed aromatic compounds, such as the amino acids, phenylalanine and tyrosine, which are usually obtained from the diet. However, aromatic compounds can also be synthesized from non-aromatic substrates. For example, specialized aromatase enzymes can generate aromatic steroids from non-aromatic precursors, as shown in Figure 7.41. Although these reactions are mediated by enzymes they are still examples of aromatic chemistry.

Aromatic amino acids, neurotransmitters, hormones and drugs

Aromatic compounds feature in the important catechol neurotransmitter/hormone family, which includes adrenaline, noradrenaline and dopamine. The importance of these compounds in the body cannot be overstated and numerous drugs interact with the biochemical systems associated with them. The biosynthesis of adrenaline (via noradrenaline and dopamine) starts with aromatic amino acid tyrosine and is shown in Figure 7.39.

Figure 7.40 shows how aromatic rings feature strongly in drugs that interact with the biochemical systems that synthesize these compounds, and the receptors on which these compounds act. Drugs that interact with the catechol system include:

- asthma treatments salbutamol and salmeterol (β_2-adrenoreceptor agonist)

- high blood pressure medication β-blocker antihypertensives e.g. Propranolol (β_1-adrenoreceptor antagonist)

- several classes of antidepressant, including tricyclics such as amitriptyline, and the noradrenalaline reuptake inhibitor, reboxetine

- the smoking cessation aid, bupropion, which is also an antidepressant

- the Parkinson's Disease treatment L-DOPA (a dopamine precursor).

Oestrogens and aromatase enzymes

Oestrogen hormones (oestrone, oestradiol and oestriol) play an important role in a class of breast cancers known as hormone dependant cancers. If a tumour is dependent on these hormones for proliferation then compounds that either block the action of those hormones (antagonists) or prevent their biosynthesis may offer treatments. The biosynthesis of the oestrogen hormones involves the action of the enzyme aromatase on andostrenedione and testosterone to form

FIGURE 7.39 The biosynthesis and deactivation of adrenaline.

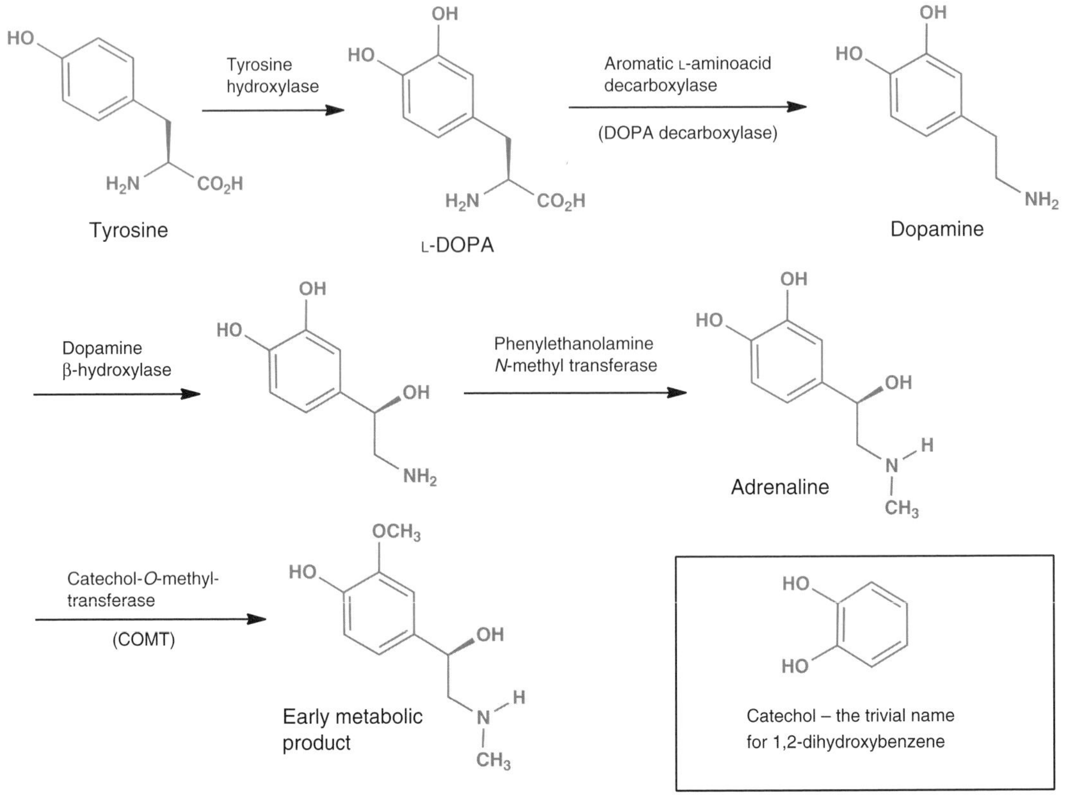

FIGURE 7.40 The structures of some drugs that interfere with the synthesis or action of aromatic compounds in the body.

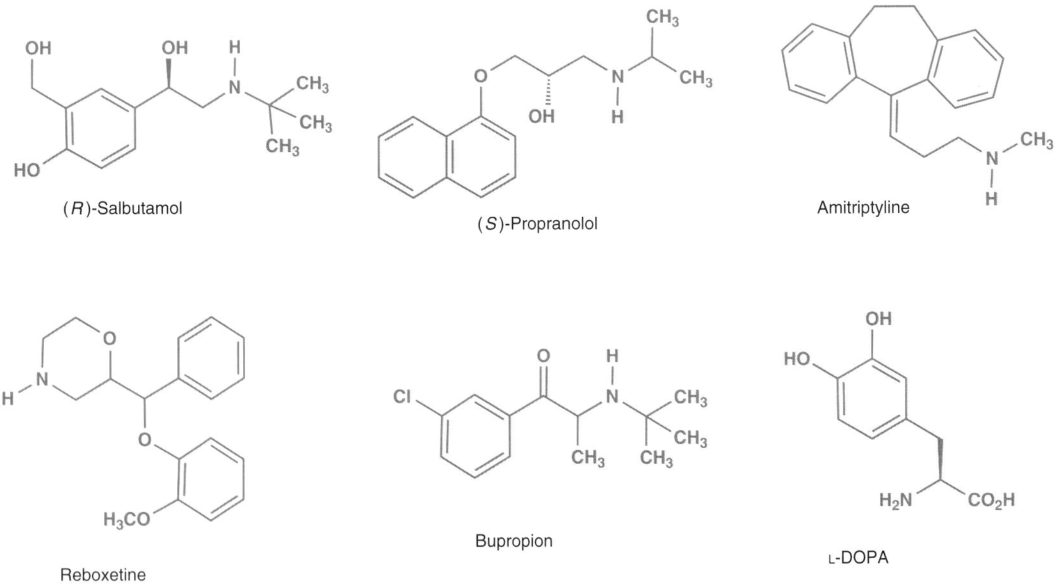

(R)-Salbutamol

(S)-Propranolol

Amitriptyline

Reboxetine

Bupropion

L-DOPA

oestrone and oestradiol respectively, as shown in Figure 7.41.

Tamoxifen and fulvestrant are antioestrogens (oestrogen antagonists), which block the action of the oestrogens directly. Anastrazole and letrozole (see Figure 7.42) are reversible aromatase inhibitors, and 4-hydroxyandrostenedione is a **suicide substrate** that reacts with the aromatase enzyme and renders it permanently

FIGURE 7.41 The action of the enzyme aromatase on steroids to form two of the oestrogen hormones.

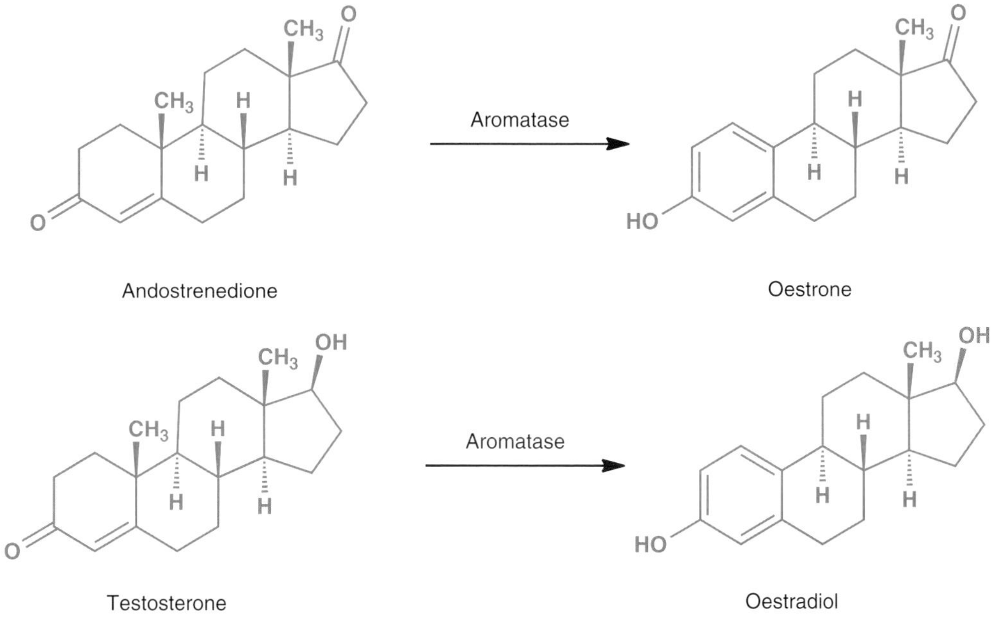

Andostrenedione

Aromatase

Oestrone

Testosterone

Aromatase

Oestradiol

FIGURE 7.42 Two antioestrogen drugs and three aromatase inhibitors.

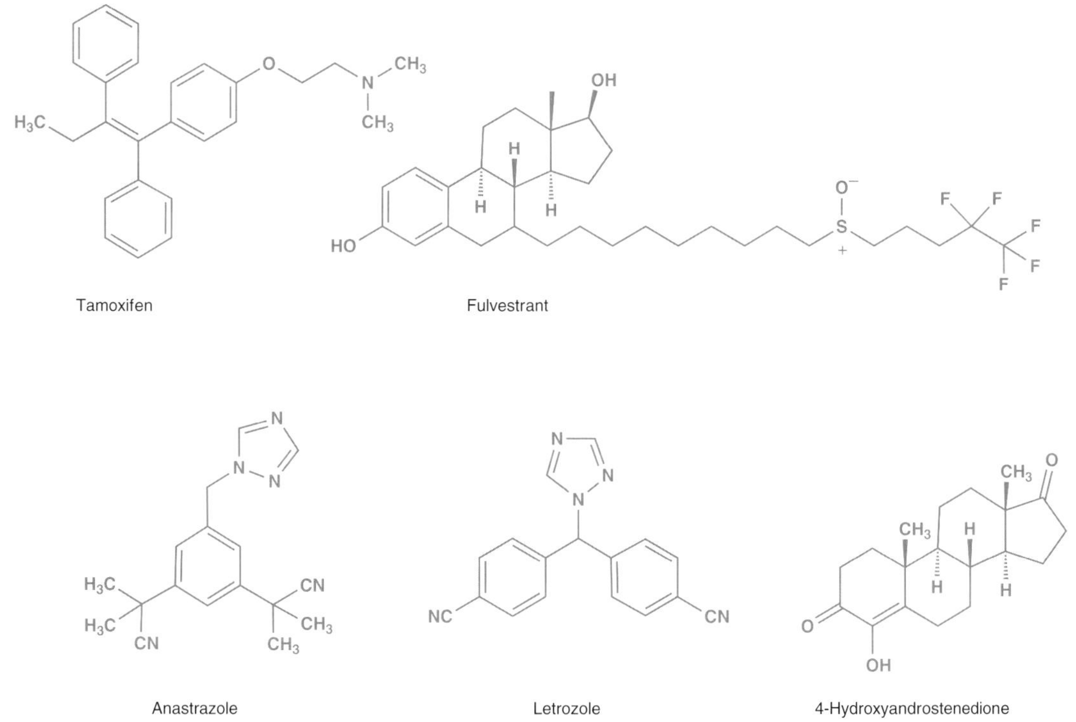

Tamoxifen

Fulvestrant

Anastrazole

Letrozole

4-Hydroxyandrostenedione

deactivated. Antioestrogens and aromatase inhibitors are extremely effective treatments for hormone sensitive breast cancer and the lives of huge numbers of breast cancer sufferers have been saved since the introduction of tamoxifen around thirty years ago.

 Tamoxifen is discussed in more detail in Chapter 3 'Stereochemistry and drug action' of this book and in Chapter 4 'Introduction to drug action' in the *Therapeutics and Human Physiology* book in this series.

Metabolism of aromatic compounds

The metabolism of aromatic compounds generally involves oxidation, usually mediated by one of the **cytochrome P450 enzymes**; the initial product is often an epoxide (a three membered ring containing two carbons and an oxygen).

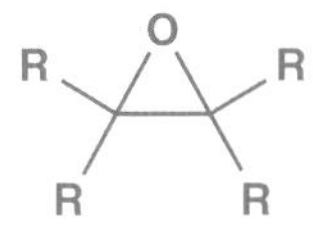

Subsequent reactions and rearrangements can take place and the metabolic pathway often involves the installation of a hydroxyl group via rearrangement of the initial epoxide. Hydroxylation often takes place at the *para*-position in singly substituted aromatic rings, but the situation is more complicated in polysubstituted systems.

In some cases, metabolism of aromatic compounds can lead to an active metabolite (for example, paracetamol from acetanilide, as illustrated in Figure 7.43), but such situations are rare and a compound is more likely to be deactivated directly or conjugated and eliminated. The body has evolved mechanisms to rid itself of foreign compounds. The formation of oxygenated compounds generally increases their water solubility and renders them more easily removed from the body – a process known as phase one metabolism. Phase two metabolism involves the formation of conjugates between the oxygenated molecules and glucose derivatives to further enhance their water solubility and further aid their removal from the body.

 Cytochrome P450 enzymes and metabolism are also discussed in Chapter 1 'The scientific basis of therapeutics' in the *Therapeutics and Human Physiology* book in this series.

 Epoxides are discussed in more detail in Chapter 5 'Alcohols, phenols, ethers, organic halogen compounds, and amines'.

When designing new drugs, it is important to consider the possible metabolism of the new compound. As a consequence, promising drug candidates are now regularly screened against a range of P450 enzymes. If troublesome metabolism is detected or expected, additional functional groups can be installed to prevent or slow those metabolic processes. One method of reducing the problem of unwanted *para*-hydroxylation is to place a fluorine atom on the aromatic ring, usually in the *para*-position. Since fluorine is the most electronegative element, its presence alters the electronic nature of the aromatic ring, making it less prone to oxidation. The fluorine atom is not much larger than hydrogen (van der Waals radius 1.47 Å vs 1.2 Å), so that a single substitution of hydrogen for fluorine will not have a significant steric effect. The use of fluorine as a metabolic blocker has been employed in the drug ezetimibe (see Figure 7.44), used to treat patients with high levels of cholesterol.

Metabolism and toxicity

The toxicity data for benzene makes worrying reading: it 'may cause cancer' and 'may cause heritable genetic damage'. The use of benzene is now restricted to pro-

FIGURE 7.43 Metabolism of acetanilide to form the analgesic paracetamol.

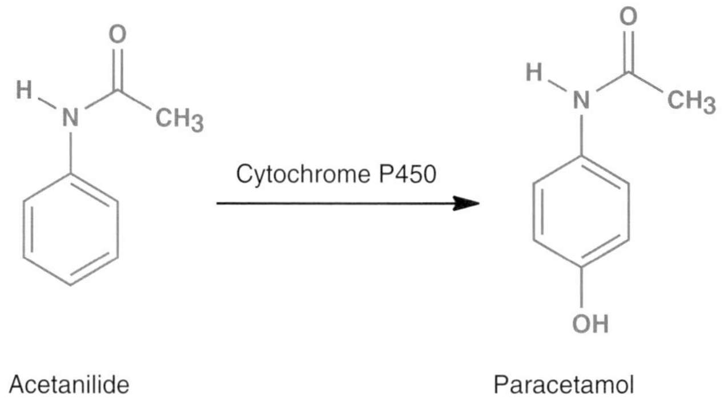

Acetanilide

Cytochrome P450

Paracetamol

FIGURE 7.44 The use of fluorine as a metabolic blocker to prevent oxidation at the positions bearing the fluorine.

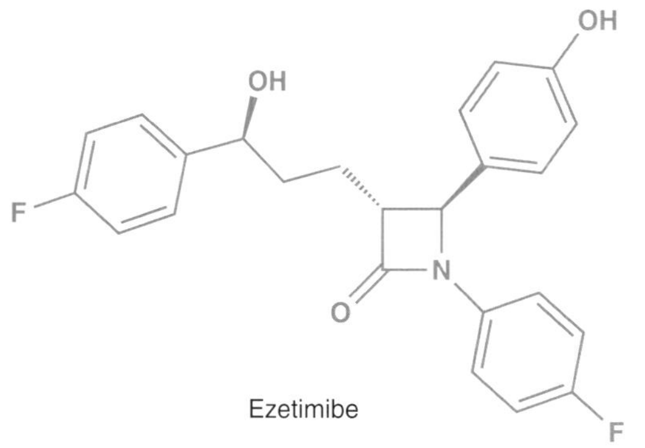

Ezetimibe

fessional chemists and similarly qualified personnel. However, we should not assume that all aromatic compounds have the same safety profile. Indeed, the fact that so many drugs contain aromatic rings is indicative of the way that the toxicity of benzene is not commonly carried over into its derivatives.

Metabolism of benzene begins with the formation of an epoxide, mediated by P450 enzymes, as shown in Figure 7.45. The epoxide can then react further to form a variety of metabolites, which can interact with biomolecules, including DNA. It is these metabolites that have significant toxicity, rather than benzene itself. Exposure to benzene is inadvisable but it is actually the body's attempts to metabolize and eliminate benzene that make it problematic. It is important to note that the epoxidation of benzene is a very difficult reaction to perform by simple chemical means, but it occurs in the body by the action of the P450 enzymes.

Doll and Hill famously published the statistical link between smoking and lung cancer in 1950. The exact cause, and therefore proof, of the connection was published 24 years later. A compound called benzo-[a]-pyrene, a member of the polyaromatic hydrocarbon family, is the villain. More specifically, it is its metabolic product that contains an epoxide and a diol that is most responsible for the carcinogenic properties, as illustrated in Figure 7.46. Reaction of the epoxydiol with DNA via guanine, for example, is responsible for DNA damage that can result in tumourogenesis.

FIGURE 7.45 The metabolism of benzene.

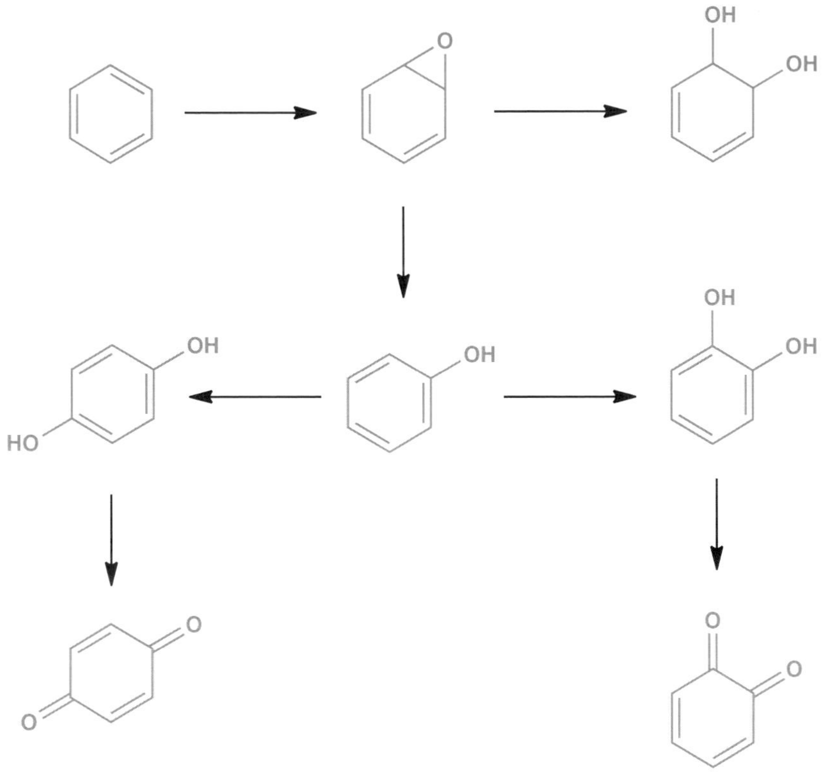

FIGURE 7.46 Metabolism of a constituent of cigarette smoke and the reaction of the metabolite with DNA.

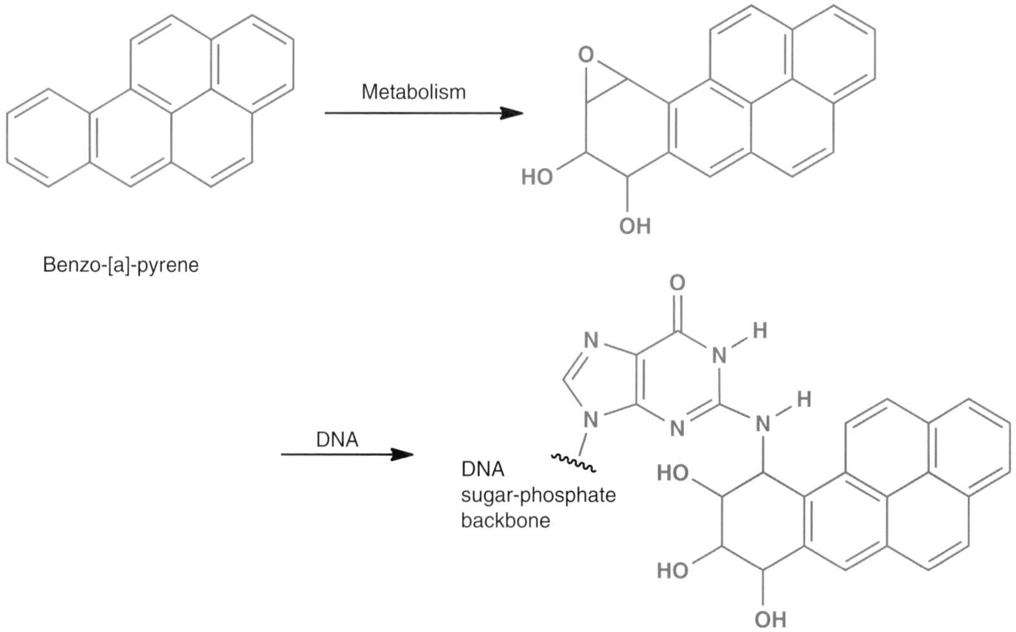

Benzo-[a]-pyrene

CASE STUDY 7.1

A customer in Dilip's pharmacy wants to purchase 100 paracetamol tablets to save her from coming to the pharmacy as often. Dilip has to explain to the customer that she is not allowed, legally, to purchase that number of paracetamol tablets.

REFLECTION QUESTIONS

1. Why is the customer not allowed to buy 100 paracetamol tablets?

2. What is the mechanism of paracetamol toxicity?

Answers

1 The restriction on the quantity of paracetamol tablets that can be sold is related to the serious consequences of paracetamol overdose. even though, at normal doses, paracetamol is a very safe drug. Overdose of paracetamol can cause irreversible liver damage and is potentially fatal.

2 The liver toxicity problem with paracetamol is not due to paracetamol itself, but due to one of its metabolites. A small portion of paracetamol (~4%) is oxidatively metabolized in the liver to a quinoneimine molecule. This is a very reactive species and could potentially cause damage to liver cells. However, if the normal doses of paracetamol are taken, this is not a problem because this reactive metabolite is deactivated by glutathione (see Chapter 5 for the mechanism). Although only 4% of the reactive metabolite is produced, in an overdose situation, this will be a significant amount and may be sufficient to deplete the liver's supply of glutathione. Any remaining reactive metabolite will then cause irreversible damage to the cells of the liver.

CHAPTER SUMMARY

We have seen that aromatic chemistry features heavily in many areas of medicinal and pharmaceutical chemistry. The aromatic ring is common in many drugs, including many of the most successful. A number of aromatic compounds are essential to our survival and we would not be here today without them.

We have also seen that the body's attempts to rid itself of aromatic compounds is sometimes unsuccessful and we end up generating dangerous metabolites. An understanding of aromatic chemistry allows us to comprehend these processes and take steps to circumvent them.

In addition, aromatic chemistry allows us to design, prepare and assess new chemical entities that can help alleviate the suffering associated with disease. Do not forget that many of the successful medicines that we take for granted (and many that we hope never to need), have to be prepared on a huge (multi tonne) scale and this requires an understanding of aromatic chemistry so that the drugs can be prepared safely. Take a minute to think about how much aspirin is consumed globally per year.

FURTHER READING

Further information on post-substitution functional group interconversion and metal-based reactions can be found in:

Clayden, J., Greeves, N., and Warren, S., *Organic Chemistry*, 2nd edn. Oxford University Press, 2012.

8 Inorganic chemistry in pharmacy

GEOFF HALL

The elements you will be most familiar with are those that are frequently encountered in the organic molecules found in biological systems: carbon, hydrogen, oxygen, and nitrogen. However, lots of inorganic chemistry occurs in the body as well, and a number of drugs are inorganic, containing no carbon. In this chapter, we shall study the chemistry of these inorganic elements. The roles of some of the elements may be familiar to you whilst others may surprise you. The chapter is organized so that after a revision of atomic structure, the periodic table, and types of bonding, we will study individual inorganic elements. For this chapter, they are organized into the alkali and alkaline earth metals, transition metals, zinc and iron, and the precious metals, silver, gold and platinum. The non-metals studied will be phosphorus and sulfur. Many other elements are also important to drug development, drug action and delivery. It is not possible to go into detail about each element, but by understanding the chemical properties of an element, it should be possible to understand its biological function.

Learning objectives

Having read this chapter you are expected to be able to:

➤ understand concepts in inorganic chemistry namely:
  ➤ types of bonding
  ➤ electron distribution in orbitals
➤ recognize the importance of alkali and alkaline earth metals for normal cell function
➤ carry out calculations associated with concentrations of ions
➤ identify how the biological role of iron and zinc can be related to their chemistry and how some drug actions and side effects can be explained by their interaction with these metal ions

➤ recognize that seemingly inert metals such as the precious metals can have biological activity
➤ understand the importance of phosphorus and sulfur in biological systems, and how this can be explained in terms of their chemical properties.
➤ identify various phosphorus and sulfur-containing functional groups in drug molecules, and be able to explain the role of these functional groups for either the formulation of the drug into a medicine or the biological action of the drug.

8.1 Concepts in inorganic chemistry

To introduce the chapter, try to complete Table 8.1 by listing two transition metal ions that are found in biological systems and two precious metals that are found in drugs. Give a short description of the function of each element.

The periodic table

Group Period	1	2	3	4	5	6	7	8	9	10	11	12	13	14	15	16	17	18
1	1 H 1.008																	2 He 4.0026
2	3 Li 6.94	4 Be 9.0122											5 B 10.81	6 C 12.011	7 N 14.007	8 O 15.999	9 F 18.998	10 Ne 20.180
3	11 Na 22.990	12 Mg 24.305											13 Al 26.982	14 Si 28.085	15 P 30.974	16 S 32.06	17 Cl 35.45	18 Ar 39.948
4	19 K 39.098	20 Ca 40.078	21 Sc 44.956	22 Ti 47.867	23 V 50.942	24 Cr 51.996	25 Mn 54.938	26 Fe 55.845	27 Co 58.933	28 Ni 58.693	29 Cu 63.546	30 Zn 65.38	31 Ga 69.723	32 Ge 72.63	33 As 74.922	34 Se 78.96	35 Br 79.904	36 Kr 83.798
5	37 Rb 85.468	33 Sr 87.62	39 Y 88.906	40 Zr 91.224	41 Nb 92.906	42 Mo 95.96	43 Tc [97.91]	44 Ru 101.07	45 Rh 102.91	46 Pd 106.42	47 Ag 107.87	48 Cd 112.41	49 In 114.82	50 Sn 118.71	51 Sb 121.76	52 Te 127.60	53 I 126.90	54 Xe 131.29
6	55 Cs 132.91	56 Ba 137.33	* 71 Lu 174.97	72 Hf 178.49	73 Ta 180.95	74 W 183.84	75 Re 186.21	76 Os 190.23	77 Ir 192.22	78 Pt 195.08	79 Au 196.97	80 Hg 200.59	81 Tl 204.38	82 Pb 207.2	83 Bi 208.98	84 Po [208.98]	85 At [209.99]	86 Rn [222.02]
7	87 Fr [223.02]	88 Ra [226.03]	** 103 Lr [262.11]	104 Rf [265.12]	105 Db [268.13]	106 Sg [271.13]	107 Bh [270]	108 Hs [277.15]	109 Mt [276.15]	110 Ds [281.16]	111 Rg [280.16]	112 Cn [285.17]	113 Uut [284.18]	114 Fl [289.18]	115 Uup [288.19]	116 Lv [293]	117 Uus [294]	118 Uuo 294

*Lanthanoids *	57 La 138.91	58 Ce 140.12	59 Pr 140.91	60 Nd [144.24]	61 Pn 144.91	62 Sm 150.36	63 Eu 151.96	64 Gd 157.25	65 Tb 158.93	66 Dy 162.50	67 Ho 164.93	68 Er 167.26	69 Tm 168.93	70 Yb 173.05
**Actinoids **	89 Ac [227.03]	90 Th 232.04	91 Pa 231.04	92 U 238.03	93 Np [237.05]	94 Pu [244.06]	95 Am [243.06]	96 Cm [247.07]	97 Bk [247.07]	98 Cf [251.08]	99 Es [252.08]	100 Fm [257.10]	101 Md [258.10]	102 No [259.10]

As you may remember from your previous chemistry studies, an element's position in the periodic table is based on its atomic number and on the distribution of its electrons. The vertical columns are called groups, and elements within the group have similar chemical behaviours. Traditionally, the main groups are designated by Roman numerals (I-VIII) with some more complicated arrangements for elements such as the transition metals. In the IUPAC system the groups are designated 1 to 18. Groups 1 and 2 are the same as groups I and II in the traditional format, with groups 13 to 18 being the same as groups III to VIII. The IUPAC system is used in this chapter. Therefore lithium is in group 1 and carbon is in group 14.

The horizontal rows of the table are known as periods. Across a period there tends to be a gradual and regular change in physical properties but the elements possess quite different chemical properties. We see an exception to this with the transition metals where adjacent elements tend to have similar chemical properties as the arrangement of electrons in the outer shell can be similar.

TABLE 8.1 Metals in pharmacy

Metal	Action/Use
e.g. Cobalt	Found in vitamin B_{12}, used to treat pernicious anaemia

Chemical bonds

In addition to covalent bonds, elements studied in this chapter take part in other forms of bonding. Both ionic and **covalent dative bonding** are important in the biological action of various elements – for example, in the interactions of ligands with biological macromolecules.

 Further information on covalent bonds can be found in Chapter 2 'Organic structure and bonding'.

An ionic bond is formed between positively and negatively charged ions. When an ionic bond is formed both electrons come from one of the ions and there is usually a large difference in electronegativity between the elements whose ions form the bond. (Consider common salt, for example, which consists of sodium and chloride ions. Na has an electronegativity of 0.93, Cl 3.16).

Dative covalent bonds were introduced in Chapter 5. Nitrogen in quaternary ammonium salts donates a pair of electrons to form a positively charged group, as shown in this tetramethylammonium ion.

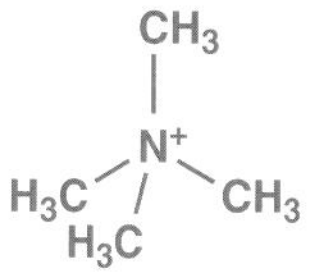

$$H_3C-\overset{\overset{\displaystyle CH_3}{|}}{\underset{\underset{\displaystyle H_3C}{|}}{N^+}}-CH_3$$

Dative covalent bonding can also occur between complex **ligands** and metal cations. Ligands can be considered to be **Lewis bases** in that they donate electrons to the metal ions; the metal ions, in turn, act as **Lewis acids** by accepting the electrons.

Dative covalent bonding can result in the formation of complexes; if the ligand can donate more than one pair of electrons to the metal then it is called a **chelating agent**. Chelating agents may be used to treat poisoning with metal ions; for example, Figure 8.1 shows the structure of desferrioxamine, which is used to treat iron poisoning.

FIGURE 8.2 The structure of pyrrole; draw the structure of histidine alongside.

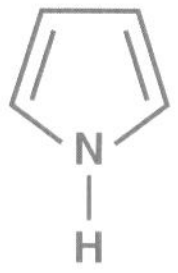

Pyrrole

Ligands require lone pairs of electrons, and the lone pair usually comes from oxygen or nitrogen. Ligands are often based on amino acids such as histidine, or the pyrrole ring of porphyrin. The structure of pyrrole is shown in Figure 8.2; it is an example of an aromatic heterocyclic compound. The lone pair of electrons on the nitrogen is available for donation to a metal ion in biological complexes.

SELF CHECK 8.1

The structure of pyrrole is shown in Figure 8.2. Draw the structure of the amino acid histidine alongside it and highlight the atoms which may act as Lewis bases when histidine is one of the amino acids of a protein.

Oxidation and oxidation states (numbers)

Oxidation is the loss of electrons. Metals are oxidized to metal ions M^{n+} when they lose electrons to a more electronegative element. Metal ions in group 1, such as sodium and potassium, lose the outer s electron very easily and tend to exist as ions e.g. Na^+ and K^+. Similarly group 2 metals, such as calcium and magnesium, lose both electrons from the outer s shells easily giving the divalent ions Ca^{2+} and Mg^{2+}.

FIGURE 8.1 The structure of desferrioxamine.

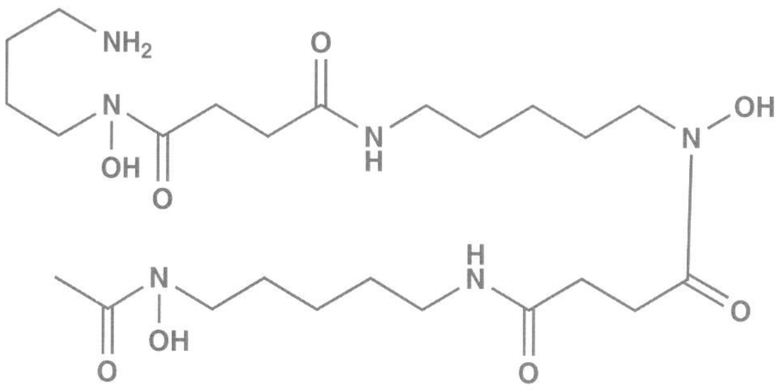

The situation with transition metals is more complicated as they can have multiple oxidation states. Iron(II) results from the loss of the two $4s^2$ electrons whilst iron(III) results from the loss of two $4s^2$ electrons and one 3d electron. As the energy difference between the 4s shell and the 3d shell is small, the transition between Fe^{2+} and Fe^{3+} can occur easily in biological situations.

The oxidation state (**oxidation number**) of an element refers to the number of electrons an element has lost or gained. Positive oxidation numbers mean that the element has lost electrons while a negative oxidation number means that the element has gained electrons. As discussed above, iron commonly exists in two oxidation states +2, as in iron(II) (Fe^{2+}), or +3 as in iron(III) (Fe^{3+}). Iron(II) is still commonly referred to as 'ferrous' (as in ferrous sulfate tablets) whilst iron(III) is 'ferric' (as in ferric chloride, which may be used in tests for phenols).

8.2 Metals: introduction

The normal activity of cells depends on the correct concentrations of certain metal ions both inside and outside the cell (see Box 8.1). Metal ions are also found in medicines. For instance, sodium and potassium ions are frequently used in salts of carboxylic acids. Benzylpenicillin sodium is used as a treatment for bacterial meningitis, and iron(II) salts are used for the treatment of anaemia.

SELF CHECK 8.2

Complete Table 8.2 by calculating the missing values. Remember: number of moles = (number of grams)/ (molar mass).

TABLE 8.2   Amounts of metals in the body and concentration of ions in serum or plasma

Metal	Average amount in the adult body (g)	Average amount in the adult body (mol)	Ion serum/plasma concentration reference range
Sodium		4	137–145 mM
Potassium	140		3.6–5.0 mM
Calcium		30	2.1–2.6 mM
Iron	4		male 11–28 µM female 7–26 µM
Zinc	2		10–24 µM

BOX 8.1

Ion serum/plasma concentration reference range

Whilst watching a TV medical drama you may have heard a doctor order 'Us' and 'Es' for a patient. The 'Es' refer to electrolyte tests, which is the determination of the concentration of specific ions in the blood. These tests are very useful for the diagnosis of a patient's disease or condition. The resulting ion concentrations can be compared with a set of values known as the ion serum concentration range. The detection of an abnormal electrolyte value may facilitate diagnosis and subsequent treatment. A range of values is quoted as a reference because normal values can vary according to age and gender.

8.3 Group 1 metals

The group 1 metals, known as the alkali metals, include lithium, sodium and potassium. All the group 1 metals react with water to release hydrogen, forming hydroxides. The reaction can be represented as below

$$2M + 2H_2O \rightarrow 2M^+ + 2OH^- + H_2$$

The effect of placing small pieces of these metals into water is quite spectacular. (There are a number of videos available on the internet, if you have not seen this before.) The reaction becomes more vigorous as you descend the group.

The relative molecular mass (RMM) and the electronic configuration of some group 1 and 2 metals are shown in Table 8.3.

Sodium and potassium ions are the most important of the group 1 metal ions for normal cell function, whilst lithium ions can be used in the prophylaxis and treatment of mania and other disorders.

Sodium

The normal adult body contains about 4 moles of sodium, of which approximately half is found in the extracellular fluid. Sodium levels can affect the volume of extracellular fluid, and excess sodium ions in the blood can lead to a number of conditions including hypertension. Sodium ions are absorbed very easily from the gastrointestinal tract and many of us ingest more than we need.

The recommended dietary intake for sodium in the UK is 1.6 g daily for adults, equivalent to 4 g of sodium chloride (salt). The target maximum daily intake is 6 g of salt, yet the average intake is approximately 9 g. Figure 8.3 shows these amounts of salt relative to a teaspoon, dessertspoon and tablespoon. The salt taste that most of us enjoy is due to sodium ions, which are able to enter ion channels in receptors on the tongue.

TABLE 8.3 Properties of some group 1 and group 2 metals

Element	Symbol	Atomic number	RMM	Electronic configuration
Sodium	Na	11	22.99	$1s^2 2s^2 2p^6 3s^1$
Potassium	K	19	39.10	$1s^2 2s^2 2p^6 3s^2 3p^6 4s^1$
Magnesium	Mg	12	24.31	$1s^2 2s^2 2p^6 3s^2$
Calcium	Ca	20	40.08	$1s^2 2s^2 2p^6 3s^2 3p^6 4s^2$

FIGURE 8.3 Amounts of sodium chloride (salt) in our diet.

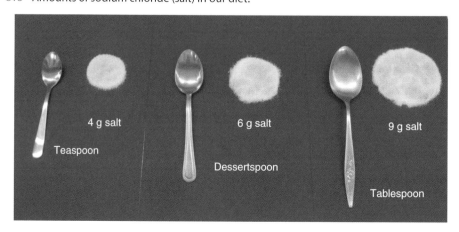

Salt substitutes contain potassium ions, which are small enough to enter these same ion channels.

Intravenous sodium chloride solution is prescribed when there are low levels of sodium in the body. Sodium Chloride Infusion is isotonic with the physiological concentration of sodium and has a concentration of 0.9% w/v (see Figure 8.4).

SELF CHECK 8.3

Calculate the concentration of Na+ ions in 0.9% w/v sodium chloride solution in g L⁻¹ and in mM. How does the latter value compare with the value in Table 8.2?

The levels of sodium in the body can be affected by the administration of medicines, including penicillin antibiotics. These compounds contain a carboxylic acid functional group and are formulated as sodium salts to give water solubility. The BNF specifically warns that high doses of injectable penicillins, or even normal doses given to patients with renal failure, can lead to the build up of an excess of sodium ions. Neonates can also be adversely affected by medicines with high sodium levels. The sodium intake is further increased if the medicine is reconstituted or infused with a sodium chloride solution.

SELF CHECK 8.4

Draw the structure of benzylpenicillin sodium, look up its relative molecular mass, and calculate the number of moles of sodium ions in a vial containing 600 mg of benzylpenicillin sodium. Use the BNF to check your answer.

 The administration of salts of carboxylic acids and amines is discussed in Chapter 6 'The carbonyl group and its chemistry'.

Potassium

Low levels of potassium can be caused by fluid loss and can cause muscle weakness. Fruit and vegetables are good sources of potassium. Many sports men and women try to combat muscle fatigue by eating bananas, which are especially rich in potassium. A medium banana contains about 420 mg potassium, equivalent to about 800 mg of potassium chloride. Figure 8.5 shows this amount of KCl relative to a banana and a two pence piece.

Digoxin, which is used to treat some cardiac problems, normally competes with K+ ions for the same binding site on the Na+/K+ ATPase pump, an enzyme essential for maintaining correct intracellular Na+/K+ ratios. Some patients are therefore treated with oral potassium supplements to try to minimize the unwanted side effects of digoxin treatment. In the oral supplements, the potassium is usually formulated as the chloride salt.

FIGURE 8.4 Ampoules of Sodium Chloride BP, Water for Infections and Sterile Potassium Chloride Concentrate BP.
Reproduced from Lankshear et al. (2005) Evaluation of the implementation of the alert issued by the UK National Patient Safety Agency on the storage and handling of potassium chloride concentrate solution. 14: 3. with permission from BMJ Publishing Group Ltd.

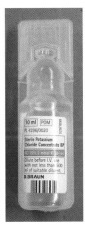

FIGURE 8.5 The amount of potassium in a banana.

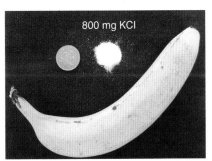

The normal dosage for oral potassium supplements is 2–4 g KCl per day, in divided doses. Calculate the weight (in g) and number of millimoles of potassium that a patient would receive in a day if they were prescribed 10 mL, three times daily, of a solution containing 7.5% w/v potassium chloride. Compare this with the amount of potassium contained in a banana.

Membrane potentials

Potassium ions inside cells work with extracellular sodium ions to maintain electrical potentials across cell membranes. The concentrations of Na$^+$ and K$^+$ ions inside cells are approximately 10 mM and 150 mM, respectively. If you compare these values with the values in Table 8.2, which refers to extracellular plasma concentrations, you will see that ratios of Na$^+$ to K$^+$ are reversed. In total, however, the concentrations of group 1 cations are high, both inside and outside cells.

Organic anions inside the cell balance the positively charged metal ions, whilst chloride ions provide the anions outside the cell. The concentration of positive ions outside the cell is slightly higher than that inside the cell, generating a small electrical potential across the cell membrane; the membrane potential for a resting neuron is about –70 mV. For information on calculating membrane potentials, see Box 8.2.

The correct balance between sodium and potassium ion concentrations both inside and outside cells is essential to maintain electrical potentials in nerves and muscles, and for these to work efficiently. Signals are transmitted along nerve fibres by the influx of sodium ions into the cell and a corresponding movement of potassium ions out of the cell. After the signal has passed the cell returns to its normal state due to the action of the Na$^+$/K$^+$ pump, which transports ions across the membrane against concentration gradients.

 For more about membrane potentials see Chapter 5 'Communication systems in the body – neural' in the *Therapeutics and Human Physiology* book in this series.

Calculation of membrane potentials

Electrical potentials across cell membranes are generated by the difference in concentrations of ions across the cell membrane. For individual ions the potential can be calculated from the Nernst Equation.

$$E = -\frac{RT}{zF} \ln \frac{[ion]_{in}}{[ion]_{out}}$$

Where R is the gas constant, T is the temperature in Kelvin, z is the net charge on the ion, [ion] is the concentration of the ion and F is the Faraday constant. For K$^+$ at body temperature this would give a value of –91 mV.

For resting membranes, the main contributor to the membrane potential is K$^+$, but there are small contributions from Na$^+$ and Cl$^-$. The overall potential can be calculated from the Goldman equation, which is beyond the scope of this book.

Lithium

Although lithium does not appear to be essential for human life, it is used to treat mania and recurrent depression. Despite being used since the 1950s, the exact mechanism of action is still unknown. There have been suggestions that it interferes with membrane transport of metal cations (including sodium ions), but this has not been proven. It has also been suggested that lithium blocks an enzyme pathway that uses magnesium ions and it has been observed that lithium ions (Li$^+$) and magnesium ions (Mg^{2+}) are of a similar size.

Lithium is usually formulated as the carbonate salt in tablets (Li$_2$CO$_3$) or as the citrate salt in liquids Li$_3$C$_6$H$_5$O$_7$.

Lithium carbonate tablets contain 200 mg of lithium carbonate. Calculate the weight of lithium citrate (Li$_3$C$_6$H$_5$O$_7$.4H$_2$O) required in a 5 mL spoonful of lithium citrate oral solution to produce the same number of mmoles of lithium as are present in a 200 mg tablet of lithium carbonate, Check your answer against the figures quoted for lithium preparations in the BNF.

8.4 **Group 2 metals**

Group 2 metals (magnesium, calcium, strontium, barium and radium) are referred to as the alkaline earth metals. They are less reactive than the alkali metals, and their salts are not always soluble in water. Group 2 metal salts with polyvalent anions – for example, carbonate, phosphate and sulfate – tend not to be soluble. The limit test for sulfates, which can be found in the British Pharmacopoeia, is based on the formation of insoluble barium sulfate under controlled conditions.

Insoluble calcium salts can give rise to medical problems and can decrease the absorption of drugs. For instance, kidney stones, which can cause severe pain, usually consist of calcium oxalate or calcium phosphate salts.

Biologically, the most important group 2 metals are calcium and magnesium. Magnesium ions tend to be concentrated inside the cells whilst calcium ions are mainly outside the cells.

Calcium

Over 95% of the calcium in the body is found in bone and teeth where it exists mainly as calcium phosphate salts (hydroxyapatite, $Ca(OH)_2.3Ca_3(PO_4)_2$). The concentration of calcium in plasma is about 2.5 mM, and, of this, 40–50% is bound to plasma proteins, particularly serum albumin. Only free calcium ions are biologically active so adjustments to the serum values determined need to be made if it is known that the patient has abnormally high or low levels of serum albumin.

 More information on phosphates can be found in Section 8.7 of this chapter.

Calcium ions are involved in blood clotting, transmission of nerve signals and muscle contraction. When muscle cells are stimulated there is a rapid influx of sodium ions, which produces an action potential. There follows an influx of calcium ions, which maintains the action potential and leads to contraction of muscle. Calcium channel blockers such as verapamil

and nifedipine prevent the entry of calcium ions into cells, preventing contraction of vascular smooth muscle and conduction of signals in myocardial cells. Both these drugs are used to treat several conditions; for example, verapamil is used to treat hypertension and nifedipine to treat angina.

Adults require a daily intake of approximately 700 mg calcium for maintenance of bone and plasma levels. This is increased for children, nursing mothers and in pregnancy. Dairy products are a good source of calcium, which is absorbed from the gastrointestinal tract. However, although calcium ions themselves are well absorbed they can interfere with the absorption of drugs. For example, they can form chelates with tetracycline and quinolone antibiotics, which cannot be absorbed. As a consequence, patients are advised to take tetracycline antibiotics one hour before food or two hours after meals to prevent the interaction, which can reduce absorption of the antibiotic by up to 80%.

The chelation reaction between tetracycline and calcium can have other serious effects. Tetracyclines should not be prescribed in the first trimester of pregnancy as they can affect bone development in the foetus. If possible, they should also be avoided later in pregnancy and in childhood, as they can cause serious discolouration of teeth.

Although chelation reactions may cause problems with drug interactions they are useful in pharmaceutical analysis. For instance, the British Pharmacopoeia uses a titration with ethylenediamine tetracetic acid (EDTA), shown in Figure 8.6, to assay calcium gluconate tablets, which are used to

FIGURE 8.6 **Ethylenediamine tetracetic acid (EDTA).**

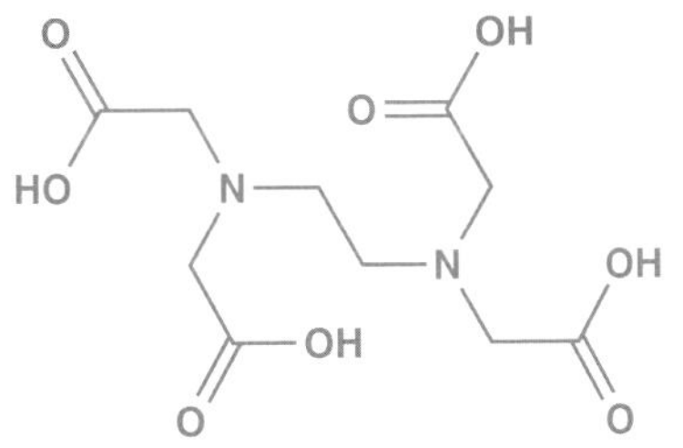

treat calcium deficiency. The reaction is a 1:1 reaction and the titration is carried out at a high pH, as the complexes are unstable at low pH.

SELF CHECK 8.7

Draw the structure of the quinolone antibiotic ciprofloxacin and identify a part of the structure that could be responsible for the interaction with Ca^{2+} ions. Are there any functional groups common to EDTA and ciprofloxacin?

Magnesium

There are approximately 25 g of magnesium in the average adult. Half of this is found in bone, with most of the remainder being inside cells. Magnesium ions are important cofactors for enzymes and are therefore required for normal body function. There are about 500 kinases in the human body, which catalyse the transfer of a phosphate group from adenosine triphosphate (ATP) to an acceptor. Kinases typically contain magnesium ions, which complex with the ATP.

Magnesium levels within the body are carefully controlled and absorption from the gut is slow. The fact that magnesium salts are not well absorbed from the gastrointestinal tract has meant that compounds such as magnesium sulfate, commonly called Epsom salts, have been widely used as osmotic laxatives.

INTEGRATION BOX 8.1

Antacids

Many of us suffer from dyspepsia (indigestion and heartburn), creating a large market for antacid preparations to relieve symptoms. Most of these medicines contain salts of aluminium or magnesium, which react to neutralize excess acid in the stomach. In general, magnesium-containing antacids tend to have a laxative effect whilst those containing aluminium may be constipating. Not surprisingly, therefore, preparations such as Co-magaldrox, which contain both magnesium and aluminium ions, are available.

SELF CHECK 8.8

Write equations for the reactions of aluminium hydroxide and magnesium hydroxide with hydrochloric acid. Use equations to explain why antacids that contain magnesium carbonate may make patients belch.

KEY POINT

Group 1 and group 2 metal ions play a vital part in maintaining normal cell functions. Imbalances in concentrations of these ions usually need to be corrected as soon as possible. Concentrations of ions in body fluids and medicines are expressed in a variety of units and you need to be very competent in working between weight per litre, percentage solutions and molarity.

8.5 Period 4 metals

In this section, we study the metals zinc and iron in more detail. These metals are frequently known as transition metals, and some of their electrons are contained in d-orbitals. Look at Table 8.4 and notice how zinc actually has a complete set of d-electrons, and its chemistry is much simpler (and less colourful) than that of iron, copper or vanadium; for this reason inorganic chemists do not usually consider it to be a transition metal. Nevertheless, zinc, like iron, has important functions in biological systems and it is the

ability to use the d-orbitals that explains many of the biological activities of these metals. Some basic information about three of the period 4 elements is shown in Table 8.4.

Period 4 chemistry is greatly influenced by d-orbitals. There are five d-orbitals, each able to hold two electrons. Four of these have dumb-bell type shapes, each with four lobes all of which lie in a plane. The fifth orbital, the so-called d_{z^2} orbital, is different in that it has a dumb-bell shaped orbital running along

TABLE 8.4 Properties of some period 4 elements

Element	Symbol	Atomic number	RMM	Electron distribution
Iron	Fe	26	55.85	$1s^2 2s^2 2p^6 3s^2 3p^6 3d^6 4s^2$
Copper	Cu	29	63.55	$1s^2 2s^2 2p^6 3s^2 3p^6 3d^{10} 4s^1$
Zinc	Zn	30	65.39	$1s^2 2s^2 2p^6 3s^2 3p^6 3d^{10} 4s^2$

FIGURE 8.7 Shapes of p and d orbitals.

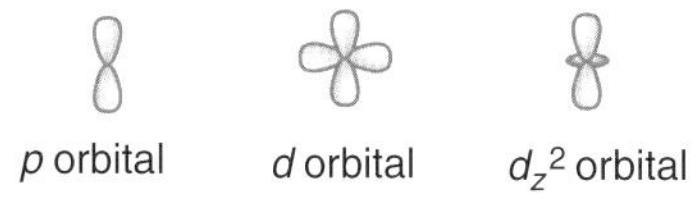

p orbital d orbital $d_z{}^2$ orbital

the *y*-axis and a doughnut shaped Taurus round the middle. Figure 8.7 shows a p orbital, the typical arrangement of the first four d orbitals and the shape of the $d_z{}^2$ orbital.

 See Chapter 2 'Organic structure and bonding' to remind yourself about s and p orbitals and hybridization of orbitals. There are also associated materials on the Online Resource Centre.

Iron

Our bodies contain about 4 g of iron, which is essential for life. It is probably best known as a carrier of oxygen in haemoglobin, but it is also found in cyto-chrome enzymes and in electron chain reactions. As iron is such an important component of the body, iron deficiency must be corrected by its administration (see Case Study 8.1).

Biological activity associated with redox reactions

About 70% of the iron found in the body is found as haemoglobin (see Figure 8.8), where it acts to bind oxygen molecules, allowing blood to transport oxygen round the body. Haemoglobin comprises four peptide chains (globins), each of which has a haem group (a porphyrin) bound. This arrangement relies on the iron (Fe^{2+}) ion being able to form six dative covalent bonds. Four of the bonds are to nitrogen atoms in pyr-role rings in porphyrin, the fifth bond is to a nitrogen in a histidine ring in the globin, and the sixth dative covalent bond is to either oxygen (O_2) or to a water molecule. The bond between oxygen and iron is weak, allowing haemoglobin to give up its oxygen easily

219

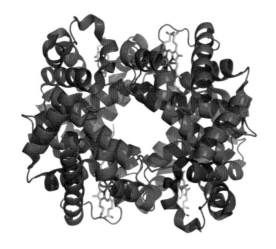

when required – hence making haemoglobin an efficient biological transporter of oxygen.

Figure 8.9 shows the structure of haem, illustrating the bonds between the pyrrole rings and Fe^{2+}. Also shown are simplified diagrams showing the overall arrangement of iron in haemoglobin and cytochrome c.

The bright red colour of blood is due to formation of a **complex** between the Fe^{2+} ion and the porphyrin ring. Oxidation of Fe^{2+} to Fe^{3+} in haemoglobin occurs when blood is exposed to air, resulting in a colour change from red to brown. Although iron is absorbed and largely used in its iron(II) oxidation state, it is stored in ferritin as the iron(III) state.

Iron is also an essential component of cytochrome enzymes, which are involved in respiration and drug metabolism. **Cytochromes** have a similar structure to haemoglobin with the iron ion being bonded to four nitrogen atoms from porphyrin. The fifth bond is to a nitrogen atom from an associated protein. The difference between cytochromes and haemoglobin is that the sixth bond can be to other proteins or to a sulfur atom, as in cytochrome c (see Figure 8.9).

The cytochromes involved in respiration belong to three groups, a, b and c. Each cytochrome has a slightly different **reduction potential** (i.e. a slightly different tendency to acquire electrons and be reduced), because the haem units are associated with different polypeptides. In the respiration cycle, these are arranged in the order b, c_1, c, a, a_3 so that the energy obtained from the oxidation of glucose is released gradually and in a controlled manner. A simplified description of the electron transfer chain follows and is shown in Figure 8.10.

Initially, electrons are passed from reduced complex Q to cytochrome b, reducing Fe^{3+} to Fe^{2+}. In the next stage of the chain the Fe^{2+} in cytochrome b is oxidized back to Fe^{3+}, reducing the Fe^{3+} of cytochrome c_1 to Fe^{2+}. These cycles are repeated through cytochrome c, cytochrome a and cytochrome a_3. The final stage of

FIGURE 8.9 The structure of haem and cytochrome and the interaction of the iron ion with the pyrrole rings of porphyrin, globin, water molecules, and sulfur.

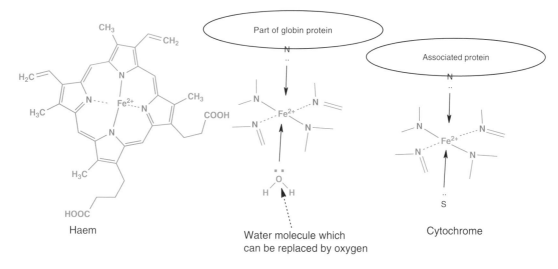

FIGURE 8.10 Electron transport chain for cytochromes.

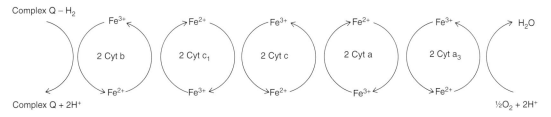

the process sees the oxidation of Fe^{2+} in cytochrome a_3 back to Fe^{3+}, with the electrons released being used in the reduction of oxygen to water. The energy released by this process can be stored by the body as ATP; see Section 8.7.

 The electron transport chain is also discussed in Chapter 3 'The biochemistry of cells' in the *Therapeutics and Human Physiology* book in this series.

Biological activity associated with ligand binding

The cytochrome P450s (CYP450) are a family of enzymes that play an important role in the metabolism of drugs. The name P450 is based on the fact that their complexes with carbon monoxide (CO) absorb light at approximately 450 nm. Cytochrome P450s are found in many tissues in the body, but the liver, kidney and intestinal enzymes are especially important. They all have a similar structure, with the iron, Fe^{2+}, coordinated to four nitrogens of haem and the thiol group of a cysteine residue, which can act as an electron donor (see Figure 8.9). The sixth coordination position is occupied by water or other ligands which can be easily exchanged.

The role of CYP450 is usually described as activation of O_2 to allow oxidation of xenobiotics. Oxidation renders these molecules more water soluble, so that they are more easily eliminated from the body. Inhibition of CYP450s is a major cause of drug interactions, because CYP450s are not very specific and many drugs are processed on the same enzyme. For example, the anti-ulcer H_2 antagonist cimetidine binds reversibly to the iron in a number of CYP450s, including CYP450 2C9, which metabolizes warfarin. This interaction can prevent the metabolism of warfarin, leading to high levels of warfarin accumulating, which can result in excessive bleeding.

 There is more about warfarin metabolism in Chapter 3 'Stereochemistry and drug action'.

SELF CHECK 8.9

The structure of cimetidine is shown below in Figure 8.11. Which parts of the cimetidine molecule do you think could be responsible for cimetidine binding to iron?

KEY POINT

The activity of iron in the body depends on the fact that it can:

- act as a Lewis acid accepting electrons from electron rich ligands
- take part in redox reactions.

Zinc

The human body contains about 2 g of zinc; many enzymes have Zn^{2+} ions in their active sites. Medicines containing Zn^{2+} ions have been used for many years. They are said to promote healing and there are frequent media reports that they can help cure the common cold.

We will consider the angiotensin-converting-enzyme (ACE) to illustrate how the chemistry of zinc relates to its biological role. ACE is a metalloprotease

FIGURE 8.11 The structure of cimetidine.

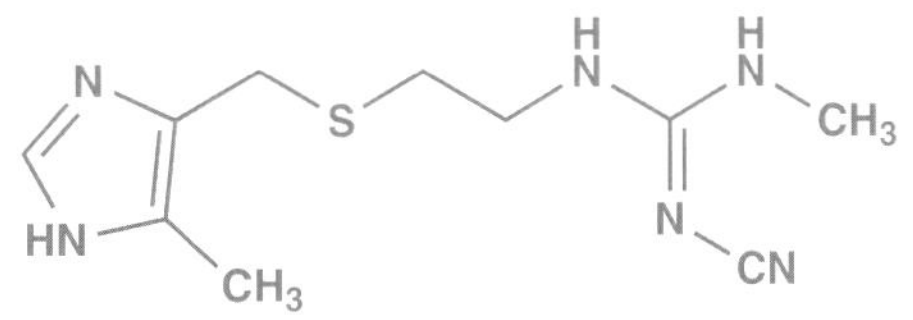

FIGURE 8.12 Diagram showing how Zn²⁺ ions are held in place in ACE.

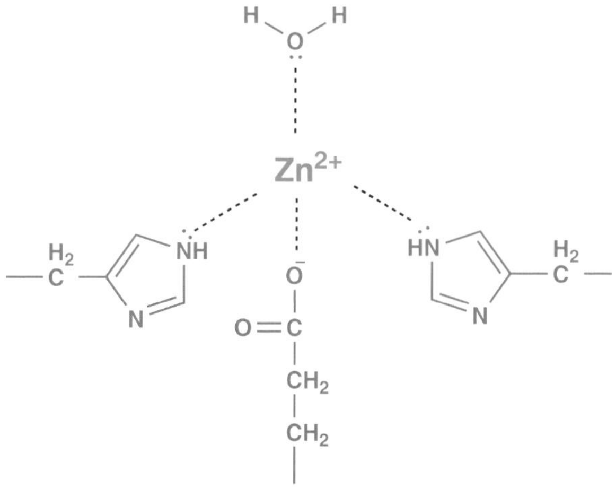

enzyme (i.e. it contains a metal ion), and is responsible for the hydrolysis of peptide (amide) links. **ACE inhibitors** are frequently used to treat hypertension and heart failure.

In biological systems, zinc exists as the Zn^{2+} ion, having lost its two $4s^2$ electrons. It is able to form complexes with four ligands in a tetrahedral shape. In ACE, the zinc ion is coordinated to the nitrogens of two histidine residues (see Figure 8.12) and the carboxylate group from an acidic amino acid. The fourth ligand is an activated water molecule. ACE is responsible for the hydrolysis of a peptide bond in angiotensin I resulting in the removal of a dipeptide to give angiotensin II. The zinc ion activates the bound water molecule so that it attacks the carbonyl group of the peptide link more readily; it is this activation that acts to catalyse the hydrolysis.

The hydrolysis of peptide bonds is discussed in Chapter 6 'The carbonyl group and its chemistry'.

All ACE inhibitors contain a functional group that is capable of interacting with the zinc ion. The thiol group (S–H) is found in captopril, the carboxylate group is found in enalaprilat and a phosphinyl group (P=O) is found in fosinoprilat.

SELF CHECK 8.10

Draw the structures of captopril, enalaprilat and fosinoprilat and highlight relevant groups that interact with the zinc ion of ACE.

The interaction between the inhibitors and the zinc ion involves ligand bonding between the lone pairs of electrons on the relevant functional group and the Zn^{2+} ion.

KEY POINT

The biological role of zinc in the body depends on the fact that zinc ions can interact with groups which have an excess of electrons.

8.6 Precious metals

You might be surprised to find that precious metals such as silver, gold and platinum have a role in medicine. They are traditionally thought of as being non-reactive, and, indeed, their use in jewellery relies on their low reactivity, however their lack of reactivity does not preclude them having a role in biological systems.

Silver

Silver nitrate sticks have been used for many years to treat warts, but silver ions (Ag^+) also possess antibacterial activity. Silver salts such as silver sulfadiazine are therefore used in dressing wounds that might be infected. The exact mechanism of action of silver ions is not yet known but may be related to their ability to interact with thiol groups in bacterial protein (see Section 8.8).

Gold

Gold in the form of sodium aurothiomalate, whose structure is shown in Figure 8.13, is used to treat active progressive rheumatoid arthritis. Once again the exact mechanism of action has not yet been fully established. It would appear that the compounds are most effective when the gold is in the monovalent form (Au^+, aurous) which is attached to a sulfur ligand.

FIGURE 8.13 Sodium aurothiomalate.

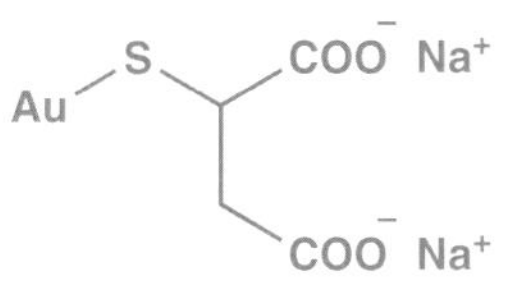

Platinum

The last precious metal to consider is platinum, which is used in drugs such as cisplatin, a cancer therapeutic. In contrast to the silver and gold compounds, the mechanism of action of platinum compounds is well understood. Cisplatin is a complex formed between oxidized platinum, chlorine and ammonia. In this form, two chlorine atoms and two ammonia molecules are bonded to the platinum in a square planar complex as shown in Figure 8.14. This arrangement leads to the possibility of geometrical isomerism; the two chlorine atoms can either be next to each other, a *cis* arrangement, or opposite each other, a *trans* arrangement. Only the *cis* arrangement (cisplatin) makes an active anticancer drug, and examination of the mechanism of action explains why.

Inside the cell the chloro groups of cisplatin are replaced by water molecules. As the water molecules are neutral, and the leaving chloride ions take both electrons from each Pt–Cl bond, the complex now carries a positive charge. This positive charge makes the complex very reactive, facilitating its binding to nitrogen and oxygen atoms in guanine residues within DNA. Cisplatin acts to form links both within and between DNA strands, with the intrastrand links predominating. Other organoplatinum compounds with the same mode of action, such as carboplatin, have also been developed.

SELF CHECK 8.11

Draw a structure for the *trans* isomer of cisplatin and explain why it would not be active.

Information on *cis* and *trans* alkenes can be found in Chapter 4 'Properties of aliphatic hydrocarbons'.

Many other metals, including vanadium and copper, are important to health and can have biological activity but there is not space to discuss them all. You should consider whether their reactions depend on redox type reactions or complex formation. The metals may function as ions or as the element itself. For instance, intrauterine contraceptive devices (IUDs) appear to have their effect through the action of copper on sperm although, once again, the exact mechanism of action is not yet known.

FIGURE 8.14 The mechanism of action of cisplatin.

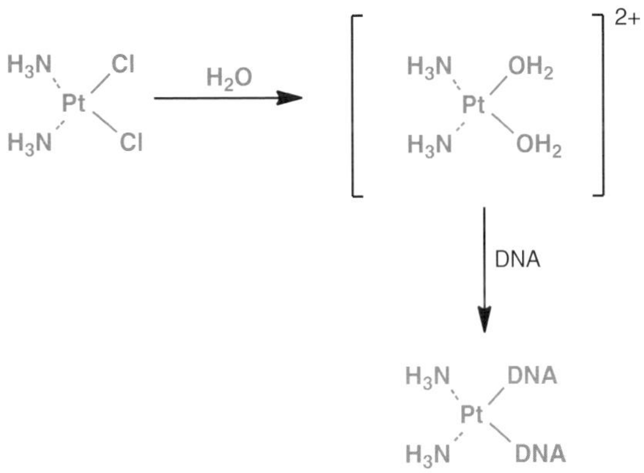

8.7 Phosphorus

Phosphorus, like nitrogen, is a group 15 element in the IUPAC system, but while nitrogen is a stable gas, phosphorus can exist in several solid forms. The most common of these, white phosphorus, is far from stable and reacts vigorously with oxygen. Consequently, white phosphorus is stored under water to prevent reaction with oxygen.

In the body, phosphates are part of DNA, ATP, the buffering system for blood, and in bone in the form of calcium hydroxide phosphate, hydroxyapatite $(Ca_{10}(PO_4)_6(OH)_2)$. Phosphorus in the form of phosphates is found widely in pharmacy. Phosphate esters can be produced to give solubility; phosphorus insecticides can be used to treat head lice; and phosphonates are used in the treatment of osteoporosis.

The electronic configuration of phosphorus is $1s^2 2s^2 2p^6 3s^2 3p^3$ and it can exist in two oxidation states; the III and the V state. In the III oxidation state only the 3p electrons are involved in bonding, and phosphorus is trivalent. By contrast, in the V state both the 3s and the 3p electrons are involved in bonding and the phosphorus is pentavalent. In order to have five valencies we need a situation in which there are five single electrons. This is achieved by using a 3d orbital as well as the 3s and 3p orbitals. This results in the production of five sp^3d hybrid orbitals each with one electron and each with the same energy. These orbitals can be used to form covalent bonds, for example in phosphorus pentachloride (PCl_5).

 Chapter 2 'Organic structure and bonding' deals with the formation of hybrid orbitals.

In pharmacy and related disciplines, we normally encounter phosphorus in derivatives of orthophosphoric acid, more widely known as phosphoric acid

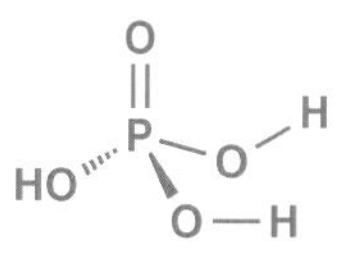

TABLE 8.5 Ionization reactions and pK_a values for the separate ionizations of acid.

Ionization	pK_a
$H_3PO_4 \rightleftharpoons H^+ + H_2PO_4^-$	1.96
$H_2PO_4^- \rightleftharpoons H^+ + HPO_4^{2-}$	6.80
$HPO_4^{2-} \rightleftharpoons H^+ + PO_4^{3-}$	~12

(H_3PO_4). This molecule consists of phosphorus that forms single bonds to three OH groups and a double bond to the remaining oxygen. The phosphoric acid molecule is tetrahedral as shown in Figure 8.15. The hydrogens attached to the oxygens are acidic such that each of the OH groups of phosphoric acid can act as an acid. Each has a separate pK_a; these values are shown in Table 8.5. Phosphoric acid is similar to a carboxylic acid in that it can produce esters, anhydrides and amides.

At physiological pH (pH 7.4), the most important ionization is the second ionization, which sets up an equilibrium between dihydrogen phosphate ($H_2PO_4^-$) and hydrogen phosphate (HPO_4^{2-}). This equilibrium enables phosphate to buffer solutions at or close to pH 7.4.

 More information about buffer solutions can be found in Chapter 6 'Acids and bases' in the *Pharmaceutics* book in this series.

When phosphoric acid is neutralized, it forms phosphate salts that can be used in medicines. Some basic drugs are formulated as water-soluble phosphate salts – for example, the analgesic, codeine phosphate (see Figure 8.16).

Phosphoesters

Phosphoric acid, like carboxylic acids, can form esters with alcohols. Each OH group can react independently to form phosphomonoesters, phosphodiesters and

FIGURE 8.15 Phosphoric acid.

FIGURE 8.16 Codeine phosphate.

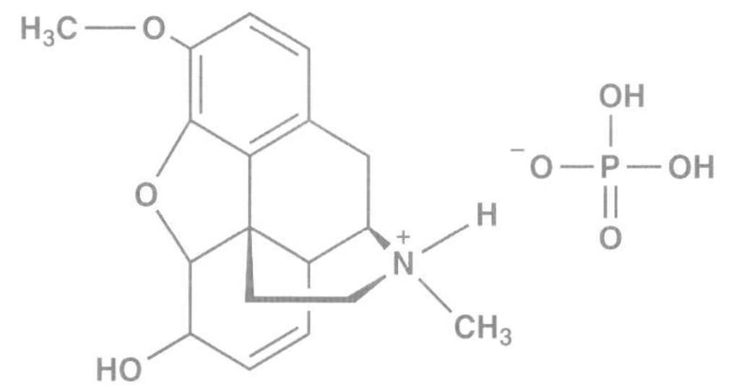

phosphotriesters. These, collectively, are also known as phosphate esters or organophosphates. In a phosphomonoester, one OH of the phosphate is replaced by OR (from an alcohol). The glycolysis intermediate glucose-6-phosphate is an ester formed between phosphoric acid and the OH in the 6-position of glucose. (Glucose is the alcohol.) The phosphoester group is shown in blue in Figure 8.17.

 Glycolysis is also discussed in Chapter 3 'The biochemistry of cells' in the *Therapeutics and Human Physiology* books in the series.

In phosphomonoesters, the remaining OH groups are still capable of ionization. This property can be put to good use in the production of ionized derivatives, which are water soluble. For example, the steroid dexamethasone, which can be used to treat cerebral oedema associated with malignancy, is formulated for injection as the sodium phosphate derivative. A structure of dexamethasone is shown in Figure 8.17; again the phosphoester group is shown in blue.

When phosphoric acid reacts with two alcohol groups, a diester is formed. A good example of a phos-

phodiester is present in the structure of DNA, as shown in Figure 8.18, where the deoxyribonucleic acid units are connected through a phosphate group. The diester part is shown in blue.

Phosphoric acids can form anhydrides as well as esters. (You might anticipate this, because of their similarity to carboxylic acids.) These derivatives are diphosphoric acid (two molecules joined together), and triphosphoric acid (three molecules joined together). These condensed phosphates are formed by the loss of water, which is why they can be considered to be anhydrides. Look at Figure 8.19 to see the structures of di and triphosphoric acid; the anhydride part is highlighted in blue.

 More details on carboxylic acid anhydrides can be found in Chapter 6 'The carbonyl group and its chemistry'.

Examples of di- and triphosphates are seen in adenosine diphosphate (ADP) and ATP. These anhydrides (like carboxylic acid anhydrides) are easily hydrolysed, leading to the release of considerable amounts of energy: 34 kJ mol^{-1} for the conversion of ATP to ADP and the same again for ADP to AMP. The body stores energy in the form of ATP and ADP and releases it when required.

 The mechanism of the hydrolysis of ATP was illustrated in Chapter 1 'The importance of pharmaceutical chemistry'.

ATP and ADP are part of the normal biochemistry of the body, but derivatives of phosphoric acid are used in medicines. One example is malathion, a phosphorus

FIGURE 8.17 Glucose-6-phosphate and dexamethasone sodium phosphate.

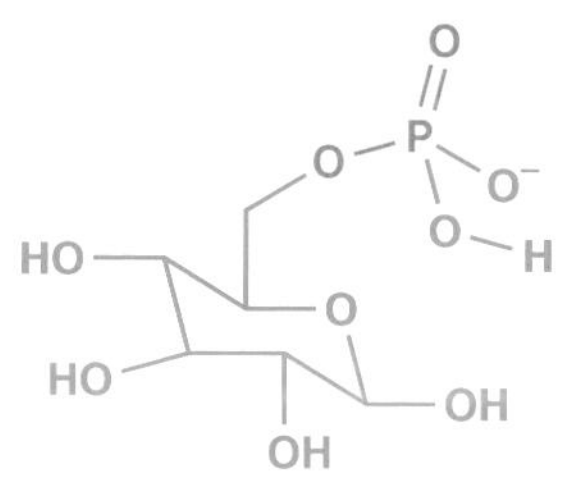

Glucose-6-phosphate

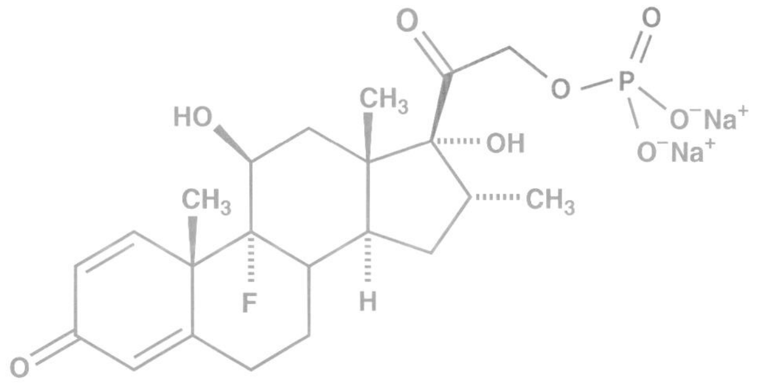

Dexamethasone sodium phosphate

FIGURE 8.18 DNA showing how a phosphodiester links two nucleosides together.

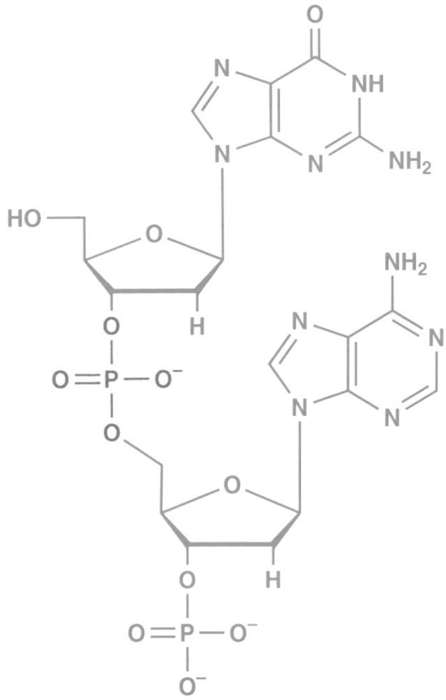

lower electron density on the phosphorus making it susceptible to attack by nucleophiles. A covalent bond is formed between the oxygen of the enzyme and the phosphorus of the insecticide, which inactivates the enzyme.

SELF CHECK 8.12

Identify the anhydride parts of the ATP in Figure 8.20. What is the normal function of ATP in the body?

At first glance, malathion may appear to be inactive because it contains a phosphorus-sulfur double bond rather than a phosphorus-oxygen double bond. Sulfur is less electronegative than oxygen so a P=S bond is less polarized than a P=O bond (and therefore less susceptible to attack by a nucleophile). Insects, however, metabolize malathion by replacing the sulfur with oxygen, whereas mammals hydrolyse the carboxylic ester to produce inactive compounds, which are excreted. The **selective toxicity** of malathion to insects, and not humans, is achieved through the different metabolic pathways in insects and mammals. The reactions showing the different metabolic routes and the reactions resulting in inhibition are shown in Figure 8.21. Nerve gases, such as sarin, attack

insecticide used to treat head lice. Malathion inhibits insect acetylcholinesterase by reacting with a serine residue at the active site of the enzyme. The reaction depends on the fact that phosphorus-oxygen double bonds (P=O) are polarized because the oxygen is more electronegative than phosphorus. There is a

FIGURE 8.19 Diphosphoric and triphosphoric acid.

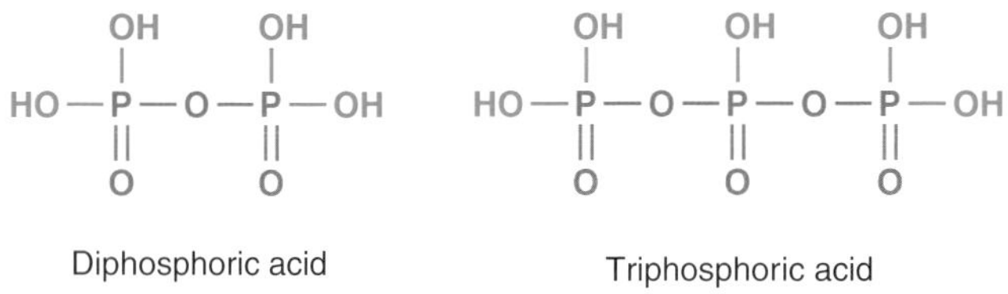

Diphosphoric acid

Triphosphoric acid

FIGURE 8.20 The structure of ATP.

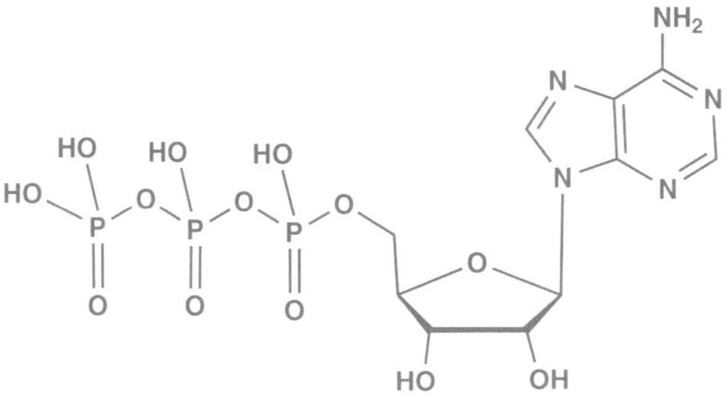

FIGURE 8.21 Metabolism and actions of malathion.

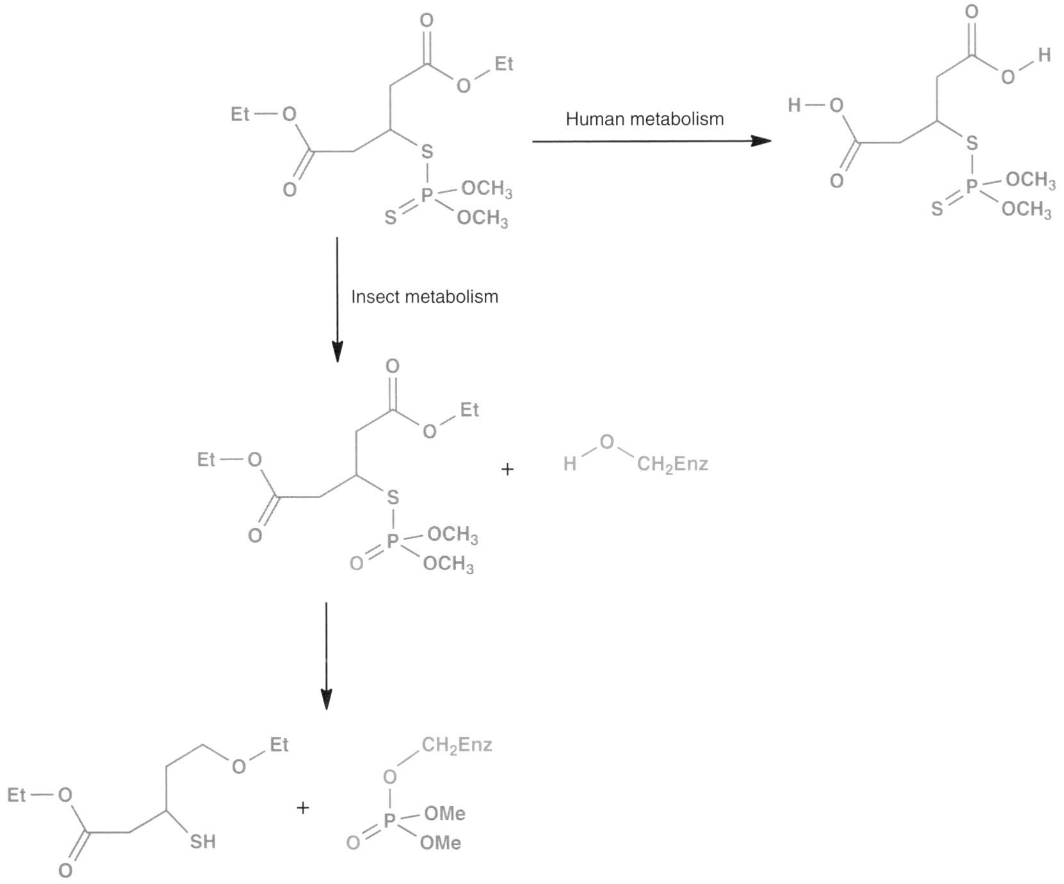

Inactivated insect acetylcholinesterase

acetylcholinesterase through a similar reaction, but in this case there is no selective toxicity.

SELF CHECK 8.13

Using the reaction between malathion and acetyl-cholinesterase shown in Figure 8.21 as an example, draw the mechanism for the interaction between the hydroxyl group of the enzyme and the phosphorus-oxygen double bond leading to inactivation of the enzyme.

Phospholipids are a major component of cell membranes; they are derivatives of glycerol in which one of the hydroxyl groups forms a phosphate ester. The other hydroxyl groups in glycerol form esters with fatty acids. The phosphate group usually exists as a diester with one of the other OH groups being esterified with a hydrophilic alcohol (for example, choline) to give lecithins, or ethanolamine to give cephalins. The other OH from the phosphate is ionized at physiological pH.

The ionized phosphate group of a phospholipid forms a polar, hydrophilic 'head', which is found on the outside of membranes (exposed to the aqueous surroundings). By contrast, the hydrocarbon chains from the fatty acids give phospholipids their hydrophobic lipid tails, which are directed towards the inside of the membrane.

A structure of a typical lecithin is shown in Figure 8.22, along with a diagram of a membrane showing the phospholipid bilayer, Figure 8.23.

Bisphosphonates

In the phosphorus compounds that we have studied so far, the phosphorus has been directly bonded to oxygen or sulfur. However, the bisphosphonates are a class of compound in which phosphorus bonds

FIGURE 8.22 A lecithin.

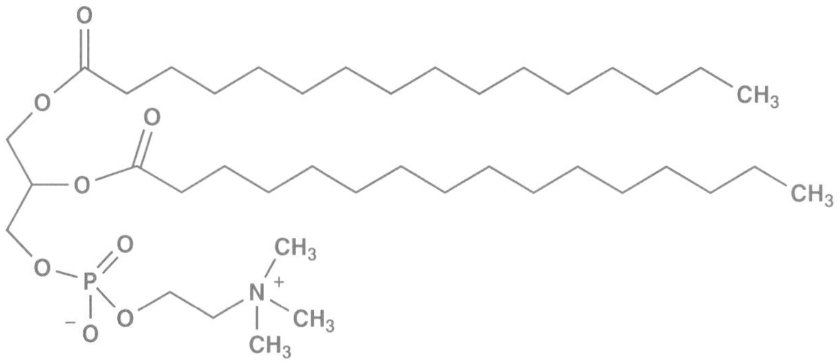

FIGURE 8.23 Phospholipid bilayer.
Adapted from Human Physiology: The Basis of Medicine. 3rd edn, by Gillian Pocock and Christopher D. Richards (2006) by permission of Oxford University Press.

to carbon. In bisphosphonates, a carbon atom bonds to two phosphorus atoms. (Take care not to confuse bisphosphonates with biphosphates, which are phosphoesters.) Alendronic acid is a bisphosphonate, which can be used to treat osteoporosis by mimicking pyrophosphoric acid in the body; both structures are shown in Figure 8.24.

Pyrophosphoric acid is another name for diphosphoric acid and is part of the hydroxyapatite component of bone. Bisphosphonates are adsorbed onto hydroxyapatite where they cannot be easily hydrolysed (because they are not esters). This leads to a decrease in the rate of bone turnover, allowing well mineralized bone to be formed. Alendronic acid is usually formulated as the sodium salt, sodium alendronate, and is taken by patients once a week.

SELF CHECK 8.14

Sodium alendronate is reported to be poorly absorbed from the gastrointestinal tract and patients are advised to take these tablets at least 30 minutes before breakfast. Looking at the structure, shown in Figure 8.24, why do you think this is?

FIGURE 8.24 Alendronic acid and pyrophosphoric acid.

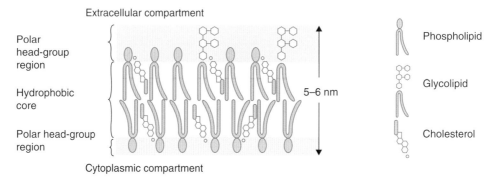

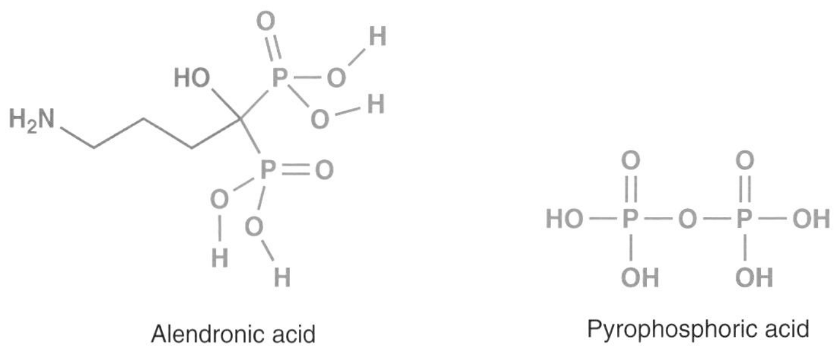

Alendronic acid   Pyrophosphoric acid

SELF CHECK 8.15

Many of the phosphorus containing substances have very similar names, or structures. Complete Table 8.6 by identifying whether the compounds listed are salts, phosphoesters, phosphorus anhydrides or bisphosphonates.

KEY POINT

In all the examples that we have looked at, phosphorus has an oxidation number of 5. Its biological activity and use in pharmacy depend on the fact that it can form a double bond with oxygen and still have O-H groups that can lose hydrogens to give salts or esters.

TABLE 8.6 **Phosphorus compounds in pharmacy**

Compound	Designation	Biological function or therapeutic use
e.g. ADP	phosphorus anhydride	mammalian energy store
chloroquine phosphate		
hydrocortisone sodium phosphate		
risedronate		
cyclophosphamide		

8.8 **Sulfur**

Sulfur and oxygen are both group 16 elements so it should be of little surprise that some of sulfur's chemistry is similar to that of oxygen. However, there are important differences: an atom of sulfur is larger, less electronegative, and has empty 3d orbitals available. Like phosphorus, sulfur is widely distributed in nature, including the human body, and is a constituent of drug molecules.

The electronic configuration of sulfur is $1s^2 2s^2 2p^6 3s^2 3p^4$ and it can exist in oxidation states ranging from −2 to +6. Table 8.7 shows the oxidation states of sulfur and gives pharmaceutical examples of these oxidation states. Sulfur's ability to exist in so many oxidation states stems from the way it can use d orbitals to share electrons from oxygen. In the rest of the chapter we shall include examples from earlier in the chapter and try to relate the biological activity to the chemistry.

SELF CHECK 8.16

Look up the structures of the compounds in Table 8.7. Draw them, highlight the relevant functional group and look up their pharmaceutical use.

Thiols

In thiols, sulfur is in the −2 oxidation state and, like oxygen in alcohols, it has accepted electrons from

TABLE 8.7 **Oxidation states of sulfur and pharmaceutical examples**

Oxidation state	Group	General structure	Biological/Pharmaceutical example
−2	thiol	R–S–H	captopril
0	sulfoxide	R–SO–R	sulindac
+2	sulfone	R–SO$_2$–R	dapsone
+4	sulfonate ester	R–SO$_2$O–R'	busulfan
+6	sulfate esters	RO–SO$_2$–OR	sodium dodecyl sulfate

other atoms. Much of the chemistry of thiols and thiophenols is similar to that of alcohols and phenols, although they are stronger acids because the larger sulfur atom is better able to spread the negative charge of the conjugate base, making it easier for a proton to be extracted. Like alcohols, thiols can react with alkyl halides to form thioethers and with acid chlorides to form thioesters, as shown in Figure 8.25.

The S–H group is sometimes called a sulfhydryl group as well as a 'thiol'. The old name for thiols was mercaptans and this is still found in the name of pharmaceutical compounds, such as the anticancer drug mercaptopurine. (Strictly, this drug is a thiophenol since the S–H group is attached directly to the aromatic ring).

Low molecular weight thiols are foul smelling – none more so than hydrogen sulfide, H_2S, which gives the characteristic smell to rotten eggs. For this reason, thiols are added to the domestic gas supply to allow us to detect leaks. Drugs containing sulfhydryl groups have an unpleasant taste.

To illustrate how the chemistry of the S–H group can explain the action of drugs and some of their side effects, we shall again consider the ACE inhibitor captopril. In Section 8.5, we saw that ACE contains a zinc ion at the active site. Thiols react readily with metal ions to form complexes, and captopril was developed by replacing a carboxylate group in a succinylproline derivative (the first inhibitor of ACE to be synthesized) with a thiol group, which has greater affinity for the Zn^{2+} ion. The structure of this derivative is shown in Figure 8.26.

Interestingly, patients who are given captopril may complain of a metallic taste in their mouth and this is attributed to the presence of a thiol group.

 See Chapter 10 'Origins of drug molecules' for more details on the development of captopril.

Thiols are able to form disulfides (R–S–S–R'), which are more stable than the analogous peroxides R–O–O–R'. Disulfides may be formed in the laboratory from mild oxidation of thiols but disulfides are also widely found in protein molecules. The tertiary structure of protein molecules is partly determined by the presence of disulfide links between the sulfur-containing amino acid residues of peptide chains. In the example shown in Figure 8.27, the amino acid residues are cysteine, which contains sulfur as a thiol. Disulfide bonds

FIGURE 8.25 Formation of thioethers and thioesters.

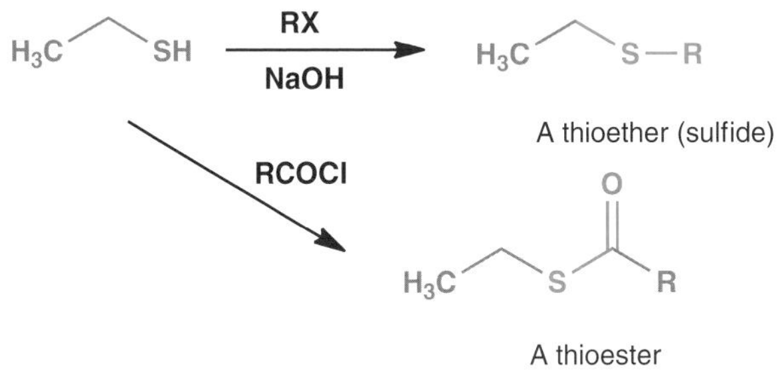

A thioether (sulfide)

A thioester

FIGURE 8.26 Succinyl derivative of proline and captopril.

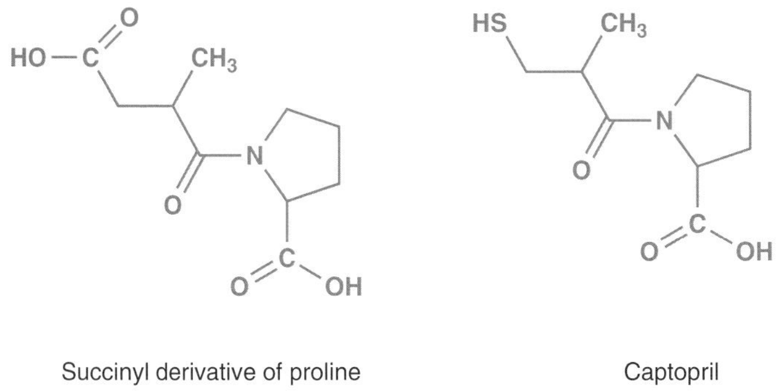

Succinyl derivative of proline

Captopril

FIGURE 8.27 (A) The structure of cysteine with the thiol group in green; (B) the structure of insulin showing the amino sequences of its composite chains, with the cross-linking of the chains via cysteine residues; (C) a diagram showing the links at a more detailed level. The disulfide bond is shown in green, the cysteine residues in pink.

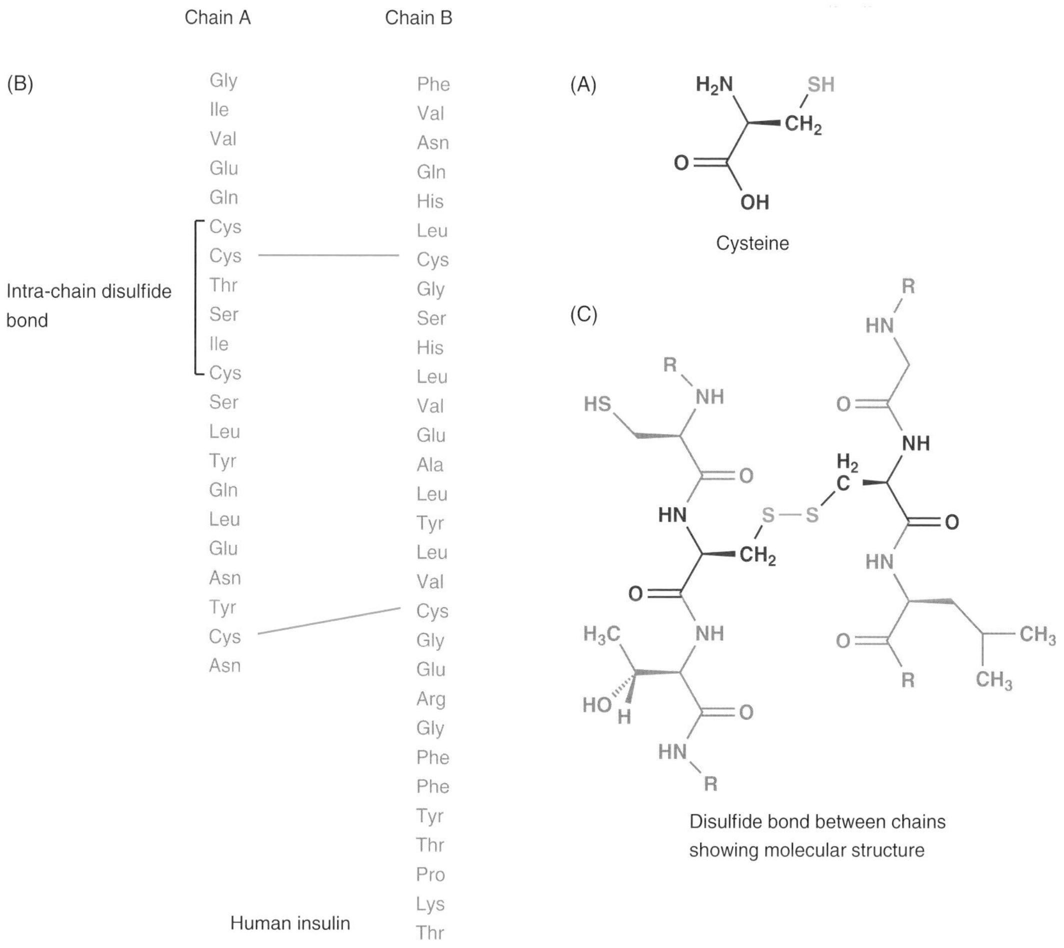

are relatively weak and can be broken and reformed relatively easily.

Thioethers (sulfides)

Thioethers are more reactive than the corresponding oxygen ethers and they act as nucleophiles: one of the lone pairs of electrons on the sulfur attacks an electrophilic centre (where there is lower electron density). In the body, the sulfur of methionine can react with ATP, as shown in Figure 8.28. This reaction results in the formation of S-adenosylmethionine (SAM), which is a very good biological methylating agent and is used by the body to add methyl groups to metabolites and to xenobiotics. For example, the enzyme catechol-O-methyltransferase transfers a methyl group onto

one of the phenolic hydroxyl groups of catecholamines, such as the neurotransmitter noradrenaline. This inactivates noradrenaline, preventing it from interacting with β_2–adrenoceptors in the lungs. Airways are dilated when these receptors are occupied.

The formation of SAM is shown in Figure 8.28 and its subsequent use to transfer a methyl group to the hydroxyl group of noradrenaline is shown in Figure 8.29. Bronchodilators such as salbutamol and terbutaline are not catechols and therefore are not metabolized by this route.

Sulfoxides, sulfones and sulfonates

Stepwise oxidation of thioethers produces sulfoxides (S=O), and then sulfones (S(=O)$_2$) as shown in

FIGURE 8.28 The formation of S-adenosylmethionine (SAM).

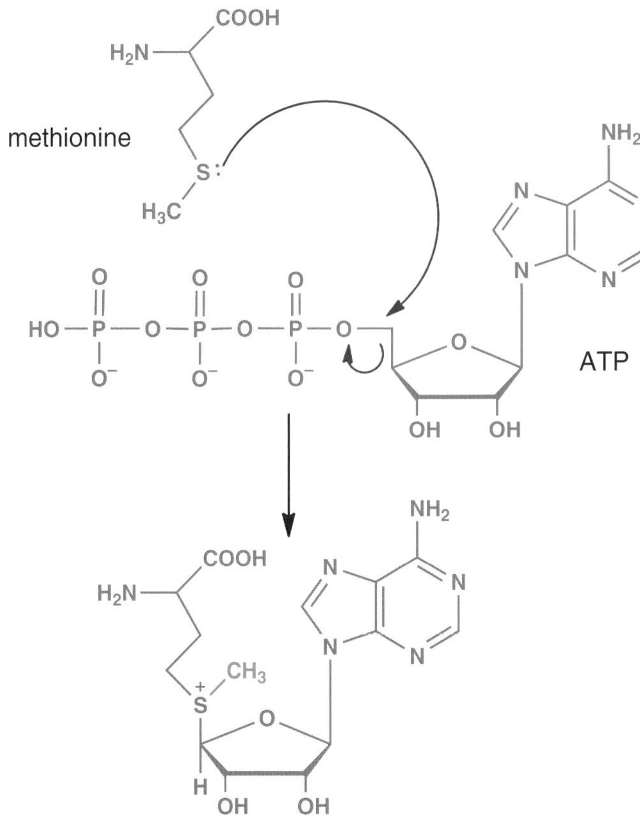

S-adenosylmethionine (SAM)

FIGURE 8.29 Transfer of a methyl group from SAM to noradrenaline.

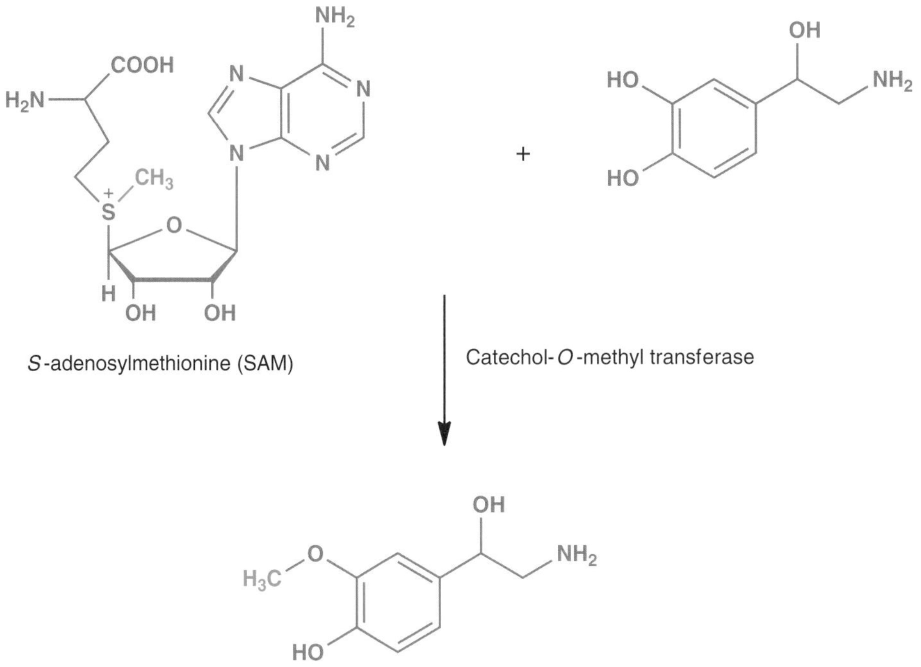

S-adenosylmethionine (SAM)

Catechol-O-methyl transferase

FIGURE 8.30 Oxidation products from thioethers.

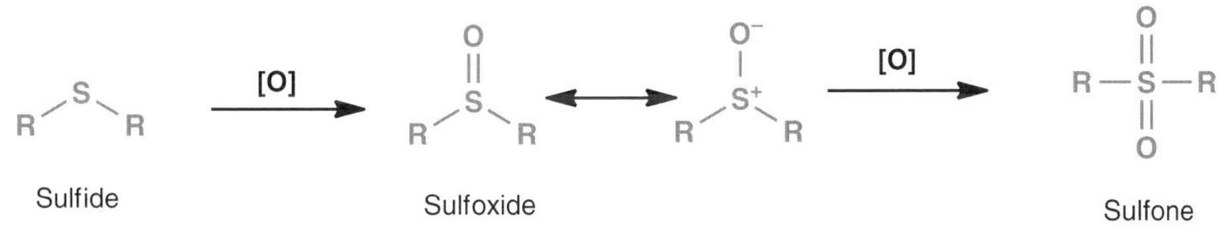

Sulfide   [O]   Sulfoxide   [O]   Sulfone

Figure 8.30. In sulfoxides, sulfur has an oxidation number of 0, and in sulfones it is +2.

Sulfoxides are often represented as having an S=O double bond, but this representation requires overlap of a 2p orbital of oxygen with a much larger 3p orbital from sulfur. Such overlap is not very effective, and the SO group is so polarized that it is often represented as S⁺–O⁻, with the greater electron density on the oxygen. This increased polarization makes sulfoxides more water soluble than sulfides and therefore easier for the body to excrete. The body can metabolize sulfides by oxidation to produce sulfoxides.

Usually the body oxidizes xenobiotics (such as drugs), but occasionally reductions do occur. For example, sulindac, which is used to treat pain and inflammation in rheumatic disease, is a sulfoxide. Sulindac is actually a pro-drug, which is reduced in the body to a lipophilic sulfide, as shown in Figure 8.31. The struc-

tural requirements for non-steroidal anti-inflammatory drugs require that they should be lipophilic so, in this case, the body reduces the hydrophilic sulfoxide group to the sulfide which is the active species.

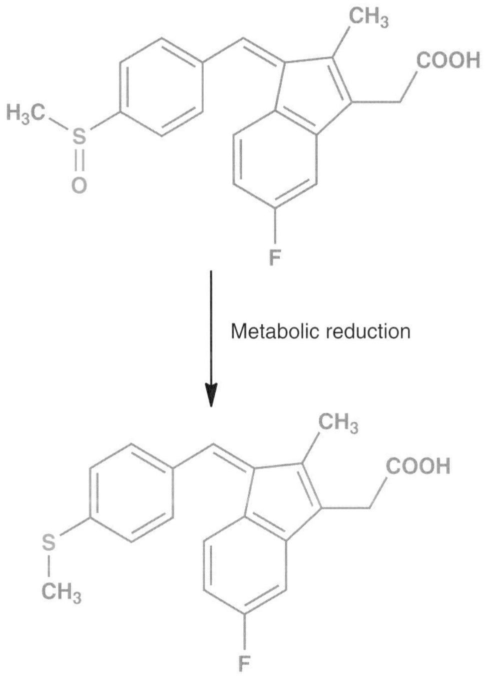

Sulfoxides are non-planar, and sulfoxides with two different substituents are chiral. The pharmaceutical significance of this is demonstrated with the proton pump inhibitor omeprazole, which is used to reduce gastric acid secretion in gastro-oesophageal reflux disease (GORD). Omeprazole, whose structure is shown in Figure 8.32, exists as both *R* and *S* isomers and is usually administered as the racemate because both isomers are active. The *S* isomer is, however, marketed as esomeprazole, which is claimed to lead to higher plasma levels of omeprazole. This is attributed to the *S* isomer being less susceptible to metabolism than the *R* isomer, such that it is removed from the body more slowly.

> *R* and *S* isomers are discussed in more detail in Chapter 3 'Stereochemistry and drug action'.

Dapsone, which has been used to treat leprosy and tuberculosis, is an example of a sulfone. Sulfones are produced by the oxidation of sulfoxides, after which sulfur has an oxidation number of +2, with double bonds to two oxygen atoms. Dapsone, like sulphonamide antibacterials, blocks the synthesis of folic acid.

Sulfonates where sulfur has an oxidation number of +4 are the next oxidation state from sulfones. Sulfonic acids, RSO_3H, where R=H or an aliphatic or aromatic carbon, are strong acids and are ionized

FIGURE 8.31 Reduction of sulindac.

Metabolic reduction

FIGURE 8.32 Omeprazole.

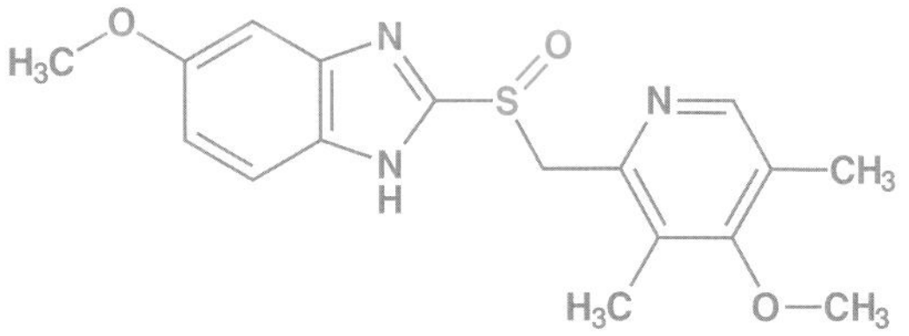

Sulphur or sulfur?

It is often said that the British and the Americans are divided by a common language. The spelling of sulfur has been standardized to the American. Several drug classes, discovered in Europe before standardization of spelling, retain the British English spelling, for example cephalosporins, sulphonamides. Individual drugs (cefalexin, sulfasalazine) are often spelt in American English. It is the duty of the pharmacist (the same spelling in British and American English) to be concerned about spelling. Drug names (digitoxin and digoxin, for example) are sometimes very similar.

at physiological pH, as depicted in Figure 8.33. They form water-soluble salts (sulfonates) with organic bases and quaternary ammonium ions. For example, the neuromuscular blocking agent atracurium is formulated as its benzene sulfonate (besylate) salt as shown in Figure 8.34.

Sulfonic acids are similar to carboxylic acids in that they can form esters and chlorides. The chlorides can be used as intermediates in the preparation of sulphonamides as shown in Figure 8.35.

The best known use of sulphonamides is as antibacterial agents. However, they also act as inhibitors of

carbonic anhydrase, a zinc-containing enzyme that catalyses the interconversion of carbon dioxide and water to carbonic acid; the carbonic acid is subsequently converted into hydrogen carbonate (bicarbonate) ions by the loss of H^+.

$$CO_2 + H_2O \rightleftharpoons H_2CO_3 \rightleftharpoons H^+ + HCO_3^-$$

Sulphonamides have an acidic hydrogen and are ionized by the loss of a proton at physiological pH. The ionized sulphonamide interacts with the zinc ion at the active site of carbonic anhydrase. Acetazolamide, whose structure is shown in Figure 8.36, is an example of a sulphonamide carbonic anhydrase inhibitor that may be given by injection. It can be used to treat glaucoma, because the inhibition of carbonic anhydrase that it promotes reduces pressure in the eye. The BNF recommends that the injection is given intravenously as intramuscular injections may be painful because of the alkalinity of the solution.

SELF CHECK 8.17

Why do you think that aqueous solutions of acetazolamide are alkaline? Hint: acetazolamide (see Figure 8.36) is formulated as the sodium salt.

The sulfonyl group readily attracts electrons. Consequently, sulfonates are good leaving groups. This characteristic is exploited in sulfonate esters such as busulfan where the C–O bond is readily broken. The reaction is an S_N2 type reaction and results in cross-linking DNA and the release of methyl sulfonate, as illustrated in Figure 8.37.

FIGURE 8.33 Sulfonic acid and sulfonate ion.

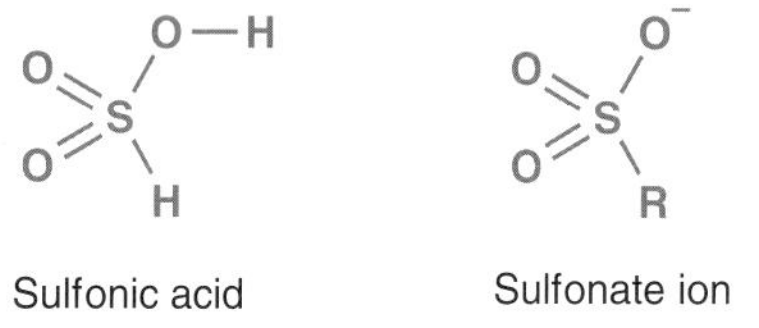

Sulfonic acid   Sulfonate ion

FIGURE 8.34 Atracurium besylate.

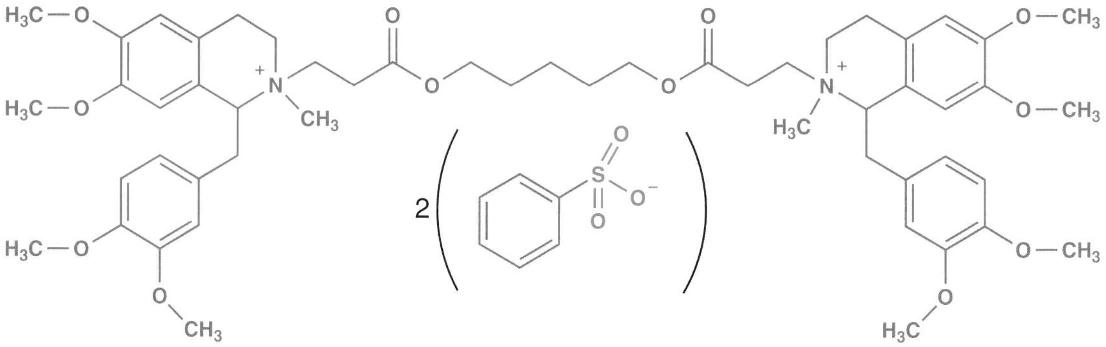

FIGURE 8.35 Synthesis and ionization of sulphonamides.

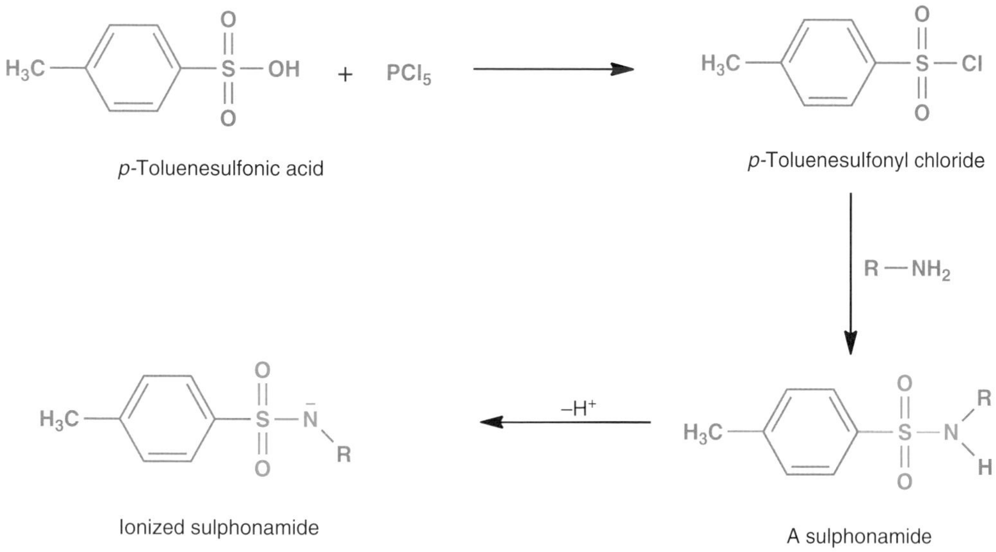

p-Toluenesulfonic acid

p-Toluenesulfonyl chloride

Ionized sulphonamide

A sulphonamide

FIGURE 8.36 Acetazolamide.

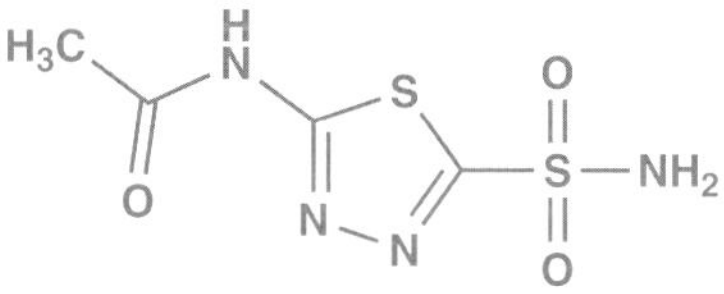

FIGURE 8.37 Mechanism of action of busulfan.

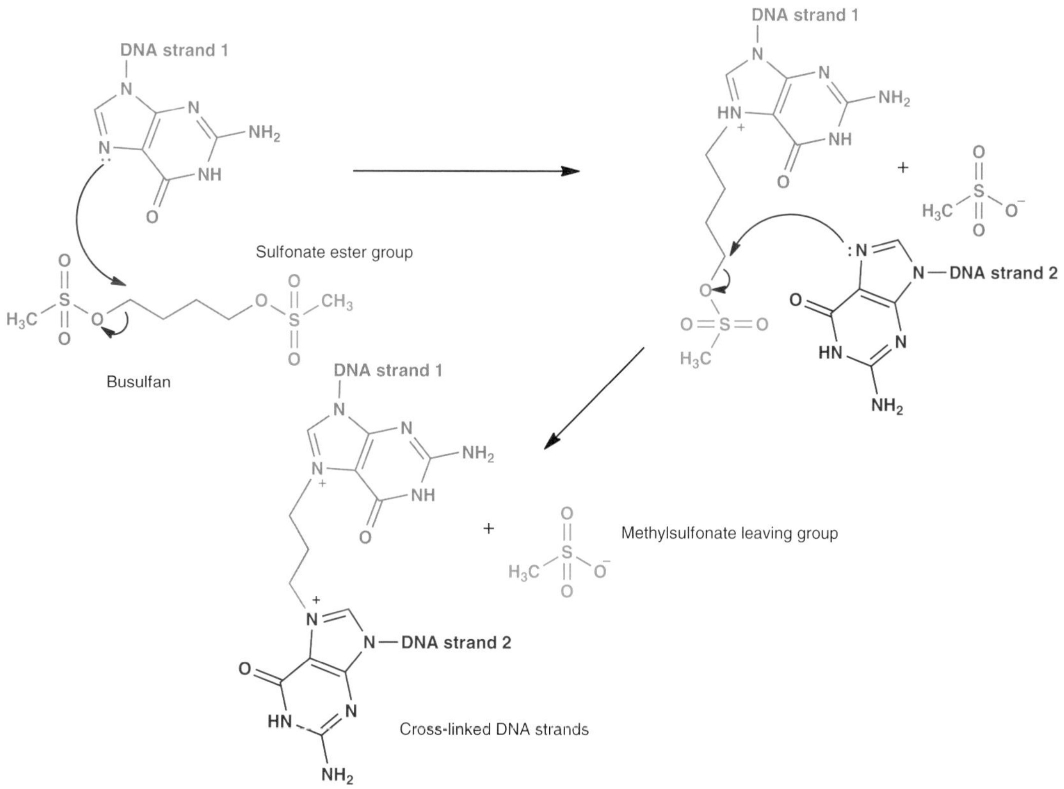

Sulfonate ester group

Busulfan

DNA strand 1

DNA strand 2

Methylsulfonate leaving group

Cross-linked DNA strands

S_N2 reactions are discussed in greater detail in Chapter 5 'Alcohols, phenols, ethers, organic halogen compounds, and amines'.

Derivatives of sulfuric acid

The highest oxidation state of sulfur is +6, as seen in sulfuric acid. The sulfur in this compound forms two double bonds with oxygen atoms and two single bonds to hydroxyl groups. Sulfuric acid is a strong dibasic acid and is capable of forming hydrogen sulfate (HSO_4^-) and sulfate (SO_4^{2-}) ions. Amine drugs are formulated as sulfate salts to give water solubility; examples include morphine sulfate and quinine sulfate.

Sulfuric acid can form sulfate esters (also called sulfoesters or organosulfates) in the same way that phosphoric acid can form phosphoesters (phosphate esters).

Sodium dodecyl sulfate is a widely used detergent. It can be formed by treating dodecyl alcohol with sulfur trioxide to form dodecyl hydrogen sulfate, which, if neutralized with sodium hydroxide or sodium carbonate, gives sodium dodecyl sulfate. The structure of sodium dodecyl sulfate is shown in Figure 8.38 with the sulfate ester group shown in green.

Dodecyl alcohol may be obtained from coconut oil, and has the trivial name lauryl alcohol. Consequently, sodium dodecyl sulfate is usually known as sodium lauryl sulfate outside the laboratory. Sodium dodecyl sulfate is amphipathic: it has a polar head and a long non-polar tail, which is capable of solubilizing oils and fats and does not form insoluble salts with calcium and magnesium ions. It is therefore very often used in the laboratory to dissociate and solubilize macromolecules such as proteins. (We say that it is an 'anionic surfactant'.) It is also used in most toothpastes – which is why they foam (see Integration Box 8.2).

Surfactants are discussed in Chapter 5 'Alcohols, phenols, ethers, organic halogen compounds, and amines' and more details about surfactants, in general, can be found in Chapter 9 'Surface phenomena' in the ***Pharmaceutics*** book in this series.

INTEGRATION BOX 8.2

Sodium lauryl sulfate in toothpaste

Not surprisingly, sodium lauryl sulfate (SLS) is an irritant, because it solubilizes proteins. Indeed, anecdotal evidence suggests that SLS in toothpastes exacerbates the pain caused by mouth ulcers. Mouth ulcers can be caused by serious disease, but they are more usually symptoms of stress or benign hormonal imbalance. Sufferers may benefit from using an SLS-free toothpaste.

Sulfate esters are also formed in the body; sulfates conjugate to alcohol and phenol functional groups. For example, the phenolic hydroxyl in salbutamol is a site for sulfate conjugation. The reaction shown in Figure 8.39 leads to the formation of a sulfate ester which is inactive and polar, and so is easily excreted in the urine. The sulfate ester is polar because only one of the available oxygen atoms is esterified; the remaining –OH acts as a strong acid (pK_a 1–2) so it will be ionized at physiological pH. Sulfotransferase enzymes, which transfer sulfate from 3′-phosphoadenosine-5′-phosphosulfate to hydroxyl groups, are found in the liver, intestine, and kidney, which are the main sites of drug metabolism in the body.

Thiocarbonyl compounds (thiones)

In the last few sections, we have been looking at the different oxidation states of sulfur. In the following examples, we will consider the replacement of oxygen in carbonyl groups, by sulfur, to give thiocarbonyl compounds, which are commonly called thiones. Thiones can have significantly different properties from their parent carbonyls, though they can tautomerize in the same way that carbonyl groups undergo keto-enol tautomerism, as illustrated in Figure 8.40.

FIGURE 8.38 Sodium dodecyl sulfate.

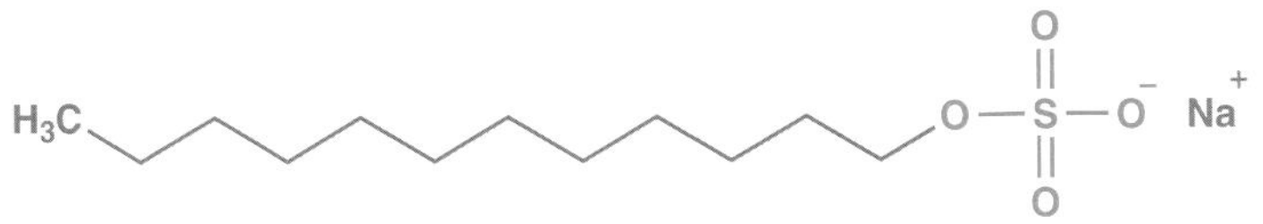

FIGURE 8.39 Metabolic sulfate ester formation.

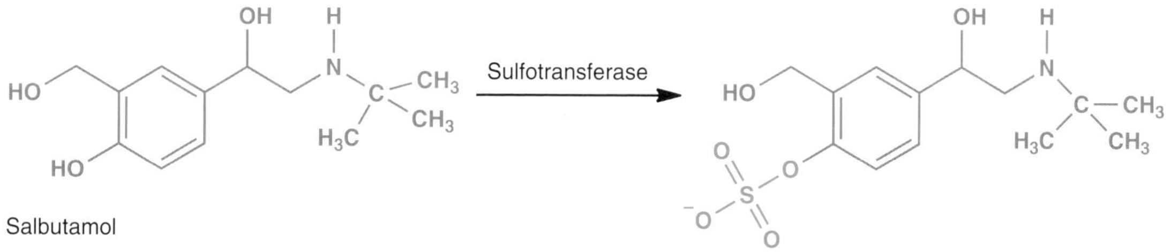

Salbutamol

Sulfotransferase

Ionized sulfate ester of salbutamol

 More details on keto-enol tautomerism can be found in Chapter 6 'The carbonyl group and its chemistry'.

The antithyroid compound propylthiouracil, whose structure is shown in Figure 8.40, illustrates how amides and the thione equivalent can tautomerize. The action of propylthiouracil may partly depend on its ability to function as a thiol.

The thione double bond is normally less polarized than the double bond of carbonyls, resulting in the compounds being less polar. The biological effect of this can be seen by considering the barbiturates, pentobarbital and thiopental. The only difference between the two compounds is that pentobarbital has three oxygen functions, whereas one of the oxygen atoms in thiopental is replaced by sulfur, as shown in Figure 8.41. The change in polarity in the compounds is reflected in the log P values – 2.1 for pentobarbital

and 2.9 for thiopental – with thiopental being more lipophilic. This change in log P has a significant effect on the biological properties of the drugs, which influences their therapeutic use.

Pentobarbital belongs to the barbiturate class of drugs, which were introduced as hypnotics in the early part of the 20th century. These drugs were very popular until the 1960s when apparently safer drugs, such as the benzodiazepines, started to replace them. In the BNF, the only barbiturates appropriate for treating insomnia are amobarbital, butobarbital, and secobarbital – and even these are only on a named patient basis. Thiopental as the sodium salt, however, is still used as an intravenous anaesthetic. Because it is very lipophilic, it quickly crosses the blood-brain barrier producing very rapid anaesthesia; it can take effect before the patient can count to 10.

The action of thiopental ends when it is redistributed into the fat deposits of the body. The resulting

FIGURE 8.40 Keto-enol and thione-lactam tautomerism.

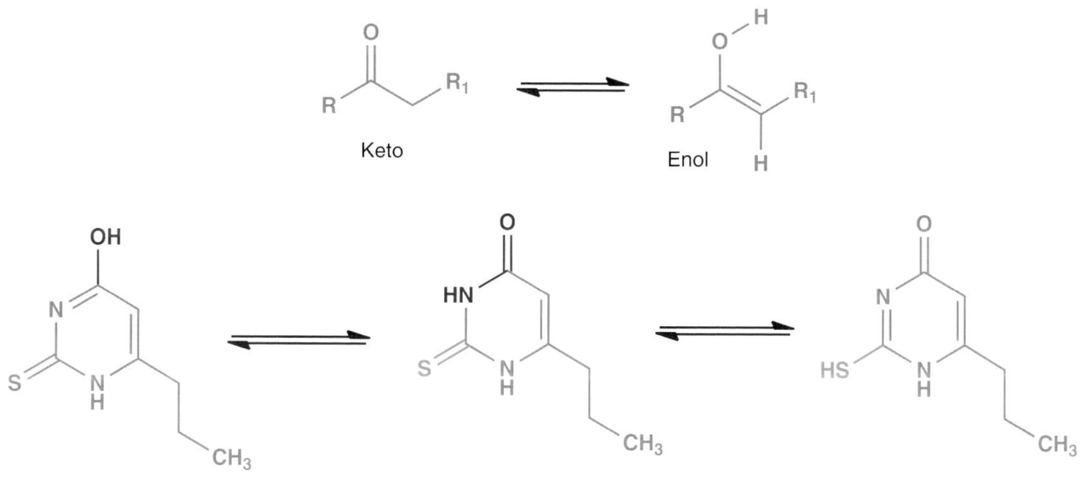

Keto

Enol

Enol type tautomer

Propylthiouracil
showing thione and lactam tautomer

Thiol tautomer

FIGURE 8.41 Pentobarbital and thiopental.

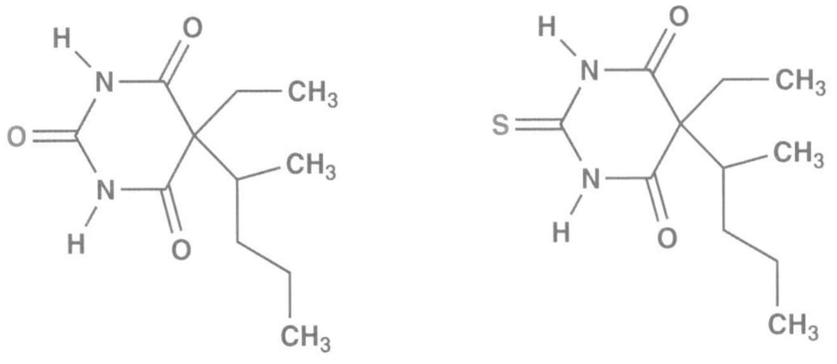

Pentobarbital

Thiopental

equilibration lowers the concentration of thiopental in the brain and the anaesthetic effect wears off. The uptake of pentobarbital into the brain is slower, so that the patient takes longer to go to sleep; the redistribu-

tion of pentobarbital into other tissues is also slower so it takes much longer for the blood level in the brain to fall and the patient to wake up.

KEY POINT

- Sulfur can exist in oxidation states from −2 to +6 and all states are found in pharmaceutical and biological molecules. The wide range of oxidation states results from its ability to use d-orbitals to accept electrons.

- The sulfur atom is bigger than the oxygen atom and is less electronegative.

- Thiols and thiophenols have similar chemistry to alcohols and phenols.

- Thiols can form complexes with metal ions.

- Oxidation of thioethers can produce sulfoxides and sulfones.

- Sulfonic acid and sulfuric acid derivatives are strong acids and can be used to produce water-soluble derivatives.

SELF CHECK 8.18

To test your understanding of the role of sulfur-containing compounds in pharmacy, look up the structures of the compounds listed in Table 8.8; for each compound identify the functional group that contains sulfur, predict the oxidation state of the sulfur and indicate the therapeutic use of the compound or its role in metabolism.

TABLE 8.8 Sulfur-containing compounds in pharmacy

Compound	Functional group	Oxidation state of sulfur	Therapeutic use or biological function of compound
e.g. gentamicin sulfate	sulfate	+6	antibiotic
thioridazine			
bretylium tosilate			
sulfamethoxazole			
dimercaprol			
carbimazole			
glutathione			

Our bodies need elements other than carbon, hydrogen, nitrogen and oxygen to allow us to function; imbalances in these elements can have serious effects on our health. Many drugs and medicines rely on the presence of both metallic and non-metallic elements to produce their biological effect and for their formulation.

For metals:

➤ Most metals usually function as the cation.

➤ Metal ions can usually interact by ionic bonding or complex formation.

➤ Iron has activity both through complex formation and through redox reactions.

➤ Ligands are molecules that contain groups capable of acting as Lewis bases by donating electrons.

Phosphorus is found in both the normal biochemistry of the body and in drugs.

➤ Phosphorus is usually pentavalent in pharmaceutical and biological molecules because it can make use of d orbitals.

➤ Phosphoric acid is tribasic and the 2nd ionization is important for regulating body pH.

➤ Phosphoric acid can form esters and anhydrides in the same way as carboxylic acids.

➤ Phosphate esters are important for biological molecules and can be used to produce water-soluble derivatives of drugs.

Sulfur can have an oxidation number from −2 to +6 and all states are found in pharmaceutically relevant molecules. It has so many oxidation states, because it can use the d orbitals to share electrons from oxygen atoms.

➤ Some of the uses of sulfur in drugs depends on the fact that it is a bigger atom than oxygen.

➤ Thiols have a similar chemistry to alcohols.

➤ Although thiones can undergo tautomerism in the same way that ketones can, they tend to be less polar because of the larger size of the sulfur atom.

➤ Sulfonate and sulfate esters are polar and can increase water solubility of drugs.

Although most drug action can be explained in terms of the chemistry of the elements, there are still a number of drugs for which the exact mechanism of action has not been determined.

8.8 Sulfur

239

FURTHER READING

More in depth coverage of inorganic chemistry may be found in books such as:

Atkins, P., Overton, T., Rourke, J., Weller, M., and Armstrong, F. *Shriver and Atkins' Inorganic Chemistry*. 5th edn. Oxford University Press, 2009.

Rayner-Canham, G, and Overton, T. *Descriptive Inorganic Chemistry*. 5th edn. W H Freeman and Company, 2010.

Dewick, P.M., *Essentials of Organic Chemistry, for Students of Pharmacy, Medicinal Chemistry and Biological Chemistry*. Wiley, 2006.
   This gives good detail on hydrolysis of ATP and ADP and release of energy.

Patrick, G., *Introduction to Medicinal Chemistry*. 5th edn. Oxford University Press, 2013.
   This gives mechanisms for the action of drugs.

Other books you might find helpful are:

The British Nation Formulary (BNF).

Kean, S., *The Disappearing Spoon*. Doubleday, 2011.
   This is a slightly different book which is full of anecdotes about the elements and their discovery and effects.

Pocket clinical pharmacy reference books such as:

Wiffen, P., Mitchell, M., Snelling, M., and Stoner, N. *The Oxford Handbook of Clinical Pharmacy*, 2nd edn. Oxford University Press, 2012.
   These are good for giving normal ranges of ions in body fluids and infusion fluids.

http://www.chemguide.co.uk/ Accessed 20/09/11.
   Gives good clear explanations and diagrams based on A-level syllabi, which may help you with revising some of your previous chemistry.

Online, you might like to research related news articles. For example:

- Zinc can be an 'effective treatment' for common colds (http://www.bbc.co.uk/news/health-12462910)

- Zinc tablets may shorten the duration of a cold (http://www.independent.co.uk/life-style/health-and-families/health-news/zinc-tablets-may-shorten-the-duration-of-a-cold-7720704.html)

Lankshear, A.J., Sheldon, T.A., Lowson, K.V., Watt, I.S. and Wright, J. Evaluation of the implementation of the alert issued by the UK National Patient Safety Agency on the storage and handling of potassium chloride concentrate solution. *Qual Saf Health Care* 2005; 14: 196–201.

The chemistry of biologically important macromolecules

ALEX WHITE

This chapter will outline the chemistry of molecules important to biology, a subject also referred to as **biochemistry**. Biologically important molecules are often large molecules, which are also called macromolecules. At first glance, the study of complex-looking large molecules may seem daunting, but hopefully this chapter will show that the principles of chemistry, presented in other chapters of this book, can be applied to the study of these molecules to help us understand their functions in the cells of our bodies. In particular, macromolecules are often drug **targets**: their biological activity is changed when a drug molecule interacts with them, so an understanding of their structure and function is an important part of drug design and development.

Learning objectives

Having read this chapter you are expected to be able to:

- ➤ recognize nucleic acids, proteins, carbohydrates and lipids by studying their structures

- ➤ discuss the biochemistry of nucleic acids, proteins, carbohydrates and lipids and outline how these molecules function at the molecular level

- ➤ explain how the structures of biological macromolecules are related to their function

- ➤ understand why studying biological macromolecules is important from a pharmaceutical perspective.

9.1 Small molecules versus large molecules

At the outset, it would be helpful for us to clarify the pharmaceutical definitions of *small* and *large* molecules. Small molecules are defined as organic, drug-like, compounds with molecular weights less than about 500. Many drugs fall into this category. Large molecules have molecular weights higher than 500, often considerably higher, in the region of thousands or tens of thousands.

Recent years have seen the emergence of large molecules used therapeutically, a class of drugs often referred to as the **biologics**. Traditionally, these molecules were not considered to be drug-like, and they are often much harder to formulate, deliver and manufacture than small drug molecules, such as ibuprofen or penicillin. Small molecule drugs still greatly outnumber biologic drugs but biologics are increasing in importance, especially economically. It is predicted that by 2016, 8 of the 10 most profitable drugs on the market will be biologics.

 The current five top-selling drugs can be found in Chapter 7 'Introduction to aromatic chemistry'.

Biology often creates large molecules from combinations of building blocks, pieced together in defined ways to create larger molecules. Large molecules can therefore be thought of as **biopolymers**, built from smaller **monomer** building blocks. It is helpful to study the monomers before seeing how they are combined to create more complex molecules.

9.2 **Nucleic acids and nucleosides**

Two nucleic acids, DNA (**D**eoxyribo**N**ucleic **A**cid) and RNA (**R**ibo**N**ucleic **A**cid), are of fundamental importance to biochemistry and all life on Earth. DNA is the carrier of genetic information, and is divided into specific sequences called genes. Genes are typically packaged into separate chromosomes, the distinctive structures found in nearly every cell, in every higher organism.

Most genes are sequences of DNA that carry instructions for making proteins. RNA, a molecule whose chemical composition is very similar to DNA, is mostly concerned with the processing of the information contained within DNA as part of the process of 'reading' the information stored in the DNA to make proteins.

Understanding the chemistry of nucleic acids is essential for many aspects of pharmaceutical science. Many important anticancer, antiviral and antibacterial drugs target nucleic acids. An understanding of nucleic acid chemistry shows us how nucleic acids work, and how drugs interact with them.

 DNA and RNA are discussed in more detail in Chapter 2 'Molecular cell biology' in the *Therapeutics and Human Physiology* book in this series.

Nucleic acid structure: the backbone

DNA (see Figure 9.1A) and RNA are relatively simple biopolymers, each composed of just four monomers, known as nucleotides. Each nucleotide has three components, a phosphate group, a ribose sugar and

a **heterocyclic** organic base. A nucleotide unit is highlighted in purple in Figure 9.1B. As nucleotides are joined together, alternating phosphate and sugar groups form the nucleic acid backbone. In DNA, the sugar component of the nucleotide is 2′-deoxyribose – a molecule of ribose lacking a hydroxyl (OH) group. In RNA, the sugar component is ribose itself.

Look at Figure 9.1B and notice how the phosphate groups are covalently bonded to two sugar groups, forming a functional group analogous to an ester. For this reason, this functional group is classified as a phospho*diester*. The acidic nature of DNA and RNA arises from the presence of this group. The P–OH group acts as an acid at physiological pH (that is, it donates its proton), making the phosphodiester negatively charged. The stability of the negatively charged oxygen group is the driving force for the loss of the proton. DNA and RNA are usually drawn with negatively charged phosphodiester groups since proton donation is always assumed to occur.

KEY POINT

DNA and RNA are made up of nucleotides that are joined together by phosphodiester groups.

SELF CHECK 9.1

Be sure you can identify the base pairs in Figure 9.1A. They are the *rungs* of the DNA *ladder*. Make sure you understand how these base pairs are related to the base pair structures shown in Figure 9.1B.

FIGURE 9.1 Nucleic acid structure. (A) A model of a short sequence of DNA (12 base pairs). The nucleic acid backbone is highlighted by a grey ribbon. The bases are shown in the ball and stick format (carbon – grey; nitrogen – blue; oxygen – red. Hydrogen is omitted for clarity). (PDB code: 423d.) (B) The structure of two single base pairs. A nucleic acid nucleotide unit is highlighted in purple. Hydrogen bonds are indicated by dashed lines.

(A)   (B)

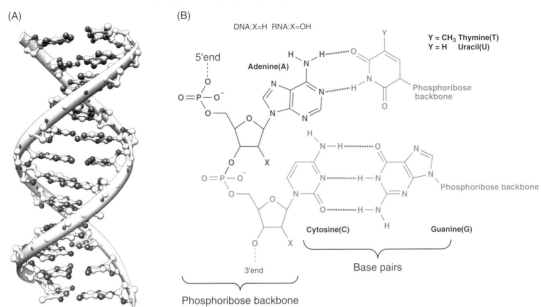

 Phosphoesters are discussed in more detail in Chapter 8 'Inorganic chemistry in pharmacy', and more details on acid/base chemistry can be found in Chapter 6 'Acids and bases' in the *Pharmaceutics* book in this series.

Nucleic acid structure: bases

The varying nature of the base distinguishes individual nucleotides and influences the composition of the nucleic acid. DNA contains four heterocyclic bases: adenine (A), guanine (G), cytosine (C) and thymine (T). RNA differs from DNA in that uracil (U) replaces thymine. The structure of each base is shown in Figure 9.1. The single-letter abbreviations (in brackets) are usually used to denote the nucleic acid bases and have widespread acceptance. They will be used in this chapter and you will encounter these abbreviations throughout the scientific literature. It is very useful to learn and remember them. Each base is bonded to the nucleic acid backbone via the sugar group.

Base pairing in nucleic acids

The interaction between bases is the key to understanding nucleic acid function. The bases are examples of heterocycles, organic ring systems that contain an element other than carbon as part of that ring. The bases come in two varieties, purines and pyrimidines. A pyrimidine contains a single six-membered ring, whereas a purine has a bicyclic structure with a six-membered ring fused to a five-membered ring. Both purines and pyrimidines are nitrogen-containing heterocycles. The presence of electron rich heteroatoms and other functional groups such as amino (NH_2) and carbonyl (C=O) groups allow for the formation of hydrogen bonds between base pairs. As we see in Chapter 2, individual hydrogen bonds are rather weak, typically 1–5 % of the strength of a covalent bond. Despite their weakness, networks of *many hydrogen bonds* within a molecule are much stronger and impart great stability to macromolecules.

The pairing between bases is very specific. In DNA, *A always pairs with T* and *G always pairs with C*. RNA does not contain T so A pairs with U instead. Note that two hydrogen bonds are formed in the A–T and A–U base pairs, whereas three hydrogen bonds stabilize the G–C base pair. Each base pair contains a purine and pyrimidine. These pairings result from the structure of each base; look at the base pairs in Figure 9.1 and notice how the functional groups complement each other perfectly. This is essential for the formation of the hydrogen bond network.

The DNA double helix

The properties outlined above give rise to the well-known and very distinctive DNA **double helix**. Looking at the model of DNA in Figure 9.1A, we see two separate DNA backbones wrapping around each other about a central axis. The structure is held together by the network of hydrogen bonds formed between the inward facing bases. It is hard to detect or visualize hydrogen bonds by common physical techniques (e.g. NMR), but their existence is strongly suggested by the orientation of the bases towards each other within the double helix.

As a consequence of this structure, the base sequence is complementary between the two strands. Recall that A always pairs with T and C always pairs with G. Therefore, if you know the sequence of one strand, you can deduce the sequence of the other. The strands run in opposite (or antiparallel) directions to each other. The direction of each strand is defined by the unbonded groups at its extreme ends: a ribose 5′-OH is located at one end (the 5′ end), with the strand running towards an unbonded ribose 3′-OH at the other (3′) end. By convention, bases are always listed in the 5′ to 3′ direction. Therefore, a particular sequence of DNA could be communicated in the following way:

5′ – AAGCGATAGCTC – 3′

This is much easier than drawing the full chemical structure! This shorthand method is used to communicate information about nucleic acids, and it ignores the common, repetitive parts of the structure. But you should always remember that there is a more complex structure behind the shorthand notation!

> **KEY POINT**
>
> The structure of DNA is a double helix of two antiparallel polynucleotide strands. The two strands are joined by a network of hydrogen bonds between purine and pyrimidine bases on the inside of the helix.

> **SELF CHECK 9.2**
>
> If 5′–AAGCGATAGCTC–3′ is the sequence of one strand of a short section of DNA, what is the sequence of the other strand? Remember that a DNA sequence is always listed in the 5′ to 3′ direction.

Ribonucleic acid (RNA)

RNA differs from DNA: it usually exists as a single strand and folds into a greater variety of three-dimensional shapes than DNA. The structure of each folded strand is stabilized by the strand interacting *with itself* by base pairing. There are three main types of RNA: messenger RNA (mRNA), transfer RNA (**tRNA**) and ribosomal RNA (rRNA). The functions of these molecules have been covered in more depth elsewhere in this series, but to summarize: single stranded mRNA is transcribed from a complementary strand of DNA and travels to the ribosome, a structure that is composed of rRNA and protein. The ribosome translates the mRNA sequence into proteins, using the amino acids attached to tRNAs.

> **KEY POINT**
>
> RNA molecules, which contain copies of the genetic code, are produced by transcription from DNA.

 Genetic code, genes, transcription and translation are discussed in more detail in Chapter 2 'Molecular cell biology' in the *Therapeutics and Human Physiology* book in this series.

Transfer RNA (tRNA) chemistry

Although its biochemical function is discussed elsewhere in the series, tRNA is an interesting molecule from a chemical perspective. Its structure, a recognizable 'cloverleaf' shape, is maintained by intrastrand base pairing (i.e. within the same strand), but it also contains loops of unpaired bases. tRNA plays an important role in protein synthesis; one of the loops contains the *anticodon*, a triplet of bases that is complementary to the codon in mRNA, which corresponds to a particular amino acid (see Figure 9.2).

During protein synthesis, the ribosome brings together the various tRNA molecules needed to synthesize a protein. A specific amino acid is bonded to the end of a tRNA molecule by an ester linkage. This is a functional group that is especially reactive towards nucleophiles. The nucleophilic amine group of an adjacent amino acid-tRNA molecule reacts as

FIGURE 9.2 A mechanism for protein synthesis within the ribosome. This scheme is highly abbreviated to focus on the reaction between two amino acids; in particular, the double-headed curly arrow represents the formation of a tetrahedral intermediate, followed by reformation of the carbonyl group and loss of −OtRNA.

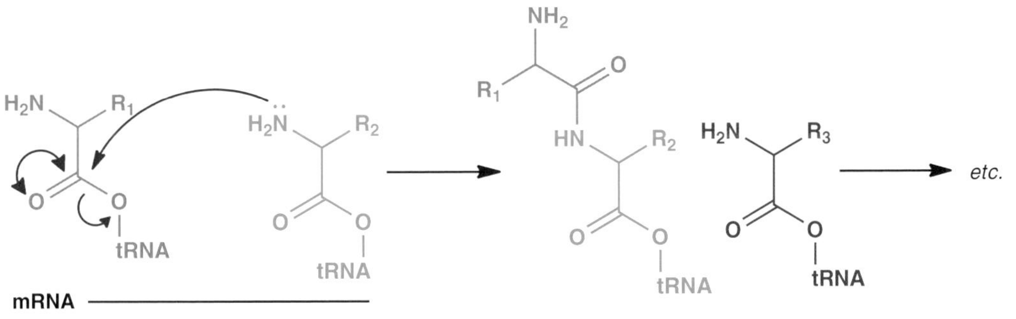

shown by the curly arrows in Figure 9.2, and forms a new amide functional group. Consequently, the protein grows by one amino acid. Another amino acid-tRNA molecule (with side chain R_3) can then enter the ribosome, and the growing protein is extended. This illustrates that even the chemistry of complex biological reactions can be understood by the use of curly arrows. The chemistry is identical to reactions we have already studied in smaller and simpler molecules. Once learnt and understood, the functional group interconversion of an ester to an amide can be applied to any example!

Several antibiotic drugs kill bacteria by interfering with specific steps in their synthesis of proteins. For example, chloramphenicol (used to treat eye infections) inhibits the transfer of the growing peptide chain between tRNA molecules, while the tetracycline class of antibiotics block the binding of tRNA to the ribosome.

SELF CHECK 9.3

Identify the ester functional group in Figure 9.2. Make sure you can follow the curly arrows and identify the nucleophile, the electrophile and the newly formed bonds in the products of this reaction.

 Carbonyl chemistry is discussed in detail in Chapter 6 'The carbonyl group and its chemistry'. Protein synthesis and translation, and antibiotic therapeutics are discussed in Chapter 2 'Molecular cell biology' of the *Therapeutics and Human Physiology* book in this series.

Nucleosides

Nucleosides, an important class of molecules closely related to nucleotides, have many important roles in biochemistry. They contain a heterocyclic base and a ribose sugar but lack the phosphate group at the 5′-position of the sugar. Nucleosides can be derived from ribose or 2′-deoxyribose as illustrated in Figure 9.3.

A nucleoside in action: SAM

S-Adenosylmethionine (SAM) is a nucleoside coupled with methionine, an amino acid (see Section 9.3). SAM is used in biochemistry as a source of the methyl group (see Figure 9.4). Adrenaline (an important hormone) is synthesized from its precursor noradrenaline by the addition of a methyl group, supplied by SAM.

This biochemical process occurs in the adrenal gland and is catalysed by an enzyme, which uses noradrenaline as the substrate and SAM as an essential cofactor. The structures might look complicated but just concentrate on the reaction itself, shown by the curly arrows, and you will see that this is a simple nucleophilic substitution. The positively charged sulfur is a good electrophile and the driving force is the conversion of SAM to a more stable, neutral molecule.

 S_N2 reaction mechanisms are discussed in greater detail in Chapter 5 'Alcohols, phenols, ethers, organic halogen compounds, and amines'. The inactivation of noradrenaline by catechol-*O*-methyltransferase also involves methylation by SAM, but at a phenolic oxygen; more details on this reaction can be found in Chapter 8 'Inorganic chemistry in pharmacy'.

FIGURE 9.3   Structure and names of some common nucleosides.

Pyrimidine nucleosides

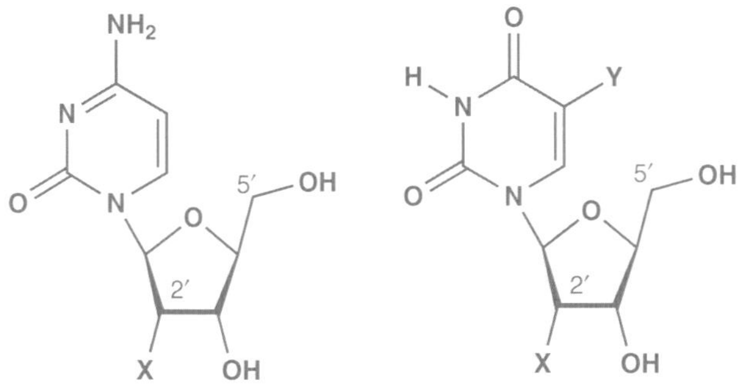

X = H   2'-Deoxycytidine
X = OH   Cytidine

X = H Y = CH₃   2'-Deoxythymidine
X = OH Y = CH₃ Thymidine
X = Y = H 2'-Deoxyuridine
X = OH Y = H Uridine

Purine nucleosides

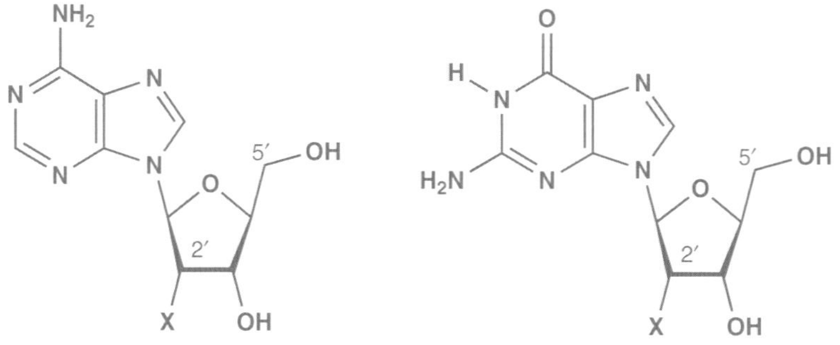

X = H   2'-Deoxyadenosine
X = OH   Adenosine

X = H   2'-Deoxyguanosine
X = OH   Guanosine

FIGURE 9.4   A mechanism showing the role of *S*-adenosylmethionine as a biological methylating reagent.

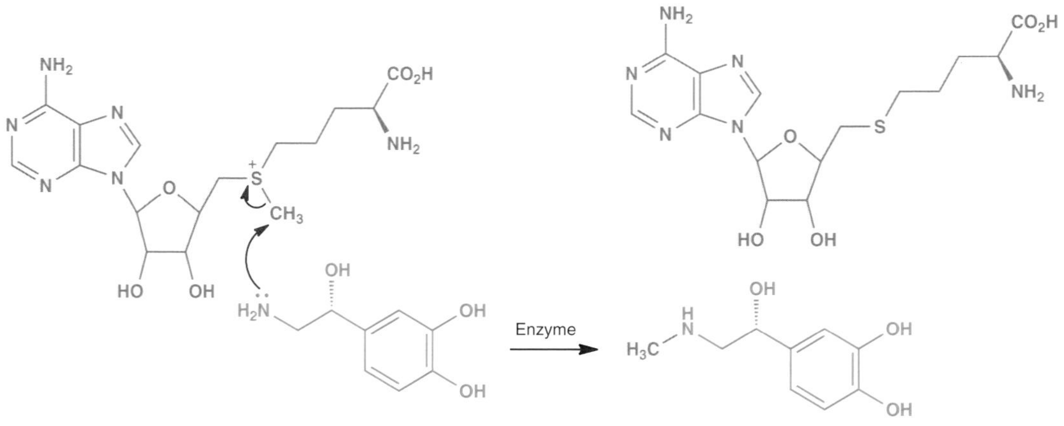

Noradrenaline (a nucleophile)   Adrenaline

A nucleotide in action: ATP

You have probably heard of ATP (adenosine triphosphate), a derivative of the nucleoside adenosine, with *three* phosphate groups attached to the 5′-hydroxyl group. ATP is used widely in biological systems as a source of chemical energy, essential to power the biochemistry of life. It has been estimated that the average human adult synthesizes and consumes 1.5 kg of ATP per hour, at rest!

Study the mechanism of action of ATP shown in Figure 9.5. Notice how this is a reaction with water and is therefore a **hydrolysis** reaction (which literally means *water-breaking*). ATP is particularly reactive because of the presence of a phosphoanhydride (i.e. P–O–P) bond. Water is a good nucleophile and reacts readily with the electrophilic phosphorus. This breaks a phosphorus-oxygen bond, liberating a large amount of energy and a molecule of phosphate (often referred to as *inorganic phosphate* or symbolized as P_i). The negative charge on the oxygen is stabilized by resonance (see box within Figure 9.5) making the phosphate a good leaving group. The release of a good leaving group in this way is a powerful driving force for this reaction.

ATP is introduced in Chapter 1 'The importance of pharmaceutical chemistry', and phosphoanhydrides are discussed in Chapter 8 'Inorganic chemistry in pharmacy'.

DNA as a drug target

When a cell divides, its DNA unwinds before being copied and distributed to new daughter cells. Cancer is a disease that is characterized by excessive cell division. Many anticancer drugs work by inhibiting cell division; some drugs achieve this by directly targeting DNA.

Cell division is discussed in more detail in Chapter 2 'Molecular cell biology' in the *Therapeutics and Human Physiology* book in this series.

For example, doxorubicin is an **intercalating drug** in common use for the treatment of cancer. It contains a planar aromatic ring system that targets DNA by sliding between two adjacent base pairs in the DNA ladder (see Figure 9.6). The binding of the drug is stabilized by hydrophobic interactions between the aromatic rings of the drug and the base pairs. This stable complex helps to prevent the unwinding of DNA and so blocks cell division; the binding of the drug to DNA ultimately leads to the death of the cancer cell.

Nucleoside derivatives as drugs

Nucleoside derivative drugs are often not naturally occurring nucleosides, but have very closely related

SELF CHECK 9.4

Redraw the structures in Figures 9.4 and 9.5. Try to add the curly arrows without looking at the book. Identify the important components of the reaction such as the nucleophile, the electrophile and the leaving group. This will help you understand why these reactions happen.

FIGURE 9.5 The structure of adenosine triphosphate (ATP) and the mechanism of its hydrolysis, a reaction that releases energy for use in biological systems; the double-headed curly arrow is once again used to combine two steps: the formation of a negatively charged intermediate and the reformation of a double bond with loss of a leaving group.

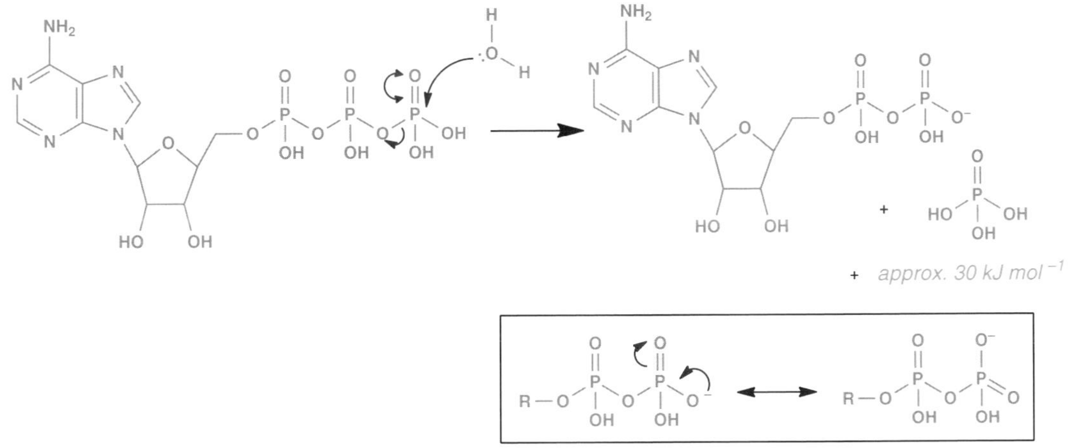

+ approx. 30 kJ mol^{-1}

FIGURE 9.6 A model of the cancer drug doxorubicin binding to DNA. This model shows doxorubicin (carbon atoms highlighted in green) inserting between two DNA base pairs (PDB code: 1D12).

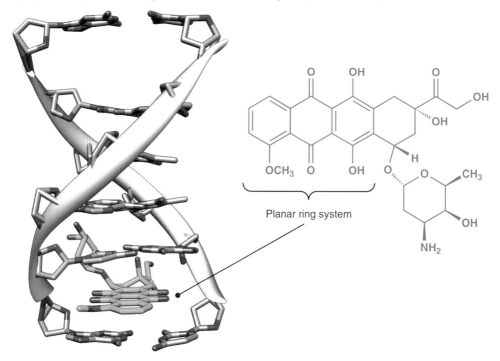

Planar ring system

structures. There are many examples of this class of drug, illustrated by two well-known anti-viral drugs.

Zidovudine (also known as AZT or azidothymidine) is a nucleoside used to treat HIV infection. It contains an unusual azido (N_3) functional group at the 3′-position. Zidovudine is added to a growing strand of viral DNA, whereupon the azide group blocks the addition of further nucleotides, and therefore the virus cannot replicate further (see Figure 9.7).

Aciclovir is often encountered as an effective treatment for cold sores (a viral infection of the lips). It has a nucleoside-like structure that is missing part of the

FIGURE 9.7 Two nucleoside derivative drugs, including an illustration of the mechanism of action of Zidovudine.

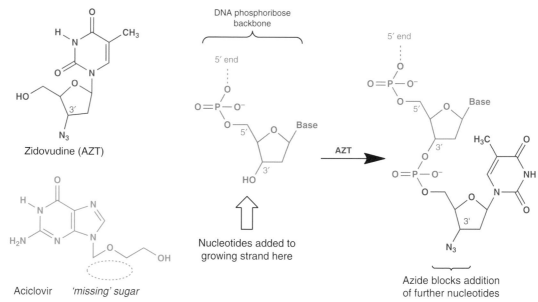

Zidovudine (AZT)

Aciclovir   'missing' sugar

DNA phosphoribose backbone

Nucleotides added to growing strand here

AZT

Azide blocks addition of further nucleotides

sugar group, including the 3′-OH required for elonga-tion of the nucleic acid chain. Once added to the chain,

aciclovir halts nucleic acid synthesis by blocking its elongation; this deactivates the virus.

SELF CHECK 9.5

Study Figure 9.7 and draw the structure that forms when a nucleic acid reacts with aciclovir.

KEY POINT

Many anticancer and antiviral drugs interfere with the replication and synthesis of nucleic acids.

9.3 Proteins

Proteins make up a highly important class of mac-romolecules in biochemistry. Early scientists named these molecules from the Greek word *Proteios*, mean-ing first or primary, since they were abundant in every biological system they studied. Today, we know that proteins have a myriad of functions in cells and organ-isms. To pick some examples: antibodies (part of the immune system), collagen (involved in physical struc-tures), membrane-bound receptors, and enzymes are all distinct types of proteins.

Proteins are biopolymers constructed from amino acid monomers. Compared with nucleic acids, pro-teins have more varied and complex three-dimen-sional structures, due in part to the greater number of monomers used to construct them. As a result, each protein has a unique three-dimensional shape, which is directly related to the function it performs.

Amino acids

All proteins are built from 20 naturally occurring amino acid monomers, which share the same general structure. An acidic carboxylic acid group and a basic amino group are both bonded to a central carbon, which is referred to as the α-carbon. The α-carbon is also bonded to a hydrogen atom and a variable side-chain group, unique for each amino acid. The amino acids are classified, as shown in Figure 9.8, by the chemical properties of the side chain. Take note of the abbreviations used for each amino acid. These are used throughout the scientific literature so it is well worth spending some time learning both the three-letter and single-letter abbreviations (as well as the structures of the 20 side chains).

Amino acid stereochemistry

With the exception of glycine, all the amino acids in Figure 9.8 have four different groups bonded to the α-carbon atom. This means that 19 of the 20 amino acids contain at least one chiral centre and can exist as pairs of **enantiomers.** In practice, all naturally occurring amino acids are single enantiomers and rotate plane-polarized light. The amino acids in proteins have the L-configura-tion. The L/D notation is an 'old fashioned' system but its use is widespread in protein chemistry. All L-amino acids have the same orientation of groups around their α-carbon as the reference compound L-glyceraldehyde (which, rather confusingly, is a simple sugar). The oppo-site enantiomers, D-amino acids, do exist in nature but they are rare and are not used in protein synthesis. (They are synthesized by microorganisms, often by isomerization from a naturally occurring L-isomer.)

 More details on stereochemistry can be found in Chapter 3 'Stereochemistry and drug action'.

Amino acids: acid-base chemistry

Amino acids are more acidic than you might expect. Although acidity varies somewhat between the 20 amino acids, the carboxylic acid (CO_2H) group has a pK_a of approximately 2. Compare this with acetic acid (CH_3CO_2H), an organic compound we normally con-sider to be quite acidic, which has a $pK_a = 4.7$. Since pK_a is a logarithmic expression of acidity, a change of one unit equals a 10-fold difference in that property. Therefore amino acids are over 100 times more acidic than acetic acid.

FIGURE 9.8 The 20 naturally occurring amino acids that form the building blocks of proteins, grouped by side-chain properties.

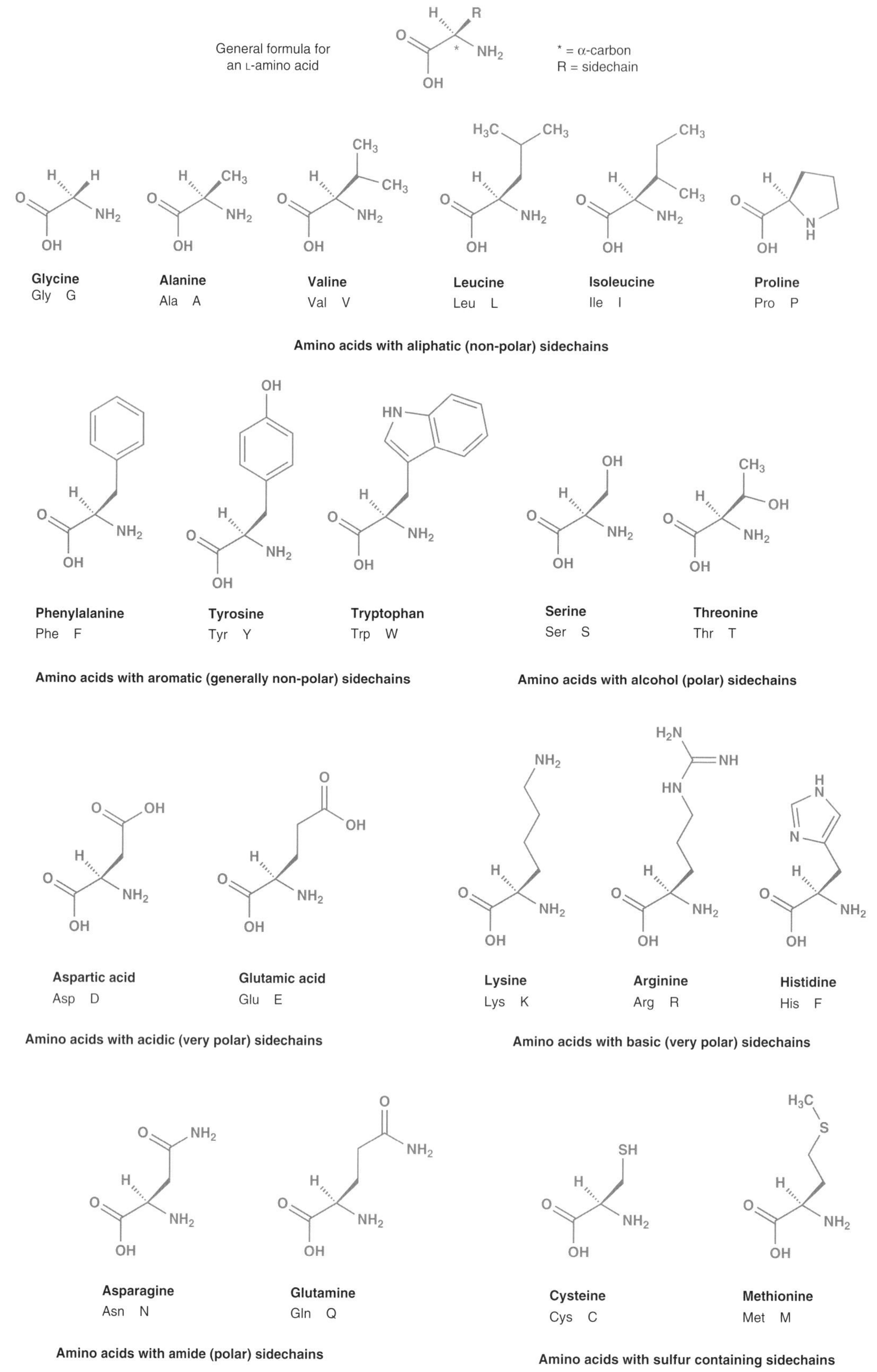

General formula for an L-amino acid

* = α-carbon
R = sidechain

Glycine
Gly G

Alanine
Ala A

Valine
Val V

Leucine
Leu L

Isoleucine
Ile I

Proline
Pro P

Amino acids with aliphatic (non-polar) sidechains

Phenylalanine
Phe F

Tyrosine
Tyr Y

Tryptophan
Trp W

Serine
Ser S

Threonine
Thr T

Amino acids with aromatic (generally non-polar) sidechains

Amino acids with alcohol (polar) sidechains

Aspartic acid
Asp D

Glutamic acid
Glu E

Lysine
Lys K

Arginine
Arg R

Histidine
His F

Amino acids with acidic (very polar) sidechains

Amino acids with basic (very polar) sidechains

Asparagine
Asn N

Glutamine
Gln Q

Cysteine
Cys C

Methionine
Met M

Amino acids with amide (polar) sidechains

Amino acids with sulfur containing sidechains

FIGURE 9.9 An amino acid zwitterion.

The pK_a of the protonated amino (NH_2) group in amino acids is approximately 9. Therefore, under physiological conditions (in other words, an aqueous solution with neutral pH), the carboxylic acid *protonates* the amino group, resulting in a neutral ion that contains two opposite charges. We refer to this type of species as a *zwitterion* (see Figure 9.9). Zwitterions are quite stable, which accounts for the high acidity, high polarity and generally good water solubility of amino acids.

> Acid-base chemistry is discussed in more detail in Chapter 6 'Acids and bases' in the Pharmaceutics book in this series.

Protein structure: primary sequence

Amino acids are linked together by amide bonds, which are also known in protein biochemistry as **peptide bonds**. A peptide bond is formed by a **condensation reaction** between a carboxylic acid and an amine, and repeated reactions can lead to a continuous *peptide backbone*, highlighted in Figure 9.10 by the continuous bold line. Note that there is an amino group not involved in a peptide bond at one end of the protein and similarly an unbonded carboxylic acid group at the opposite end. Consequently, the ends of a peptide are called the *N*-terminus (where the free –NH_2 group is found) and the *C*-terminus (where the free –CO_2H group is located).

By convention, the primary sequence (or primary structure) of a protein is described by listing the amino acids in the correct sequence, starting from the *N*-terminal amino acid and finishing at the *C*-terminus. Either three-letter or single-letter abbreviations can be used for individual amino acids.

Figure 9.11 shows three different ways of representing the same short protein. The full structural formula is given on top, and below are two abbreviated methods. It is common, and far easier, to describe proteins in an abbreviated form, but do not forget the chemical structure that underlies this shorthand.

FIGURE 9.10 Protein primary sequence. Amino acid monomers, linked by peptide bonds, form an extended chain. The peptide backbone, highlighted by bold bonds, runs unbroken from the *N*-terminus to the *C*-terminus of the protein. The sidechains, abbreviated as R_1, R_2 etc point outwards from the backbone.

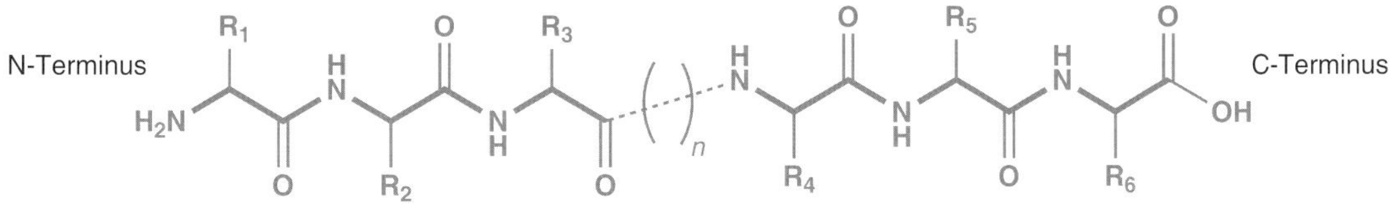

FIGURE 9.11 Three different methods to illustrate the same small protein. A structural formula; the three-letter amino acid abbreviations; and the single-letter amino acid abbreviations are all shown. In each case, the amino acids are always drawn/listed from the *N*- to the *C*-terminus.

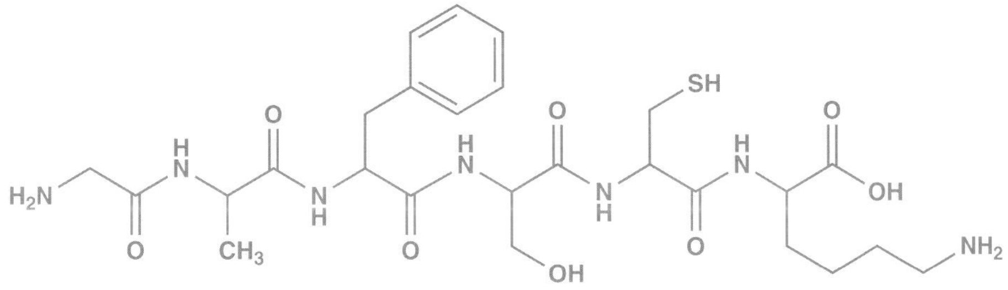

Gly-Ala-Phe-Ser-Cys-Lys

GAFSCK

KEY POINT

Proteins are made from the 20 naturally occurring L-amino acids, joined together in linear chains by peptide bonds. The order of amino acid residues in a protein is its primary structure.

SELF CHECK 9.6

Test your knowledge of amino acids and protein structure. Pick a word made up from just the letters used as single-letter abbreviations for amino acids (only J, O, U and X are not used). Draw the structure of the protein resulting from that sequence of letters. Try PHARMACY for starters!

Peptides

Small proteins with less than 50 amino acids are usually classified as peptides. Peptides comprising just two amino acids are dipeptides; those with three amino acids are tripeptides and so on. Therefore, Figure 9.11 illustrates a hexapeptide.

Protein folding

In practice, proteins do not exist as linear peptide chains. Immediately after synthesis, proteins fold into more complex structures. Large proteins even start to fold up before their synthesis is complete. There are several factors that influence the folding of a protein:

- The peptide bond. The peptide bond has approximately 40% double bond character due to resonance. This can be explained by the curly arrows in Figure 9.12. Recall that double bonds are rigid and cannot rotate. *This restricts the movement of the peptide backbone*. Furthermore, almost all peptide bonds in proteins exist as *trans*-amides to avoid the unfavourable steric interactions that arise from *cis*-peptide bonds (e.g. between R and R_1 in Figure 9.12).
- Rotation of single bonds. Since the peptide bond is generally rigid, a protein folds by movement of the remaining single bonds in the peptide backbone. As shown in Figure 9.12, these are known as the Φ and Ψ bonds.
- Interactions. **Bonding interactions** occur between different parts of a protein chain. These might be hydrogen bonds, ionic bonds, van der Waals (hydrophobic) interactions or covalent bonds. These are intramolecular interactions (i.e. within the same molecule) and are vital for maintaining a protein's shape.

For more details on stereochemistry, see Chapter 3 'Stereochemistry and drug action'; cis and trans-isomers are discussed in greater detail in Chapter 4 'Properties of aliphatic hydrocarbons'.

Protein structure: secondary structure

Secondary protein structure can be described as the repetitive three-dimensional arrangement of parts of the primary structure. There are two main types of secondary structure, the α-helix and the β-sheet.

- The α-helix results from the folding of a *single backbone strand* of the peptide backbone into a coil around a central axis to form a **single helix**. This must not to be confused with the double helix of DNA which contains *two backbones* – a common mistake! The coil is tight: just 3.6 amino acids are required for each complete turn. The structure is

FIGURE 9.12 Factors affecting protein folding.

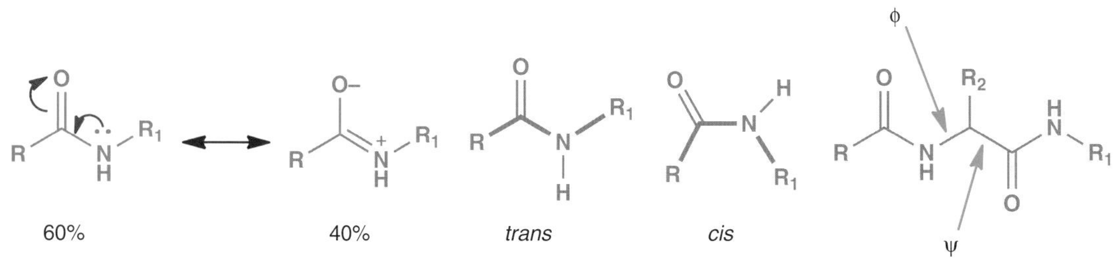

stabilized by a network of hydrogen bonds between the amide carbonyl (C=O) and NH groups along the length of the helix.

- The β-sheet forms when two or more adjacent segments of a peptide backbone interact with each other. Usually several sections of the backbone, called β-strands, interact with each other to form a relatively flat, sheet-like structure. Hydrogen bonding between strands is the most common stabilizing interaction, although hydrophobic interactions can occur between adjacent non-polar amino acid side chains.

The β-sheet exists in two varieties: parallel and antiparallel. The antiparallel sheet is more common and arises when β-strands run in opposite directions to each other. The opposite is seen in the parallel sheet, where β-strands run in the same direction.

Even though Figure 9.13 shows diagrammatic representations of the β-sheet structure, study the arrangement of the stabilizing hydrogen bond interactions. Throughout chemistry, bonding is more efficient when the groups involved are favourably aligned. You can see the hydrogen bonds are regularly aligned in the antiparallel β-sheet. This forms a more stable structure than the parallel β-sheet. This observation is confirmed in nature; antiparallel β-sheets are far more common, and exist in sheets of 2–15 strands (although the average number of strands is six). Parallel β-sheets are far less common; the largest discovered contains just five strands. This is an excellent illustration of the stabilizing ability of a network of relatively weak hydrogen bonds.

Protein structure: tertiary structure

Proteins with more than about 50 amino acids adopt complex three-dimensional shapes, which we call tertiary structures. Each protein folds into a very specific shape, known as the native conformation, which is often the biologically active form of the protein.

Protein tertiary structures usually contain two or more elements of secondary structure. The following examples, an enzyme and collagen, illustrate how the structure of a protein relates to its biological activity (see Figure 9.14).

Enzymes are excellent examples of proteins with tertiary structure. The structures of many enzymes contain both α-helices and β-sheets. The three-dimensional arrangement of these structures creates an overall shape that is unique for every protein. Most enzymes have a roughly spherical native conformation and contain a cleft in their surface known as the active site. The active site is the location within the enzyme at which the reaction it catalyses, occurs.

Collagen is one of the most abundant proteins in nature. It has a very simple tertiary structure (in fact some texts may classify collagen as just having secondary structure). It consists of three independent peptide backbones wound together in a triple-coil. This forms a very rigid fibre-like structure that provides mechanical support in tissues such as skin, tendons and bones. α-Keratin is a related protein with a similar structure. It has a double coil of α-helices and is a principal component of hair.

 Enzyme biochemistry is discussed in more detail in Chapter 3 'The biochemistry of cells' in the *Therapeutics and Human Physiology* book in this series.

Protein structure: quaternary structure

Some proteins have more than one tertiary peptide chain in their structure. Haemoglobin, the protein that transports oxygen in the blood, is composed of four distinct tertiary protein subunits. There are two identical α-subunits and two identical β-subunits. Proteins of this type are described as having quaternary structure.

Proteins as drug targets

Proteins, especially enzymes and receptors, are important molecular targets for many drugs. Here we will focus on the example of a drug targeting an

FIGURE 9.13 Secondary protein structure. (A) Two representations of the α-helix showing a partial structure highlighting the backbone (bold) and a cartoon single helix; (B) antiparallel β-sheet; (C) parallel β-sheet. For clarity, all amino acid side chains are abbreviated as R.

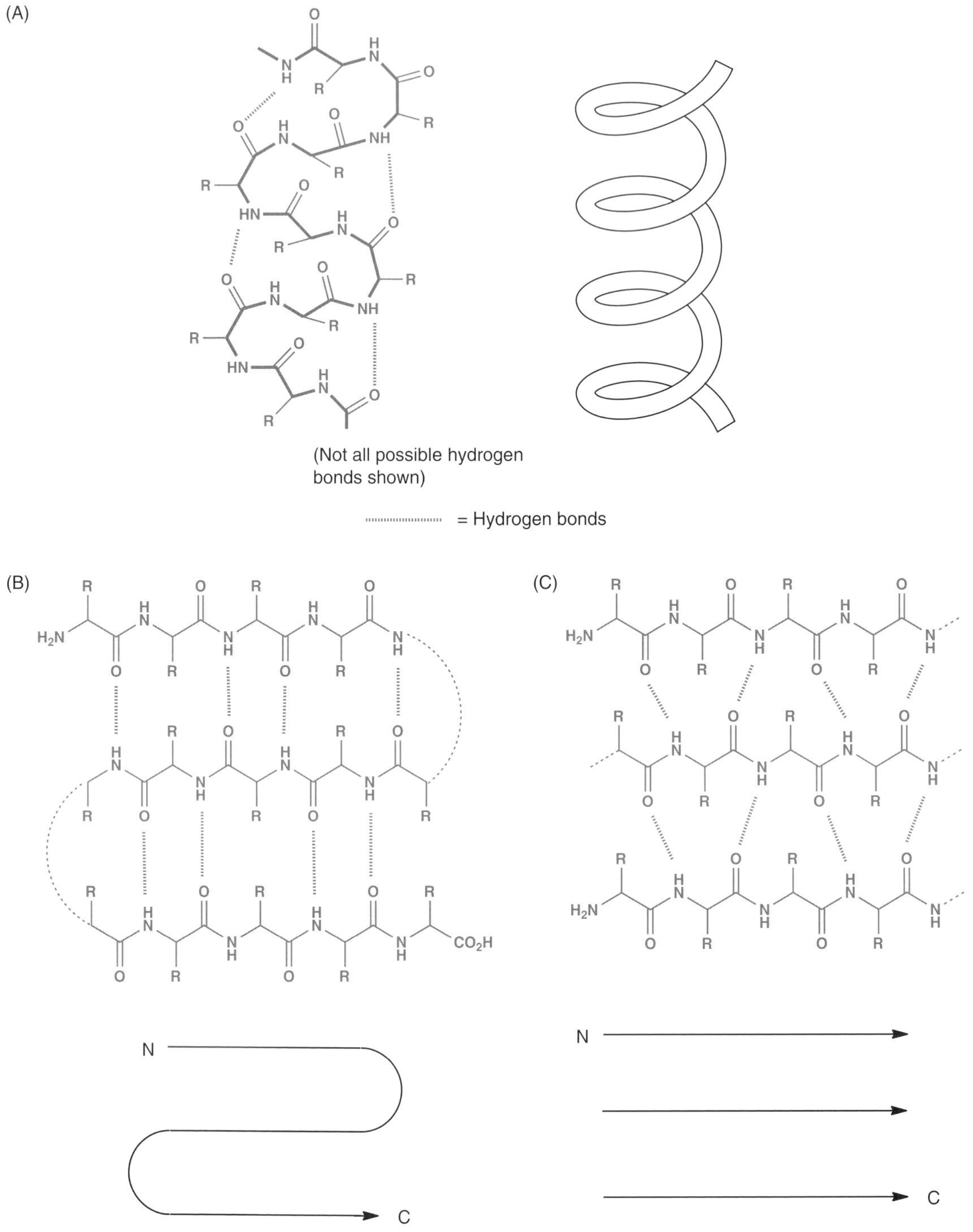

(A)

(Not all possible hydrogen bonds shown)

⋯⋯⋯⋯ = Hydrogen bonds

(B)

(C)

enzyme, although membrane-bound receptors are also very important drug targets.

The unique tertiary structure of an enzyme often leads to the formation of a very specific active site. Enzymes are biological catalysts that regulate a wide range of biochemical reactions and their active sites bind specific substrates. Although other mechanisms of action are possible, a drug that targets an enzyme often binds to the active site, blocking the binding of the substrate, and hence inhibiting the function of the enzyme. This is commonly referred to as *enzyme inhibition*.

FIGURE 9.14 The structure of a typical enzyme, carbonic anhydrase (A) and collagen (B). Note the regular elongated structure of collagen (each of the three peptide chains is coloured separately), and the compact, spherical, but irregular structure of the enzyme (PDB code: 1CA2).

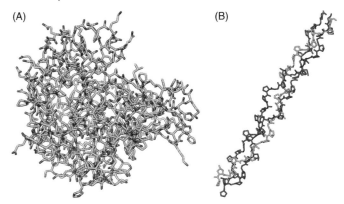

(A)   (B)

FIGURE 9.15 A detailed view of atorvastatin bound to the active site of HMG-CoA reductase. The shape of the drug complements the shape and amino acid functional groups of the enzyme active site. The various **bonding interactions** between the drug and protein are highlighted as black solid lines (hydrogen bonds) and black dashed lines (non-polar interactions) (PDB code: 1HWK).

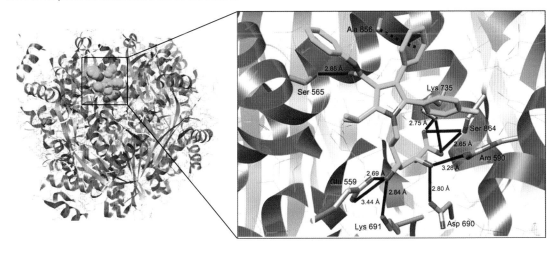

 Hydrogen bond   Hydrophobic interaction

A good example of a best-selling drug that causes enzyme inhibition is atorvastatin, which is used to treat high cholesterol levels. It works by inhibiting HMG-CoA reductase, an enzyme involved in the production of cholesterol. As illustrated in Figure 9.15, the drug has been designed to bind perfectly in the active site, forming a number of interactions with specific amino acids.

The wide variety of protein structures means that usually a drug is very specific for a certain protein target (see Chapter 1). This is in contrast to drugs targeting DNA where many planar compounds can potentially intercalate with DNA. DNA targeting

drugs are therefore much less specific for their target and serious side effects are common for DNA targeted therapeutics (see Section 9.1).

> Enzyme and receptor pharmacology is discussed in Chapter 3 'The biochemistry of cells' and the lock and key hypothesis is covered in Chapter 4 'Introduction to drug action' of the *Therapeutics and Human Physiology* book in this series.

KEY POINT

Protein molecules, particularly enzymes, are often drug targets.

9.4 Carbohydrates

Carbohydrates (literally meaning *carbon* and *water*) are molecules composed almost exclusively of carbon, hydrogen and oxygen. Carbohydrate monomers are called *monosaccharides* (trivially called *sugars*) and are found throughout nature. Carbohydrates are synthesized in plants during the process of photosynthesis, with their carbon atoms being obtained from atmospheric carbon dioxide. The best-known role of carbohydrates in humans is nutritional (that is, they provide energy), but here we will focus on other aspects of their biochemistry. The carbohydrates are a huge class of organic molecules, so we must limit the discussion here to selected examples.

Simple carbohydrates – monosaccharides

Monosaccharides are the simplest carbohydrates and are classified by the number of carbons they contain. Most common monosaccharides contain five carbons (classified as *pentoses*) or six carbons (known as *hexoses*). The general structure of these carbohydrates features a carbon chain functionalized with

several hydroxyl groups and one or two more oxidized groups; aldehydes, ketones or carboxylic acids are all encountered.

Monosaccharide structure

Monosaccharides exist in two isomeric forms: open-chain and cyclic. In aqueous solution, these structures are in equilibrium. (Remember that biological systems are aqueous and that all biochemistry occurs in water.) The cyclic form is, generally, more stable and therefore the more likely structure.

Understanding carbohydrate structure is complicated by the large number of stereoisomers that are possible. This issue is addressed by the use of Fischer projections, a convenient way to represent three dimensions on the page of a book. In Figure 9.16, Fischer projections are used to show the structure of the open-chain forms of two common carbohydrates, glucose (a hexose) and ribose (a pentose).

 More details on Fischer projections can be found in Chapter 3 'Stereochemistry and drug action'.

FIGURE 9.16 Monosaccharide structure. Different ways to represent the structures of: the hexose sugar, glucose; and ribose, a pentose sugar.

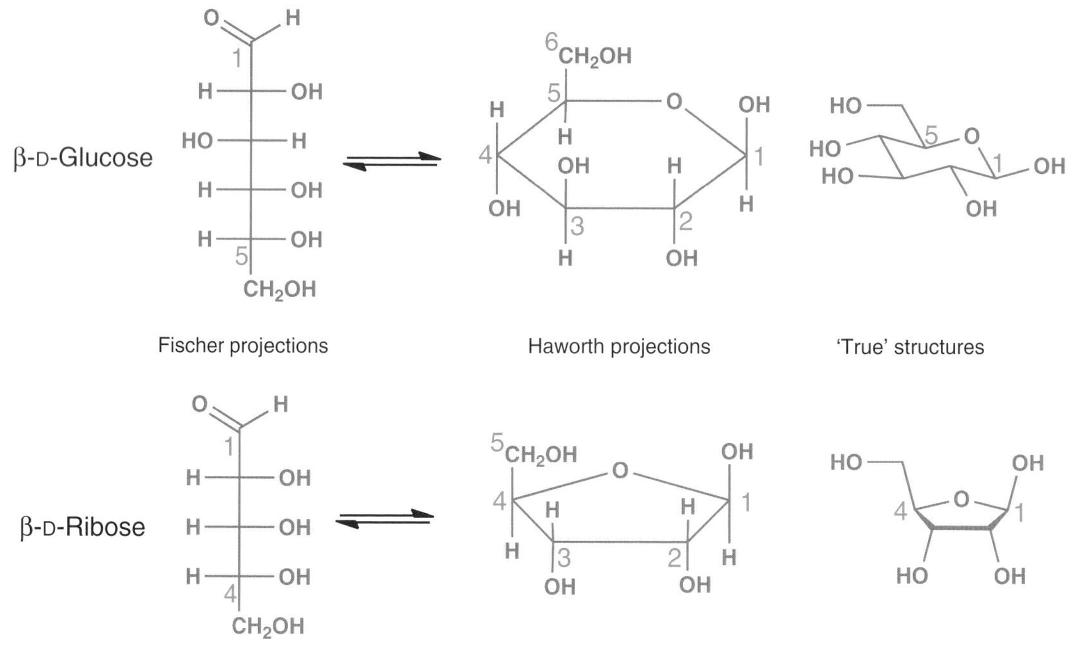

β-D-Glucose

Fischer projections   Haworth projections   'True' structures

β-D-Ribose

SCHEME 9.1

Monosaccharide stereochemistry

The D- and L-notation is used to describe carbohydrate stereochemistry, as well as amino acid stereochemistry. The three-carbon monosaccharide glyceraldehyde is the reference compound. This sugar contains a single chiral centre, resulting in two mirror image enantiomers, assigned D (*dextro*) and L (*levo*).

D-Glyceraldehyde   L-Glyceraldehyde

Monosaccharides matching this configuration at the highest numbered chiral carbon are named D and those matching the configuration of L-glyceraldehyde are labelled L. It is important to note that the D/L system is independent of both the R/S system and of optical activity. For example, L-monosaccharides may rotate plane-polarized light in either direction. Although the R/S system is arguably more logical than the D/L method, most common carbohydrates are found in the D form and so the D/L system has become the accepted standard in carbohydrate chemistry.

Stereochemistry is discussed in more detail in Chapter 3 'Stereochemistry and drug action'; amino acid stereochemistry is discussed further in Section 9.3.

SELF CHECK 9.7

Redraw the Fischer projections of glucose and ribose and draw the full mechanism for hemiacetal formation. Make sure you understand how the cyclic structure is derived from the open-chain isomer.

Hemiacetal formation

The cyclic forms of glucose and ribose arise from a nucleophilic reaction between one of their hydroxyl groups and the aldehyde functional group present in both these monosaccharides. Compare the structures in Figure 9.16 and the generalized mechanism shown (see Scheme 9.1):

Another type of drawing can be used to represent the cyclic form of carbohydrates. Known as *Haworth projections* they also help us see the three-dimensional shape of the molecule, especially the chiral centres; this type of projection is particularly favoured in biochemistry textbooks. This is still a diagrammatic (easy to draw) way of showing a structure. For comparison, more accurate structures are also shown in Figure 9.16. Note that hexoses such as glucose normally have stable 'chair' conformations.

Included on the Haworth projections in Figure 9.16 is the numbering system used in carbohydrates. By convention, the hemiacetal carbon (also called the anomeric carbon) is labelled C1; the remaining carbons then follow in order.

We might think that any of the hydroxyl groups within an open-chain carbohydrate could react with the highest oxidized group (carbon-1) to yield a hemiacetal. In principle this can and does happen, but at equilibrium the most stable (that is, energetically favoured) cyclic structures are more likely to form. Therefore, in Figure 9.16, the cyclic structures shown for glucose and ribose are those most commonly encountered at equilibrium.

 More information on chair and boat conformations can be found in Chapter 3 'Stereochemistry and drug action'. Hemiacetals and acetals are discussed in greater detail in Chapter 6 'The carbonyl group and its chemistry'.

Monosaccharide isomerism: anomers

D-Glucose

Hemiacetal

α–anomer

β–anomer

D-Ribose

Hemiacetal

The formation of cyclic monosaccharides can result in two isomers, called the α- and β-anomers. In cyclic D-glucose and D-ribose, the hemiacetal carbon (the aldehyde carbon in the open-chain isomer) is known as the anomeric carbon. The hemiacetal functional group and the *anomeric carbons* are highlighted below as a 'C' with related bonds in green.

The stereochemistry of the anomeric centre is assigned on the basis of stereochemical priorities. In most cases, including D-ribose and D-glucose, the α-anomer has the hydroxyl group in the axial (or down) orientation, whereas in the β-anomer the hydroxyl group is equatorial (or up). Remember that carbohydrates are in equilibrium between the open-chain and cyclic forms. Therefore, in aqueous solution, D-glucose and D-ribose exist as a mixture of the α- and β-anomers. In both these examples, the more stable β-anomers are the major form.

KEY POINT

Cyclic monosaccharides can form two isomers called anomers.

Disaccharides

As we have already observed, nucleic acids (see Section 9.1) and proteins (see Section 9.2) are complex biochemical molecules, built from simple monomer building blocks. This is also the case for polysaccha-rides, which are large carbohydrates constructed from simple monosaccharides.

Oligosaccharides are small polysaccharides con-taining fewer than ten carbohydrate monomers. A good example is sucrose, a molecule familiar to us as table sugar. Sucrose is a **disaccharide** of D-glucose and D-fructose joined by a linkage known as a **glyco-sidic bond**. These bonds are formed by loss of water (another condensation reaction), between two adja-cent hydroxyl groups and are highlighted in bold on the structures in Figure 9.17.

Study the structure of sucrose in Figure 9.17 carefully. The D-glucose monomer is the α-anomer (the O-group at carbon 1 is 'down'), and bonds directly to the anomeric carbon of β-D-fructose. This is the first time we have met fructose. Note that it contains six carbon atoms, so is a hexose carbohydrate, but forms a five-membered ring, because the carbonyl group in the open-chain form was at C2, rather than C1. Therefore carbon-1 is outside the ring system and the anomeric carbon is at position 2.

Glycosidic bonds are described using a conven-tion that describes both the number of each carbon involved in the bond and its anomeric form. In sucrose, the glycosidic bond is between two anomeric carbons and is described as: D-glucose-(α1→β2)-D-fructose.

Lactose is a disaccharide of D-galactose (a hexose) and D-glucose. It is present in the milk of many mam-mals, including cows and humans. The glycosidic linkage between the β-anomeric (C1) hydroxyl group of galactose and the hydroxyl group of C4 in glucose

FIGURE 9.17 Sucrose and lactose, two disaccharides (oligosaccharides composed of *two* carbohydrate monomers). The glycosidic bonds are highlighted in bold.

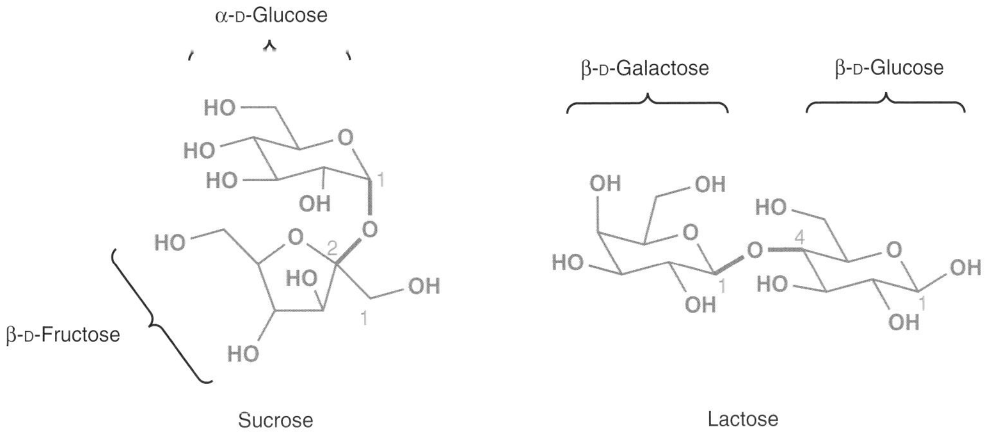

Sucrose

Lactose

(not an anomeric carbon) is described as D-galactose-(β1→4)-D-glucose. (See Integration Boxes 9.1 and 9.2 on lactose intolerance and the use of lactulose in laxatives.)

The advantage of this system to describe glycosidic bonds is hopefully obvious – with so many variations possible this simple method helps anyone easily determine the nature of the bond between two monosaccharides.

SELF CHECK 9.8

Using the structures within this chapter, draw the structure of the disaccharide D-galactose-(α1→6)-β-D-fructose.

Complex carbohydrates – polysaccharides

Large carbohydrates contain greater numbers of monosaccharide building blocks. Known as **polysaccharides**, they can be composed of repeats of a single monosaccharide monomer or a mixture of different sugars. A distinctive feature of these macromolecules is their ability to form *branched structures*; this is never seen in nucleic acids or proteins.

Glycogen is a common animal polysaccharide used to store glucose. When energy is required, glycogen is rapidly degraded to release glucose monomers. Glycogen is a polymer composed entirely of glucose

INTEGRATION BOX 9.1

Lactose intolerance

Lactose is an important dietary carbohydrate, especially for infant mammals who consume large quantities of milk. Young children express an intestinal enzyme, β-D-galactosidase, (commonly known by the trivial name *lactase*) that breaks down (or *digests*) lactose into the two constituent monosaccharides which are readily absorbed into the bloodstream. However, most mammals do not consume milk beyond infancy, so the production of this enzyme decreases into adulthood.

Some humans, notably those living in the West, continue to express lactase into adulthood and are able to consume milk and dairy products throughout life. By contrast, most South-east Asian adults do not express lactase and cannot digest lactose. Milk and dairy products are rarely consumed in these parts of the world.

A minority of Western adults do not express lactase and are said to be *lactose intolerant*. Eating dairy products causes abdominal discomfort as bacteria in the lower gut ferment the lactose. Fortunately this condition is easily treated. A variety of preparations of the enzyme lactase are available that can be added to milk, or taken during a meal, to digest lactose and help prevent symptoms occurring.

Lactulose in laxatives

β–D-Fructose

β–D-Galatose

Lactulose is a disaccharide of D-galactose and D-fructose linked by a β1→4 glycosidic bond. Lactulose is used widely as a laxative for the treatment of constipation; it does not occur naturally and is manufactured synthetically. It is taken orally as a syrup, and remains undigested until it reaches the colon. There it causes water to concentrate by osmosis; this softens the accumulated stools, making them easier to pass.

 Osmosis is discussed in more detail in Chapter 11 'Colligative properties' of the *Pharmaceutics* book in this series.

monomers joined by (α1→4) bonds. The structure of glycogen, shown partially in Figure 9.18, contains many branch points. Additional (α1→6) glycosidic bonds start a new branch of the polymer.

Polysaccharides have numerous additional functions in biology. The ABO blood group system is a good example. Oligosaccharides of four or five monosaccharides attached to the surface of the cell membranes of red blood cells act as antigens that determine the blood group of any individual.

 Blood groups are discussed in more detail in Chapter 9 'Haematology' of the *Therapeutics and Human Physiology* book in this series.

FIGURE 9.18 The structure of glycogen. Note the branch point (only one is shown) from carbon-6 of one of the glucose monomers.

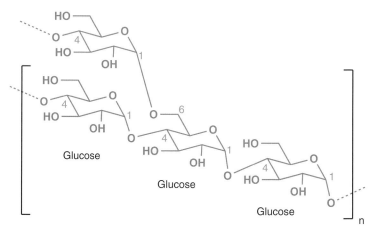

9.5 Lipids

The final class of macromolecules we will study in this chapter are the lipids. They have a greater variety of structural types than nucleic acids, proteins, or carbohydrates and can be broadly defined as *water insoluble molecules from biological systems that are readily soluble in non-polar organic solvents*. In other words, they are hydrophobic. This property is unusual; the molecules we have studied so far are hydrophilic, and dissolve readily in water. We shall see how the non-polar properties of lipids are vital to several important aspects of biochemistry.

Lipids differ further from the other molecules in this chapter in that they do not form large polymeric structures. They are also somewhat smaller and simpler molecules.

Fatty acids

Fatty acids are the best-known examples of lipids although they are usually found in biochemistry as components of larger molecules. Fatty acids have a single carboxylic acid functional group attached to an unbranched hydrocarbon chain. The length of this chain, and the degree of saturation, characterizes individual fatty acids. The chain length can be classified as short (less than 6 carbons), medium (6 to 12 carbons) or long (greater than 12 carbons). These chains may be fully saturated (no double bonds) or unsaturated (containing one or more double bonds). Almost all naturally occurring fatty acids contain an even number of carbons.

Medium- and long-chain fatty acids are **amphiphilic**, a term used in biochemistry to describe molecules with both polar and non-polar characteristics. This property arises because the carboxylic acid is a very polar functional group but the long hydrocarbon chain is very non-polar.

 Properties of amphiphilic molecules are discussed in Chapter 5 'Alcohols, phenols, ethers, organic halogen compounds, and amines'.

You have met acetic acid (ethanoic acid), the most important short-chain fatty acid, many times in this book. Most important fatty acids in human biochemistry are long-chained, with up to about 20 carbon atoms. Figure 9.19 illustrates three typical long-chain fatty acid structures. Each is a *free fatty acid*, meaning that the carboxylic acid functional group is not bonded to another group. Stearic acid is an 18-carbon, saturated fatty acid. Oleic acid is an unsaturated fatty acid with a *Z* (*cis*) double bond at carbon-9.

Arachidonic acid is a *poly*unsaturated fatty acid – in other words, it contains *many* double bonds. All four double bonds have the *Z* (*cis*) geometry. Arachidonic acid is notable as the precursor for the 20-carbon eicosanoids, a group including the prostaglandins, thromboxanes and the leukotrienes, which are all important molecules in cellular signalling. Fatty acids with such a high degree of unsaturation are otherwise unusual.

 Polyunsaturated fatty acids and prostaglandins are discussed in more detail in Chapter 7 'Communication systems in the body – autocoids and hormones' of the *Therapeutics and Human Physiology* book in this series.

Study Figure 9.19 and note the effect that double bonds in the hydrocarbon chain of each fatty acid have on the structure of the molecule. Stearic acid has a straight and elongated structure. Oleic acid has a distinctive kink due to the double bond. Arachidonic acid, with four double bonds, has a very different three-dimensional shape. Since double bonds do not rotate, the geometry of the double bonds maintains these fatty acids in distinctive and fixed shapes.

> **KEY POINT**
>
> Fatty acids contain one carboxylic acid group, almost always have an even number of carbons and, if the alkyl chain is unsaturated, the double bonds will be *cis*.

FIGURE 9.19 Examples of common fatty acids. Compare each drawn structure against the accompanying molecular model showing the precise three-dimensional structure.

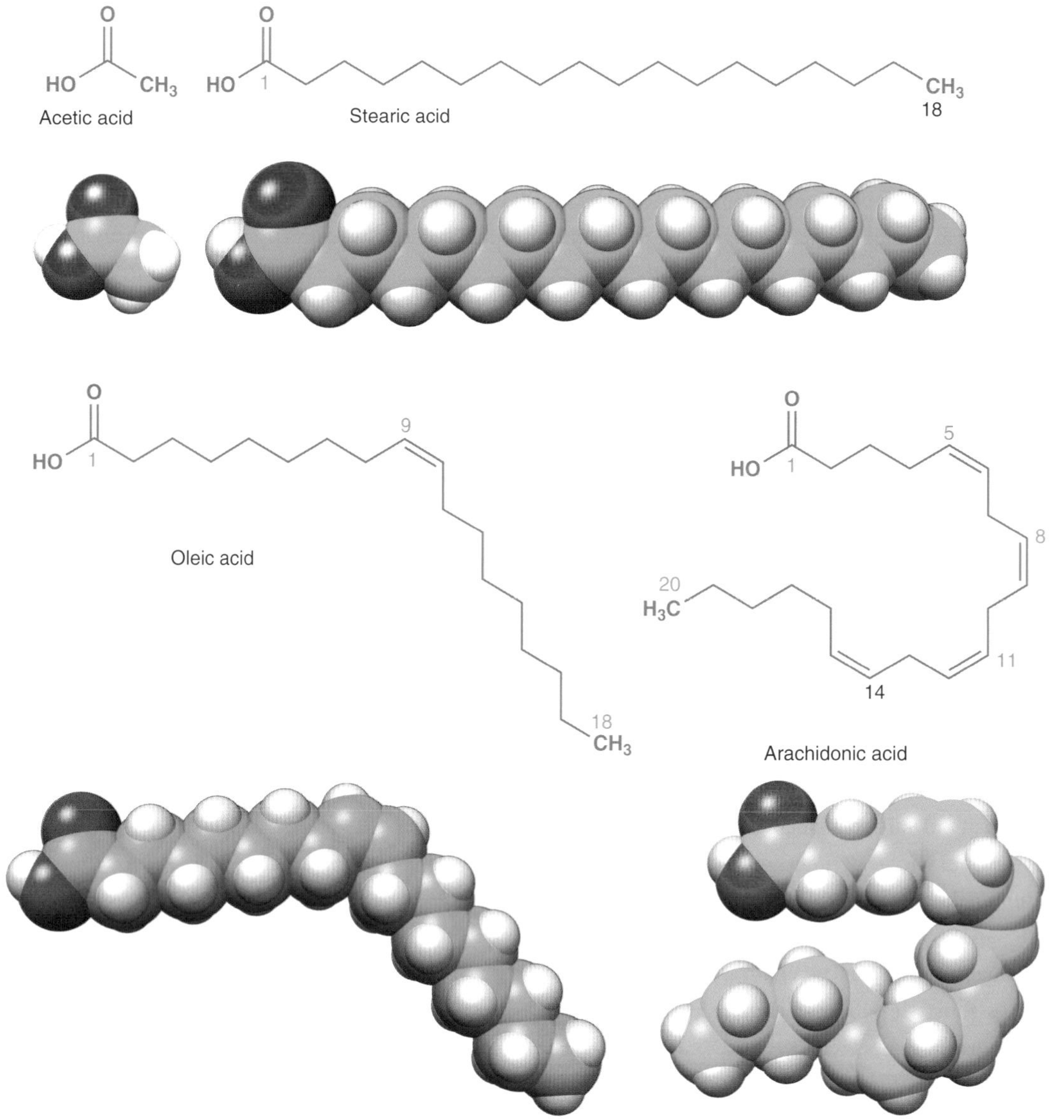

Acetic acid

Stearic acid

Oleic acid

Arachidonic acid

Triglycerides

Triglycerides are a class of lipids commonly referred to as fats, and are an important part of our diet. A triglyceride is derived from a molecule of glycerol that is bonded to three fatty acids by ester functional groups, as illustrated in Figure 9.20. Triglycerides are even less polar than free fatty acids. This is because the polarity of the free carboxylic acid groups has been masked by conversion into non-polar fatty acid esters.

 Carboxylic acids and esters are discussed in Chapter 6 'The carbonyl group and its chemistry'.

SELF CHECK 9.9

Study the triglyceride structure and identify the ester functional group and the central glycerol molecule. Think about *why* esters are less polar than carboxylic acids rather than just learning this as a fact.

SELF CHECK 9.10

You should also remind yourself about the reactivity of esters. Are esters more or less reactive to water than (a) acyl chlorides (b) amides?

FIGURE 9.20   The structure of a typical triglyceride.

Palmitic acid

CH_3 16

CH_3 14

Myristic acid

Palmitoleic acid

9

CH_3 16

HO — OH, OH

Glycerol

In Figure 9.20, the structure of glycerol alone is included for reference, next to a typical triglyceride. The glycerol part of this large molecule is highlighted in bold. Two saturated fatty acids, palmitic acid (16 carbons) and myristic acid (14 carbons) and one unsaturated fatty acid, palmitoleic acid, are bonded via esters to glycerol forming a triglyceride. The three fatty acids used in this example are for illustration; any combination of fatty acids can be used to form a triglyceride. In nature, mixed triglycerides (glycerol esterified with three different fatty acids) are very common. By contrast, triglycerides with three identical fatty acid side chains are rare.

Phospholipids

The most significant role of lipids in biology is their function as a major component of cell membranes. Cell membranes are formed from a **bilayer** of a particular type of lipid known as a phospholipid. Looking at the chemistry of these lipids in more detail will help us understand how cell membranes work. Figure 9.21 illustrates the structure of phosphatidylcholine. This is the most common phospholipid in the mammalian cell membrane, so we will focus on the chemistry of this example.

Phospholipids are related to triglycerides; the only difference is that one fatty acid is replaced by a polar

head group. The two fatty acids constitute the non-polar *tail*. In phosphatidylcholine this consists of two fatty acids, joined by ester linkages, to carbon-1 (C1) and carbon-2 (C2) of glycerol. The fatty acid chain at C1 is usually saturated and 16 or 18 carbons in length. Attached to C2 of glycerol is a fatty acid that is often mono-unsaturated and is 16, 18 or 20 carbons in length. A choline molecule is bonded to glycerol C3 by a phosphodiester linker (we have met this functional group already this chapter in the backbone of DNA and RNA). The polarity of the head group arises from the two charged functional groups, a positively charged quaternary amino group and a negatively charged phosphodiester group. By contrast, the tail of the molecule is extremely non-polar, being composed of two long-chain fatty acids.

 Membrane structure is discussed in more detail in Chapter 2 'Molecular cell biology' of the *Therapeutics and Human Physiology* book in this series.

This is another example of an amphiphilic molecule. The head group is soluble in polar solvents (i.e. water, the *only* available solvent in a biological system), whilst the tail has an affinity for other non-polar groups. When large numbers of amphiphilic phospholipids are packed together, they form the cell membrane lipid bilayer, illustrated as a cartoon in Figure 9.21. The hydrophilic, polar head groups on both sides of the

FIGURE 9.21 Phospholipids and the cell membrane.

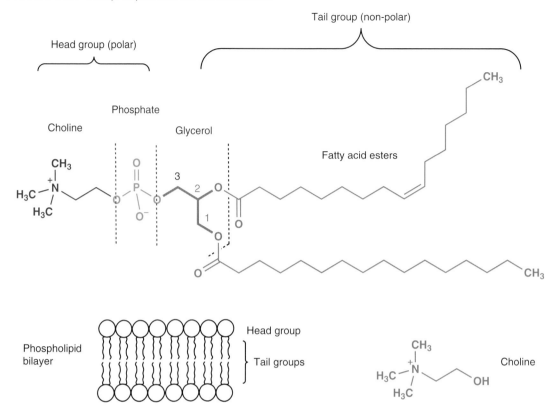

membrane (i.e. the extracellular/outside and cytoplasmic/inside) are in contact with water. In the interior of the bilayer, the hydrophobic fatty acid hydrocarbon chains strictly exclude water. Hydrophobic interactions between tail groups further stabilize the bilayer.

So we see how the structure of the membrane is directly related to the strongly amphiphilic properties of the phospholipid. It is important to remember that other phospholipids, with differing head and tail groups, but very similar physical properties, are also found in a typical cell membrane.

The existence of a lipid bilayer is essential for forming very stable cells, but it often provides the pharmaceutical scientist with a significant challenge when a drug needs to cross a cell membrane and enter a cell to find its biological target!

 More information on drug delivery and biological barriers can be found in Chapter 13 'Drug development and delivery' of the *Pharmaceutics* book in this series.

KEY POINT

Phospholipids possess very polar or ionic sites in addition to long hydrocarbon chains. This makes them ideal for forming the cell membrane lipid bilayer.

Fats: the ultimate energy source

Fatty acids are an important source of energy, yielding more energy than the equivalent quantity of carbohydrate. They are stored as triglycerides in specialist tissue (adipose cells), and broken down by lipase enzymes to free fatty acids and glycerol when energy is needed. Each free fatty acid molecule is then metabolized by the addition of a hydroxyl group, a process known as β-oxidation, since oxygen is added specifically to the second carbon from the carboxylic group. Since fatty acids are less oxidized than carbohydrates (which have large numbers of oxygen functional groups), they require further oxidation to be degraded and are therefore a very efficient energy store (see Scheme 9.2).

Fats and diet

The popular press is full of stories about fat in our diet. Newspapers often use *cis/trans*-terminology (e.g. 'trans-fats') to describe unsaturated fats instead of the scientifically correct convention of *E/Z* so it is important that you are familiar with both nomenclatures.

Saturated fatty acids are generally obtained from animal fats, whereas unsaturated fats are obtained from plants. Unsaturated fatty acids tend to be liquids or oils. This property arises because of the distinctive 'kink' in their structure, caused by the double bond. This kink causes the hydrophobic interactions between the non-polar side chains in adjacent fat molecules to be less effective. Saturated fatty acids tend to be solids; their more regular structure promotes efficient non-polar interactions between adjacent molecules.

In the fatty acids we have seen so far, all the unsaturated examples have contained *cis*-double bonds. This is no accident; most double bonds in naturally occurring fatty acids have this geometry.

In the food industry, liquid polyunsaturated plant oils are partially hydrogenated, by a chemical process, to manufacture solid oils. This process is used in the manufacture of margarine. In processed foods of this type, some of the *cis*-fatty acids are isomerized to *trans*-fatty acids; although unsaturated, these *trans*-fatty acids have a similar shape to their saturated equivalents. *Trans*-fatty acids are linked to an increased risk of cardiovascular disease whereas polyunsaturated (*cis*) fatty acids are considered to be much better for you. A minor change to the chemistry and shape of a molecule can have significant effects on its properties!

Acetyl CoA is released by this process and enters the citric acid (or Krebs) cycle, leading to the production of energy in the form of ATP. More acetyl CoA is synthesized from fat than from the comparable amount of carbohydrate. Therefore you might expect fats to be the body's primary energy source. However, carbohydrates are preferred because they can enter the citric acid cycle more quickly than fats and thus release their energy more rapidly.

Thus dietary carbohydrate is used for energy in preference to fat stores, which explains why a diet low

SCHEME 9.2

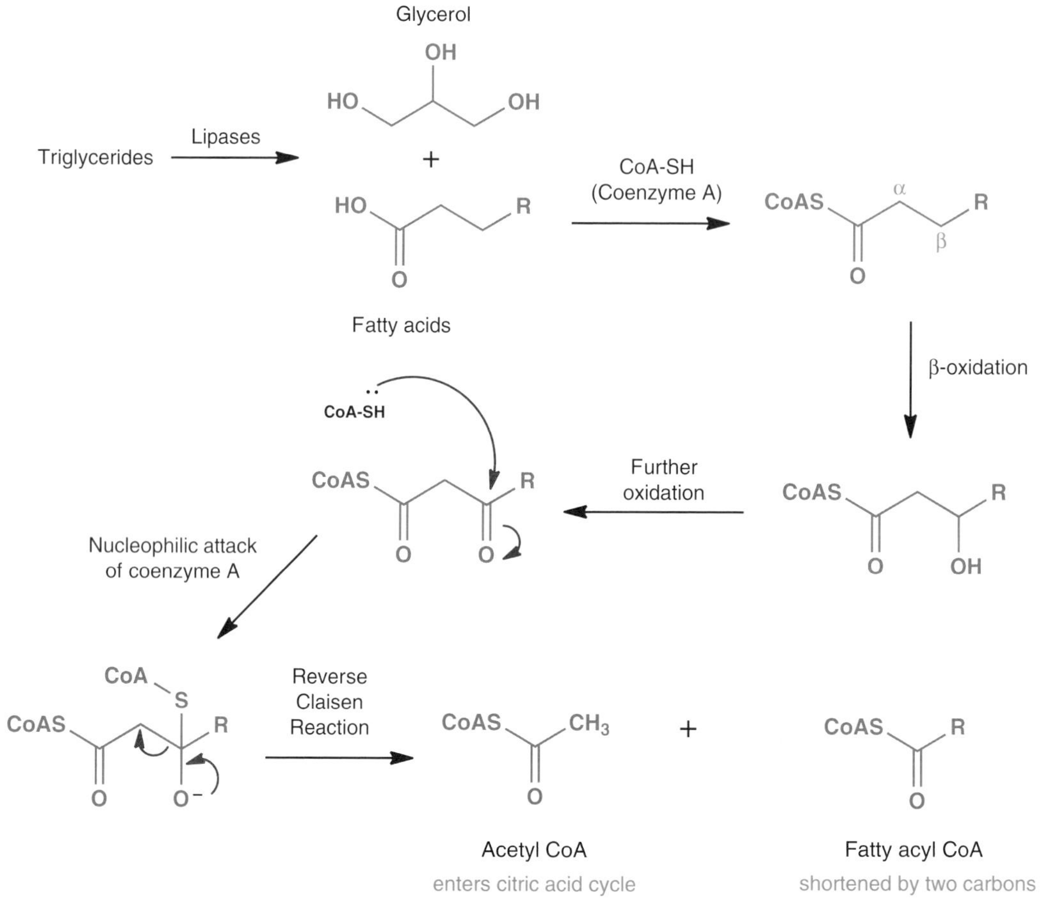
Glycerol
OH
HO
OH
Triglycerides
Lipases
+
HO
R
O
Fatty acids
CoA-SH
(Coenzyme A)
CoAS
α
R
β
O
β-oxidation
CoA-SH
CoAS
R
O
O
Further
oxidation
CoAS
R
O
OH
Nucleophilic attack
of coenzyme A
CoA
S
CoAS
R
O
O⁻
Reverse
Claisen
Reaction
CoAS
CH3
O
+
CoAS
R
O
Acetyl CoA
enters citric acid cycle
Fatty acyl CoA
shortened by two carbons

in carbohydrate promotes fat metabolism – and why people choose low-sugar diets to reduce body fat and weight.

Steroids

Steroids are an important class of lipids that share the characteristic hydrocarbon skeleton shown in Figure 9.22. They contain four fused, usually saturated, hydrocarbon rings, named, by convention, A, B, C, and D. The A, B and C rings are six-membered and the D ring is five-membered. All steroids have this basic structure.

Cholesterol, the most common steroid in animals, is a molecule that you may well have heard of, and is an important component of the cell membrane. It is a typical steroid that illustrates some common features of this class of lipids. Steroids often contain an oxygen containing functional group at C3; in the case of cholesterol this is a hydroxyl group. The presence of this group allows further classification of this molecule as a *sterol*, the –ol suffix denoting an alcohol. Steroids often contain methyl groups at C10 and C13 and larger hydrocarbon groups at C17. This is further illustrated by studying the structure of ergosterol (see Figure 9.22). You can tell immediately from its structure that it is a steroid (and a sterol) and has hydrocarbon groups in the characteristic positions.

Steroid structure

You will find steroids frequently drawn as shown in Figure 9.22, because it is a convenient way to illustrate a complex structure on a flat page. However, it is important to remember that *all* molecules have three-dimensional structures. To emphasize how different three-dimensional structures can be from flat

FIGURE 9.22 A general steroid structure (showing the ring naming and steroid numbering system) and two examples of typical steroids.

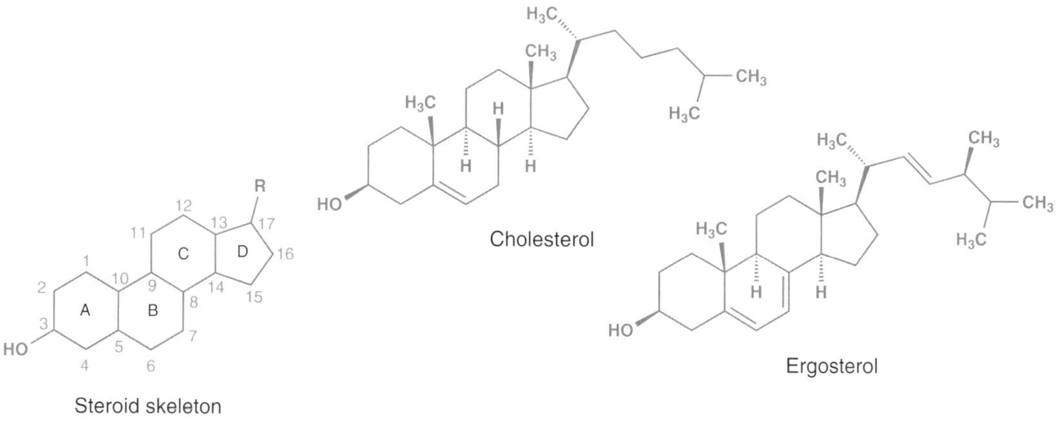

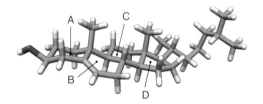

Cholesterol

Ergosterol

Steroid skeleton

representations, Figure 9.23 shows a computer generated model of cholesterol. Note how the structure is far from flat. The model is orientated to highlight the shape of the steroid A-ring. Note how it adopts the 'chair' conformation typical of six-membered saturated

FIGURE 9.23 A computer generated model of the three-dimensional structure of cholesterol. Carbon atoms are shown in green, hydrogen atoms are grey and the oxygen atom is red. The A, B, C, D rings are labelled.

rings. Ultimately, this arises from the tetrahedral conformation that most of the saturated, sp³-hybridized, carbons in the structure adopt.

 Conformational isomerism is discussed in more detail in Chapter 3 'Stereochemistry and drug action'.

Steroid biochemistry

Steroids have many important roles in the body and are used widely as drugs (see Figure 9.24). For example, cholesterol is a major component of animal cell membranes. Because it is non-polar, cholesterol interacts with the tail groups of phospholipids, and together these two types of lipids determine the overall properties of a membrane.

FIGURE 9.24 Some steroid hormones and drugs. Despite the similarity of these structures (they all contain the steroid ring system) they have very different biological activities.

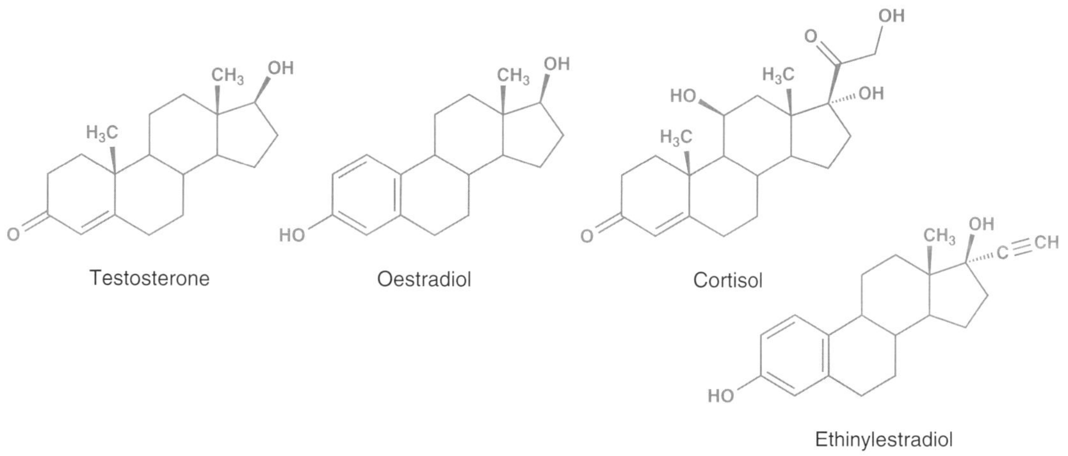

Testosterone

Oestradiol

Cortisol

Ethinylestradiol

Cholesterol is also important as the **biosynthetic precursor** for the steroid hormones, molecules vital to human physiology. For example, cholesterol is converted into the sex hormones testosterone and oestradiol, responsible for male and female sexual development and characteristics, respectively. Cortisol is a hormone synthesized in the body from cholesterol that regulates the metabolism of lipids, proteins and carbohydrates. It is also used as a therapeutic drug (under the name hydrocortisone) for a variety of uses, ranging from ulcerative colitis of the bowel to skin conditions such as eczema.

However, many steroid therapeutics are not naturally occurring molecules. Ethinylestradiol is a syn-

thetic derivative of oestradiol containing an unusual carbon-carbon triple bond functional group. It is widely used in many oral contraceptives ('the pill') and possibly one of the most widely prescribed drugs in the world.

 Ethinylestradiol is also discussed in Chapter 4 'Properties of aliphatic hydrocarbons'.

KEY POINT

Cholesterol is one of the most important steroid molecules. It is an important component of cell membranes and is a precursor of steroid hormones.

CASE STUDY 9.1

A teenage girl attends Dilip's pharmacy requiring emergency hormonal contraception. Although worried about needing this treatment, she is also anxious about taking the drug. How does Dilip do his best to reassure her?

REFLECTION QUESTIONS

1. How does the drug work?

2. Are there any risks associated with the drug?

Answers

1 Emergency hormonal contraception (EHC), also known as the 'morning after pill', utilizes the drug levonorgestrel. This drug is a synthetic progestin, mimicking the activity of the natural steroid hormone progesterone. Progesterone is biosynthetically derived from cholesterol. Levonorgestrel works by preventing the ovaries from releasing an egg and altering the lining of the womb so an egg cannot be embedded.

2 Whilst the patient may experience some minor side effects, such as nausea, there are no significant risks associated with this drug, except for patients with serious liver disease.

CHAPTER SUMMARY

This aim of this chapter has been to highlight some aspects of biochemistry that are essential to understand before progressing to a more advanced study of pharmacy. Although many facts have been presented that need to be committed to memory, it is hoped that you have also learnt to apply concepts of chemistry you already know to more

complex molecules. Hopefully, you now agree that this can be achieved with relative ease. The concept of molecules as three-dimensional entities has also been an important theme throughout this chapter. It is important to think about the shape of a molecule, as well as its chemistry, when studying its role in a biological system.

Voet, D and Voet, J.G. *Biochemistry*. 4th edn. John Wiley and Sons, 2011.

> An excellent comprehensive textbook that offers alternative and far more detailed descriptions of all the material presented in this chapter.

Dewick, P.M. *Essentials of Organic Chemistry: For Students of Pharmacy, Medicinal Chemistry and Biological Chemistry.* Wiley-Blackwell, 2006.

> A textbook focusing on chemistry, especially structures and mechanisms. Its writing is directed to Pharmacy students and it is well illustrated with pharmaceutical examples.

Patrick, G.L. *An Introduction to Medicinal Chemistry*. 5th edn. Oxford University Press, 2013.

> A very accessible textbook that expands upon the chemistry of drugs and their mechanisms of actions against target macromolecules.

Origins of drug molecules

TIM SNAPE

Our ancestors chewed tree bark and drank herbal tea to relieve their illnesses, whereas today we are more likely to visit the doctor and take prescribed medication. The medication generally appears in tablet or capsule form, and the active ingredient will have first appeared in pure form on a chemist's bench. This chapter addresses the origins of the active ingredient.

Some of today's drugs are still derived from tree bark and herbal teas, others from more sustainable (easily replaced) biological sources, such as moulds and soil bacteria. Some drugs are synthetic (they are made wholly by chemists using chemical reactions), and some are a mixture (semi-synthetic). You will learn to recognize the structures of natural products based on their complexity and to understand that fully synthetic drugs are usually a lot simpler than natural products.

Despite the structural complexity of natural products, they can be used by chemists as a source of inspiration for the development of new drugs, not only to optimize their biological activity, but also to prepare novel compounds for **patent** protection. Genetic engineering is of ever-increasing importance in the production of drug molecules, enabling us to use whole proteins as drugs. The chapter ends with a discussion of how compounds prepared in the laboratory or isolated from nature (or a mixture of the two) can be purified ready for biological testing.

Learning objectives

Having read this chapter you are expected to be able to:

➤ explain the main origins of drugs, including drugs which originate from nature and those which are totally synthetic, and those with hybrid origins

➤ identify those drugs whose structures originated in or were inspired by nature and those that were not

➤ state the basic principles of organic synthesis and explain how new drugs are prepared and purified.

10.1 Drugs, dyes, and cleaning fluid: similarities and differences

Most drug molecules are small and organic, made up of carbon and a small number of other elements, most of which are commonly found in living things. Drug molecules are typically based on carbon and contain hydrogen, oxygen, nitrogen, and perhaps chlorine, sulfur or phosphorus.

There is one element frequently found in drug molecules that is hardly ever found in living organisms. What is it?

These drug molecules typically have molar masses of a few hundred grams per **mole**, and can be found as clinically-approved medicines, (for example aspirin, antibiotics, anti-angina drugs) or as illicit substances used to alter the mind (for example cocaine, heroin, LSD etc). Both types of 'drug' exert their effects by altering the biochemical processes in our bodies.

From a molecular point of view, there is no difference between a legal and an illicit drug.

 For more details on the distinction between legal and illicit drugs, see Chapter 3 'Legal and ethical matters' of the *Pharmacy Practice* book in this series.

Figure 10.1 shows the structure of diamorphine, alongside morphine and codeine, which are closely related. Diamorphine is a strong analgesic, used clinically to treat severe, including cancer-related, pain. However, the same substance is also known as heroin, an illegal recreational drug. The distinction between diamorphine and heroin is not a molecular distinction, but a social and legal distinction.

Organic molecules are found in food, drugs, plastics, all living things, household cleaning products, fabrics, dyes and more. For the most part, they are made up of a small number of different chemical elements, arranged in a limited number of ways, yet these quite subtle variations in structure can lead to major differences in properties. For example, Figure 10.2 shows the chemical structure of the local anaesthetic lidocaine, alongside the most bitter compound yet discovered, denatonium benzoate (Bitrex®). Bitrex® was discovered by chance in 1958, by chemists trying to develop novel local anaesthetics, and was approved for use in the early 1960s as an additive to ethanol to make it undrinkable. (Ethanol for industrial use is not subject to the same taxes as alcoholic beverages, and may also contain harmful impurities.) Since then, denatonium benzoate has found many applications. For example, it is added to cleaning products and antifreeze to prevent accidental poisoning, and is used in nail biting preventions. The chemical structures of denatonium benzoate and lidocaine are almost identical (shown in blue); a small difference in molecular structure (in this case the addition of a benzyl group) may cause a hugely different biological effect.

271

Four of the compounds shown below are medicinal drugs; the other is a potent poison. Can you identify it?

(A)

(B)

(C)

(D)

(E)

FIGURE 10.1 The analgesics morphine, codeine and diamorphine that originate from poppies. Poppy photo source: Photodisc.

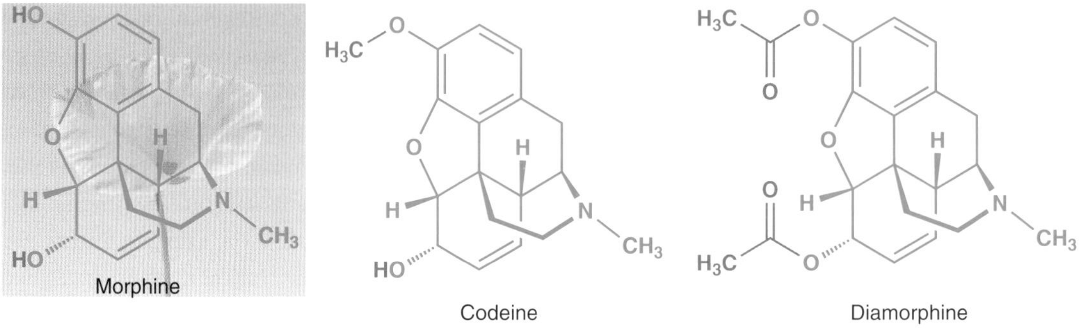

Morphine

Codeine

Diamorphine

The main difference between drugs and other small organic molecules is cost. A new drug not only requires some desirable medicinal effect, but also low toxicity, and no serious unwanted side-effects – it is far from easy to make such a drug. In fact, roughly, for every 10,000 structures made or discovered, 500 will reach animal testing, 10 will reach phase I clinical trials and only one will get to market, usually at a cost of about £450 million. Before any new drug is approved for the market it must be shown to be an improvement on existing therapies. Extensive legislation requires that lengthy clinical trials are carried out, and, at the end these, a new drug must be able to generate a profit so that the costs of its production can be recouped.

The drug discovery process raises numerous ethical questions to which there are no easy answers. Like any industry, the pharmaceutical industry must pay its employees (a very large and skilled workforce), offset its costs and reward its shareholders (such as pension funds). It is undoubtedly more profitable to invest in long-term illness of the rich nations (hypertension, cancer, diabetes) than short-term illness of the poor nations (tuberculosis, trypanosomiasis, malaria). Charities, notably the Bill and Melinda Gates Foundation, and the World Health Organisation work alongside the pharmaceutical industry to fill these gaps.

Thorough and expensive drug testing ensures that our medicines (with some well-documented exceptions) are safe and efficacious (they work); however, the regulation of alternative therapies is very weak, as herbal teas are classed as food, not medicines. This

FIGURE 10.2 The chemical structures of denatonium benzoate (Bitrex®) and the label used to denote its presence in products, and the local anaesthetic lidocaine. The structural similarity is shown in blue.

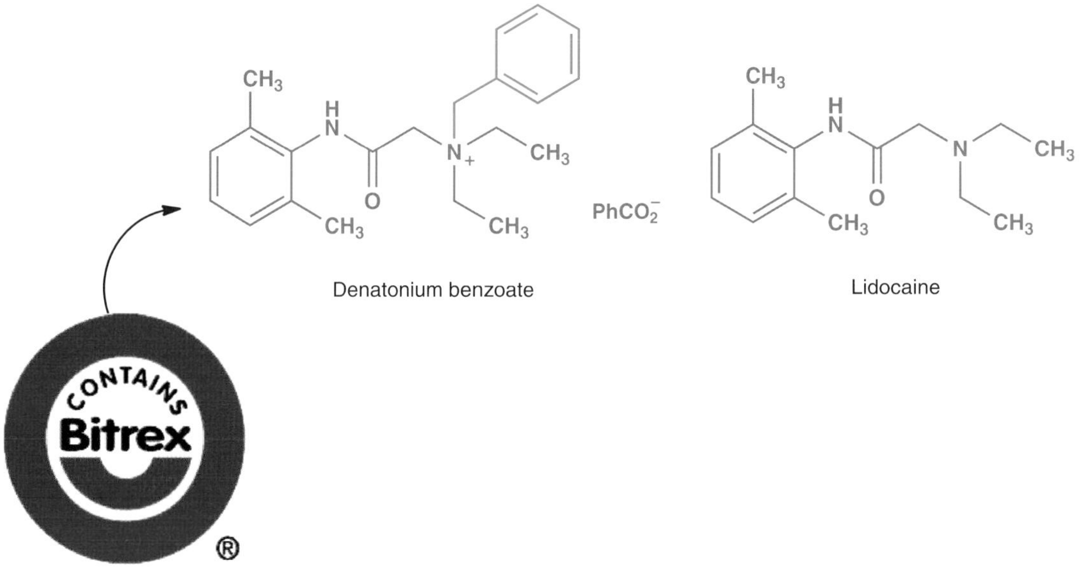

Denatonium benzoate

Lidocaine

same testing means that new drugs are so expensive that even a rich nation cannot always afford them. In the UK, the National Institute for Health and Clinical Excellence is charged with determining whether the country should pay for a particular therapy, or whether the money would be better spent on a different therapy. Governments decide how much money to spend on healthcare overall.

Ethics is now an important component of Pharmacy courses, and although the ethics of dealing with patients is the main focus of most of this teaching and learning, you should be aware that there are ethical considerations underlying all aspects of the production and use of medicines.

Let us now take a look at some of the drugs on today's market that are products from nature.

10.2 Natural products as drugs and medicines

A hundred years ago almost all the medicines in the pharmacopoeia were of natural origin, and even today, it is estimated that a third of the compounds we use as drugs are derived from nature. Our complex relationship with nature has developed over tens of thousands of years. It is almost inevitable that products of nature, found in animals and plants, will have some kind of medicinal effect on our bodies. Whether this effect is beneficial or not will depend on a number of factors, including the amount taken (the dose) as well as the nature of the chemicals (see Case Study 10.1).

In the following sections, we will look at several well-known medicines, where the active drug is made from, or based on, a natural product.

Aspirin

Aspirin (chemical name – acetylsalicylic acid), the well-known painkiller, was developed from a natural source, the bark of the willow tree. You will have seen aspirin in the form of over-the-counter medicines such as Anadin®, Aspro® and Disprin® but have you ever wondered how aspirin became a drug with a worldwide production of several thousand tonnes per year?

Aspirin itself is not a natural product. In fact it was the first drug to be prepared synthetically. Its structure is, however, very similar to salicin, the active product isolated from willow tree bark. Nature was an inspiration for aspirin, rather than its source, a distinction that will be made in more detail in the following

---

Dilip, a pharmacist, catches his mother using some dried leaves brewed in hot water to treat her aches and pains. He is worried about her and asks what they are. She says it is just herbal medicine she got from a friend and tells Dilip not to worry because they are natural, so they must be ok.

REFLECTION QUESTION

Dilip advises his mother to stop using the dried leaves immediately. Why is this?

Medicines that come from natural products have been rigorously tested to make sure they are safe.

Lots of very toxic chemicals come from nature.

There is a common misconception that natural products cannot do any harm. This, however, is simply incorrect. Just because something is natural does not necessarily mean it is safe to eat.

Answer

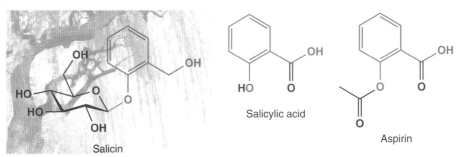

Salicin

Salicylic acid

Aspirin

sections of this chapter. Shown in Figure 10.3 are the structures of salicin, salicylic acid and aspirin. The areas of structural similarity are highlighted in blue. Aspirin, salicin and methyl salicylate are all pro-drugs and are converted to salicylic acid in the body.

For more details about salicin, salicylic acid and aspirin, see Chapter 7 'Introduction to aromatic chemistry'.

SELF CHECK 10.3

Find the active ingredient of the topical painkiller wintergreen oil and draw its structure alongside the structures in Figure 10.3. Mark the areas of structural similarity.

SELF CHECK 10.4

Draw the mechanisms for the conversion of aspirin to salicylic acid and the conversion of methyl salicylate to salicylic acid.

SELF CHECK 10.5

Draw the mechanism for the conversion of salicin to 2-hydroxybenzyl alcohol. Hint: salicin is an acetal.

If you find these questions difficult, look at the reactions of carbonyl compounds in Chapter 6.

Willow bark is still sold as an alternative therapy. The sales literature claims that it is slower to act than aspirin but that the effects last longer. As you can see from Self Check 10.5, salicin in willow bark requires

two chemical reactions to convert it into salicylic acid; the acetal function is hydrolysed and then the drug is oxidized in the liver (see Chapter 1). The active ingredient is released more slowly than in the case of aspirin.

Opioid analgesics

Opioids are molecules which interact with the opioid receptors in the central nervous system. The body makes its own opioids, of which endorphins (produced in response to injury or exercise) are the most well-known, and their effects can be mimicked by other compounds. Opioid drugs are used to relieve moderate to severe pain.

Naturally occurring opioids are extracted from opium (hence their name), which comes from the seed head of the opium poppy. Opium contains morphine (5–20%), and smaller amounts of related compounds, including codeine; it is one of the oldest herbal medicines known. So sought after were the extracts of the opium poppy that wars were fought in its name (Opium Wars 1839–42 and 1856–60), and such were the problems associated with its use that the Opium Commission and later the League of Nations were set up to regulate its use and production. This goal was not wholly met, and illicit trade in opium continues in certain parts of the world today.

In Figure 10.1, you can see the structures of morphine and codeine, which are both natural products, together with the structure of diamorphine (heroin), which is prepared in a chemical laboratory from morphine. The structural differences between the molecules have been highlighted in blue. As you can see, these three chemical structures are extremely similar. Both codeine and diamorphine are metabolized to morphine in the body.

SELF CHECK 10.6

Describe the differences between the structures of morphine, diamorphine and codeine. Would you expect codeine and diamorphine to be more or less hydrophobic than morphine?

Opioid analgesics are some of the most effective pain-killers available to medicine; the compounds act in the brain and appear to work by elevating the pain threshold of the patient thus decreasing the brain's awareness of pain. The side-effects of morphine are quite broad and include: nausea, constipation, drowsiness, respiratory depression, euphoria, **tolerance** and **dependence**. While side-effects are not usually desirable, euphoria can actually be helpful for treating pain in terminally ill patients. You may wish to think of the consequences of euphoria as a side-effect in those people taking drugs for recreational use! Tolerance and dependence are of course very negative side-effects of morphine, but the most dangerous side-effect is depression of breathing. The most common cause of death from a morphine overdose is suffocation.

One person's unwanted side-effect can be another person's treatment, and there are numerous examples in medicine of careful observation of side-effects being put to good use. Opioid drugs, including morphine and codeine, which can have the side-effect of causing constipation, are now used for the treatment of diarrhoea and of irritable bowel syndrome. Conversely, derivatives of erythromycin (whose side-effects include diarrhoea) are used to treat constipation in intensive care patients.

KEY POINT

The observation of side-effects with a particular drug can often lead to drugs being used to treat conditions other than the one for which they were initially used.

SELF CHECK 10.7

Can you draw the mechanism for the conversion of diamorphine to morphine? (Think about the hydrolysis of aspirin.)

SELF CHECK 10.8

We make diamorphine from morphine, and then diamorphine acts as a pro-drug for morphine. At first sight, this looks like a waste of time and money, yet diamorphine is actually more effective than morphine. Why?

Antibiotics

There is a large selection of naturally occurring antibiotics available for the treatment of bacterial infections. These include: penicillins, cephalosporins, tetracyclines and macrolides. Examples of these antibacterial drugs can be seen in Figure 10.4. As you can see from their chemical structures, they are structurally very different, yet they are all antibacterial drugs. This is quite different from the opioid analgesics, which all have similar chemical structures and thus have similar biological activities. It takes a little thought to appreciate that there is no contradiction here.

There is more than one way of killing a bacterium. The bacterial cell contains many different biological targets, and these are targeted by different drugs. A crude way of looking at this is to say:

Different biological target = Different chemical structure of drug

This makes sense; we would expect proteins and nucleic acids with different functions to be structurally different and to bind differently-structured drugs.

 See Chapter 9 'The chemistry of biologically important macromolecules' to refresh your memory on biologically important drug targets. Also for more information about drug targets, see Chapter 4 'Introduction to drug action' of the *Therapeutics and Human Physiology* book in this series.

Macrolide antibiotics

Macrolides are molecules which contain a large carbon-based ring containing an ester. Cyclic esters are known as **lactones**.

Erythromycin (see Figure 10.4) belongs to the macrolide class of antibacterial agents. It was isolated from a soil organism in the 1950s in the Philippines, and is one of the safest antibacterial agents in clinical use. Erythromycin works by binding to the 50S subunit of the bacterial ribosome, which prevents the bacteria from synthesizing essential proteins. Its specific shape and the position of its functional groups enable it, and a few derivatives, to bind to a particular binding site on the ribosome.

See Chapter 1 'The importance of pharmaceutical chemistry' for information on the formulation and drug delivery of erythromycin.

FIGURE 10.4 Common antibiotics. Note the structure of erythromycin and its ball-and-stick and space-filling models which provide more realistic insights into the shape of this molecule.

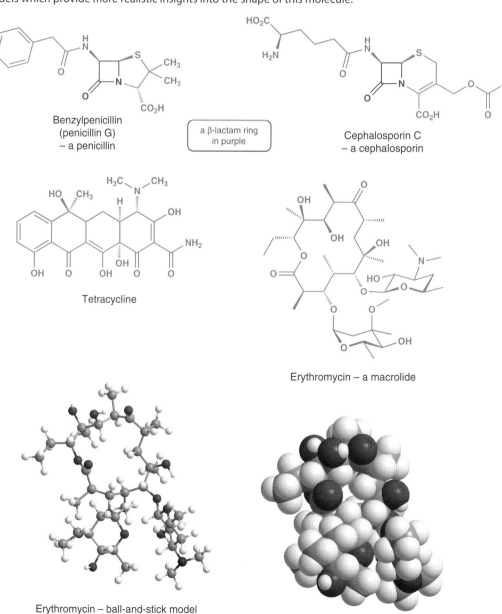

Benzylpenicillin (penicillin G) – a penicillin

a β-lactam ring in purple

Cephalosporin C – a cephalosporin

Tetracycline

Erythromycin – a macrolide

Erythromycin – ball-and-stick model

Erythromycin – space-filling model

β-Lactam antibiotics

The β-lactam class of antibiotics include the penicillins and cephalosporins, plus a couple of other related structures (see Figure 10.4). β-Lactam antibiotics work by inhibiting the enzymes responsible for bacterial cell wall synthesis. They all possess a β-lactam ring (shown in purple in Figure 10.4), which is essential for their activity. Without this structural feature these compounds are inactive. Perhaps surprisingly, this essential β-lactam ring is partly responsible for the allergic reaction to the penicillins experienced by some patients. When the β-lactam ring is attacked by specific cellular proteins, it creates an antigen which can stimulate an immune response, resulting in hypersensitivity to the drug. This also accounts for the cross-sensitivity to cephalosporin antibiotics, which share the β-lactam ring structure. If you are allergic to the penicillin antibiotics, the similar chemical structures should also alert you to the fact that you may be allergic to the cephalosporins and other β-lactam antibiotics too. An understanding of chemical structures could save your life (see Case Study 10.2)!

Penicillin G was famously discovered by Alexander Fleming as the product of a mould, *Penicillium notatum*. Moulds, like soil bacteria, can be cultured and developed to produce very high yields of drugs. Cephalosporins are also produced by fungi, of the *Acremonium* species.

SELF CHECK 10.9

Look at the chemical structures to decide why the penicillin and cephalosporin antibiotics may trigger a similar allergic response.

CASE STUDY 10.2

6-year-old Maya has an upper respiratory tract infection. The doctor has prescribed her amoxicillin. Before

Glycopeptide antibiotics

Glycopeptide antibiotics work by targeting the building blocks of the bacterial cell wall preventing cell wall synthesis and leaving the cell vulnerable. A common glycopeptide antibiotic is vancomycin (see Figure 10.5). Vancomycin is important because of its ability to combat deadly methicillin-resistant strains of *Staphylococcus aureus* (MRSA). Unfortunately, we are beginning to see vancomycin resistant strains of bacteria now too.

The structure of this antibiotic is hugely complex. It was first isolated in 1956 from the fermentation broth of the soil bacterium *Streptomyces orientalis*. Remarkably, given its complexity, it was synthesized chemically in 1998 (see Box 10.1). However, as is usually the case with complex molecules, the chemical synthesis is not economic, and the drug continues to be produced from the soil bacterium.

Look at the structure of vancomycin and note the number of chiral centres present: 18 in a single molecule! Each one of these chiral centres is fixed and exists in nature as drawn. Such complexity is only ever found in natural products. Chemists making new drugs would never dream of designing such a complex molecule from scratch unless they were being inspired by a natural product.

KEY POINT

You should be able to look at the chemical structures of the drugs that are spread throughout this chapter and identify which ones are synthetic and which ones are natural products based on their structures alone.

 Refer back to Chapter 3 'Stereochemistry and drug action' for a refresher on stereoisomers, if you need to.

she did so, she checked that Maya was not allergic to penicillin.

REFLECTION QUESTIONS

1. Why did the doctor need to check Maya's allergies?

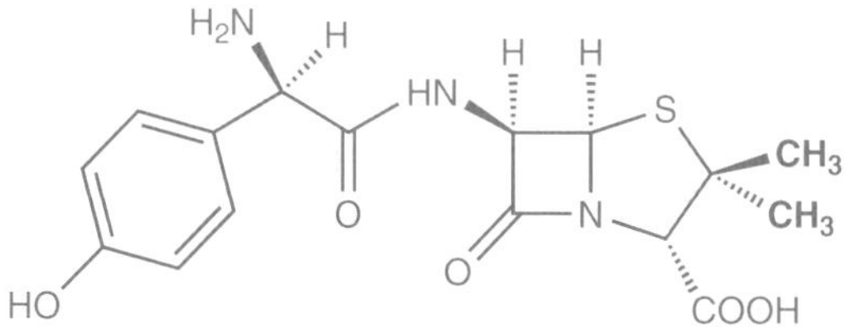

2. The original penicillin was benzylpenicillin. As you can see, the structures of amoxicillin and benzylpenicillin are very similar:

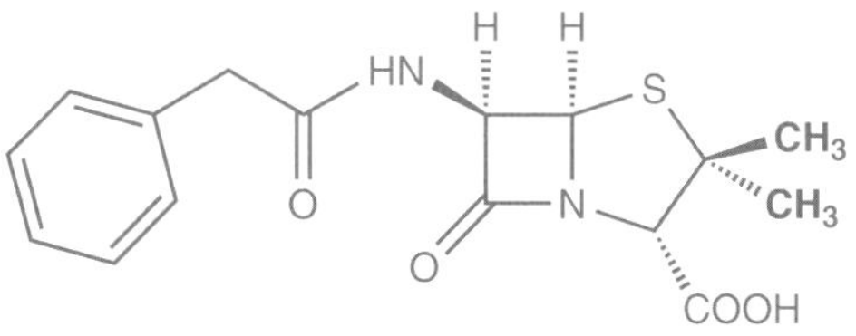

Amoxicillin

Benzylpenicillin (Penicillin G)

Why was benzylpenicillin not prescribed to Maya?

3. Maya quite enjoys taking her medicine. Why could this be?

Answers

1 Amoxicillin is a type of penicillin. Since many people have an allergy to penicillin, it is important to check that anyone that anyone prescribed this medication is safe to take it.

2 Benzylpenicillin is broken down by stomach acid, so it has to be injected. It can also be broken down by lots of bacteria (they produce enzymes called β-lactamases). Amoxicillin is less susceptible to these.

3 Amoxicillin does not taste very nice, but, thankfully, it is easily masked by other flavours. Paediatric amoxicillin suspension is banana flavoured.

FIGURE 10.5 Vancomycin.

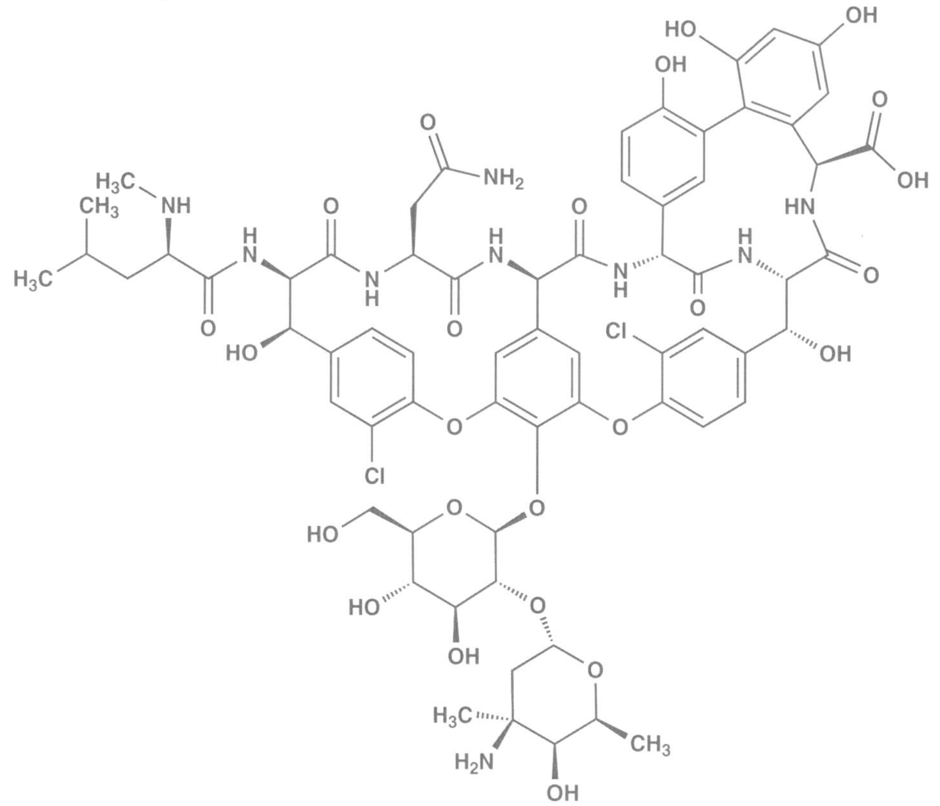

Vancomycin

Stereoisomers

For each chiral centre present in a molecule there are 2^n isomers possible (where n=the number of chiral centres). Vancomycin is a single isomer, not $2^{18} = 262144$ different molecules. Making a single compound of such complexity means making one out of 262144 possible variations (very low odds indeed) – however, chemists achieved this in 1998, a remarkable feat!

There are many different classes of antibiotics and they kill bacteria by a variety of mechanisms.

Anticancer agents

Cancer is the second most common cause of death in the western world, exceeded only by heart disease. There are more than 200 different types of cancer as a result of different cellular defects, and so a treatment for one particular cancer may not be effective in controlling another type of cancer. The continuing battle we have against cancer means that we continue, as ever, to look for new drugs and treatments for the disease. Since 1980, Cancer Research UK has advanced over 100 novel anticancer agents through various stages of preclinical and early phase clinical trials. A comparison of the BNF from September 2001 to March 2011 shows an increase in the number of **cytotoxic** drugs available from 57 to 86. Some of the natural product drugs used to combat cancer are shown in Figure 10.6.

The first thing that should strike you is the sheer size and complexity of the compounds in Figure 10.6. As previously discussed, these complex compounds are natural products. For example, vincristine is found in the Madagascan periwinkle.

Madagascan periwinkle
Licensed under the Creative Commons Attribution-Share Alike 3.0 Unported, 2.5 Generic, 2.0 Generic and 1.0 Generic license. Source: Arria Belli.

Can you identify the number of chiral centres and types of functional groups in the anti-cancer compounds in Figure 10.6? Can you identify the different hybridization states of the atoms in those molecules and predict which parts of the molecules are planar?

 See Chapter 2 'Organic structure and bonding' and Chapter 3 'Stereochemistry and drug action' for a more in-depth coverage of chiral centres and hybridization.

Anticancer drugs which interact with DNA

The structures of bleomycin (which is dispensed as a mixture of bleomycin A_2 and B_2), doxorubicin, and mitomycin C, are all very different and complex; they are natural products after all! Despite their structural differences, they all act as anticancer agents by interacting with DNA.

Bleomycin is a very complex three-dimensional molecule isolated from the soil bacterium *Streptomyces verticillus*. It causes breaks in the strands of DNA, which ultimately leads to cell death – a good thing for a cancer cell! It is used intravenously or intramuscularly to treat, amongst other things, certain types of skin cancer.

Doxorubicin is also derived from a soil bacterium, *Streptomyces peucetius,* but, in contrast to bleomycin, doxorubicin has four fused rings in which 14 of the 18 carbon atoms are sp^2 hybridized, meaning they are planar (i.e. flat). The planar structure means that doxorubicin is able to slip in between the base pairs of DNA. This process is called intercalation and it prevents DNA replication, and ultimately leads to cell death. Doxorubicin is used to treat acute leukaemias, lymphomas, and a variety of solid tumours.

Mitomycin C, another *Streptomyces* product, has yet another way to interact with DNA and kill cancer cells. The mechanism by which mitomycin C works is rather complex and beyond the scope of this book; its interesting structure enables it to alkylate DNA. It is used for the treatment of upper gastrointestinal cancer, as well as breast cancer. It is one of the most toxic anticancer drugs in clinical use, causing delayed bone-marrow toxicity and it can result in permanent bone-marrow damage if used for long periods of time.

FIGURE 10.6 The chemical structures of a few anticancer drugs.

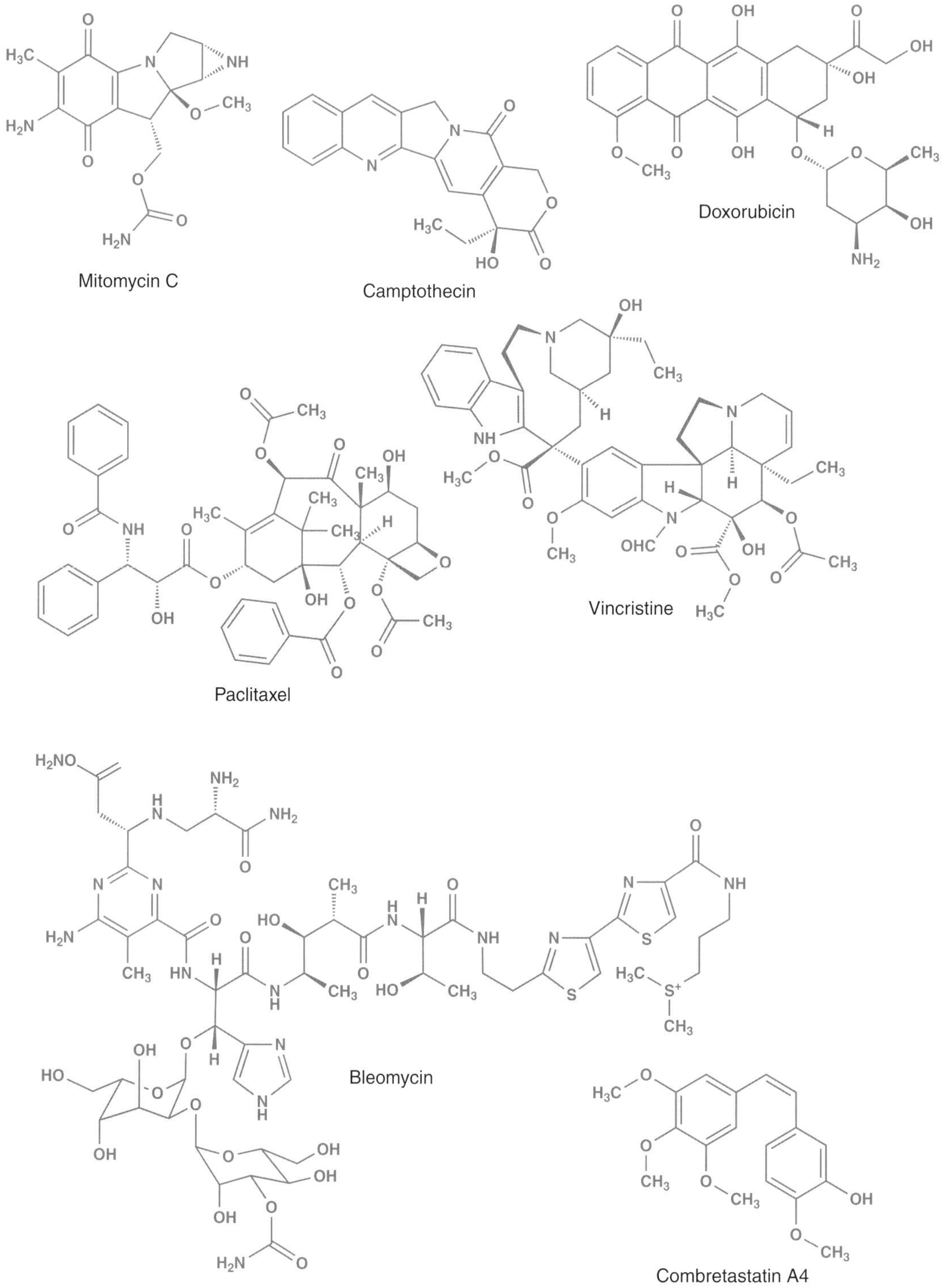

Mitomycin C

Camptothecin

Doxorubicin

Paclitaxel

Vincristine

Bleomycin

Combretastatin A4

 For the structure of DNA, refer back to Chapter 9 'The chemistry of biologically important macromolecules'. Chapter 5 'Alcohols, phenols, ethers, organic halogen compounds, and amines' gives details of some simpler alkylating agents. Refer back to Chapter 8 'Inorganic chemistry in pharmacy' for an account of cisplatin – an inorganic anticancer agent that interacts with DNA.

Anticancer drugs which act on important proteins

There are huge numbers of important proteins in our bodies (see Box 10.2). Enzymes and receptors are (normally) proteins and both are used as targets for a range of therapies in a number of therapeutic areas. Paclitaxel, combretastatin A4 and vincristine (see Figure 10.6) are naturally occurring anticancer agents which induce cell death by interfering with a crucial structural protein called tubulin. Paclitaxel (Taxol®) is a member of the taxane group of drugs and was originally derived from the Pacific yew tree, but is now produced from sustainable sources (see Section 10.3). It can be used to treat ovarian, breast, and non-small cell lung cancers and is administered by intravenous infusion. Combretastatin A4, from the South African Bush Willow, is in clinical trials for the treatment of thyroid cancer. Vincristine and related compounds from the Madagascan periwinkle plant also interfere with tubulin and are used to treat a variety of cancers including leukaemias, lymphomas and some solid tumours such as breast and lung cancer. Like paclitaxel, they are administered intravenously. Inadvertent intrathecal (injection into the spinal cord) administration of vincristine sometimes occurs because the drug is often given in combination with intrathecal medicines. This is usually fatal.

KEY POINT

Anticancer agents treat cancer using a number of different mechanisms and, hence, they are structurally quite different.

BOX 10.2

The Human Genome Project

The Human Genome Project has shown that humans have about 27,000 genes. It used to be thought that one gene was responsible for one protein, but we now know that proteins can be modified in so many ways that one human gene may give rise to perhaps 10 different protein molecules. So we do not know how many different proteins there are in the body, but it is more than 27,000.

Hormones

Many naturally occurring hormones are used as medicines; Figure 10.7 shows some of these. In many cases the hormone (or a simple derivative of it) simply 'tops up' the patient's own reserves of the naturally occurring molecule and restores normal function to the body. A dose of the hormone in a level similar to that found in the body is called a physiological dose, whilst a dose that exceeds that found in the body is a pharmacological dose. The effects on the body of a physiological dose and a pharmacological dose may be quite different and this difference has been exploited in the development of useful therapies.

The most commonly prescribed hormones are the oestrogen and progestagen steroids, which are prescribed for contraception and for menstrual disorders (see, for example, estradiol and progesterone in Figure 10.7A and B). Other steroids produced by our bodies are used for treating a range of inflammatory and allergic disorders such as autoimmune diseases, skin disorders and respiratory problems, including asthma. For example, prednisolone (see Figure 10.7C) may be taken orally as an immune-suppressant, as a cream for the topical treatment of skin disorders, or by inhaler for the treatment of asthma.

SELF CHECK 10.11

Can you identify the structural similarities between the three steroids in Figure 10.7? What shape do you think these molecules will have? Hint: refer back to Chapter 3.

The common hormone thyroxine (see Figure 10.7D) is sold with the name levothyroxine. The prefix *levo* indicates that the drug is enantiomerically pure and rotates the plane of polarized light anticlockwise. Levothyroxine is used as a maintenance treatment for hypothyroidism (underactive thyroid). Hypothyroidism typically leads to tiredness, weight gain and a general slowing down of metabolism. Thyroxine is the major hormone produced by the thyroid gland and patients taking levothyroxine do so in order to 'top up' the levels of thyroxine in their bodies to ensure that normal function is maintained.

 Enantiomers are discussed in more detail in Chapter 3 'Stereochemistry and drug action'.

FIGURE 10.7 Naturally occurring hormone drugs.

(A)

Estradiol, a steroid

(B)

Progesterone, a steroid

(C)

Prednisolone, a steroid

(D)

Levothyroxine

Insulin (a peptide containing 51 amino acids; see Figure 10.8) plays an important role in the regulation of carbohydrate, fat and protein metabolism. Insulin is used as a therapy for the treatment of Type 1 diabetes, a disease where a patient's own body destroys the insulin-producing cells of the pancreas. The subsequent lack of insulin then leads to an increase in blood and urine glucose, and if untreated the condition is usually fatal. Insulin is sensitive to both stomach acid and digestive enzymes in the intestine, so patients with Type 1 diabetes have to inject the hormone regularly.

 For more information about insulin and diabetes, see Chapter 7 'Communication systems in the body – autocoids and hormones' and Chapter 10 'This is just the beginning' of the *Therapeutics and Human Physiology* book in this series.

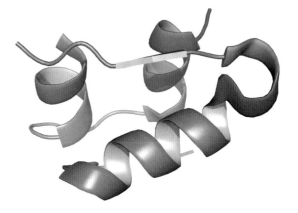

FIGURE 10.8 The structure of the human insulin monomer. Insulin monomers tend to aggregate as hexamers, but the monomer is the active form.
Copyright Jonathan Crowe.

10.3 Semi-synthetic drugs

As we have seen in Section 10.2, many drugs can be isolated from natural sources. These compounds, however, did not evolve in nature to benefit humans, but to provide the plants and micro-organisms that produce them with a competitive advantage. We would therefore expect there to be a limit to their efficiency as drugs, both in terms of activity and toxicity. Luckily, with the aid of chemistry, we are able to take the natural product and alter its chemical structure appropriately to make similar compounds (**analogues**) to the natural structure with the aim of improving activity, reducing toxicity or introducing some other desired property.

SELF CHECK 10.12

Think back again to the sections on the analgesics and antibiotics. Why do similar structures have similar activities?

The process of manipulating the structure of a naturally occurring compound in the laboratory is called semi-synthesis. Semi-synthesis may produce

FIGURE 10.9   Semi-synthetic analogues of natural products.

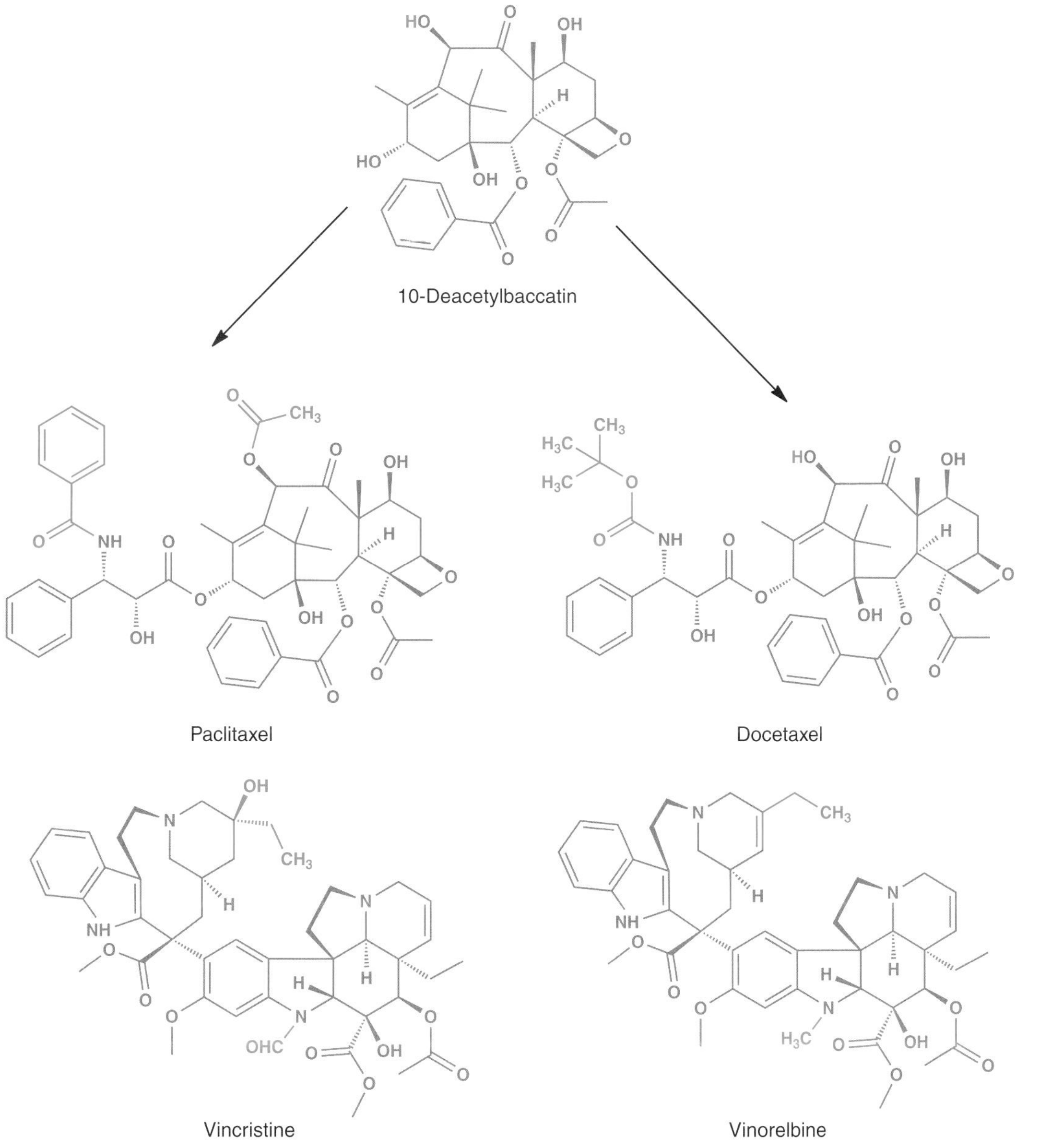

10-Deacetylbaccatin

Paclitaxel

Docetaxel

Vincristine

Vinorelbine

the natural product itself. Usually, however, the process will produce new compounds in the hope that they are better drugs than the parent natural product. There is a lot of trial and error in this process. Lots of analogues are prepared and tested before the best analogue is progressed further. The 'best analogue' is judged on its overall activity, toxicity, aqueous solubility and many other factors. The examples in Figure 10.9 show some of these analogues alongside their naturally occurring 'parent' structure.

Semi-synthetic paclitaxel and its analogues

Paclitaxel, mentioned in Section 10.2, was originally isolated from the bark of the Pacific Yew tree during a project aimed at finding new anticancer agents. Intense research ultimately led to paclitaxel in minute quantities, but unfortunately, the extraction of 300 mg (0.3 g) of the compound (about a single dose for a cancer patient) required an entire 100-year-old yew tree. Obviously, there are simply not enough trees to meet this demand, and it would be far too slow to plant and grow more trees. One potential solution to this problem came about when it was found that the needles and twigs of the European Yew tree contained a molecule called 10-deacetylbaccatin (see Figure 10.9), which could be converted into paclitaxel relatively easily in the laboratory using semi-synthesis. Needles and twigs can be harvested without killing the tree; they are a renewable source that allowed relatively large quantities of paclitaxel to be produced. Nevertheless, to meet the huge demand on this new drug, chemists set about trying to synthesize this huge molecule 'from scratch'. This required many synthetic steps but was achieved, by two independent research groups, in 1994. As is often the case, the synthetic route was not economical on a large scale, and paclitaxel is still manufactured by semi-synthesis from deacetylbaccatin.

Now that a semi-synthetic paclitaxel had been developed, analogues could be made to see if its biological properties as an anticancer agent could be improved. Docetaxel is one such analogue, and its structure can be seen in Figure 10.9 alongside pacli-

taxel. Both compounds inhibit tubulin depolymerization, but docetaxel has greater water solubility than paclitaxel. Both compounds are given by intravenous infusion. The small difference between these two structures (shown in green) means that although both docetaxel and paclitaxel can be used for the treatment of metastatic breast cancer and non-small cell lung cancer, of the two, only paclitaxel can be used to treat ovarian cancer.

SELF CHECK 10.13

Vinorelbine is a semi-synthetic derivative of vincristine. What do you think are the clinical indications for vinorelbine?

Semi-synthetic penicillins

The penicillin antibiotics are quite reactive. The four-membered ring (the square containing nitrogen, see Figure 10.10) in the middle of the structure is primed to pop open, and this reactivity enables it to act as an antibiotic. Unfortunately, the same reactivity means that it is difficult to make in the laboratory. In 1957, John Sheehan's persistence paid off when he managed to make a small amount of penicillin V (see Figure 10.10). His synthesis also enabled him to make 6-aminopenicillanic acid (see Figure 10.10), the precursor that would ultimately enable chemists to make a range of penicillin analogues.

Once again, however, a brilliant piece of organic synthesis did not prove to be an economical way of producing a compound with several chiral centres. In 1958, the Beecham company announced the isolation of 6-aminopenicillanic acid from *Penicillium* fungi. The free amine function is a good nucleophile and can be used to make numerous amides. This technology was developed and optimized to allow the synthesis of new penicillins at will, and the limitations of the only existing penicillin on the market in 1958 – penicillin G (see Figure 10.4) – could be tackled. You only have to consider the enormous number of penicillin antibiotics on the market today to see what a huge discovery this was (see Figure 10.10).

FIGURE 10.10 Semi-synthetic penicillins.

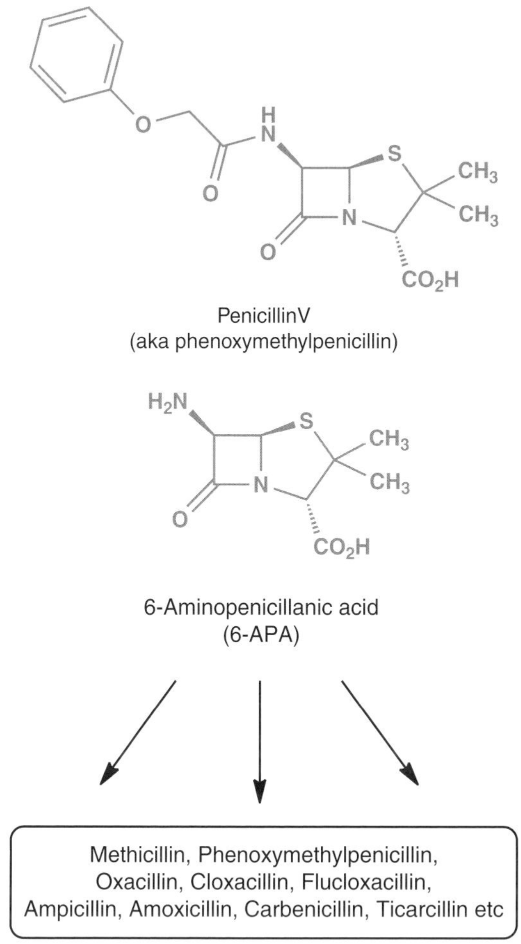

PenicillinV
(aka phenoxymethylpenicillin)

6-Aminopenicillanic acid
(6-APA)

Methicillin, Phenoxymethylpenicillin,
Oxacillin, Cloxacillin, Flucloxacillin,
Ampicillin, Amoxicillin, Carbenicillin, Ticarcillin etc

SELF CHECK 10.14

Look up the structures of the penicillins named in Figure 10.10. Can you identify their structural similarities?

SELF CHECK 10.15

How would you synthesize penicillin G from 6-amino-penicillanic acid? Draw the reagents and mechanism for the amide formation. (Hint: if you are struggling, look back at Chapter 6).

KEY POINT

Semi-synthetic drugs are made by chemical modification of natural products. Semi-synthesis is often used to improve the properties of a drug, but it may also be used to produce a natural product in an economical or sustainable way.

10.4 **Synthetic drugs**

As you have seen throughout this chapter, natural products can be very complex structures indeed; you should be able to spot one at a thousand paces by now! Some of these complicated structures possess amazing properties that can lead to diseases being cured.

Natural products do not, however, provide for all our medicinal needs. A huge number of drugs have been designed and made entirely in the chemical laboratory. Such drugs can be termed 'synthetic', and as you will see their chemical structures are much simpler than those of many natural products. Figure 10.11 shows some of today's synthetic drugs. These compounds are completely novel, so how did the chemists think them up? What inspired their thinking?

SELF CHECK 10.16

Count the chiral centres in each of the molecules in Figure 10.11. Whenever you meet a drug for the first time, you should also look up its BNF entry.

The strategy of simplification

Natural products are very complex and the biological activity of a natural product may reside in just a small part of the molecule. Drugs can be prepared by simplifying natural products; identifying the important part of the structure and synthesizing it, thus stripping away all of the molecular fragments that

FIGURE 10.11 Some examples of the drugs made by total synthesis.

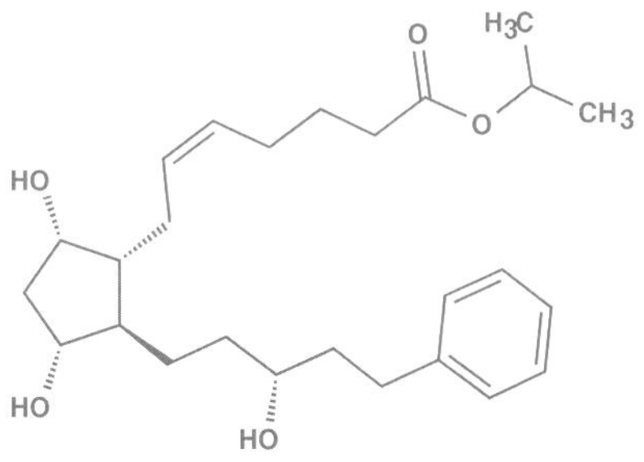

Latanoprost
a glaucoma treatment

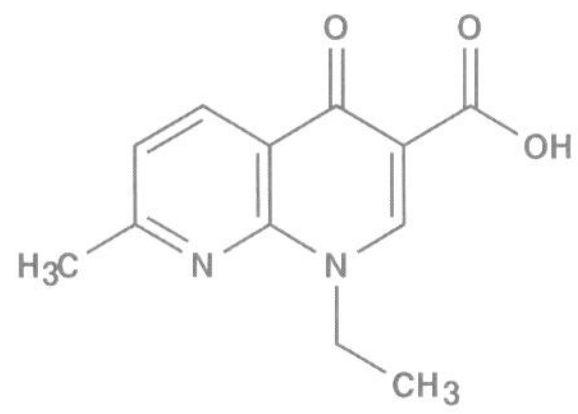

Nalidixic acid
an antibiotic

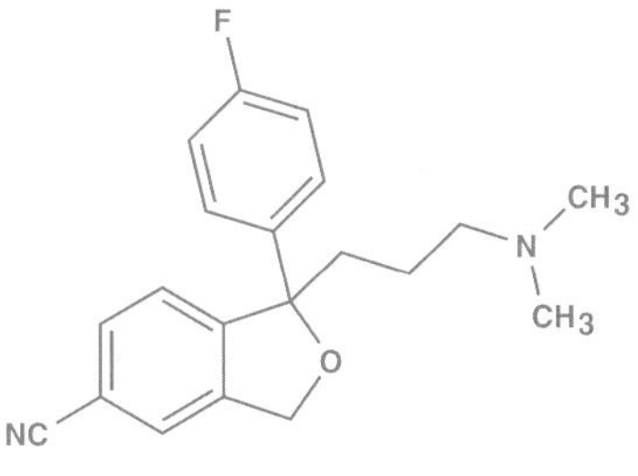

Citalopram
an antidepressant

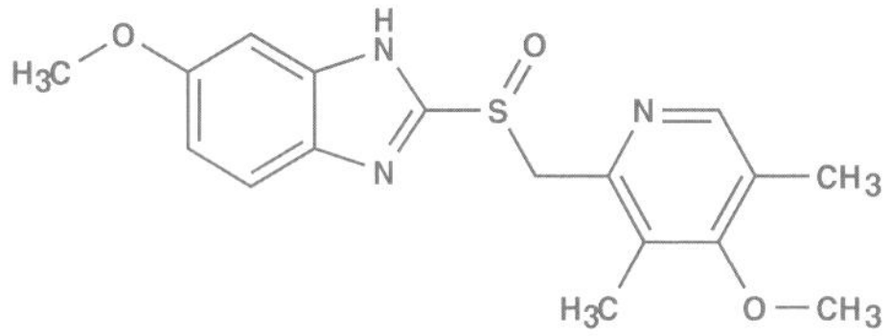

Omeprazole
a proton pump inhibitor

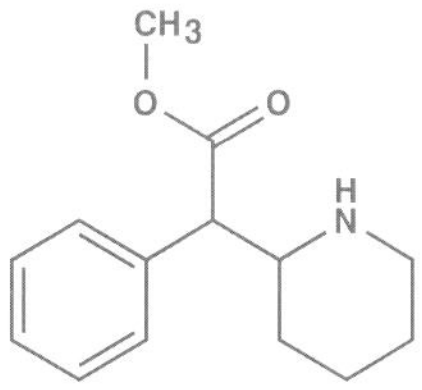

Methylphenidate
(ritalin)
for attention deficit hyperactivity disorder

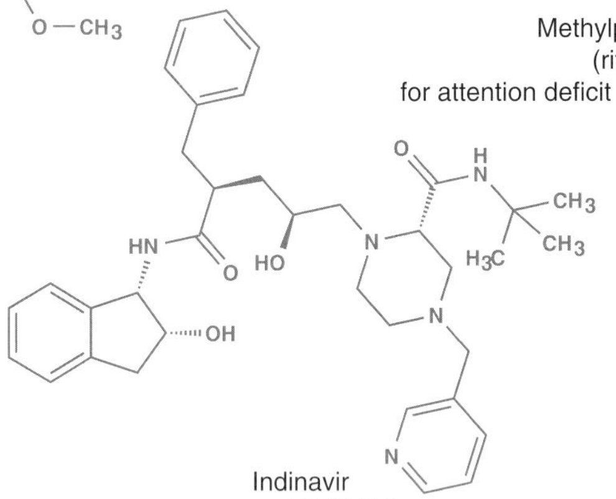

Indinavir
to treat HIV/AIDS

are not required for biological activity. This essential core can be optimized further as some complexity is added back to the core structure with the aim of making a new drug. These drugs are totally synthetic molecules. Figure 10.12 depicts a number of drugs which have been prepared by such a 'simplification' strategy.

Methadone and procaine were developed from morphine and cocaine respectively. Figure 10.12 shows how procaine and cocaine overlay one another in three dimensions. The structural requirements that give cocaine its good anaesthetic properties (i.e. the benzene ring, ester and amine shown in blue) are also present in procaine *in the same orientation in space* thus giving it similar properties.

Atorvastatin is discussed in Chapters 1, 3 and 9. At first sight, it does not look simpler than the natural product lovastatin, but actually, it has many fewer chiral centres. Statins are used to lower cholesterol levels.

Synthetic drug structures inspired by nature

If a new drug cannot be developed from a natural product itself and if a simplification strategy cannot be used, nature can still help to *inspire* new drug structures.

Figure 10.13 shows a number of drugs (in blue) and the natural inhibitor that inspired their development (in purple). The drug and the natural product differ substantially because the drugs have undergone a huge amount of development before they made it on to the market. Such development includes:

- simplification of the natural product, if possible

- optimization of the number, type and orientation of functional groups remaining once simplification has taken place

FIGURE 10.12 Drugs designed by a simplification strategy. (A) Methadone, derived from morphine; (B) simplification of cocaine to give procaine (the insert shows how the two drugs can be overlaid – they have the same shape); (C) Lovastatin is found in oyster mushrooms.
Oyster mushroom source: Photodisc.

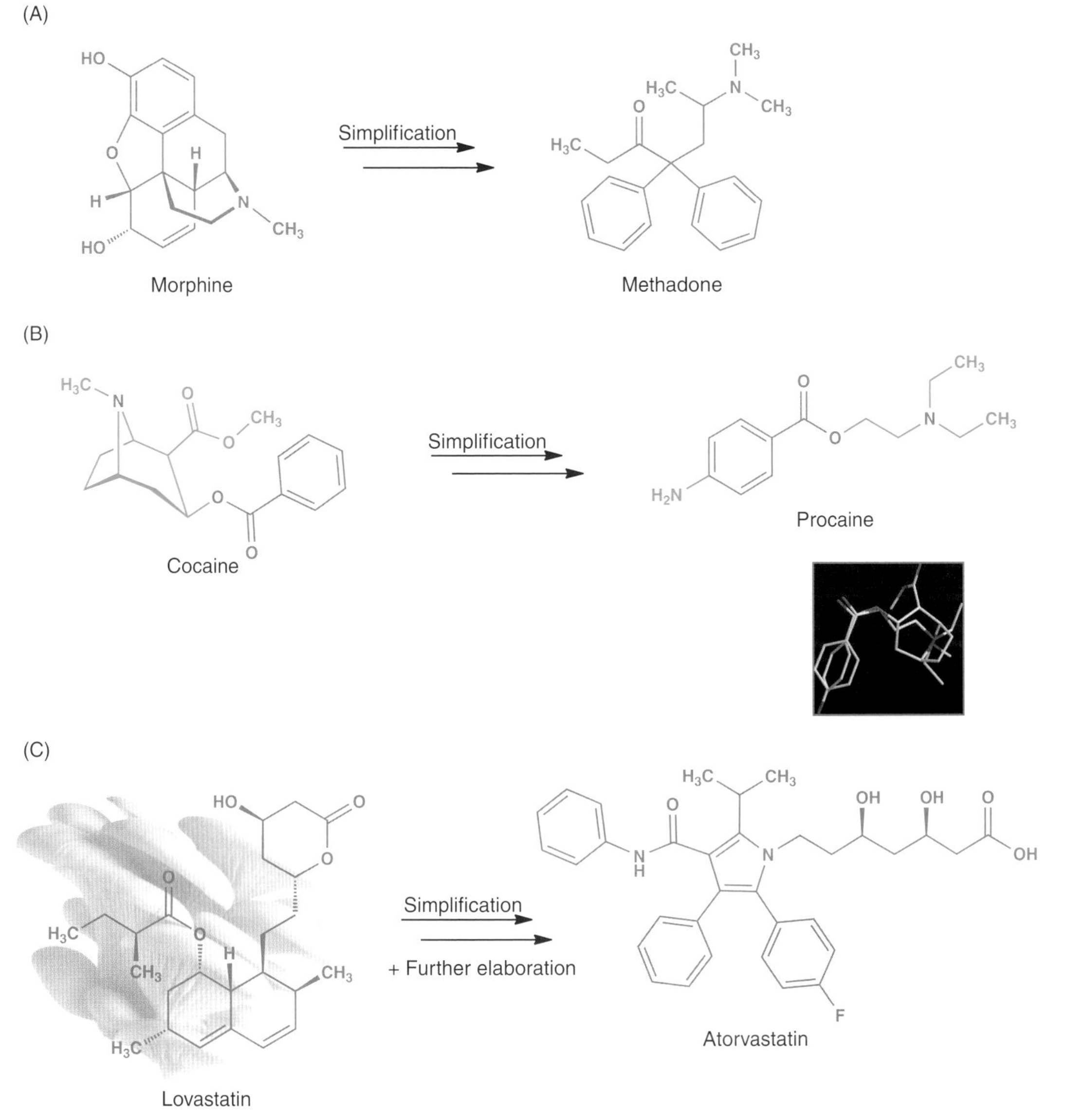

(A)

Morphine → Simplification → Methadone

(B)

Cocaine → Simplification → Procaine

(C)

Lovastatin → Simplification + Further elaboration → Atorvastatin

- optimization of aqueous or lipid solubility

- minimization of toxicity.

Why would these development steps be carried out? There may be a number of reasons:

- simplification to remove unnecessary parts of the molecule

- optimization of the functional groups to give the best fit into the drug target

- modification of drugs that are either too hydrophobic or too hydrophilic to easily get through membranes and biological fluids to their sites of action

- the minimization of toxicity to avoid harming the patient.

Here we will briefly consider some examples of drugs inspired by nature. Drugs, such as captopril and enalapril, were inspired by a very toxic xenobiotic, whereas both cimetidine and salbutamol were inspired by compounds found in the body.

FIGURE 10.13 Some drugs inspired by natural products: captopril and enalapril were inspired by teprotide, a constituent of Brazilian pit viper venom.

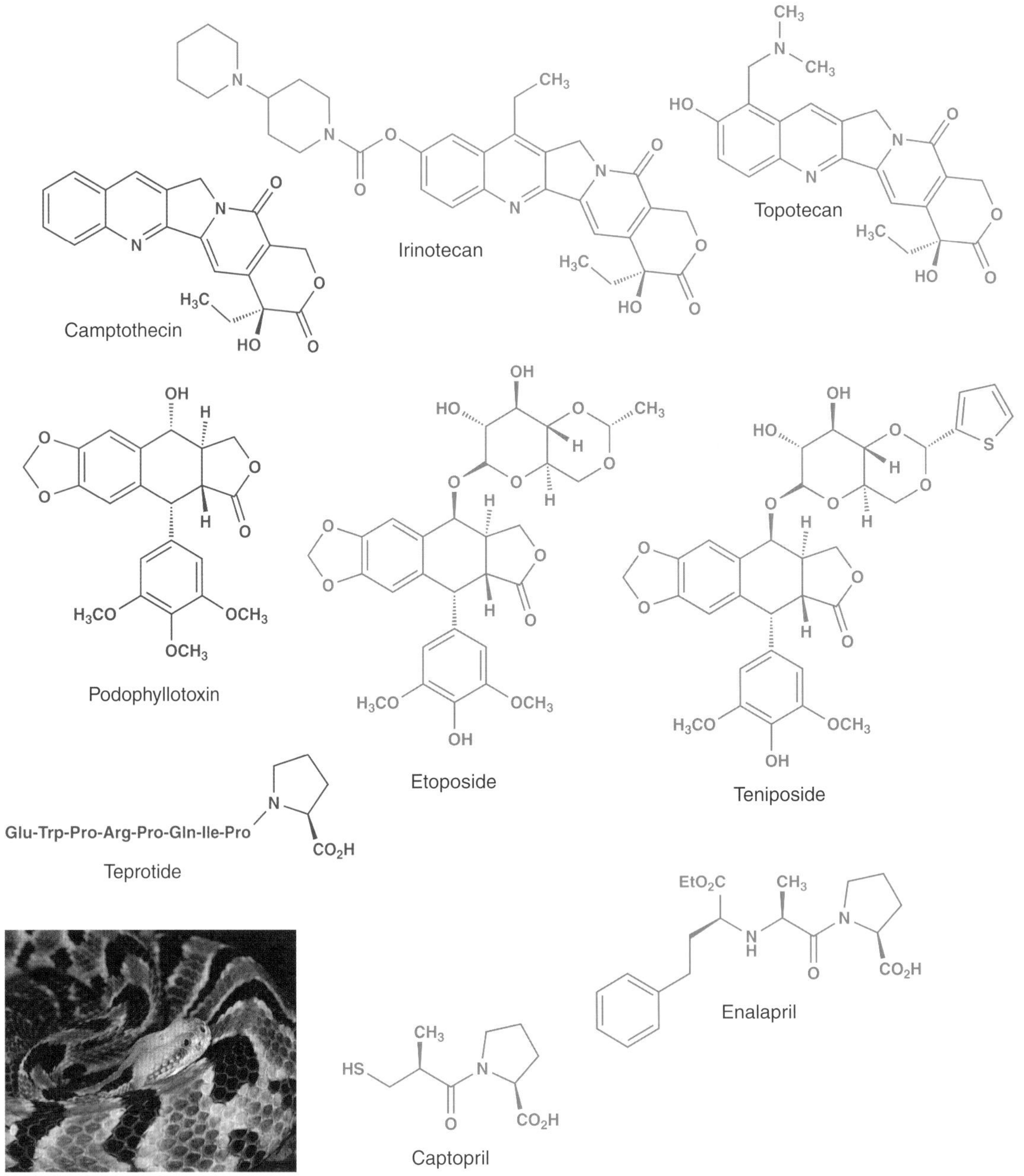

Irinotecan

Topotecan

Camptothecin

Podophyllotoxin

Etoposide

Teniposide

Glu-Trp-Pro-Arg-Pro-Gln-Ile-Pro

Teprotide

Enalapril

Captopril

Xenobiotic-inspired drugs

The venom from the Brazilian pit viper was found to inhibit angiotensin converting enzyme (ACE), reducing blood pressure. Uncontrolled lowering of blood pressure leads to death and pit viper venom was histor-ically used as an arrow poison. But the understanding of the mode of action to this poison, led to the development of ACE inhibitors like captopril and enalapril, which can be used as drugs to combat hypertension.

You can read more about the development of captopril in Chapter 6 'The carbonyl group and its chemistry' and Chapter 8 'Inorganic chemistry in pharmacy' of this book.

Drugs inspired by compounds found in the body (endogenous compounds)

Figure 10.14 shows the chemical structures of salbutamol, the asthma medication, and cimetidine, used for the treatment of peptic ulcers. Alongside their structures are the natural compounds from which they were developed.

Cimetidine was inspired by a natural compound found in the stomach – histamine (see Figure 10.14). Histamine stimulates the production of gastric acid, which is a problem for patients with peptic ulcers, causing pain and inflammation. Nothing was known about the physiological site of action of histamine, so how could scientists design a drug to interact with it and stop the production of the problematic acid? The answer was ingenious. Histamine binds to the site, so

chemists used the structure of histamine as a starting point. Their aim was to produce a new drug that would compete with histamine for the active site but would bind more strongly to it, and would suppress, rather than trigger acid secretion. The structure of cimetidine was developed after much experimentation, both by computer (*in silico*) and in the laboratory. For several years cimetidine was the number one selling prescription product in several countries with worldwide annual sales of $1,000,000,000.

 Chapter 7 'Introduction to aromatic chemistry' describes the development of salbutamol from adrenaline. If you cannot remember why salbutamol is a better asthma treatment than adrenaline, you should revise that material now.

SELF CHECK 10.17

Look at the structures in Figure 10.14. Can you imagine how the natural ligand led to the development of the popular drugs?

FIGURE 10.14 Drugs inspired by natural ligands.

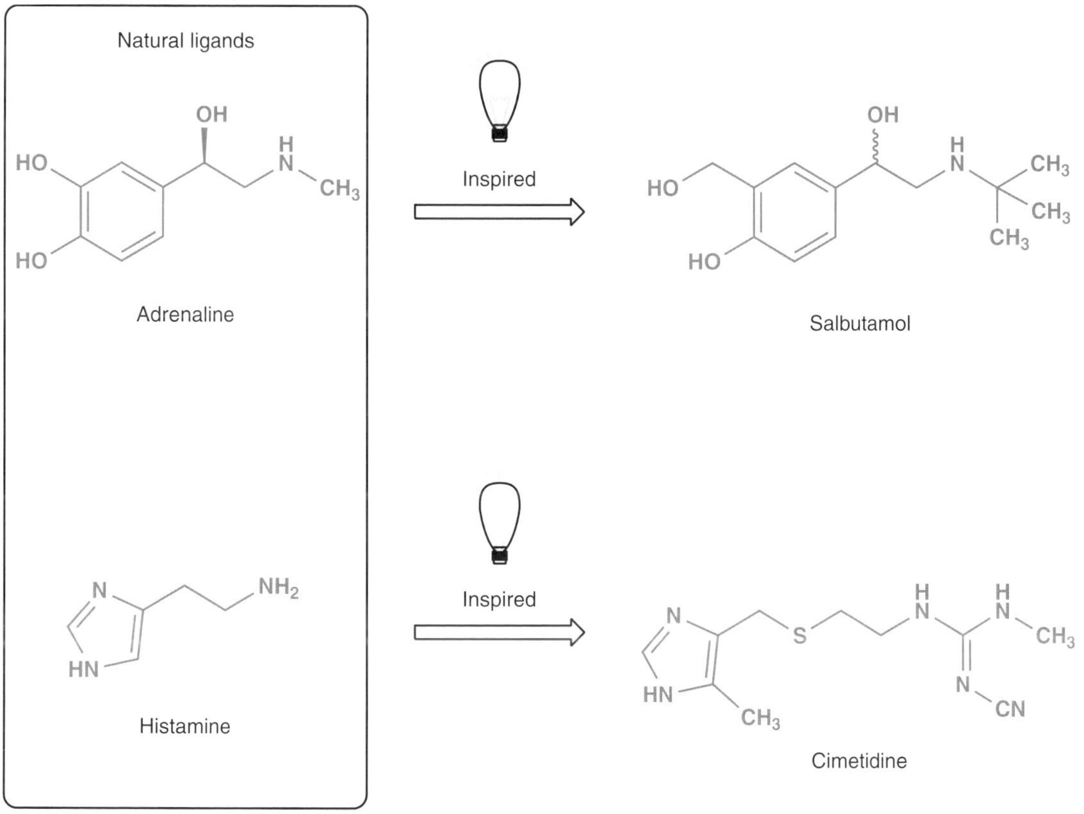

KEY POINT

Totally synthetic drugs have often been inspired by the biological activity of a complex natural product.

Computer aided drug design

Some drugs are designed entirely on a computer, using what we call 'computational chemistry'. Provided they start with good quality information provided by competent scientists, computers can make a huge contribution to the design of drugs, by designing or selecting compounds for testing against a desired biological target (see Box 10.3). We can use computers to predict molecular properties such as aqueous solubility and shape. Sophisticated programs now enable us to predict the ability of a molecule to bind to biomolecules, including proteins, lipids and nucleic acids. This can then help to speed up the drug discovery process.

Captopril, a clinically important antihypertensive drug inspired by the Brazilian snake venom was one of the first drugs to be designed logically with the aid of computers in the 1970s. During its development, scientists at the Squibb Research Institute in the USA made use of a hypothetical model of the active site of the ACE to guide the design and synthesis of specific inhibitors. Using the model, they were able to propose which amino acids in the active site of ACE were important in binding inhibitors, leading to the preparation and biological testing of a **lead compound**,

BOX 10.3

The Lifesaver Screensaver Project

The Lifesaver Screensaver project (2003–2009) harnessed 3.5 million home computers in more than 200 countries to help with synthetic drug design. The project built a database of billions of small drug-like synthetic molecules, and tried to fit them into important biological macromolecules. The main diseases under study were cancer, anthrax and smallpox. Up to 10% of the molecules that the computers predicted to have activity, were verified by scientists under experimental conditions.

which supported their hypothesis. Optimization of this lead compound ultimately led to the development of captopril (see Figure 10.13).

'Me-too' drugs

'Me-too' is a peculiar, but useful name. One definition of a 'me-too' drug is: a drug which follows an initial therapy and works through essentially the same mechanism. Once a new drug has proved itself so that it passes all the biological tests, clinical trials and regulatory issues, it is relatively easy for other companies (or the same company even) to make similar molecules that do a similar thing. Ideally, these new molecules will have improved properties compared with the existing drug. Patent protection will be in place to stop unscrupulous copying of structures, but patent protection can only cover so many analogues and it is often possible to develop new compounds that have similar or better biological profiles without infringing the original patent. Figure 10.15 shows some examples of 'me-too' drugs. We have already met the proton pump inhibitor omeprazole, the antidepressant citalopram, and cimetidine, used to treat peptic ulcers. Esomeprazole, escitalopram, and ranitidine are 'me-too' drugs resembling these.

More information on patents can be be found in Chapter 3 'Legal and ethical matters' of the *Pharmacy Practice* book in this series.

Library screening

Some drug companies possess large banks of compounds (often up to several hundred thousand), which they can potentially screen against any disease (see Box 10.4). To develop a new drug they screen the entire library to see if any of the molecules in it has any activity, no matter how small, against their disease of choice. If a 'hit' is identified, the chemists then make hundreds of analogues of it to optimize its activity. Several rounds of analogue preparation and optimization then takes place to see if these new structures have the potential to become the next marketable drug.

FIGURE 10.15 'Me-too' drugs.

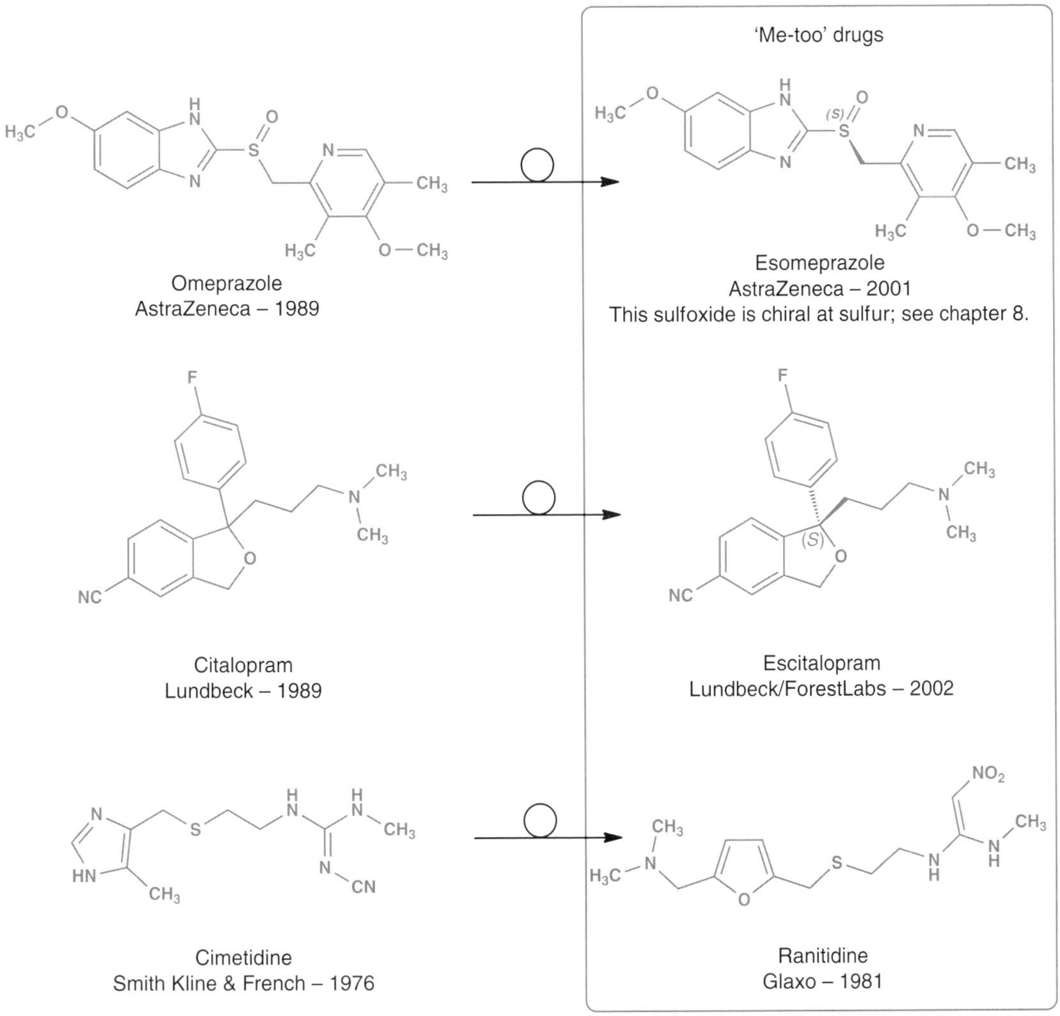

'Me-too' drugs

Omeprazole
AstraZeneca – 1989

Esomeprazole
AstraZeneca – 2001
This sulfoxide is chiral at sulfur; see chapter 8.

Citalopram
Lundbeck – 1989

Escitalopram
Lundbeck/ForestLabs – 2002

Cimetidine
Smith Kline & French – 1976

Ranitidine
Glaxo – 1981

BOX 10.4

High-throughput screening

High-throughput screening is a method that enables a large number of compounds (typically thousands) to be **screened** in an automated fashion and on a small scale. Activity needs to be easily detected: for example, a colour change caused by enzyme action. While high-throughput screening itself may not result in a blockbuster drug, it can help find new lead compounds for further optimization by medicinal chemists.

10.5 Genetic engineering and fermentation (biotechnology) for the production of drugs

Huge advances in molecular biology, over the past few years, have meant that scientists can use genetic engineering to help produce important molecules, particularly important proteins. By cloning specific genes and incorporating them into the DNA of fast-growing cells (such as bacteria or yeast), we can use the cells as a kind of 'factory' to produce large quantities of proteins and other molecules. Genetic engineering is used for the mass-production of numerous drugs, including insulin, human growth hormones, human albumin, monoclonal antibodies and vaccines.

Genetic engineering is a flexible method of drug preparation, because it is possible to manipulate the genetic information so that bacteria or yeast produce not just natural compounds but analogues of the natural compounds for us too. The first genetically-engineered drug made in this way, and approved, was human insulin for the treatment of diabetes.

It is very easy to purify genetically-engineered proteins, and this substantially reduces the risk of contamination by viruses, such as HIV, or by disease-causing proteins, such as those causing Creutzfeldt-Jakob disease.

Since the introduction of genetically-engineered insulin, the insulin gene has been manipulated to pro-

FIGURE 10.16 Genetically-engineered insulins. Differences between insulin analogues and normal human insulin are shown in red.

```
Human Insulin

GIVEQCCTSICSLTQLENYCN
          |
FVNQHLCGSHLVEALYLVCGERGFFYTPKT

Insulin lispro

GIVEQCCTSICSLTQLENYCN
          |
FVNQHLCGSHLVEALYLVCGERGFFYTKPT

Insulin glargin

GIVEQCCTSICSLTQLENYCG
          |
FVNQHLCGSHLVEALYLVCGERGFFYTPKTRR
```

duce analogues of insulin. Figure 10.16 shows normal human insulin together with two clinically useful analogues, which help diabetics maintain near-normal insulin activity. Insulin lispro is a fast-acting insulin and insulin glargin is a long-acting insulin, useful for maintaining constant insulin levels.

10.6 The principles of organic synthesis

Over half the drugs in clinical use require synthetic chemistry at some stage in their production. Despite the growth of biotechnology as a source of drugs, it seems unlikely that chemical synthesis will go out of fashion. So the question is: how do chemists make molecules? The section sets out the basic principles to give you some idea as to how organic molecules, such as drugs, are made.

In Chapters 2–7, you have learnt a basic toolbox of chemical reactions and we are now in a position to consider the best strategy to make molecules. We can build them up using a number of different chemical reactions. The difficult part is deciding what order to do the reactions in. Below is a summary of the related key principles involved in making molecules of any kind:

• make the core framework of your target molecule, (i.e. get all the carbon atoms in place)

• introduce any additional functional groups, remove unwanted ones, or change the existing ones to get the functionality you need. Some functional groups may need to be masked to make them unreactive

• make sure the selectivity you need is present. Can you make the single enantiomer you want (R or S) or the desired C=C double bond (E or Z)?

• make your pure compound as efficiently as possible, as cheaply as possible and with minimal environmental impact. On an industrial scale, this also means that your synthetic route should be as short as possible, robust, and provide enough of the desired products with minimal waste products.

Since there are a huge number of different theoretical synthetic routes to any target molecule, an important skill is to ensure that the best route is chosen.

This route must contain the best reactions so that the *overall* process is as efficient as possible. For all but the simplest of molecules, this is usually achieved by starting (on paper) from the target molecule and working *backwards* until you arrive at cheap, commercially available starting materials. This process is called **retrosynthesis**. The details of this are beyond the scope of this book, but can be found in an organic chemistry textbook (see Further Reading).

10.7 **Purification methods**

Once an organic compound has been isolated, either from nature or after a chemical synthesis in the laboratory, the compound needs to be purified and the identity of the pure compound needs to be established beyond any doubt, so that we are certain exactly which compound is having the biological effects we are seeing. Chapter 11 outlines the common methods to identify compounds unequivocally once they have been made, however, before this can be done, it is best to have the compounds pure. This section outlines a few of the more common methods for purifying compounds from chemical reactions. Some of the methods for small scale reaction monitoring and purification are also discussed.

Aqueous work up

This technique is a crude purification step that works especially well for acids and bases. A synthetic drug is normally made in an organic solvent; a natural product may be isolated from plant material in an organic solvent, or found in a micro-organism growth medium in aqueous solution. At neutral pH, most small drug molecules dissolve better in organic solvents, such as ethyl acetate or chloroform, than in water. A drug in aqueous solution can therefore be extracted into an organic solvent (one that does not mix with water).

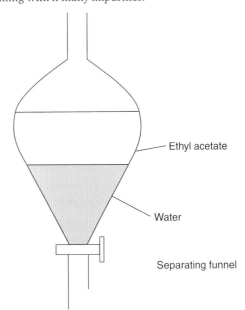

Ethyl acetate and chloroform

Whatever the source, we now have an organic solution of a drug that can be placed in a separating fun-nel, shaken with water and the water layer removed, taking with it many impurities.

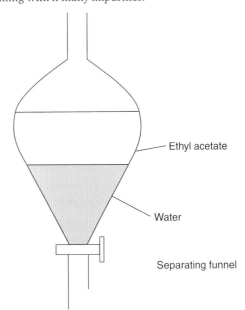

Ethyl acetate

Water

Separating funnel

If the drug is acidic and stable to base (salicylic acid is a good example), you can purify it further by shaking the remaining organic solution with a basic solution, such as saturated sodium carbonate (pH 11). The drug deprotonates and dissolves in the water layer, leaving more impurities behind in the organic layer, which is removed. You can now acidify the water layer and extract the drug back into organic solution. The organic solvent is removed by evaporation, giving the purified drug.

SELF CHECK 10.18

Azithromycin is a basic drug that is stable to acid below pH 3. By analogy with the purification of salicylic acid how could you use an aqueous work up of an ethyl acetate solution of impure azithromycin to remove many of the impurities? (Note that the methyl groups are not explicitly represented on this structure because of crowding.)

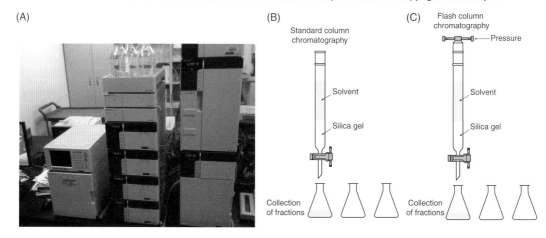

Azithromycin

Liquid chromatography

Liquid chromatography is a method of separating compounds (which may be small or large molecules), based on how they partition between a solid and a liquid. In normal phase chromatography, the solid is polar and the liquid is a non-polar organic solvent. In reversed phase chromatography, the solid is very hydrophobic and the liquid is typically based on water and is polar. A concentrated solution of the mixture to be separated is applied to the top of a column containing the solid phase, and the liquid phase is passed through. Compounds that bind poorly to the solid phase reach

the bottom of the column first, and are collected. Compounds that bind more tightly to the solid phase (polar compounds in normal phase chromatography, non-polar compounds in reversed phase chromatography), reach the bottom (are **eluted**) slowly. All the compounds are collected for analysis or further purification.

Liquid chromatography includes HPLC (high performance liquid chromatography), which is highly mechanized and boxed up to look rather generic (see Figure 10.17A). It also includes methods, such as gravity and flash chromatography, that are based on very simple pieces of equipment that are operated manually (see Figure 10.17B and 10.17C).

Recrystallization

Recrystallization may be used instead of, or as well as, column chromatography. It is especially suitable for purifying large quantities of material, but is only likely to be successful when the desired product is the major component of the crude mixture. This technique requires the crude product to be dissolved in the minimum amount of a carefully chosen hot solvent before being allowed to cool slowly. The idea is that the impurities will remain dissolved in the solvent after cooling and the product will crystallize out in a pure form from which it can be isolated by filtration. This method is used on an industrial scale to purify solid compounds.

FIGURE 10.17 Chromatography: (A) an HPLC system, (B) standard gravity chromatography and (C) flash chromatography (in which solvent is under pressure).
10.17A: licensed under the Creative Commons Attribution-Share Alike 3.0 Unported license. Copyright: Nandostyle.

CHAPTER SUMMARY

➤ The structure of the drug is important. It must have the correct functional groups in the correct orientation to interact with its biological target and initiate a response.

➤ Drugs can originate from a range of sources, including: nature, chemical laboratories or a combination of the two.

➤ Even drugs from nature are sometimes made in a chemical laboratory so that enough of the active ingredient is available.

➤ Nature can be used as an inspiration to chemists when developing new drugs.

➤ Many analogues of new drugs have to be made before the best candidate is found.

➤ Organic chemistry is used to make most small-molecule drugs.

➤ Drugs usually need to be pure before biological testing can take place.

FURTHER READING

Patrick, G.L. *An Introduction to Medicinal Chemistry*. 5th edn. Oxford University Press, 2013.

Presents a very in-depth account of medicinal chemistry and includes an excellent section on drug discovery, design and development.

Mann, J. *Murder, Magic and Medicine*. 2nd edn. Oxford University Press, 2000.

A very readable book outlining the evolution of modern drugs from their origins in nature and their uses as poisons, hallucinogenics and medicines.

Clayden, J., Greeves, N., and Warren, S. *Organic Chemistry*. 2nd edn. Oxford University Press, 2012.

An excellent reference work about all aspects of organic chemistry.

Newman, D.J., Natural Products as Leads to Potential Drugs: An Old Process or the New Hope for Drug Discovery? *J. Med. Chem.* 2008; 51: 2589–99.

Presents a concise account of the influence of natural products on drug discovery.

Nicolaou, K.C., Yang, Z., Liu, J.J., Ueno, H., Nantermet, P.G., Guy, R.K., Claiborne, C.F., Renaud, J., Couladouros, E.A.,

Paulvannan, K., and Sorensen, E.J. Total synthesis of taxol. *Nature* 1994; 367: 630–4.

One of the first total syntheses of taxol.

Fu, Y., Li, S., Zu, Y., Yang, G., Yang, Z., Luo, M., Jiang, S., Wink, M. and Efferth, T. Medicinal chemistry of paclitaxel and its analogues. *Current Medicinal Chemistry* 2009; 16: 3966–85.

An in-depth account of the state-of-the-art medicinal chemistry of paclitaxel and its analogues.

Elander, R.P. Industrial production of ß-lactam antibiotics. *Appl. Microbiol. Biotechnol.* 2003; 61: 385–92.

An excellent review on the production of ß-lactam antibiotics.

Richards, W.G. Virtual screening using grid computing: the screensaver project. *Nature Reviews Drug Discovery* 2002; 1: 551–5.

A description of how massively distributed computing using screensavers has allowed databases of billions of compounds to be screened against protein targets in a matter of days.

Introduction to pharmaceutical analysis

LARRY GIFFORD AND TONY CURTIS

The deliberate contamination (adulteration) of food and drugs with poisons has been practised for hundreds of years. In 1820, Friederich Accum published a treatise describing the chemical analysis of commonly adulterated substances. To draw attention to the seriousness of the problem he adorned the front cover with a skull and crossbones and the text 'There is death in the pot'.

He highlighted such practices as the colouring of jelly, and blancmange, red by the addition of mercury(II) sulfide mixed with red lead! Legislation now makes these activities illegal. Today, however, we are still at risk from unscrupulous criminals wishing to make profit from the sale of fake medicines. The World Health Organisation (WHO) warns about SFFC medicines – Spurious/Falsely labelled/Falsified/Counterfeit medicines. The use of SFFC medicines can result in treatment failure or even death.

The quality of medicines intended for human use is controlled by the analytical department of a drug company, which comprises scientists from a range of disciplines. They are required to provide data that can answer very important questions, such as: Does the medicine contain the right drug in the right amount? What is its shelf life? How should the drug be taken?

In order to obtain the necessary data, they use different instrumental methods of analysis, the most popular of which are discussed in this chapter.

Learning objectives

Having read this chapter you are expected to be able to:

➤ understand the purpose of the qualitative and quantitative tests described in pharmacopoeial monographs

➤ calculate the pharmaceutical content of formulated products from ultraviolet-visible spectrophotometric data

➤ obtain structural information from infrared, mass spectrometric and nuclear magnetic resonance data

➤ select an appropriate gas or liquid chromatographic method for the analysis of a mixture of substances in formulated products or clinical samples.

11.1 Quality control of pharmaceuticals and formulated products

Reference texts detailing how to make healing remedies have been known for thousands of years. **Pharmacopoeias** (literally meaning 'drug making') are reference works containing specifications for drug substances and formulated medicines. Early English pharmacopoeias contained recipes for medicines, containing such things as animal excrement, cobwebs, moss found growing on the human skull and virgins' milk! Remedies containing such ingredients would clearly cause more harm than good. Today manufacturers of medicines must meet more exacting standards.

Two types of standards are used to control the quality of pharmaceutical products.

1. **Product licence standards** are manufacturing standards laid down by the Licencing authority, which in Britain is the Medicines and Healthcare products Regulatory Agency (MHRA). The MHRA publish a book entitled 'Rules and Guidance for Pharmaceutical Manufacturers and Distributors', often referred to as 'The *Orange* guide' because of its colour. The specifications set to obtain a product licence for any product are usually confidential between the manufacturer and the MHRA.

2. **Published standards** (or Pharmacopoeial monographs) for users of medicinal products are to be found in the **British Pharmacopoeia** (BP), European Pharmacopoeia (Pharm Eur) and the United States Pharmacopoeia (USP). These standards not only take into account the required quality during manufacture but also possible degradation during the shelf life of the product.

Sources of impurities in pharmaceutical products

Impurities can arise from a range of different sources. These include raw materials, products of the manufacturing process, degradation products, packaging materials and microbiological contaminants. The complexity surrounding the manufacturing process is illustrated by the synthesis of aspirin.

Figure 11.1A shows the well-known synthesis of aspirin from salicylic acid. At first glance, you might expect that the only impurities would be excess starting materials. But things are not always as they seem. The British Pharmacopoeia identifies six impurities (referred to as related substances) that may be found

FIGURE 11.1 The synthesis of aspirin. (A) The conversion of salicylic acid to aspirin. (B) The synthesis of salicylic acid.

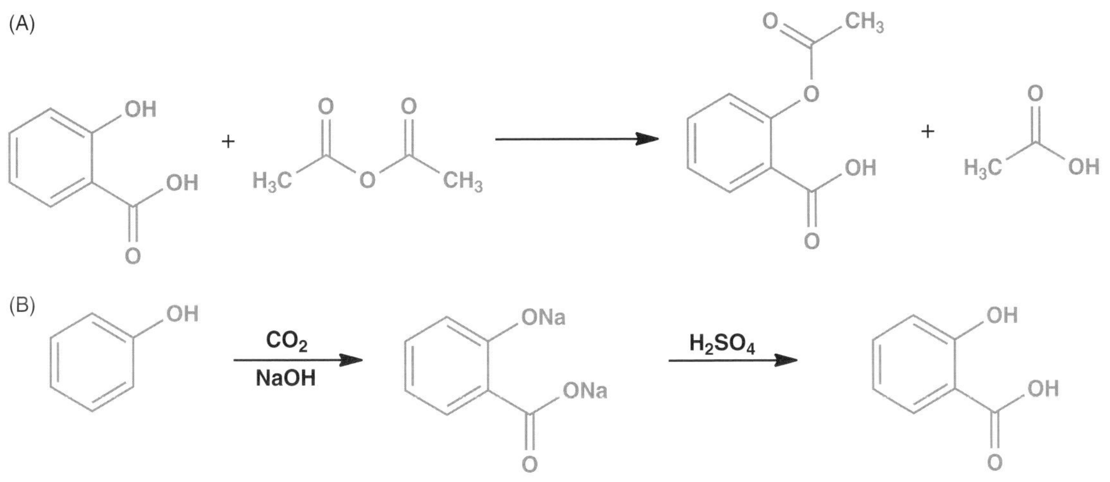

in aspirin intended for the manufacture of medicines. This raises questions about how they are identified, how much of each is present, and how they got there.

When you make aspirin in the laboratory, you start from salicylic acid, and side reactions may lead to impurities. In an industrial setting, however, salicylic acid itself is synthesized by the process shown in Figure 11.1B from phenol, and this synthesis leads to a number of side-products, as well as the desired salicylic acid.

SELF CHECK 11.1

Suggest a mechanism for the first step of the reaction pathway shown in Figure 11.1B.

The six impurities found in aspirin preparations are shown in Figure 11.2. 4-Hydroxybenzoic acid (impurity A) and 4-hydroxybenzene-1,3-dicarboxylic acid (impurity B) arise during the synthesis of salicylic acid, if the conditions of temperature and pressure are not carefully controlled. The third possible impurity (impurity C) is salicylic acid itself. This is, of course, a synthetic precursor of aspirin, but can also arise from the hydrolysis of aspirin, if the drug is stored in damp conditions. Impurity D (acetylsalicylsalicylic acid) arises from the reaction of aspirin and salicylic acid, and impurity E (salicylsalicylic acid) is formed by the reaction of two salicylic acid molecules. Finally, impu-

SELF CHECK 11.2

Using your knowledge of aromatic chemistry (Chapter 7), can you say why 4-hydroxybenzoic acid (impurity A) and 4-hydroxybenzene-1,3-dicarboxylic acid (impurity B) are formed during the synthesis of salicylic acid?

SELF CHECK 11.3

Draw the mechanism for the hydrolysis of aspirin to give salicylic acid.

SELF CHECK 11.4

Look carefully at the new (purple) portion of these molecules in Figure 11.2. What functional group is formed?

rity F is 2-(acetyloxy)benzoic anhydride (acetylsalicylic anhydride) and comes from the condensation of two aspirin molecules.

It is clear that competing reactions during synthesis can result in several products being present. It is important that there are limits put on the amounts of these to prevent adverse or toxic reactions being suffered by the patient. The British Pharmacopoeia sets out a **high-performance liquid chromatography (HPLC) assay** for the identification and quantification of these impurities in aspirin. (The word 'assay' means

FIGURE 11.2 The six impurities found in aspirin preparations.

(A) 4-Hydroxybenzoic acid

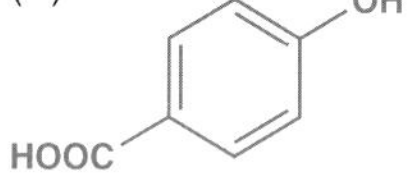

(B) 4-Hydroxybenzene-1,3-dicarboxylic acid (4-Hydroxyisophthalic acid)

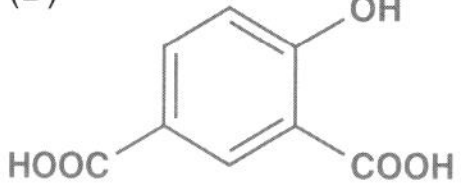

(C) Salicylic acid

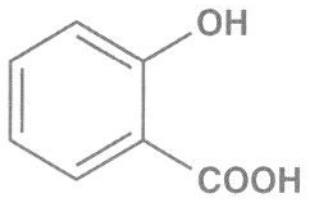

(D) 2-[[2-(Acetyloxy)benzoyl]oxy]benzoic acid (Acetylsalicylsalicylic acid)

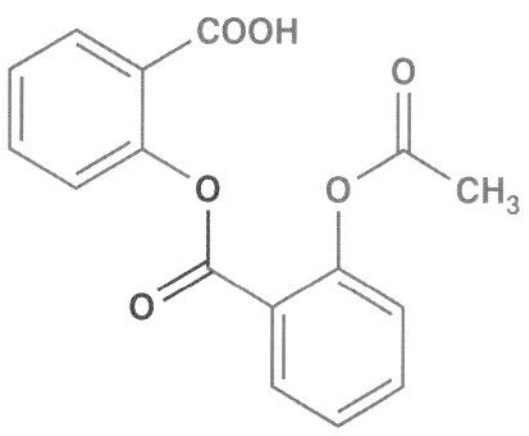

(E) 2-[(2-Hydroxybenzoyl)oxy]benzoic acid (Salicylsalicylic acid)

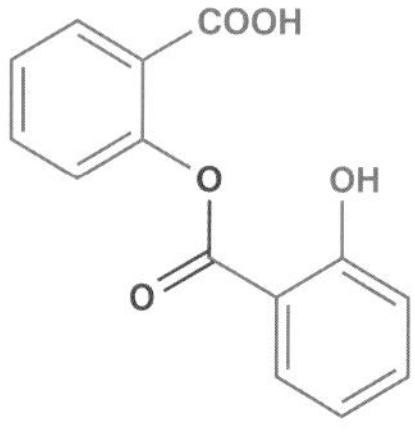

(F) 2-(Acetyloxy)-benzoic anhydride

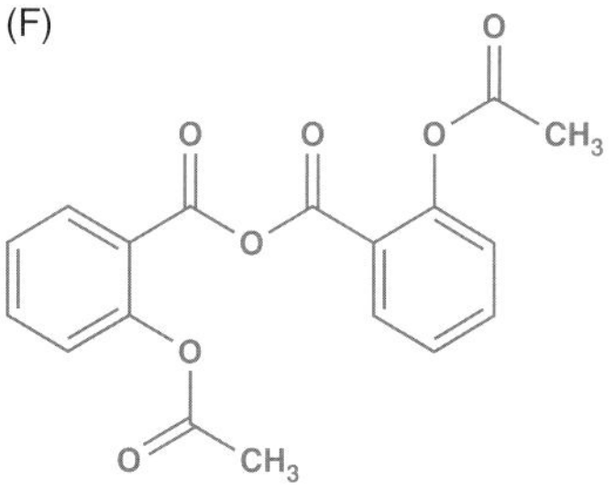

method of analysis). This analysis is quite straightforward, and involves comparing peak areas on a **chromatogram**. A simple assay is sufficient for this routine check because the possible impurities are well-known.

 High Performance Liquid Chromatography is discussed in more detail in Section 11.7.

Suppose, however, that you wished to assay aspirin from a new manufacturing process, in order to check for unexpected impurities. This requires several steps which typically include isolation of each impurity from within the reaction mixture. The unknown compound (impurity) may be separated from the other compo-

nents using chromatography and, once pure, can be analysed using spectroscopy. This is done using **spectrometry** and spectroscopy techniques including:

- **ultraviolet-visible spectrophotometry** (see Section 11.3)
- **infrared spectroscopy** (see Section 11.4)
- **nuclear magnetic resonance (NMR) spectroscopy** (see Section 11.5)
- **mass spectrometry** (see Section 11.6)

Measurements of **physicochemical properties**, such as melting point and solubility should also be taken.

11.2 **The electromagnetic spectrum**

Our eyes can only detect electromagnetic radiation with a wavelength between about 400–740 nm; we refer to this radiation as visible light. This places an immediate limit on the size of what we can see. The best light microscope in the world cannot enable our eyes to detect objects smaller than about 1 μm in diameter, because a light wave will strike such objects only once or not at all.

There is a lot more to the electromagnetic spectrum than visible light (see Figure 11.3). At short wavelengths, electromagnetic radiation is very ener-

getic and, as it can destroy cells, is often dangerous. For example, ultraviolet (UV) radiation (wavelength 10–400 nm) causes sunburn and can cause skin cancer, however, used carefully it can also be very useful. X-rays (wavelength 0.01–10 nm) are used in medicine and in x-ray crystallography, and gamma rays (wavelength below 0.01 nm) are used in cancer treatment and to sterilize hospital equipment.

Radiation with longer wavelength is also very useful. Infrared (IR) (wavelength 0.74 μm–1 mm) is

FIGURE 11.3 The electromagnetic spectrum.

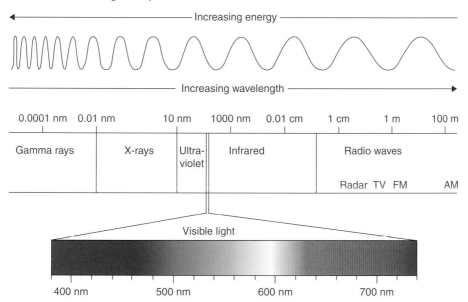

used in infrared lamps and in barcode scanners, and microwave radiation (wavelength 1 mm–1 m) is used in cookery and astronomy. Radiowaves (wavelength 1 mm–100 km) are, of course, used in radios, as well as radiotelescopes.

Most parts of the electromagnetic spectrum are used in spectroscopy in laboratories. Different radiation elicits different behaviours in atoms and molecules depending on the wavelength of the radiation.

For example, UV radiation causes electronic transitions in materials, which are measured using ultraviolet-visible spectrophotometry; IR radiation makes molecules vibrate and this is exploited in infrared spectroscopy; and radiowaves produce nuclear spins, which are used to analyse molecules in NMR spectroscopy. These techniques are discussed in more detail in the following sections.

11.3 Ultraviolet-visible spectrophotometry

Ultraviolet–visible spectrophotometry is an analytical technique used extensively in pharmacopoeial **monographs**. It can be used as a qualitative method (to answer the question: *What* is in these tablets?) and as a quantitative technique (to answer the question: *How much* of the drug is in these tablets?).

Making spectrophotometric measurements

A diagram of the basic instrument required for making spectrophotometric measurements is shown in Figure 11.4. It has the following components:

- a light source (or sources) capable of producing visible and ultraviolet light
- a monochromator capable of splitting the light into its constituent wavelengths
- a cell (called a cuvette) containing the sample

- a photocell (also called a photodetector) capable of measuring light intensity
- an electronic display module (output).

In this simple system, light of intensity I_0, shines on a sample solution contained in the cuvette. Some of the light is absorbed by the solution and the remainder passes through the sample cell and emerges with the lower intensity I_t. (We say that the light that passes through the sample is *transmitted*.) The intensity of the light falling on the photodetector is displayed, in absorbance units, on the electronic display.

For qualitative analyses the absorbance of the sample solution is measured at different wavelengths by adjusting the monochromator. The resultant spectrum is a plot of absorbance vs wavelength; it shows how the absorbance of the sample changes with the wavelength of the light source. Each drug molecule has a characteristic spectrum – essentially, a chemical 'fingerprint'.

FIGURE 11.4 Diagram of a simple spectrophotometer.

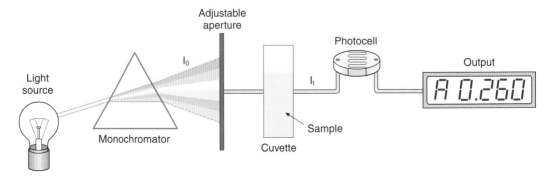

FIGURE 11.5 UV-visible spectra for paracetamol in an acid solution (blue) and an alkaline solution (red).

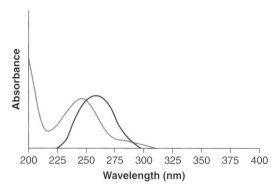

In order to absorb radiation in the UV-visible region, a molecule requires a **chromophore** (a number of conjugated π-bonds and/or lone pairs of electrons). For example, paracetamol can absorb radiation in the UV-visible region because it has a benzene ring, and the lone pair of electrons on the oxygen can donate electron density into the π-system.

Figure 11.5 shows two UV–visible spectra for paracetamol. The blue line shows paracetamol in an acid solution and the red line shows the spectrum of paracetamol in an alkaline solution. The difference between the two plots demonstrates how the technique can be used, not just to identify chemicals but also to monitor chemical changes – in this case ionization.

 You met conjugated double bonds in Chapter 7 'Introduction to aromatic chemistry'; they are double bonds separated by one single bond.

The Beer-Lambert Law

The sample in the cuvette is most often a liquid solution, although UV-visible spectrophotometry can also be used for gases and solids. The amount of light passing through a coloured solution depends on two things: (i) the depth of the solution (i.e. the distance the light has to travel to pass through the solution), and (ii) the concentration of the solution.

The relationship between how much light will pass through a coloured solution and the depth of the solution is easily visualized. This is because when the light has to travel further through the solution, more light is

absorbed. This is easily visualized by comparing the colour of fruit juice viewed through the narrow part of a glass (a champagne flute works especially well) and the wide part of the glass. This relationship has been understood for several hundred years, with Bouguer (1729) and later Lambert (1768) deriving a mathematical relationship between the intensity of light travelling through a solution and how far through the solution the light travels.

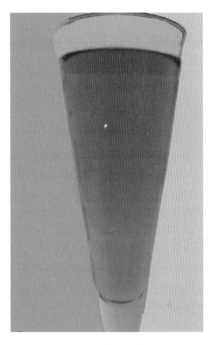

©Nicola Wood

In 1852, Beer also established the relationship between the amount of light passing through a solution and the concentration of the absorbing species in that solution. The higher the concentration of the solution, the more light will be absorbed and the darker the solution will appear.

Beer and Lambert's mathematical relationships are usually combined and expressed as the Beer-Lambert Law, which can be expressed in grammes or moles.

Beer-Lambert Law expressed in grammes

This expression of the Beer-Lambert Law is written as:

$$A = a\,b\,c$$

where, A is the absorbance (this value is obtained from the spectrophotometer), a is the absorptivity (this is the absorbance of a solution having a concentration of 1 g L^{-1}; it is a constant), b is the path length of the sample cell in cm (usually 1 cm), and c is the concentration in g L^{-1}.

Many BP monographs use standard values of A (1%,1 cm) in spectrophotometric analyses. The equation then becomes:

$$A = A \, (1\%, 1cm) \, bc$$

A (1%,1 cm) is a constant value at a given wavelength and is the absorbance of a 1% w/v (1 g/100 mL) solution. A cell of 1 cm path length is used, and c the concentration is expressed in g/100 mL.

Beer-Lambert Law expressed in moles

This expression of the Beer-Lambert Law is written as:

$$A = \varepsilon \, bc$$

where, A is the absorbance (this value is obtained from the spectrophotometer), ε is a constant known as the molar extinction coefficient (the absorbance of a 1 M solution of the substance being analysed), b is the path length of the sample cell in cm (usually 1 cm), and c is the concentration in moles L^{-1}.

It is very important to take care over units. The Beer-Lambert Law is ancient and is sometimes expressed in very old-fashioned units!

Example calculation

Labetalol Hydrochloride injection is used to treat hypertension. It is supplied in 20 mL ampoules at a strength of 5 mg/mL.

Labetalol Hydrochloride

The BP monograph for Labetalol Hydrochloide injection is as follows:

'Dilute a volume containing 50 mg of Labetalol Hydrochloride to 100 mL with water. To 10 mL of the solution add 10 mL of 0.05 M sulfuric acid and dilute to 100 mL with water. Measure the absorbance of the resulting solution at the wavelength 302 nm in a 1 cm pathlength cell. Calculate the content in the injection taking 86 as the value of A(1%, 1 cm) at 302 nm.'

First, let's go through the method:

• In the 20 mL Labetalol Hydrochloride ampoule there is 5 × 20 mg = 100 mg. The BP monograph states that a volume containing 50 mg is needed for the analysis, so only 10 mL of the ampoule is needed. This is diluted with water to make a 100 mL solution (50 mg/100 mL).

• To 10 mL of this solution, 10 mL of sulfuric acid is added and enough water to produce a final volume of 100 mL. The original solution contained 50 mg Labetalol Hydrochloride, so this solution now contains 5 mg Labetalol Hydrochloride in 100 mL (5 mg/100 mL).

• By measuring the absorbance of this solution we can perform a calculation to determine if the sample ampoule contained the expected amount of Labetalol Hydrochloride and thereby confirm the strength of the original injection solution.

Using the method described, an analyst found the absorbance of the final solution to be 0.420 at 302 nm. The BP monograph states that the injection should contain between 90 and 110% of the stated amount.

Using the equation:

$$A = A \, (1\%, \, 1cm) \, bc$$

where A = 0.420, A (1%, 1 cm) = 86, b = 1 cm, we can calculate c, the concentration of the solution measured in the spectrophotometer. (Remember this is NOT the concentration of the injection solution since we have performed several dilution steps during the process.)

$$c = \frac{A}{A(1\% \, 1 \, cm)b}$$

$$\therefore c = \frac{0.420}{(86 \times 1)}$$

$$\therefore c = 0.0049 \, \text{g}/100 \, \text{mL}$$

(Remember to use the correct units of concentration.)

To calculate the percentage of the stated dose in the sample we need to do the following calculation:

$$\frac{concentration\ found\ in\ the\ sample}{concentration\ expected\ in\ the\ sample} \times 100$$

$$\frac{0.0049}{0.005} \times 100 = 98\%$$

Therefore the sample meets the BP standard as it is within the limits of 90 and 110% of the stated amount.

You might wonder why the method requires so many dilutions and why we do not measure the absorbance of the injection solution directly. The answer is that the sample would be too concentrated and the measurements would be inaccurate. In modern spectrophotometers absorbance is proportional to concentration over a range from 0 to 3 absorbance units. Most analysts aim for readings of about 0.4 absorbance units to get the best results.

SELF CHECK 11.5

Lidocaine hydrochloride injection is supplied in 2 mL ampoules containing 20 mg/mL. If a 1 mL sample is diluted to 100 mL with water and its absorbance at 263 nm found to be 0.380, calculate the concentration of the injection if the A (1%,1 cm) is known to be 19.

KEY POINT

UV spectrophotometry can be used either qualitatively or quantitatively.

KEY POINT

The Beer-Lambert law relates absorbance to concentration of analyte and allows UV spectrophotometry to be used quantitatively.

11.4 **Infrared spectroscopy**

Pharmacists most usually encounter infrared spectroscopy in pharmacopoeial monographs, but it is also used extensively by synthetic chemists, as an aid to determining molecular structure. If you compare the infrared spectrum in Figure 11.6 with the UV-visible spectrum, you will see that the infrared spectrum contains a wealth of detail, which can be interpreted to give structural information.

Recording infrared spectra

Infrared spectra are recorded on an IR spectrometer. The design of the IR spectrometer and the sample preparation techniques employed are somewhat different from those used in UV–visible spectrophotometry, which makes use of the fact that many materials, including glass, quartz, some plastics, water, and a range of other solvents are transparent to light in the UV-visible region of the spectrum.

The same situation is not true for infrared radiation. Extremely few materials are transparent to IR radiation; even water deeper than a few centimetres absorbs IR radiation. Therefore, in order to obtain good resolution in the spectrum, the sample must be finely dispersed. In pharmaceutical analysis, sample preparation most usually involves a so-called KBr disc. Potassium bromide is mixed with a finely ground solid sample of the substance being studied, and the mixture is then compressed under high pressure to produce a disc about the size of a 5p coin. The KBr disc is scanned in the spectrometer. Specialist materials are used in the construction of the optical components of the spectrometer, to prevent absorption of infrared radiation by the instrument.

You might have noticed that the IR spectrum for ibuprofen (see Figure 11.6) is displayed a bit differently from its UV spectrum. The peaks are negative (they could be called troughs) because the y-axis is '%Transmittance' rather than 'Absorbance'. The x-axis also seems to have been designed to confuse the unwary. Instead of wavelength in μm, we have wavenumber (1/wavelength) in cm^{-1}. This is a historical

FIGURE 11.6 (A) UV-visible spectrum of ibuprofen; (B) IR spectrum of ibuprofen (ATR). Note: IR spectra are displayed with 0% transmission displayed at the bottom of the spectrum, whilst in UV spectra zero absorbance is displayed at the bottom. UV spectra are a plot of absorbance (A) or log (I_0/I_t) vs wavelength (λ) in nm. IR spectra are a plot of $I/I_0 \times 100$ (% Transmission) vs wavenumber, which is the reciprocal of wavelength or $1/\lambda$ in cm^{-1}.

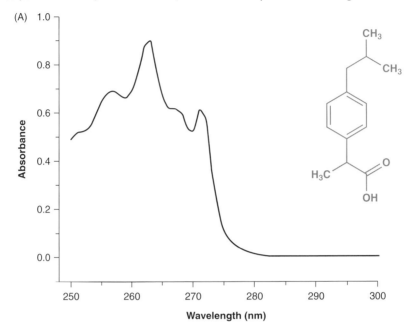

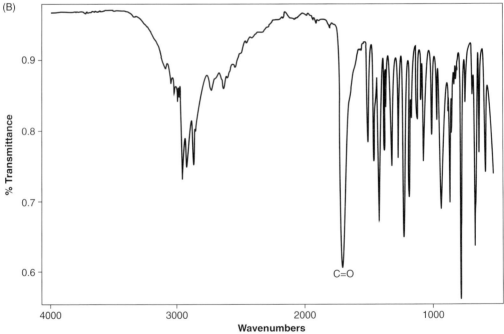

accident. The two techniques were developed independently and nobody has ever felt it important enough to standardize!

(Note: some modern instruments are capable of producing a spectrum directly from the sample without the need to make a KBr disc. This uses a technique known as Attenuated Total Reflectance or ATR.)

Spectral information

Pure samples of drugs, like other organic compounds (but unlike formulated medicines, as these are mixtures of different materials) produce spectra that are unique to themselves and can therefore be used as a means of identification. Collections of spectra are

FIGURE 11.7 Stretching and bending vibration modes. The arrows indicate the direction of the vibration; the + and – signs represent, respectively, vibrations above and below the surface of the page.

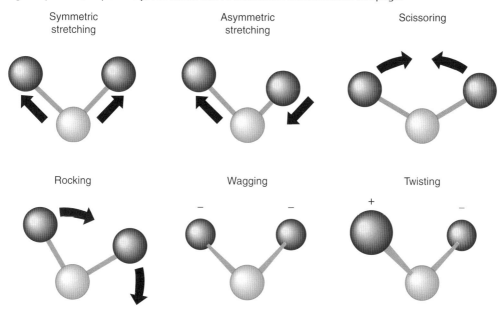

Symmetric stretching   Asymmetric stretching   Scissoring

Rocking   Wagging   Twisting

published for this purpose and The British Pharmacopoeia contains a reference collection of IR spectra. Thus it is possible to differentiate between chemically similar drugs by studying their IR spectra.

Although the infrared region of the electromagnetic spectrum extends further in both directions, spectra are normally recorded in the range 4000–750 cm⁻¹, as in Figure 11.6. For practical purposes, we can divide this into the mid-infrared region (4000–1400 cm⁻¹) and the fingerprint region (1400–750 cm⁻¹):

- The mid-infrared region contains frequencies that are indicative of specific functional groups (group frequencies). Signals in this region can be used to aid the structure determination of unknown compounds.

- The fingerprint region is used rather differently. Every molecule has its own distinctive fingerprint spectrum, so, provided standards are available, it can be used for precise identification.

The absorption of infrared radiation sets up different mechanical vibrations of atoms or functional groups in a molecule, which give rise to absorption bands (or peaks) in the spectra. For example, a methylene group can exhibit both stretching and bending of its bonds. This can be represented as six different types of vibration: symmetric stretching, asymmetric stretching, scissoring, rocking, wagging, and twisting (see Figure 11.7).

In the region from 4000–1400 cm⁻¹ absorption bands characteristic of many common functional groups can be identified. Stretching frequencies of bonds to hydrogen (OH, NH, CH) appear in the region 4000–2500 cm⁻¹, triple bonds 2500–2000 cm⁻¹, double bonds 2000–1500 cm⁻¹ and single bonds 1500–750 cm⁻¹ (the fingerprint region) (see Figure 11.8).

An infrared spectrum is often plotted so that the scale of the 2000 cm⁻¹ to 750 cm⁻¹ region is bigger than the rest of the spectrum. This is because there are more bands below 2000 cm⁻¹.

FIGURE 11.8 The principal stretching frequencies in the infrared spectrum.
Redrawn with permission from Clayden et al. *Organic Chemistry*, 2nd edn, (2012), OUP.

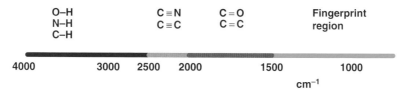

O–H	C≡N	C=O	Fingerprint
N–H	C≡C	C=C	region
C–H			

4000   3000   2500   2000   1500   1000

cm⁻¹

Can you find a drug molecule with an infrared signal between 2500 cm⁻¹ and 2000 cm⁻¹? (Hint: see Chapter 4).

The carbonyl stretching vibration at about 1715 cm⁻¹ is characteristically strong and sharp. Different types of carbonyl appear at slightly different frequencies and this band is often diagnostic of particular types of carbonyl (see Table 11.1).

TABLE 11.1 Principal stretching vibrations of some carbonyls

Functional group	Structure	C=O stretch (cm⁻¹) usually ± 10 cm⁻¹
Acyl chloride	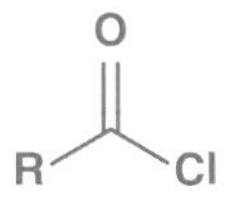	1800
Acid anhydride	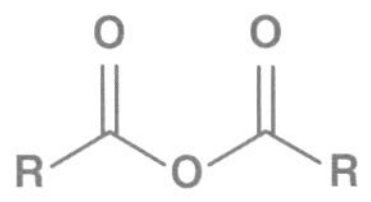	1815, 1765; two bands
Ester	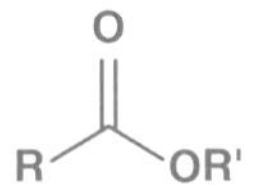	1745
Aldehydes (saturated)	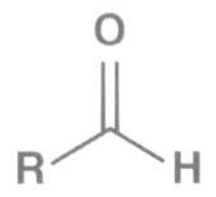	1730
Ketones (saturated)	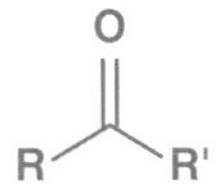	1715
Carboxylic acid symmetric stretch	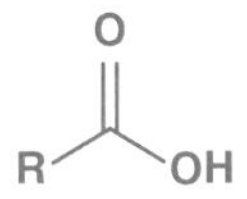	1715
Amide	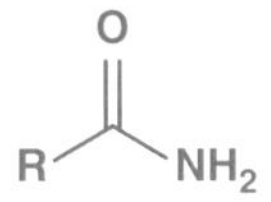	1650
Carboxylic acid asymmetric stretch	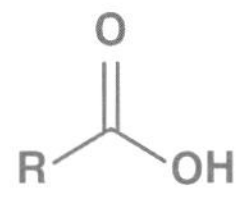	1580

Detailed correlation tables for functional groups are readily found in the literature, for example (see Further Reading). These tables enable scientists to look up a functional group and determine where to look for its infrared signals.

In the spectrum shown in Figure 11.9 for paracetamol the following distinctive signals can be identified:

C=O amide stretch – 1650 cm⁻¹
C=C aromatic stretch –1608 cm⁻¹
N–H amide bend – 1568 cm⁻¹
C=C aromatic stretch – 1510 cm⁻¹

Taken with data from NMR and mass spectrometry the IR data usually provides medicinal chemists with sufficient information to determine the structures of new compounds.

The IR spectrum of phenacetin shows peaks at 1655, 1555, 1513 cm⁻¹. By comparison with the data in the text for the principal IR bands in the spectrum of paracetamol, identify the functional groups associated with these signals.

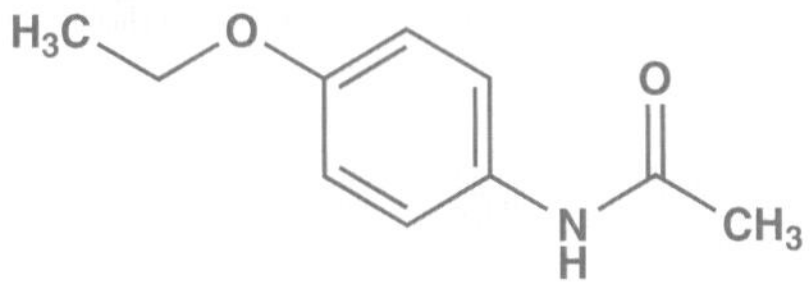

Phenacetin

IR spectra can be used as a means of identification of an unknown substance by comparison with known reference samples.

IR spectroscopy is useful for obtaining structural information about the functional groups in an unknown substance.

FIGURE 11.9 IR spectrum of paracetamol (ATR).

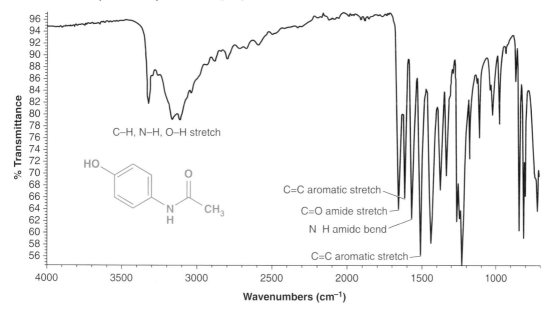

11.5 Nuclear magnetic resonance (NMR) spectroscopy

NMR spectroscopy was discovered by physicists, Felix Bloch and E.M. Purcell, who shared the 1952 Nobel Prize in Physics. It was not immediately taken up by chemists, but by the 1970s simple NMR spectrometers could be found in most university chemistry departments.

NMR spectroscopy is uniquely powerful in that every hydrogen atom (or sometimes carbon, phosphorus, nitrogen or other atoms) gives rise to a signal, and that each signal contains information about the position of that atom in a molecule. For the identification of new compounds or unexpected contaminants in solution, it is thus the most important tool available to the scientist. At some stage in its development, almost every drug you dispense will have been analysed by NMR spectroscopy. However, as it is both expensive and time-consuming, community pharmacists may never use an NMR spectrometer or even see an NMR spectrum. Hospital pharmacists are quite likely to see a spectrometer, but in a rather special setting; **magnetic resonance imaging (MRI)** uses nuclear mag-

netic imaging to allow us to see inside the body (see Figure 11.10). Unlike x-rays, which are mainly used to look at bone damage, MRIs can also be used to view soft tissue, including muscles and the brain.

FIGURE 11.10 Magnetic resonance image of a knee. Source: Jennifer Sheets/iStock.

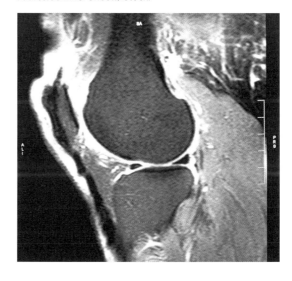

In NMR spectroscopy, a solution of the compound of interest is irradiated with long wavelength, radio frequency radiation within a strong magnetic field, causing some atomic nuclei to behave essentially like bar magnets.

Protons and neutrons are paired in a similar way to electrons, and nuclei with unpaired particles have **nuclear spin**. ^1H (proton), ^{13}C, ^{15}N, ^{19}F and ^{31}P, each of which has a nuclear spin of ½, are by far the most important NMR nuclei. ^2H (deuterium), which has a nuclear spin of +1, is occasionally used as an NMR nucleus. Most importantly though, it is used in NMR solvents; deuterated solvents are used to prepare samples for ^1H NMR analysis because they do not absorb radiation at the frequencies used for ^1H NMR.

The isotope with spin ½ is not always the most abundant. Although ^1H makes up more than 99% of total hydrogen, and ^{19}F and ^{31}P are the only stable isotopes of their respective elements, the most abundant isotope of carbon is ^{12}C, with ^{13}C having a natural abundance of just over 1%. This, together with other factors beyond the scope of this book, means that 5 mg of a small organic molecule is more than enough to run a ^1H spectrum in seconds, but the corresponding ^{13}C spectrum is run over a much longer time period.

In an applied magnetic field, NMR active nuclei align themselves with the field: in the low-energy ground state the nuclei are aligned with the direction of the magnetic field, whereas in the higher energy excited state the nuclei are opposed to the direction of the magnetic field. When the sample is irradiated more of the nuclei become opposed to the magnetic field, but when irradiation stops these high energy nuclei relax and release their energy as radiofrequency that makes up the NMR signal. NMR spectroscopy works by detecting the energy emitted by the sample as it relaxes, representing it as a spectrum that shows the frequency of radiation absorbed by each of the nuclei within the compound. Three key pieces of information can be used to analyse any simple NMR spectrum: **chemical shift**, **relative integration** and **spin-spin coupling** (discussed further, later in this section).

The technique is far more versatile and powerful than we can illustrate here. It can be used to determine the shapes of small and large molecules, the dynamics of binding of small molecules to large molecules, the diffusion rates of small molecules and much more.

The instrumentation required

An NMR spectrometer consists of a digital radio transmitter and receiver, coupled to a very powerful magnet (see Figure 11.11). The superconducting magnet is a hugely sophisticated piece of engineering, requiring liquid helium cooling. The spectrometer does not work properly unless the magnetic field is of the same strength (i.e. is homogeneous) across the entire sample, and this requirement is responsible for much of the sophistication and cost of the instrument.

NMR spectrometers and their magnets are usually described by the frequency of radiation required to excite a proton within the magnetic field generated by the superconductor. The notation '300 MHz NMR spectrometer' implies that a proton contained within

FIGURE 11.11 A schematic diagram of an NMR spectrometer.

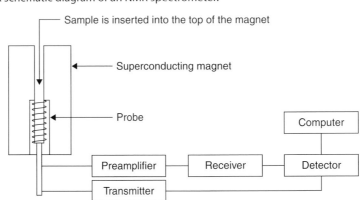

the superconducting magnet attached to that particular NMR machine will become excited by radiowaves at a minimum frequency of 300 MHz.

300 MHz NMR machines are very commonly found in analytical facilities, are frequently automated, and provide spectra that are of sufficient resolution for routine ^{1}H and ^{13}C NMR analyses of small organic molecules. A much stronger magnetic field is required to obtain meaningful information about proteins and even larger organic molecules and more energy is required to raise the nuclei to the excited state. Magnets that generate strong magnetic fields (up to 1000 MHz, or 1 GHz) are large, expensive, and very costly to maintain so these powerful machines are typically located at specialist facilities.

Like modern infrared spectrometers, NMR spectrometers use complex mathematics to derive the spectrum: the powerful pulse of radiation excites all of the nuclei within the compound, but the consequence of this is that all of the nuclei relax and release their energy together as soon as the transmitter is turned off. This extremely convoluted radio signal, detected over a short period of time (up to a few seconds), is converted by computer using a **Fourier transformation** into the NMR spectrum. The NMR spectrum represents the irradiation frequency emitted by each of the nuclei.

Analysing an NMR spectrum

NMR spectra are normally run using *solutions* of compounds. ^{1}H NMR spectra are most usually acquired using deuterated chloroform ($CDCl_3$) or deuterated water (D_2O) as solvent, because deuterium is not observed in the ^{1}H NMR experiment. The solution is placed in a long, narrow tube made of high-quality glass (poor glass disturbs the homogeneity of the magnetic field), which is lowered into the NMR magnet. The solution is then irradiated at the resonant frequency of the nuclei under study; as highlighted above, this frequency is dependent upon the strength of the magnetic field generated by the superconducting magnet.

Each ^{1}H nucleus signal has a position (chemical shift), a size (integral), and a shape (spin-spin coupling) and each parameter is informative. This is best illustrated with an example: here we will consider the 300 MHz ^{1}H NMR spectrum of aspirin.

Chemical shift

In a 300 MHz NMR magnet, all protons are excited to a higher energy level (**resonate**) at approximately 300 MHz. The electrons associated with covalent bonds generate small local magnetic fields acting with or against the applied magnetic field, so that each type of proton has its own characteristic resonant frequency. We reference resonant frequencies to the 12 identical ^{1}H nuclei of tetramethylsilane, $Si(CH_3)_4$, either by adding a drop of this volatile liquid to the sample or by calculating its position from another signal (which may be the solvent deuterium). The chemical shift, δ, in parts per million (ppm) of a nucleus is then defined as:

$$\delta = \frac{frequency\ (Hz) - frequency\ TMS\ (Hz)}{frequency\ TMS\ (MHz)}.$$

δ is therefore independent of the operating frequency (magnetic field strength) of the instrument, and spectra of the same compound taken on different instruments can be directly compared. An NMR signal may be described as having a chemical shift of, for example, 2 ppm or δ 2. They mean the same thing.

Let us now consider a ^{1}H NMR spectrum of aspirin, with chemical shift but no spin-spin coupling information. It is possible to obtain such a spectrum, called a pure shift spectrum, using NMR methodology published in 2010 (see Further Reading) but the one shown in Figure 11.12 is simulated.

In the ^{1}H NMR spectrum of aspirin, the signal at δ 2.35 is assigned to the methyl group of the ester. There is also a group of four signals at δ 7.14, 7.36, 7.63 and 8.12, in the region that is normally associated with protons attached to a benzene ring.

With experience you quickly learn to recognize that the benzene ring protons resonate at about δ 7 and that methyl protons next to carbonyls resonate at about δ 2. But correlation tables and charts in NMR are very useful for interpreting spectra, because hundreds of spectra have gone into their compilation. Figure 11.15 gives a simple guide to where you would expect various proton chemical shifts. Use this chart

FIGURE 11.12   ¹H NMR spectrum of aspirin with no spin-spin coupling information.

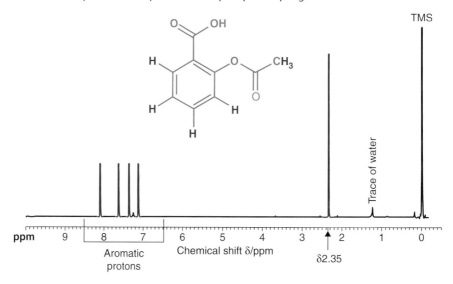

to confirm that the ¹H NMR signals of aspirin are in the expected places.

Finally, there are various ways of describing chemical shifts in relation to one another. Small chemical shifts are found on the right of the spectrum, with large chemical shifts on the left. This corresponds to low and high frequency. You may also come across the terms 'shielded' and 'deshielded'. Each hydrogen nucleus is surrounded by electrons that shield it from the applied field. Adjacent electronegative atoms attract electron density away from the nucleus and it becomes deshielded, causing it to resonate at higher chemical shift. It is sometimes easier to cut through all this and use terms like east and west, but they should be regarded as purely colloquial. Figure 11.16 is a summary.

Relative integration

The area of each peak, known as the integral, is directly related to the number of protons in that particular environment. By comparing the integrals in a spectrum, it is possible to deduce the relative number of protons in each environment in a molecule. If the number of protons giving rise to a particular signal is known, it is possible to deduce the absolute number of protons corresponding to each signal. Integrals normally appear in an NMR spectrum as **sigmoid curves** above each signal, as shown in Figure 11.13.

In the ¹H NMR spectrum of aspirin, the integral for the signal at δ 2.35 is three times as big as the integrals associated with each of the four signals arising from the benzene ring. This provides further evidence, in addition to the chemical shift, that the signal at δ 2.35 is due to the methyl group (three protons) of aspirin; each of the signals within the aromatic region is associated with only one proton.

Spin-spin coupling

Spin-spin coupling, also known as multiplicity, arises from the interaction of neighbouring nuclei within the applied magnetic field. In ¹H NMR spectroscopy, coupling between protons on adjacent carbons (three-bond coupling) can be used to determine the number of protons adjacent to the particular proton of interest. Coupling between protons on the same carbon (two-bond coupling) and coupling between protons on next-but-one carbons (four-bond coupling) are also seen, but are generally less important.

In the NMR experiment, adjacent excited nuclei may both be opposed to the applied field, or only one of them may be opposed to the field whilst the other remains in the lower energy alignment. These two situations give rise to very slightly different chemical shifts, and the NMR signal splits in two and becomes a

FIGURE 11.13 An expansion of the aromatic region in the ¹H NMR spectrum of aspirin.

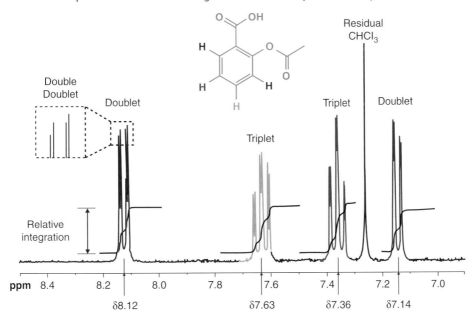

doublet. Consider the red signal in the expanded aromatic region of the ¹H NMR spectrum of aspirin (see Figure 11.13). The red proton has one neighbour (the orange proton) and therefore appears as a doublet. The orange proton has two neighbours (red and lilac) and is split twice, giving a 1:2:1 triplet that is characteristic of signal split by two similar protons. The lilac signal is also a triplet, and the blue signal a doublet. Table 11.2 gives a guide to the most common patterns.

 The Online Resource Centre (www.oxfordtextbooks. co.uk/orc/ifp) gives you a guide to predicting the shapes of ¹H NMR signals from first principles.

There are two remaining points in the spectrum of aspirin (see Figure 11.13) that may concern you. Firstly, each line in the blue doublet (and indeed in all the other signals) is split by a tiny amount. This is due to a very small spin-spin coupling with a next-door-but-one neighbour

FIGURE 11.14 The ¹H NMR spectrum of aspirin – integrals are shown in green.

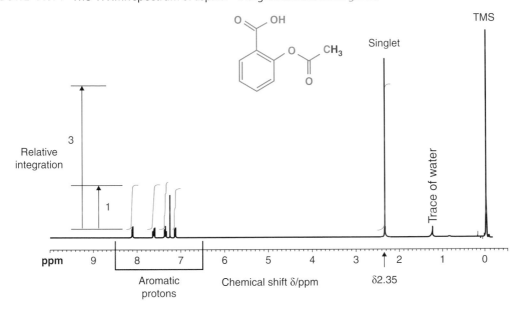

FIGURE 11.15 A simple guide to NMR chemical shifts. R=H, alkyl group or aromatic ring; X=halide, nitro, OR, NR₂, etc.

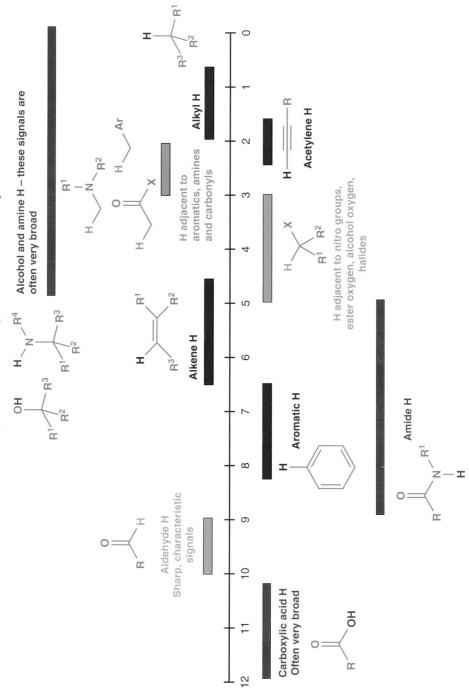

FIGURE 11.16 Different ways to compare chemical shifts.
Redrawn with permission from Clayden et al. *Organic Chemistry*, 2nd edn, (2012), OUP.

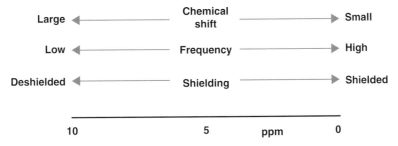

(the four-bond coupling mentioned at the beginning of the section). Secondly, we can easily see which protons give rise to doublets and which to triplets, but how do we know which proton resonates at δ 7.14 and which at δ 8.12? The answer, in brief, is that the strongly electron donating oxygen of the acetyl group is not only *ortho*, *para*-directing in terms of reactivity but is *ortho*, *para*-shielding, so the red proton is at a lower chemical shift

(see Figure 11.14). The carboxylic acid function also has an effect, and sophisticated NMR correlation tables enable you to predict the chemical shift of protons in substituted aromatic rings.

 For more information on *ortho*, *para*-directing groups, see Chapter 7 'Introduction to aromatic chemistry'.

TABLE 11.2 Common coupling patterns in ¹H NMR spectra

Number of neighbours	Spin coupling (shape) of signal	Example
1	Doublet	(1,4)-disubstituted aromatic rings – *para*-aminobenzoic acid
		Note that the red hydrogens are identical and each has a blue neighbour.
2 (in the same or very similar environment)	Triplet (1:2:1)	C**H**₃ next to CH₂ – diethyl ether, see Figure 11.17
2 (in different environments)	Double doublet (1:1:1:1)	C**H** next to two CHs – cinnamaldehyde
		The red H gives a double doublet.
3 (in the same environment)	Quartet (1:2:2:1)	C**H**₂ next to CH₃ – diethyl ether, see Figure 11.17

Diethyl ether

Organic chemists sometimes extract compounds into diethyl ether, and it can be surprisingly hard to evaporate the last traces. Diethyl ether is therefore a common impurity in NMR spectra. An expansion of two regions in the ^{1}H NMR spectrum of diethyl ether (Figure 11.17), highlights the characteristic ethyl group signature of a triplet at low δ and a quartet at higher δ. Even more common is ethyl acetate (ethyl ethanoate), which has largely replaced diethyl ether because it is less hazardous. Its ^{1}H NMR spectrum is quite similar to that of ether; can you sketch it?

SELF CHECK 11.8

Sketch the ^{1}H NMR spectrum of ethyl acetate. (Hint: Use the correlation chart in Figure 11.15, and the spectrum of diethyl ether (see Figure 11.17) to help you.)

Cinnamaldehyde

Cinnamaldehyde is the compound that gives cinnamon its flavour. It has biological activity and has been tested both as an antibacterial and an anticancer agent, but has not yet found any use as a drug.

SELF CHECK 11.9

Sketch the ^{1}H NMR spectrum of cinnamaldehyde assuming that the five aromatic protons are identical to one another (they are nearly).

SELF CHECK 11.10

You should by now be very familiar with the pain killer paracetamol. Draw its structure and then sketch its ^{1}H NMR spectrum. (Hint: phenol OH protons resonate around δ 8 – 10, and amide NH protons at around δ 7 – 8.)

FIGURE 11.17 Part of the ^{1}H NMR spectrum of diethyl ether. Note that because diethyl ether is a symmetrical molecule, the two methyl groups are identical to one another and the two CH$_2$ groups are also identical to one another.

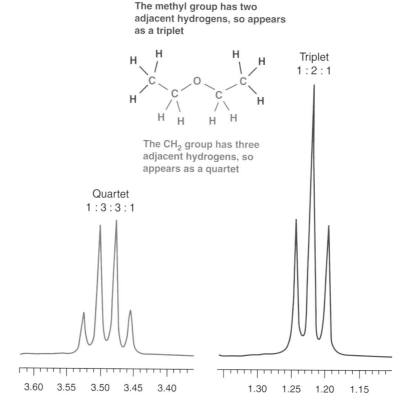

The methyl group has two adjacent hydrogens, so appears as a triplet

The CH$_2$ group has three adjacent hydrogens, so appears as a quartet

Triplet
1 : 2 : 1

Quartet
1 : 3 : 3 : 1

¹³C NMR spectroscopy and other nuclei

For synthetic organic chemists ^{13}C NMR is no less important than ^1H NMR. The main advantage of ^{13}C NMR is that signals are very widely spaced, with almost no overlap. Because ^{13}C is present at only 1.1%, there is no ^{13}C–^{13}C coupling seen and ^1H–^{13}C coupling is 'decoupled'. ^{13}C spectra are, therefore, uncluttered and often very diagnostic. The disadvantage of ^{13}C NMR spectroscopy is that it is very insensitive, about 6000 times less sensitive than ^1H NMR spectroscopy, so only practical when milligrams of material are available.

For physiologists, the most exciting nucleus is ^{31}P. All the phosphorus in the body is ^{31}P and it is possible to take ^{31}P NMR spectra of working muscle, visualizing phospho-esters, such as ATP (see Chapter 8) and making it possible to diagnose diseases of phosphorus metabolism.

^{19}F, ^{15}N, ^2H and ^{14}N NMR are all important in their own fields. Even the radioactive nucleus ^3H (tritium) is used for NMR spectroscopy!

Two-dimensional NMR spectroscopy

NMR spectra can get very cluttered, with signal overlap. Two-dimensional spectra can be used to spread out spin-spin coupling information, which can be ^1H–^1H or ^1H–^{13}C or other combinations. They can also be used to determine whether two nuclei are close in space. DOSY (diffusion ordered spectroscopy) NMR allows the spectra of several components in a mixture to be separated, without separating the components themselves.

Advanced NMR spectroscopy is a huge subject and is beyond the scope of this book. However, people devote their entire working lives to its applications, which extend beyond pharmaceutical analysis into organic chemistry, biochemistry, medicine and the food and drinks industry.

KEY POINT

NMR allows us to detect atomic nuclei and their environment within a molecule. It is the most powerful spectroscopic technique available for determining chemical structure.

315

11.6 **Mass spectrometry**

Mass spectrometry (MS), in contrast to spectroscopy, does not involve irradiating the compound to produce an excited state. MS requires the compound of interest to be ionized under high vacuum and the spectrometer provides information concerning the mass of the ions produced. This ionization usually requires relatively energetic conditions and the molecular ion, the initial species formed, often breaks up into smaller fragments. Whereas spectroscopy is typically non-destructive so that the compound can be re-isolated for further analysis, MS is a destructive technique. Fortunately, the technique does have redeeming features, in that only an extremely small amount (as little as 10^{-15}–10^{-18} moles) of the compound is required for analysis. In addition, a mass spectrometer is often coupled directly to a gas chromatography (GC) or high-performance liquid chromatography (HPLC) machine to give 'hyphenated' analytical techniques, Gas chromatography-mass spectrometry (GC-MS) for example, allows the analysis of mixtures containing hundreds of different molecules (see later).

The instrumentation required

Mass spectrometers (see Figure 11.18) employ several different ionization methods, and several different methods of separating the ions. Some ionization techniques require the sample to be mixed with another compound, known as a matrix, then introduced into the spectrometer in the solid state. The matrix assists the ionization process. For example, the matrix in MALDI (Matrix Assisted Laser Desorption Ionization), has a chromophore capable of absorbing energy produced by a laser. Most analytical spectrometers

FIGURE 11.18 A schematic diagram of a mass spectrometer.

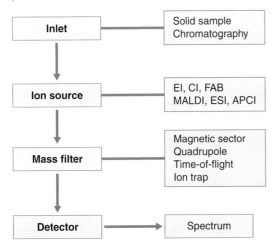

receive the sample as a gas or solution directly from a GC or HPLC instrument; the GC or HPLC separates the components of the analyte and the mass spectrometer measures a spectrum for each component.

Methods for ionization can be broadly described as 'hard' or 'soft', and the method used can dictate what is observed in the mass spectrum. Electron ionization (EI) is a hard method and involves bombarding the sample with electrons, causing the compound to shatter into smaller fragment ions; in chemical ionization (CI), the sample is bombarded with a stream of ions which causes the compound to fragment due to chemical reaction. In the majority of cases, hard ionization methods result in extensive fragmentation and the molecular ion, the ion formed before fragmentation occurs, may not be observed. Electrospray ionization (ESI), matrix assisted laser desorption ionization (MALDI), and atmospheric pressure chemical ionization (APCI) are all soft methods and result in much less fragmentation; ESI and MALDI are very much suited to analysis of very large peptides and proteins, whereas APCI is more useful for small organic molecules.

The resultant mixture of ions is then filtered according to their **mass/charge (m/z) ratio** to produce the spectrum. Ions with different m/z ratios move at different speeds under the influence of a magnetic field and are separated from one another. Small organic molecules and their fragments nearly always acquire a charge of +1 in a mass spectrometer, so it is not necessary to worry about the value of z, the charge.

The simplest, and very accurate, instruments use a magnetic sector to filter the beam of ions by measuring their deflection within an applied magnetic field; the ions are focussed on to the detector by varying the field. Magnetic sector instruments are large and inconvenient for bench-top applications. Modern instruments, such as those coupled to GC apparatus, normally use a quadrupole, time of flight (TOF) or an ion trap filter, even though the resolution is less good; when very high resolution is required, specialist instruments (orbitrap or Fourier transform ion cyclotron resonance) are used, but they are beyond the scope of this book. Magnetic sector instruments can easily determine molecular weights to within 0.001 mass unit. This can be very useful. Aspirin has a molecular formula of $C_9H_8O_4$ and therefore a formula weight of 180. $C_9H_{12}N_2O_2$ also has a formula weight of 180, when calculated to the nearest mass unit. When you calculate to the nearest 0.001 units, however, $C_9H_8O_4$ has a formula weight of 180.042 and $C_9H_{12}N_2O_2$ has a formula weight of 180.090. This is because C = 12.000, H = 1.0078, O = 15.9949 and N = 14.0031.

Time of flight instruments can be used to determine the molecular weights of large proteins to within 1 mass unit.

SELF CHECK 11.11

A paracetamol derivative has a molecular ion of m/z 165.079 in the mass spectrometer. Its infrared spectrum shows strong signals at 1653, 1605, 1565 and 1508 cm⁻¹. Its ¹H NMR spectrum is shown in Figure 11.19. Can you deduce its structure?

Analysing a mass spectrum

Twenty or thirty years ago, pharmaceutical scientists would spend a great deal of time trying to understand the complex gas-phase ion chemistry that goes on inside an electron impact mass spectrometer. Two developments have made this a more specialized activity: the invention of soft ionization techniques, and the use of superconducting magnets and Fourier transforms in NMR spectroscopy.

Using soft ionization techniques such as electrospray ionization, it is almost always possible to obtain

FIGURE 11.19   The ¹H NMR spectrum of a derivative of paracetamol.

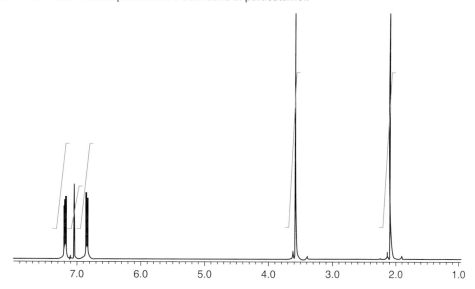

a molecular ion – (M+H)⁺ because electrospray ionization involves protonation – even when the molecule is very large (molecular weight up to millions!). Complementary to this development are the advances in NMR spectroscopy enabling detailed structural information to be obtained with milligram quantities of material. The contribution of MS to structural determination of small molecules may be limited to determining the molecular weight, but the technique has found many other uses in protein chemistry, drug metabolism, forensic science, to take just three examples.

The use of mass spectral fragmentation patterns in structure determination is described in the web resource. Here we discuss the issues common to hard and soft ionization techniques.

Each of the methods for mass filtering presents the detector with a stream of ions that have been sorted based upon their mass-to-charge ratio, m/z. The mass spectrum is a plot of m/z against intensity but it is important to note that MS is not a quantitative technique and the intensity of a particular peak is related to the stability of that particular ion, as well as to its abundance.

Figure 11.20A shows an ESI mass spectrum of aspirin. Note the strong molecular ion at m/z 181. There is also a signal at m/z 182. This is because ¹³C is present at 1.1% of total carbon. There are nine carbon atoms in aspirin so the peak at m/z 182 is approximately 9 × 1.1% (10%) of the height of the peak at m/z 181.

Brivudine (see Figure 11.20B) is an antiviral drug, used to treat *Herpes simplex*. Molecules containing chlorine and bromine give characteristic mass spectra, because both contain two common isotopes. Two stable isotopes of chlorine, ³⁵Cl and ³⁷Cl, have a natural abundance in the ratio of 3:1 in favour of ³⁵Cl; if a particular ion is accompanied by another ion that is heavier by two mass units and the intensities of the peaks are in the ratio 3:1 then this is a good indication that there is one chlorine atom in the compound. Similarly, the isotopes of bromine, ⁷⁹Br and ⁸¹Br, have roughly equal natural abundance so an ion that is accompanied by another that is heavier by two mass units and the intensities of the peaks are equal then this is a good indication that there is one bromine atom in the compound. The ESI mass spectrum of brivudine shows the characteristic 1:1 pattern.

Finally, the 'Nitrogen rule' can be used to predict the presence of that element within the molecule being analysed: if a compound has an even molecular weight value there is likely to be an even number of nitrogen atoms, or no nitrogen atoms in the molecule, whereas a molecule with an odd molecular weight probably has an odd number of nitrogen atoms.

FIGURE 11.20 (A) ESI mass spectrum of aspirin, and (B) brivudine.

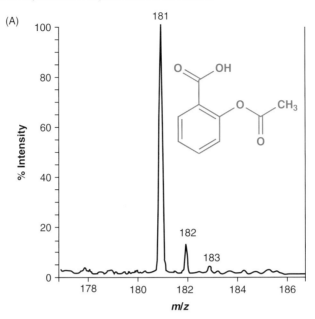

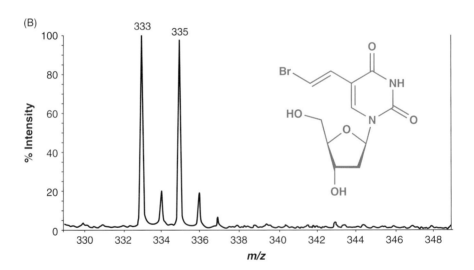

Sketch the ¹H NMR spectrum and the ESI mass spectrum for ibuprofen.

The main things to know about MS are:

- The technique can be applied to the analysis of a wide range of compounds dependent upon the type of instrument used.

- The sample is often inserted into the mass spectrometer as a gas or solution purified by some coupled chromatography apparatus, leading to hyphenated techniques, e.g. GC-MS or LC-MS.

- It is very sensitive requiring **picomole** or even **attomoles** of sample, and very selective, giving resolution measured in parts per billion.

11.7 Chromatographic methods of analysis

Chromatographic methods are used to analyse mixtures of substances particularly when other techniques, for example, spectroscopic methods, cannot be employed because of interfering impurities. These techniques also have the advantage that several components in a sample may be measured in a single test. In 1906, the Russian botanist, Michail Tswett, separated different coloured plant pigments by pouring a leaf extract down a glass column filled with calcium carbonate. He coined the word 'chromatography' meaning 'coloured writing'. Today over 60% of all analyses are chromatographically based.

The principles of preparative chromatography (chromatography for purifying compounds) are outlined very briefly in Chapter 10. In this chapter, we are concerned with analytical chromatography (small scale chromatography for the purpose of understanding what a mixture contains).

The components in a mixture can be separated by employing differences in their chemical or physical properties (e.g. solubility, charge, size, volatility or affinity to a stationary phase).

There are several different types of chromatography in common use. These include thin layer chromatography (TLC), gas-liquid chromatography (GLC), and high-performance liquid chromatography (HPLC).

All chromatographic methods are based on the distribution of the compounds to be separated between two immiscible phases. The *mobile phase* is a solvent for TLC and HPLC applications or a gas for GLC. The *stationary phase* is a specialist packing material contained in a steel column for HPLC or a glass column for GLC. The stationary phase in TLC can be made of a specialized paper or a thin layer of absorbing material coated on to the surface of a plate of glass, aluminium or plastic.

Thin layer chromatography

TLC is a simple, quick and inexpensive method of analysis. It is especially useful for monitoring chemical reactions. The equipment consists of a glass or plastic plate coated in silica gel or aluminium oxide, a chromatography tank in which the TLC plate is placed, and an appropriate solvent. A volume of about 20 µl of a dilute (0.1% w/v sample solution is a good guide) solution is applied to the chromatography plate, using a capillary tube, at a position about 1.5 cm from the bottom edge. The sample is applied in several applications keeping the diameter of the spot confined to about 2 mm. The solvent is allowed to evaporate between successive additions.

An appropriate solvent (this does not have to be the same solvent as the solvent used for applying the sample) is added to the tank to a depth of no more than 1 cm. The solvent, or mixture of solvents, is chosen in combination with the type of material used on the thin layer plate to give the best separation of components in the sample mixture. Two sides of the tank are lined with an adsorbent filter paper. The paper absorbs the solvent and keeps the atmosphere inside the tank saturated with solvent. This prevents the sample plate from drying out during the chromatographic process. The plate is then inserted into the chromatography tank. A lid is then placed on top of the tank to create a sealed atmosphere (see Figure 11.21).

FIGURE 11.21 The basic TLC experiment.

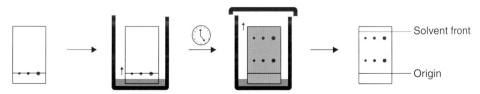

Solvent front

Origin

The solvent moves up the plate by capillary action carrying the sample with it. Components in a mixture travel up the plate at different rates due to the combined effect of their different affinities for the stationary phase and solubility in the mobile phase. If a silica gel coated plate is used with a non-polar solvent such as hexane or chloroform, polar compounds in the sample will stick longer to the polar silica than non-polar components. The non-polar compounds will move up the plate most quickly. Once the mobile phase has travelled up to a distance of about two-thirds the height of the plate, the plate is removed from the tank. The solvent on the plate is allowed to evaporate. In order to see the spots on the plate, (since most compounds are colourless), plates can be impregnated with a fluorescent indicator during manufacture. If the plate is viewed under ultraviolet light the spots are easily seen. Alternatively plates can be sprayed with chemical reagents, which produces coloured products.

The distance a compound travels up a plate, using a particular solvent system, can be used to identify a substance when compared with reference substances. This movement is defined by the retention factor or R_f value. R_f is the ratio of the distance moved by the substance (or spot on the plate) to the furthest point moved by the solvent (solvent front). Look again at the TLC plate from Figure 11.21.

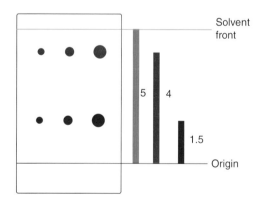

The distance from the origin to the solvent front is 5 cm. The distance from the origin to the centre of the red spot is 4 cm. So the R_f is 4/5=0.8. Similarly the distance from the origin to the centre of the blue spot is 1.5 cm; the R_f is therefore 1.5/5=0.3.

A mixture of morphine, strychnine and cocaine (Figure 11.22) was chromatographed by TLC using a silica gel plate. The solvent system was ethyl acetate-methanol-30% ammonia (in the ratio 17:2:3). Three spots were resolved with the following Rf values 0.25, 0.44 and 0.85. Match the compounds to their Rf values. Hint: silica gel has polar surface OH groups which bond most strongly to the most polar molecule – this will be the slowest to move up the plate and will have the smallest Rf value.

High-performance liquid chromatography

HPLC is one of the most frequently used analytical methods, having great sensitivity and specificity, especially when coupled with a mass specific detector. An HPLC instrument is capable of detecting **nanogram** or even smaller amounts, allowing studies to be made about how drugs function within the body, which otherwise would be impossible.

The basic instrument consists of a column packed with the stationary phase, an injector to apply samples, a high pressure pump and solvent reservoir, a detector and a data station. The stationary phase consists of very tightly packed particles of small, consistent diameter; these give the technique its very high resolution compared with simple gravity-based systems. High pressure is required to force the solvent through the tightly packed column material, so columns are encased in steel. Figure 11.23 shows a diagram of an HPLC apparatus.

The sample, contained in typically 20 µl of mobile phase, is injected onto the column containing small particles of packing material. Most commonly 5 µm silica particles, with hydrocarbon chains bonded to the surface are used and the most commonly used hydrocarbon chains are octadecyl (ODS or C_{18}) and octyl (C_8).

HPLC is therefore usually a reversed phase technique. As mentioned in Chapter 10, normal phase chromatography involves a polar stationary phase and a non-polar solvent, whereas in reversed

FIGURE 11.22 Morphine, strychnine, and cocaine

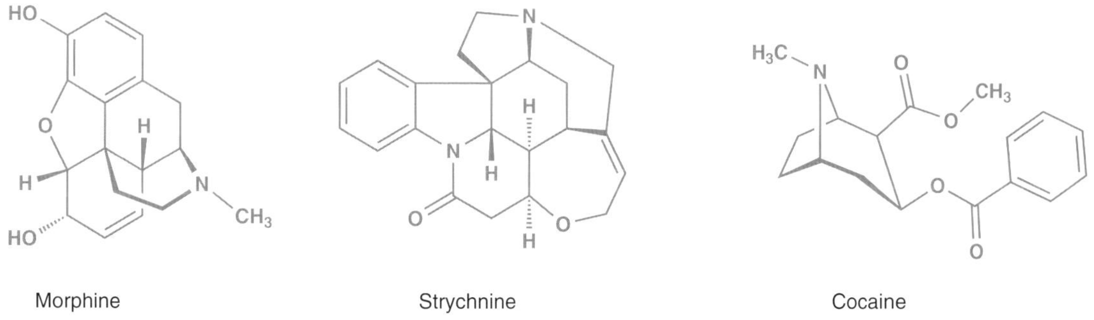

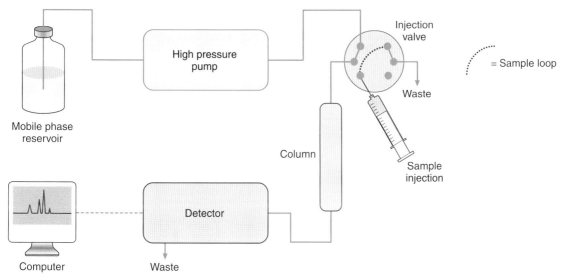

Morphine   Strychnine   Cocaine

phase chromatography the stationary phase is non-polar and the solvent is polar (usually based on water).

The mobile phase, in this instance a polar solvent system, is pumped through the column under high pressure. Polar molecules have a greater affinity for the mobile phase than for the non-polar octadecyl (or octyl) side chains of the stationary phase and therefore move rapidly through the column. Non-polar components are attracted to the stationary phase by van der Waals interactions with the hydrocarbon side chains. Their progress through the column is slowed and they are washed through (eluted) later than the polar components. Once eluted from the column the components of the sample need to be detected and quantified and a wide range of sensitive detectors are available. These include detectors based on measuring UV-Vis absorption, changes in refractive index,

fluorescence, mass spectra or radiochemical emissions. The output from these detectors is in the form of a chromatogram. As with TLC, the time taken for elution from a column (retention time) under a fixed set of experimental conditions can be used to identify a compound. This time is measured from the solvent front to the point at which the height of the peak is at its maximum. The concentration of a substance in a sample is determined by constructing a calibration graph from samples *of the same compound* of known concentration. You cannot quantify an unknown compound using HPLC.

Figure 11.24 shows an HPLC separation of the closely related painkillers naproxen (prescription only) and ibuprofen (OTC). Ibuprofen is a significant contaminant of rivers near centres of population and HPLC is used to monitor its levels.

FIGURE 11.23 Basic diagram of an HPLC.

FIGURE 11.24 An HPLC separation of a mixture of naproxen and ibuprofen.

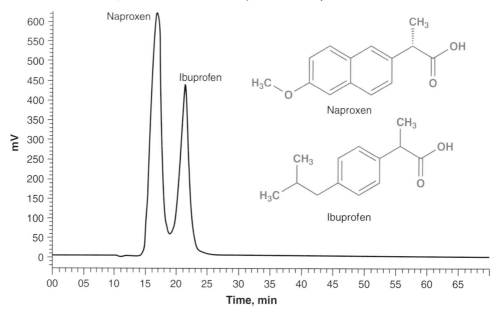

SELF CHECK 11.14

Morphine, strychnine and cocaine (see Figure 11.22) were chromatographed in a reversed phase HPLC system using a C_{18} ODS column and a mobile phase of phosphate buffer pH 3.0 and acetonitrile (50:50). Predict the order of elution of these compounds.

Hyphenated techniques have been mentioned before, and one of the most important of these is HPLC-MS. Bench-top ESI mass spectrometers are superb detectors for HPLC separations, because they are both extremely sensitive and extremely specific (related compounds seldom have the same molecular weight). Researchers and clinicians studying drug metabolism are major users of HPLC-MS (see Figure 11.25).

FIGURE 11.25 The anti-anxiety drug buspirone is metabolized by oxidation to give more water-soluble hydroxylated compounds. Hydroxyl groups can be inserted in several different places on the molecule (as indicated by the red arrows), giving several metabolites, detectable by HPLC-MS.

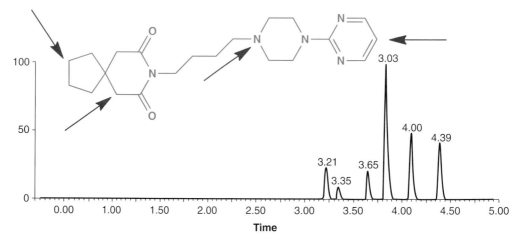

11.8 Gas-liquid chromatography

Both TLC and HPLC involve the use of a solid stationary phase and a liquid mobile phase. In GLC, the mobile phase is an inert gas such as nitrogen or helium, and the stationary phase is a high boiling liquid coated on to the surface of an inert particulate material or onto the surface of a fine glass capillary column. The sample is introduced into a hot oven where it becomes a vapour. The constituents of the sample then partition themselves between the stationary phase and the mobile phase carrier gas. Separation occurs and the individual components elute from the column. An extremely versatile detector, known as a flame ionization detector, is commonly used in GLC. The carrier gas eluting from the column is mixed with a stream of hydrogen gas and air, which once ignited produces a small, continuously burning, flame. A pair of electrodes is located within the flame and a voltage is applied. Any organic compounds eluting from the column will burn in the flame and will in the process of burning produce ions. Under the influence of the applied potential a small current can be measured between the electrodes. This is amplified and produces the chromatogram for the sample under investigation. A range of more specific and ultrasensitive detectors are available for use in GLC but are beyond the scope of this chapter. Figure 11.26 shows a schematic diagram of a simple GLC system.

GLC works exceptionally well for the analysis of volatile compounds such as fixed oils of plant origin, many of which have medicinal use. Trace amounts of volatile substances exhaled in breath can be analysed by GLC and the results used to indicate various clinical disorders. Drugs which are volatile at temperatures below 300 °C, and remain thermally stable, can be analysed by GLC, but HPLC has become the first line method for the analysis of pharmaceutical samples.

Once again, a mass spectrometer can serve as a detector for a GLC, leading to a hyphenated technique GLC-MS (also known as GC-MS). Figure 11.27 and Table 11.3 show a GC-MS analysis for a sample of lavender oil. The analytical report provides information about the identity of the constituents in the oil and quantitative information about the relative amounts of each present. This technology allows comparisons of lavender oils obtained from different sources.

FIGURE 11.26 Diagram of a gas chromatograph.

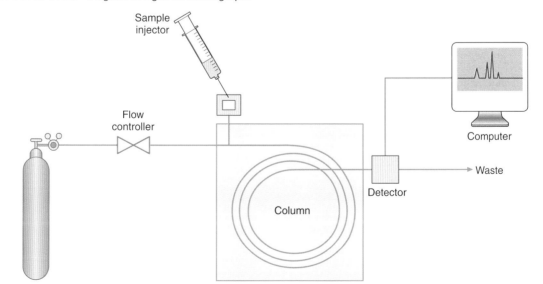

FIGURE 11.27 Capillary gas chromatogram of a sample of lavender oil. The principal components of the oil are labelled

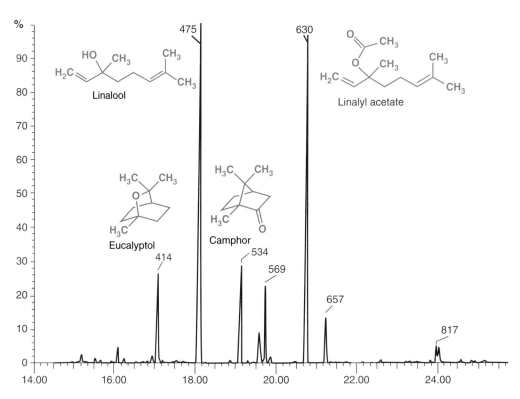

Lavender oil provence

TABLE 11.3 GC-MS analysis report for lavender provence (organic)

Peak area	Peak time	Peak scan	Marker text
1.064	0:16:06	357	Pinene or beta-myrcene
7.432	0:17:04	414	Eucalyptol
31.909	0:18:07	475	Linalool
7.518	0:19:07	534	Camphor
3.282	0:19:35	561	Borneol
5.317	0:19:43	569	CAS 562-74-3
27.504	0:20:46	630	Linalyl acetate
3.585	0:21:14	657	Geraniol acetate
1.362	0:23:58	817	Farnesene
1.085	0:24:01	820	Caryophylene

KEY POINT

Chromatography employs differences in chemical and physical properties to separate the components of a mixture.

KEY POINT

The main types of chromatography are TLC, GLC and HPLC.

CASE STUDY 11.1

John Vickers is prescribed the statin, simvastatin BP, for high blood pressure. He visits his doctor because he thinks that the medicine makes his legs ache. The doctor sends a blood sample for a lipid test, and discusses changing the statin or the dose.

REFLECTION QUESTIONS

1. What does the term BP in 'simvastatin BP' mean?

2. What is the most appropriate analytical technique to confirm:

   a) the identity of the active ingredient to be used in the manufacture of the simvastatin tablets

   b) that the tablet contains the correct amount of simvastatin

   c) that the level of impurities in the tablets is acceptably low.

3. Which analytical techniques would be appropriate for the analysis of lipids in blood plasma?

Answers

1 It means that the medicine complies with a set of standards defined in the British Pharmacopoeia. Compliance with the BP monograph confirms that the medicine contains the right drug, in the right dose and that it is free from impurities. The BP also contains a test to determine the dissolution rate of tablets. This is used as a measure of how quickly tablets will dissolve once taken and therefore how efficiently the medicine gets absorbed and will begin to take effect.

2 a) IR spectroscopy – an infrared spectrum of a sample of active ingredient should be compared with the infrared spectrum of an authenticated reference sample of simvastatin.

   b) A representative number of the tablets should be weighed and crushed and the active ingredient extracted into a suitable solvent. This solution should then be injected onto a HPLC column and the resulting peak area converted into an amount of simvastatin using a calibration graph previously prepared using an authenticated standard sample of simvastatin.

   c) the BP requires an HPLC analysis by comparison with a range of potential impurities at specified concentrations.

3 There have been a number of techniques used over the years but the most recent technique uses is MS.

CHAPTER SUMMARY

➤ Pharmaceutical analysis is essential to keep medicines safe. If a medicine is to be safe and effective, we need to know what impurities are present and in what quantities.

➤ Various forms of spectroscopy, using different parts of the electromagnetic spectrum are vital in pharmaceutical analysis. Ultraviolet, infrared and NMR spectroscopy are all used.

➤ Mass spectrometry is an especially sensitive and selective technique, and for this reason it is often coupled to chromatographic instrumentation.

➤ Various types of chromatography, especially HPLC, are widely used to detect impurities in pharmaceutical preparations.

FURTHER READING

Detailed correlation tables for functional groups are readily found in the literature, for example:

http://www.chemistry.ccsu.edu/glagovich/teaching/316/ir/table.html.

For spin coupling information, it is possible to obtain such a spectrum (called a pure shift spectrum) using NMR methodology published in 2010, but the one shown in Figure 11.12 is simulated.

http://nmr.chemistry.manchester.ac.uk/?q=node/24.

Many Pharmacy schools introduce ^{13}C NMR in year 2. If you cannot wait until then, read Clayden *et al.* pp.52–59.

Clayden, J., Greeves, N., and Warren, S. *Organic Chemistry.* 2nd edn. Oxford University Press, 2012.

British Pharmacopoeia Commission. *British Pharmacopoeia 2011.* The Stationary Office, 2011.

Moffat, A.C., Osselton, M.D., Widdop, B., and Watts, J. *Clarke's Analysis of Drugs and Poisons,* 4th edn. Pharmaceutical Press, 2011.

Watson, D.G, *Pharmaceutical Analysis: a textbook for Pharmacy Students and Pharmaceutical Chemists,* 2nd edn. Churchill Livingstone, 2005.

Joie Power. http://www.dreamingearth.com/gas-chromatographs.html.

Zhu M, Zhao W, Jimenez H, Zhang D, Yeola S, Dai R, Vachharaiani N, Mitoka J, Cytochrome P450-3A-mediated metabolism of buspirone in human liver microsomes. DMD 33: 500-507, 2005.

Pure Shift NMR: http://nmr.chemistry.manchester.ac.uk/?q=node/24.

The molecular characteristics of good drugs

JILL BARBER

We hope that this book has helped you to understand and appreciate the fundamentals of Pharmaceutical Chemistry, and that you have developed some of the skills needed in the practice of this subject. This chapter shows how these skills have been used to help us understand the molecular properties of a good drug. It also illustrates the limitations of what we know.

Learning objectives

Having read this chapter you are expected to be able to:

➤ have an understanding of Lipinski's rule of five and its exceptions

➤ have an understanding of the fundamental principles of structure-activity relationships.

You will have revised:

➤ the importance of chirality in drug design and discovery

➤ the importance of purity in drug preparations.

You will, we hope, appreciate that there is still plenty of work to be done in drug design and discovery, and plenty of fun to be had doing that work.

12.1 Rules in chemistry

Scientists and mathematicians like rules. Everyone learns Pythagoras's Theorem; many GCSE students know Newton's Laws of Motion; and most people know that $E = mc^2$, even if they do not know what E, m and c are. Chemists like rules as well, but generally the rules of chemistry are less authoritative than those of mathematics and physics. There are lots of exceptions, and the rules get rewritten as time goes by. You have already come across Markovnikov's rule in this book, but in the same chapter we discussed how to achieve anti-Markovnikov addition to an alkene. Almost immediately Markovnikov stated his rule, an important class of exceptions was discovered. This is what tends to happen to rules in chemistry.

 To reread Markovnikov's rule, see Chapter 4 'Properties of aliphatic hydrocarbons'.

Thus, there was great excitement when in 1965 Woodward and Hoffmann published *The Conservation of Orbital Symmetry*, which described what became known as the Woodward–Hoffman rules (The Woodward–Hoffman rules define the course and stereochemistry of certain types of chemical reaction.) There was a chapter entitled 'Exceptions' and the text for this chapter unusually read 'There are none'.

Ian Fleming's book *Frontier Orbitals and Organic Chemical Reactions*, in many respects an appreciation of the Woodward–Hoffmann rules, also had a chapter called 'Exceptions' but it began 'There are many'. Woodward and Hoffmann's insights changed the way organic chemistry is understood and taught, but the Woodward–Hoffmann rules have gone the way of rules in chemistry: they remain very useful guidelines, but without the authority of (say) Pythagoras's Theorem.

The most famous rules in medicinal chemistry were published in 1997 and are known as Lipinski's Rules or Lipinski's Rule of Five. There are many exceptions to these rules, but, remarkably, Lipinski and co-workers identified most of them in their original publication.

12.2 **The Rule of Five – an empirical rule**

Lipinski and co-workers compiled a database of about 2500 drug-like compounds. Although not all these compounds were used in the clinic, they had all entered phase II clinical trials, meaning that they had not shown early toxicity. Compounds likely to be administered by injection, such as peptides, were not considered. The chemists used this database to find rules for identifying the characteristics of *orally-active* drugs.

They reasoned that an orally-active drug has to have certain physicochemical properties. It has to pass from the lumen of the gut (an aqueous environment), through the gut wall (that is through hydrophobic membranes), and beyond, into the aqueous environment of the blood stream. To do this, it must not be too big, nor too hydrophilic, nor insoluble. If it hydrogen bonds too strongly to itself, it will become stuck inside lipid membranes, such as cell membranes. Lipinski and co-workers therefore looked for indicators of size, solubility and hydrogen bonding in the database of successful compounds.

They did not therefore devise The Rule of Five on a piece of paper and then test it. They were experimentalists, not theoreticians. They devised their rule based on data. This is what is meant by an empirical rule.

Size matters

In the database used by Lipinski and co-workers only 11% of compounds had molecular weights of more than 500. In order for orally-active drugs to do their job, they have to be able to pass through the gut wall. Drugs that are successful for oral delivery tend to be small, since big molecules are less good at crossing membranes. Figure 12.1 shows some of the orally-delivered drug molecules we have discussed in this book.

 SELF CHECK 12.1

Calculate the molecular weights of each of atorvastatin, metronidazole and ibuprofen.

Water and lipid solubility matter

If a compound is to cross a membrane it must be soluble both in water and in lipid, but it must not be *too* soluble in either. The parameter log$P$ (or Clog$P$) is used to determine whether a compound has appropriate solubility. Consider a compound in a vessel containing equal amounts of water and octan-1-ol. The two solvents do not mix but form distinct layers; water

FIGURE 12.1 Atorvastatin, metronidazole and ibuprofen.

Atorvastatin

Metronidazole

Ibuprofen

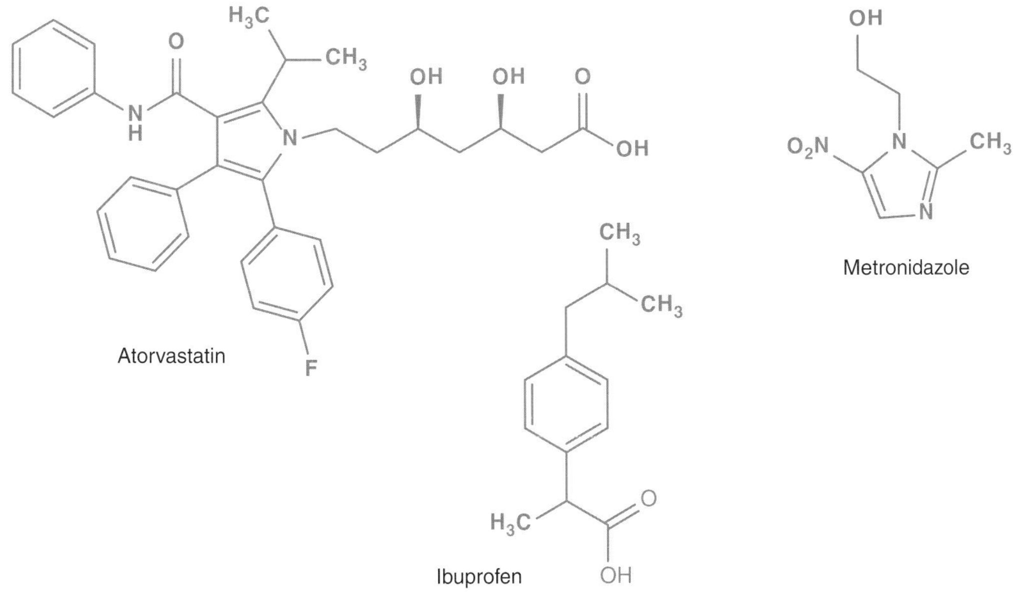

sits beneath the octan-1-ol because it is more dense. (Most organic solvents are less dense than water, but chlorine-containing solvents are more dense).

P, the octanol-water partition coefficient, is given by:

$$P = \frac{C_{octanol}}{C_{water}}$$

where C is the concentration of the drug in water or octanol at equilibrium. The partition coefficient (*P*) is usually converted to a logarithmic value (log*P*) because of the large range of values obtained for different compounds resulting from their differing solubilities. Lipinski and co-workers found that only 10% of the compounds in their database had log*P* values above 5. Log*P* can either be measured (and you may have a practical class involving the measurement of log*P*) or calculated using an on-line tool such as http://www.molinspiration.com/cgi-bin/properties. A compound with a log*P* above 5 will be very hydrophobic, and it will tend to lodge in a membrane, rather than entering an aqueous environment, such as a cell or the blood stream.

 For more about log*P*, see Chapter 8 'Partitioning and hydrophobicity' of the *Pharmaceutics* book in this series.

SELF CHECK 12.2

Use an on-line calculator to estimate log*P* for atorvastatin, metronidazole and ibuprofen.

 The importance of hydrophobic groups in contributing to drug-like characteristics was discussed in Chapter 4 'Properties of aliphatic hydrocarbons'.

Hydrogen bonding matters

Many drug molecules can form hydrogen bonds. A hydrogen bond donor (an electrophilic atom attached to a hydrogen) is required, as is a hydrogen bond acceptor (an electrophilic atom). Lipinski and co-workers suggested just counting OH and NH groups to give the hydrogen bond donors and counting O and N atoms to give hydrogen bond acceptors. If a drug molecule hydrogen bonds too strongly to itself (that is, numerous hydrogen bonds form between neighbouring molecules of the drug), it cannot cross lipid membranes, such as cell membranes. Lipinski and co-workers found that 12% of compounds in their database had more than 10 hydrogen bond acceptors and 8% of compounds had more than five hydrogen bond donors.

Count the hydrogen bond donors and acceptors in atorvastatin, metronidazole and ibuprofen.

 Hydrogen bonding was introduced in Chapter 2 'Organic structure and bonding'.

The Rule of Five

Using their observations, Lipinski and his team devised the 'Rule of Five'. It states that poor absorption (or permeation) is more likely when:

- the molecular weight is over 500
- the $\log P$ is over 5
- there are more than five H-bond donors (expressed as the sum of OHs and NHs)
- there are more than 10 H-bond acceptors (expressed as the sum of Ns and Os).

It is well worth learning the Rule of Five. It might be helpful to notice that all the numbers are multiples of five, giving the origin of the rule's name.

Exceptions

There are many, but Lipinski and his co-workers have never claimed anything else. The main classes of exception are:

- **Drugs for delivery via non-oral routes**: The Rule of Five was always intended as a guide to drugs that would be appropriate for oral delivery – it says nothing about those that are delivered via non-oral routes, such as injectable or **topical** drugs.
- **Drugs to treat disorders of the gut**: In these circumstances the drug does not need to cross the gut wall – indeed it is much better if it does not. So the Rule of Five does not apply. An example of this is vancomycin, which is normally given by injection, but can be given orally to treat infections of the gut.
- **Antibiotics, antifungals, vitamins and cardiac glycosides**: Many of these compounds do obey the Rule of Five, but some do not. Erythromycin

(molecular weight 731) has been discussed several times in this book; it is one of the safest and most effective antibiotics in clinical use. Many compounds in these categories, which are often natural products, are **actively transported** across cell membranes. They are natural products and have evolved in nature to take advantage of nature's weaknesses.

Many natural products that do not obey Lipinski's Rule of Five are suitable for use in orally-delivered drugs.

Using Lipinski's Rule of Five in drug design

The Rule of Five has proved enormously valuable to medicinal chemists involved in the design of drugs. The most important class of antifungal drugs are the triazole antifungals, which are used to treat a range of serious infections. They inhibit the formation of ergosterol, an important component of the fungal cell membrane. Mammals have cholesterol, instead of ergosterol, in their membranes so the triazole antifungals are highly selective in their action.

Two important members of this class of drug are fluconazole (see Figure 12.2A) and itraconazole (see Figure 12.2B). Itraconazole is useful because it has activity against the important Aspergillus class of fungi that causes lung infections, especially among farmers. Fluconazole does not have this activity, but despite this it has not been superseded by itraconazole. Why do you think this is? Lipinski's Rule of Five gives us an insight.

Fluconazole has:

- a molecular weight of 306
- a $\log P$ of 0.5
- one hydrogen bond donor
- seven hydrogen bond acceptors.

Itraconazole has:

- a molecular weight of 706
- a $\log P$ between 5 and 7

FIGURE 12.2 Triazole antifungals; (A) fluconazole and (B) itraconazole (which is popular in veterinary medicine).
© Nicola Wood

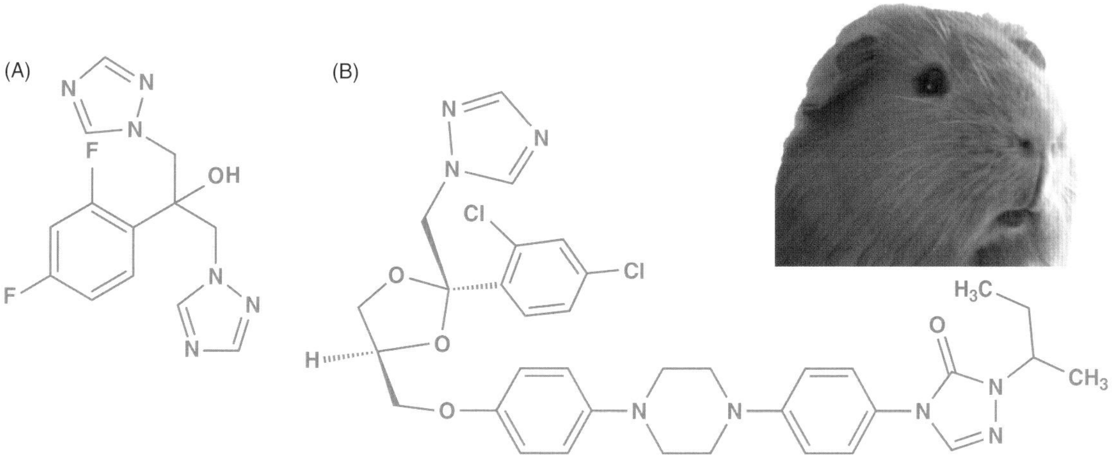

- no hydrogen bond donors

- twelve hydrogen bond acceptors.

Itraconazole violates three of the four Rules of Five, and so, not surprisingly, is poorly absorbed when given orally. If it is taken with food, absorption can reach 55%, but this is a maximum. Interestingly, it is used in veterinary medicine; it is licensed in cats, and given also to guinea pigs, which are susceptible to fungal infections. Veterinary preparations are, it seems, very tasty, and this raises another important point. If a drug tastes good, it may not all be absorbed, but the starting point is that it is swallowed. Fluconazole obeys all the Rules of Five and is absorbed well when given orally.

> **KEY POINT**
>
> The Rule of Five has proved enormously valuable to medicinal chemists involved in designing drugs. If you are trying to design a new synthetic compound that treats a disease that is very long term (like hypertension), or a disease which is usually manageable (such as an upper respiratory tract infection), it is far preferable to have oral therapy. Remember though, that ultimately it is more important to treat the disease than to deliver the drug orally and that natural products make up about a third of the drugs in clinical use. Injectable drugs and many natural products do not need to obey Lipinski's Rule of Five.

12.3 Structure-activity relationships

Paul Ehrlich (1854–1915) is described as the Father of Chemotherapy, and is particularly famous for the 'Magic Bullet'. (A magic bullet is what all those involved in drug design aim to develop – a drug that selectively kills an invading organism while leaving the patient's own cells unharmed.) Ehrlich is also credited with a number of insights into how to design or discover a drug. One of these was the importance of structure-activity relationships.

A structure-activity relationship (SAR) is the relationship between the chemical or 3D structure of a molecule and its biological activity. To define an SAR requires the preparation of analogues in a logical, systematic way. To take a simple example: butylbenzene (see Figure 12.3A) has a calculated logP of 3.8. Suppose I am looking for a molecule that is more polar (hydrophilic) than butylbenzene. It would be logical to shorten the butyl chain and also to add an –OCH$_3$ in the *para* position. I could try each of those changes. Sure enough, propylbenzene (see Figure 12.3B) has a logP of 3.2, but 1-butyl-4-methoxybenzene (see Figure 12.3C) has a calculated logP of 3.9.

FIGURE 12.3 Calculated logPs of some analogues of butylbenzene.

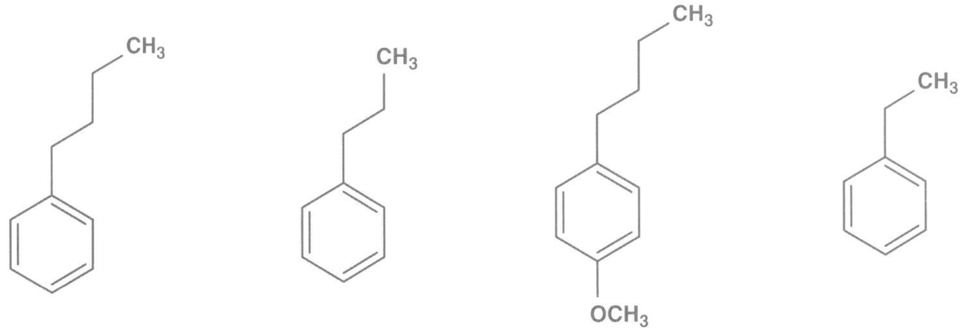

(A) Butylbenzene
LogP = 3.8

(B) Propylbenzene
LogP = 3.2

(C) 1-butyl-4-
methoxybenzene
LogP = 3.9

(D) Ethylbenzene
LogP = 2.8

This tells me that shortening the alkyl chain is the more appropriate of the two potential changes, and if propylbenzene is still too hydrophobic, I could try ethylbenzene (see Figure 12.3D). This has a calculated logP of 2.8.

Let us now consider some real examples.

Salvarsan

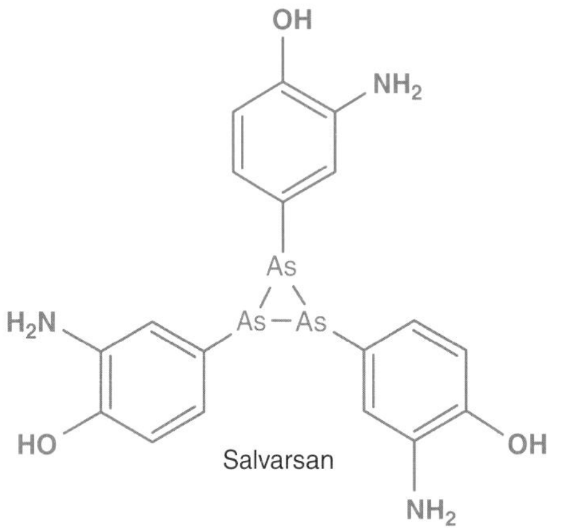

Salvarsan

Salvarsan was the first effective drug to treat syphilis and was discovered by Ehrlich's research group while they were trying to find a cure for African trypanosomiasis. Ehrlich was probably the first scientist to acknowledge the importance of luck in drug discovery. In order to create the drug, Ehrlich's group synthesized over 900 compounds, changing the structure a small amount each time and monitoring whether activity improved or reduced; salvarsan was number 606 in Ehrlich's research programme. This approach is especially effective when aromatic compounds are used. Many different substitutions can be made to aromatic rings in different positions. In Chapter 7, we learnt how these different substitutions can be manipulated, and how they can serve to make an aromatic compound more or less reactive. We also understand quite well how different substituents make an aromatic compound more hydrophobic or hydrophilic; this important point is discussed in Chapter 5 with respect to aromatic halogen compounds. Nevertheless, to find the most active compound, it is necessary to synthesize many, many molecules.

> The effect of different substitutions in the reactivity of an aromatic ring is covered in more detail in Chapter 7 'Introduction to aromatic chemistry'.

When carrying out such a huge synthetic programme, it is important to have a simple *in vitro* screening method available. The syphilis-causing bacterium, *Treponema pallidum*, can be grown in culture and used in initial screens. Each compound can be tested for its ability to kill the bacteria.

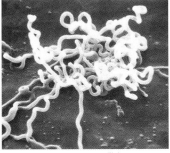

Treponema pallidum

FIGURE 12.4 Formation of a carbon-arsenic bond, using a Grignard reagent.

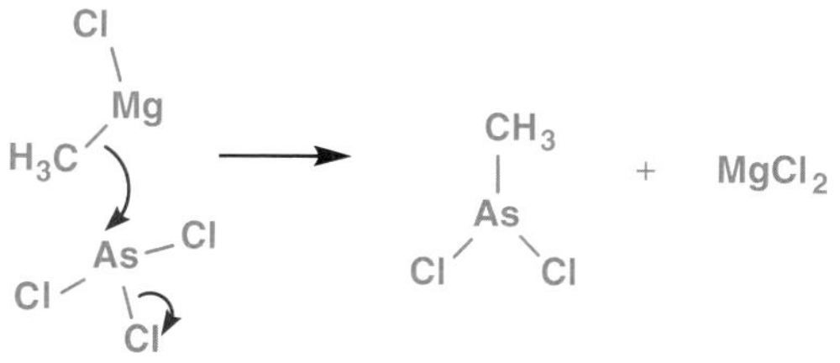

SELF CHECK 12.4

Chapter 6 describes the use of Grignard reagents in forming carbon-carbon bonds. How does the methyl Grignard reagent in Figure 12.4 react with acetaldehyde?

In the structure of Salvarsan, As represents arsenic. It can be introduced into organic compounds by quite standard chemistry – for example, by reacting Grignard reagents with $AsCl_3$ (see Figure 12.4). These days arsenic is still present in a small number of drug molecules, but phosphorus, the element immediately above arsenic in the periodic table is especially important in pharmacy. Together with sulfur, it is discussed in Chapter 8.

 Inorganic pharmaceutical chemistry is discussed in detail in Chapter 8 'Inorganic chemistry in pharmacy'.

Flucloxacillin

6-Aminopenicillanic acid (6-APA) is a natural product with little or no antibacterial activity. It can, however, be coupled with numerous carbonyl compounds, particularly aromatic carbonyl compounds, to make a range of penicillins (see Figure 12.5). In the late 1950s, when 6-APA was discovered, penicillin G (benzylpenicillin) was already in use. However, it was known to be degraded by stomach acid and to have no activity against Gram-negative bacteria. The

FIGURE 12.5 6-Aminopenicillanic acid (6-APA) and some of the penicillins that can be made from it.

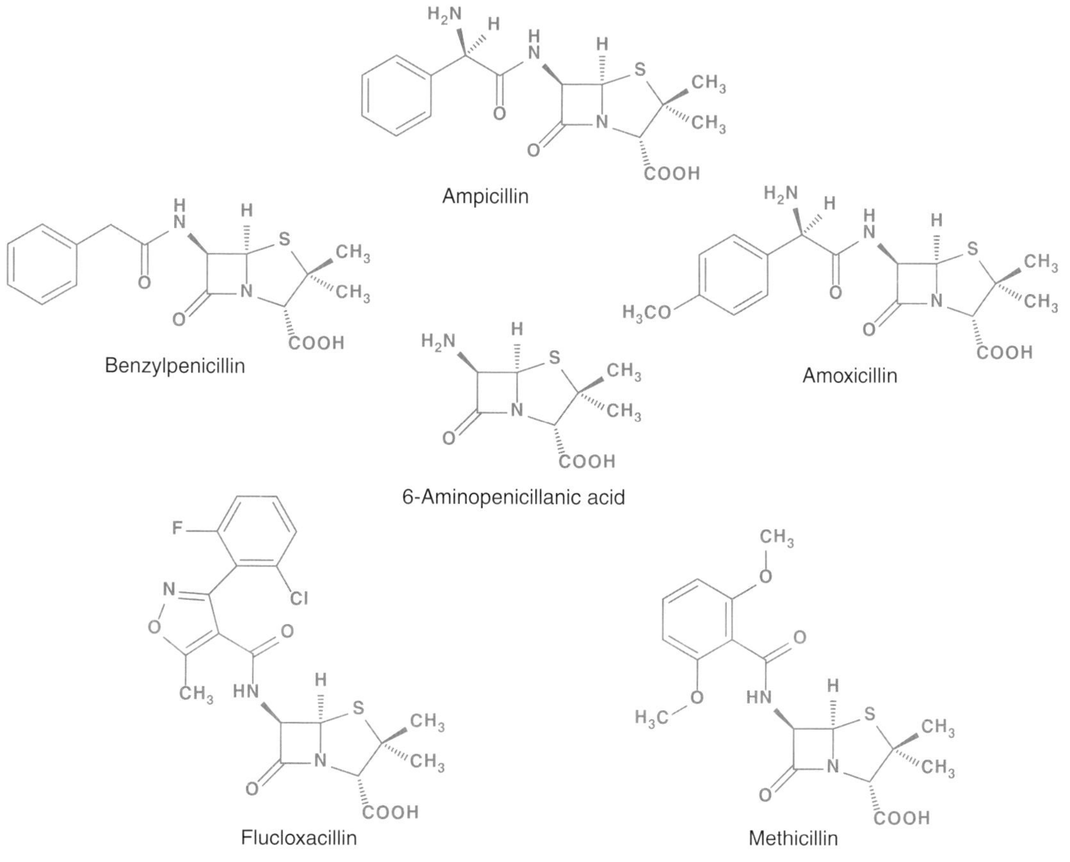

Ampicillin

Benzylpenicillin

Amoxicillin

6-Aminopenicillanic acid

Flucloxacillin

Methicillin

search was on for a broad-spectrum drug, capable of oral administration.

The difficulty was that penicillins work by reacting with a serine residue of a transpeptidase, an enzyme critical for bacterial cell wall manufacture. The side chain of serine contains an –OH group. This is activated by neighbouring groups and reacts with the cyclic amide of penicillin, which ring-opens. You cannot normally make an ester from an amide so easily, but the opening of the four-membered ring drives the reaction (see Figure 12.6). As shown in Chapter 5, water (H–OH) acts in a similar way to alcohols (R–OH) in many reactions. The concentration of water in the human body is high, and in the acidic environment of the stomach, the four-membered ring amide (lactam) is activated to ring-open.

A great many penicillins were made in an effort to overcome the acid-instability of penicillins and also to broaden their spectrum of action. One of the first successes was ampicillin, although this has largely been superseded by amoxicillin, which is particularly easy to taste-mask, making it suitable for paediatric suspensions.

SELF CHECK 12.5

Calculate the concentration of H_2O in water (H_2O has a molecular weight of 18 and a density of 1 g/mL).

SELF CHECK 12.6

Methicillin and flucloxacillin have bulky side-chains, so in Figure 12.5 they have been drawn at a different orientation from the side-chains of the other three penicillins. Redraw the structure of ampicillin, with the side chain at the top as shown for methicillin. (It really is important to keep drawing chemical structures!)

FIGURE 12.6 Reaction of a penicillin with a transpeptidase enzyme in a bacterial cell membrane. If you are drawing a mechanism like this, it can be very useful to replace part of the molecule with 'R'. Make sure you define 'R' though.

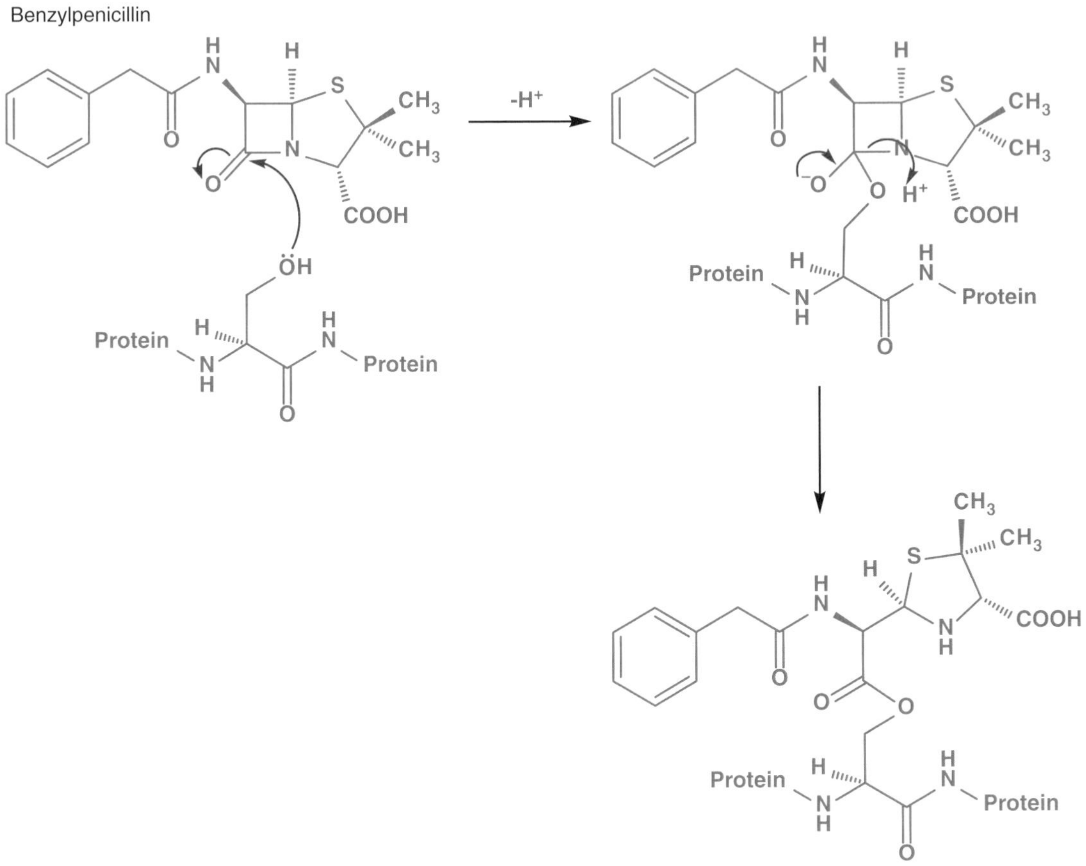

Once the acid sensitivity had been overcome, a second problem soon became apparent – the rise of penicillin resistance. Bacteria carrying **β-lactamases** (enzymes which deactivate penicillins and related drugs), became more and more common in clinical isolates as the use of the penicillins increased. A β-lactamase enzyme has a very similar active site to a transpeptidase, and its function is to catalyse the hydrolysis of penicillins and related drugs. Although β-lactamases are similar to transpeptidases, they are not, of course, identical. For one thing, transpeptidases are bound to the cell membrane whereas β-lactamases are soluble proteins. Many more penicillins were made using the semi-synthetic route, and it became clear that compounds with bulky side-chains tended to be resistant to β-lactamases. Flucloxacillin was reported in 1970. Note the typical pharmaceutical use of the chlorine and fluorine atoms to modify the properties of the aromatic ring, not to take part in any chemical reaction directly (see Figure 12.5).

 Organic halogen compounds are covered in more detail in Chapter 5 'Alcohols, phenols, ethers, organic halogen compounds, and amines'; carbonyl groups are discussed in Chapter 6 'The carbonyl group and its chemistry'; and more information on semi-synthetic drugs can be found in Chapter 10 'Origins of drug molecules'.

The quinolone antibiotics

The first quinolone antibiotic, nalidixic acid, was discovered in 1962 and was used exclusively to treat Gram-negative urinary tract infections. Quinolones work by inhibiting **bacterial DNA gyrases**, which unwind circular bacterial DNA during cell replication. Mammals do not have circular DNA (they have enzymes called gyrases, but these are different in both structure and function). Consequently, these antibiotics do not harm mammalian cells.

Nalidixic acid is a simple, synthetic molecule whose structure is shown in Figure 12.7; this structure has been varied in almost every imaginable way. There are eight positions in the bicyclic ring system that can be modified (by adding different groups or changing the groups that are there); each of them has, indeed,

been modified and the resulting molecules tested for antibacterial activity. The second generation quinolones are all fluorine-containing and are known as fluoroquinolones. They include ciprofloxacin and ofloxacin.

 The rationale for introducing fluorine atoms into drug molecules is discussed in Chapter 7 'Introduction to aromatic chemistry'.

These compounds have enhanced **Gram-negative** activity, and some activity against **Gram-positive** bacteria. The third generation of quinolones, which resulted from further attempts to improve the antibacterial activity, includes levofloxacin, and grepafloxacin, and they have improved activity against Gram-positive bacteria and against **anaerobes**. The fourth generation, produced in a search for a broader spectrum of antibacterial activity, includes gatifloxin and moxifloxacin, and shows further improvements against Gram-positive bacteria, although their clinical use is limited by increased toxicity to humans. Figure 12.7 shows the structures of several quinolone antibiotics; note their similarities.

SARs in the quinolones have been so extensive that something is known about the importance of each of the eight available positions in the bicyclic ring system shown.

$$R^5 \quad O \quad O$$
$$R^6 \quad \overset{4}{\underset{N}{\bigcirc}} \quad 3 \quad OH$$
$$R^7 \quad \quad 2$$
$$R^8 \quad R^1$$

- Position 1 controls overall antibacterial activity: *N*-cyclopropane is especially effective.
- Positions 2–4 are responsible for binding. C2 can bear H or a small group; C3 and C4 cannot be changed.
- C5 carries H or CH_3 and is important for Gram-positive activity.
- C6 carries H or (better) fluorine. F is of similar size to H, but is *much* more electronegative. Substitution of F for H in an aromatic ring is often beneficial for drug action, as discussed in Chapter 7.

FIGURE 12.7 Some quinolone antibiotics.

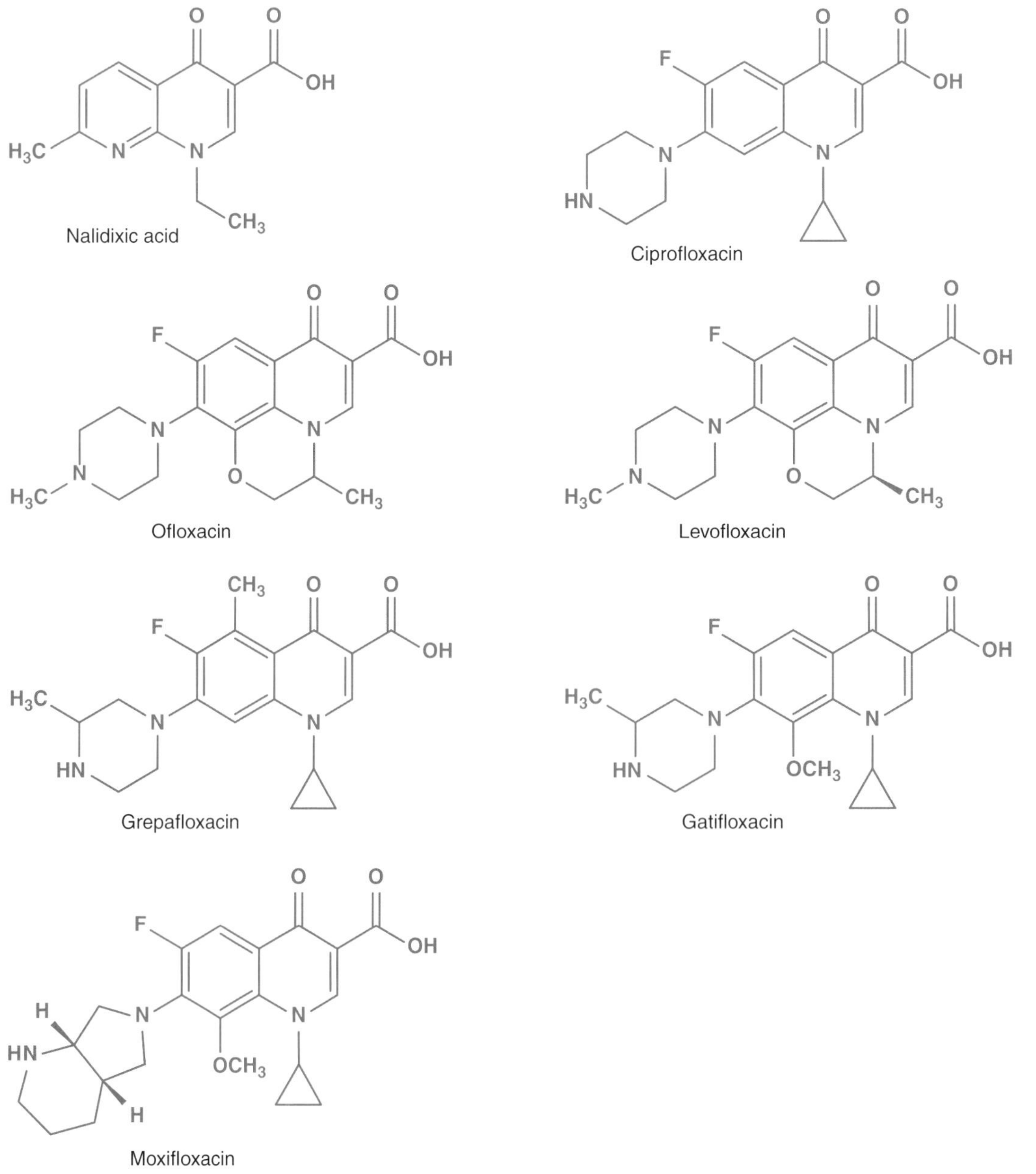

Nalidixic acid

Ciprofloxacin

Ofloxacin

Levofloxacin

Grepafloxacin

Gatifloxacin

Moxifloxacin

- Considerable variation is possible at C7, but most of the successful quinolones have a tertiary amine at this position.

- Variation is also possible at C (or N)8.

As a result of all this experimentation, a total of 20 quinolones are used worldwide.

Aspirin

Aspirin is a pro-drug. Acetylsalicylic acid is converted in the body to salicylic acid, as shown in Figure 12.8.

How do we know that the acetyl ester is the best? How many variants were tested?

There are no published reports of SARs of aspirin analogues as painkillers, and it is quite possible that this ancient medicine, that dates from 1897, is not the most effective salicylic acid ester possible. Yet no pharmaceutical company has tried to get rich by developing an alternative.

The probable reason is that aspirin is hydrolysed in the body to give a lot of acetic acid (ethanoic acid) and its salts. Acetate is a very benign substance; it enters

FIGURE 12.8 The hydrolysis of aspirin to salicylic acid. Note that both acids are deprotonated at neutral and basic pH.

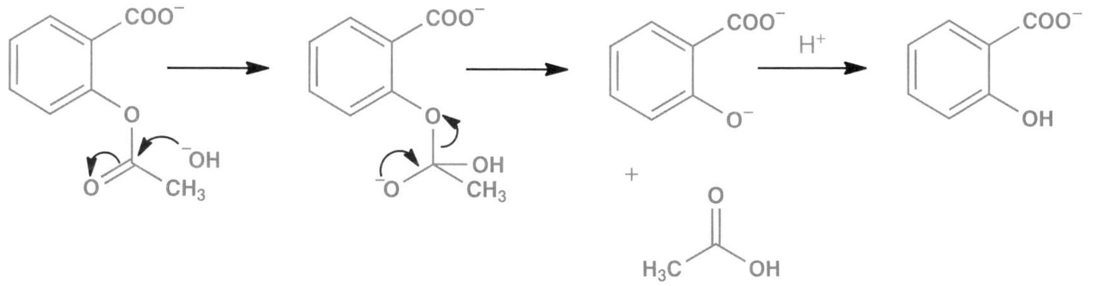

the citric acid cycle and is safely metabolized to carbon dioxide. Aspirin is taken in large doses yielding large amounts of hydrolysed carboxylic acid; the most important characteristic of this carboxylic acid is, therefore, that it must be safe. It may well be possible to make a faster-acting painkiller than aspirin; but it would be very difficult to achieve a better balance of effectiveness and safety.

The citric acid cycle (also known as the Krebs Cycle) is of central importance in mammalian biology, so both biology and pharmacy students are usually expected to learn about it. It has been mentioned already in this book (Chapters 1 and 5), because the reactions in this cycle can all be understood in terms of chemistry. This cycle is also a rich source of carboxylic acids, which are found in the body in quite large amounts. If you make a pro-drug using one of these acids, you simply add to the pool of these harmless compounds and you have one less concern about safety.

One aspirin analogue we could consider is fluoro-acetylsalicylic acid. We have seen that fluorine is often a good replacement for hydrogen. The trouble is that fluoroacetic acid is one of the most potent toxins known. It is used as a pesticide in circumstances where the pests are mammals, and is lethal to humans at a dose of about 5 mg kg^{-1}.

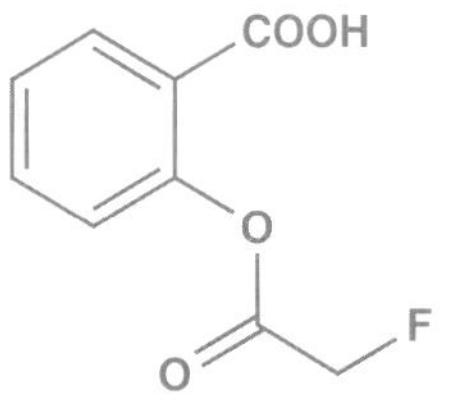

SELF CHECK 12.7

Look up the maximum daily dose of aspirin and calculate the amount of fluoracetic acid that would be generated by this amount of fluoroacetylsalicylic acid. What effect would this have on a small (50 kg) woman?

KEY POINT

Structure-activity relationships can be studied in synthetic and semi-synthetic drugs. Small changes are made to a molecule and the effects determined, usually in a simple screen. Molecules can be optimized for effectiveness, safety, solubility, spectrum of activity, stability etc.

12.4 Stereochemistry

In considering the molecular characteristics of a good drug, we have seen that several physical properties are important, especially if the drug is to be taken orally, and we have seen that SARs can be used to improve the properties of drugs. Another essential property to consider is stereochemistry.

Some time long ago, something happened in our world to make it chiral. Our amino acids have

L-stereochemistry and our sugars have D-stereochemistry. Both the DNA double helix and the α-helix in proteins are right-handed, and we might describe our world as right-handed.

 The structures and functions of biological macromolecules are described in Chapter 9 'The chemistry of biologically important macromolecules'.

'Everything that can happen does happen', we are assured by Brian Cox and Jeff Forshaw, and we can imagine a world that is identical to ours, but where something happened in the depths of history to make it left-handed. A left-handed world would be an exact mirror image of ours at the molecular level. Sugars would have L-stereochemistry, amino acids would have D-stereochemistry and DNA and proteins would have left-handed helices. If you were transported to left-handed Earth, what would you notice?

You would be able to breathe the air and drink the water with no ill-effects. If you were very observant, you might notice that your friends looked very slightly different, mirror images of themselves. My best guess, however, is that the first thing you would notice would be the smell. β-Bisabolene is just one of the enticing odours contributing to the smell outside an Indian or Chinese takeaway. In our world, it has *S*-stereochemistry, but in a left-handed Earth, it would have *R*-stereochemistry and would not interact with our receptors in the same way. A city street at night would smell quite different, although people from left-handed Earth would perceive their street in the same way as we perceive ours.

CH₂   CH₃

β-Bisabolene

A real problem for a person trapped in a mirror image world would be food. It would not smell right, it would not taste familiar and, most importantly, you would not be able to digest it. Our enzymes come from a right-handed world and they digest chiral molecules evolved in that world. D-amino acids and L-sugars do not fit into our enzymes and they cannot be digested.

To avoid starvation, you could try eating non-chiral food, but there is very little of that.

SELF CHECK 12.8

Can you think of a non-chiral food?

SELF CHECK 12.9

What advice should a practising Pharmacist give to a visitor from the left-handed Earth about contraception?

Thankfully, quite a lot of drugs would work. For example, some of the quinolones would be as effective in the left-handed Earth as they are at home.

Returning to our own planet, if a drug (such as ciprofloxacin) has no chiral centres, there is no need to worry about adverse effects of the wrong isomer, because there is no wrong isomer. The thalidomide tragedy persuaded many chemists that drugs without chiral centres (**achiral molecules**) are always preferable. If there can be no wrong isomer, there is less chance of a tragedy.

There are now some doubts about this idea and it is easy to see why. My achiral socks are size 4-7. They bind quite non-specifically to medium female feet. In a house with six medium female feet, they go missing quite a lot. This is an undesirable side-effect of socks being achiral and not fitting very closely. My chiral shoes, size 5, width medium, each fit only one of these six chiral feet. They never go missing. Similarly, it is reasoned, a chiral drug is likely to bind more specifically and more tightly to a chiral receptor.

Close-fitting chiral climbing shoes fit very specifically

Did you notice that the third generation quinolone levofloxacin is just a single enantiomer of ofloxacin? This hardly seems like a change worthy of being called a new generation, except that when the unwanted dextro-isomer is removed, the effective dose of levofloxacin is reduced to below that of ciprofloxacin (the achiral drug) for many purposes.

As you learnt in Chapter 10, the drugs made by nature are usually chiral, and sometimes have many chiral centres. This enables them to fit receptors very accurately and specifically. The price that we normally pay for this specificity is that we cannot easily make analogues of compounds with many chiral centres. Penicillins, chiral molecules with many analogues, are quite exceptional, because of the discovery of 6-aminopenicillanic acid.

 Stereochemistry is covered in greater depth in Chapter 3 'Stereochemistry and drug action'.

KEY POINT

Human beings are full of chiral molecules and these are usually stereochemically pure. Molecules made in chemical laboratories are seldom stereochemically pure. The wrong stereoisomer of a drug may be harmful, even lethal, so many drugs do not have chiral centres. This may compromise their ability to bind tightly and specifically to receptors.

12.5 **Purity**

'There is death in the pot' is the headline from Chapter 11, and impurities in drugs have been the cause of many deaths. You may have noticed reference to the taste of drugs in this book. Few parents will hold their children down and force revolting medicine into them, and it is important that medicines do not taste too dreadful.

In 1937, the introduction of sulphanilamide made bacterial pneumonia and meningitis treatable diseases, but before long, children in Oklahoma were dying in agony. The addition of diethyleneglycol to sulphanilamide, with no safety testing at all, made the drug taste sweet, but introduced a deadly poison that claimed 107 lives. The 1938 Federal Food, Drug, and Cosmetic Act was the USA government's response to this tragedy. This act enabled Dr Frances Kelsey, working at the Food and Drug Administration, to refuse to authorize the use of thalidomide in the USA. She reviewed the safety literature for thalidomide and was not satisfied. Despite pressure from the manufacturers, who pointed out that the drug was in use in Europe, she held her ground. At least 4000 children in Europe suffered birth defects as a result of this drug. Frances Kelsey was awarded the President's Award for Distinguished Federal Civilian Service by John F. Kennedy for preventing a similar tragedy in

the USA. She was only the second woman to receive this honour.

Sulphanilamide

Diethylene glycol

 The metabolism of glycols is discussed in Chapter 5 'Alcohols, phenols, ethers, organic halogen compounds, and amines'; pharmaceutical analysis is covered in more detail in Chapter 11 'Introduction to pharmaceutical analysis'.

You have learnt a great deal about the penicillin group of drugs, especially in Chapter 10. A strange fact about these drugs is that many people report allergies to them; the BNF estimates that 1% of the population experiences penicillin allergy, to the point of having experienced an anaphylactic reaction. This really should not happen. Penicillin is a small molecule that should provoke very few allergies.

On the thirtieth anniversary of the 1982 Falklands conflict, an army medic Steven Hughes spoke about his experiences treating the wounded: 'These were all guys who were gungy, their smocks are covered in whatever they had picked off the ground. So you have got contaminated wounds so everyone's having penicillin'. Mr Hughes said he was 'absolutely gobsmacked' that everyone treated at the field hospital survived (BBC News 28 May 2012). In 1982, there was no incidence of penicillin allergy.

Nobody now knows what else was in the penicillin preparations given to children in the 1950s and 1960s, but it seems increasingly likely that many who experienced a severe allergic reaction to what should have been a life-saving medicine were actually allergic to something else, possibly an impurity in the penicillin preparation.

These stories should convince you of the importance of pharmaceutical analysis and of the importance of good quality clinical and pre-clinical trials.

CHAPTER SUMMARY

It still costs millions of pounds to bring a drug to market. A great many potential drugs fail at pre-clinical testing and in clinical trials. We can try to minimize the waste by understanding the molecular characteristics that make a good drug.

➤ The Rule of Five has proved enormously valuable to medicinal chemists involved in designing drugs for oral delivery. However, injectable drugs and many natural products do not need to obey Lipinski's Rule of Five.

➤ Structure-activity relationships can be studied in synthetic and semi-synthetic drugs.

➤ Stereochemistry is important in making a good drug, although it is not yet clear whether enantiomerically pure chiral drugs are better than achiral drugs.

➤ It is essential that we understand what else (other than drug) is in a drug preparation (a medicine). Pharmaceutical analysis is essential to patient safety.

FURTHER READING

Denton, P. and Rostron, C. *Pharmaceutics*. Oxford University Press, 2013.

Gaskell, E. and Rostron, C. *Therapeutics and Human Physiology*. Oxford University Press, 2013.

Hall, J. *Pharmacy Practice*. Oxford University Press, 2013.

A short biography of Frances Kelsey can be found at:
http://www.nlm.nih.gov/changingthefaceofmedicine/physicians/biography_182.html accessed 3 January 2013

Lipinski, C.A., Lombardo, F., Dominy, B.W., Feeney, P.J. Experimental and computational approaches to estimate solubility and permeability in drug discovery and development settings. *Adv Drug Discovery Rev* 1997; 23: 3–25.

Fleming, I. *Frontier Orbitals and Organic Chemical Reactions*. Wiley, 1978; 29–109.

Woodward, R.B. and Hoffmann R. *The Conservation of Orbital Symmetry*, Vch Pub, 1970.

Glossary

Absorption (verb to absorb)   The taking up of electro-magnetic radiation by matter.

ACE inhibitor   Angiotensin-converting enzyme inhibitors inhibit the conversion of angiotensin I to angiotensin II, they may be used to treat heart failure and hypertension.

Acetylcholinesterease   An enzyme that catalyses the hydrolysis of acetylcholine.

Achiral molecules   An achiral molecule is a non-chiral molecule. It is a molecule without chiral centres. Glycine is the only achiral amino acid in nature.

Acid dissociation constant   An acid dissociation constant is the equilibrium constant for the dissociation of an acid in aqueous solution. It is a measure of acid strength.

Activating   Effect of substituents that stabilize the cationic intermediate in electrophilic substitution and hence increase the rate of reaction.

Active transport   Active transport of a metabolite involves transporting the molecule against a concentration gradient, using energy in the process. The term is used more loosely with drugs, to describe the movement across a membrane by hijack of an active transport system. Natural products can sometimes do this.

Acylation   A reaction in which an acyl group is added to a molecule.

Addition reaction   A reaction where a π bond is broken and two new σ bonds are formed.

Agonist   An agonist is a molecule that activates a particular receptor to produce a specific response.

Aliphatic   An organic compound whose structure does not contain benzene or a similar structure.

Alkanes   Saturated hydrocarbons, containing only single bonds.

Alkenes   Hydrocarbons containing one or more carbon-carbon double bonds.

Alkynes   Hydrocarbons containing one or more carbon-carbon triple bonds.

Alkylation   A reaction in which an alkyl group is added to a molecule.

Amphiphilic or Amphipathic   The property (attributed to a molecule) of both polar and non-polar characteristics (literally loving both hydrophobic and hydrophilic environments).

Anaerobes   Organisms that do not require oxygen to live. Metronidazole is an effective drug for treating anaerobic bacteria.

Analgesic   An analgesic reduces, or even eliminates, pain through a pharmacological action.

Analogue   A compound that differs from another in a small, carefully considered way. Analogues of an active compound are often made in order to optimize activity.

Antagonist   An antagonist is a molecule that binds to a receptor, but does not trigger the usual response, and can block the binding of, and activation by, an agonist at the same receptor.

Angstrom (abbreviation Å)   An angstrom is a unit of length equivalent to 0.1 nm, or 10^{-10} m. This non-standard unit is very convenient because it is of the same order of magnitude as a bond length. A C–C bond is about 1.5 Å, a hydrogen bond about 2.7 Å, for example.

Antipyretic   An antipyretic relieves a fever (which is sometimes called pyrexia), by reducing a higher than normal body temperature, through a pharmacological action.

Assay   A method of quantitative analysis.

Attomole 10^{-18} mole.

API   Active pharmaceutical ingredient.

ATP   Adenosine triphosphate is the universal currency of energy in biological systems. Its hydrolysis to ADP (adenosine diphosphate) and inorganic phosphate can yield up to about 50 kJ mol^{-1}.

Axial   In a chair conformation of a six membered ring the axial substituents stick up or down, as shown by the red and blue positions in the diagram.

Bacterial DNA gyrases   Enzymes responsible for introducing supercoils into the DNA double helix, so that it can replicate.

Bilayer   A physical structure, typically a cell membrane, that consists of two layers of molecules facing each other.

Bioavailability   The fraction of the dose of a drug that reaches the general circulation.

Biochemistry   The study of biology from a chemical/molecular perspective.

Biologics   An emerging class of drugs consisting of high molecular weight biological molecules such as proteins or nucleic acids.

Biopolymer   A polymeric molecule produced within a biological system.

Biosynthesis   The synthesis of a compound by a biological system.

Biosynthetic precursor   A substance that occurs early in the production pathway of a molecule prepared within a biological system.

Boat conformation   The boat conformation of a six-membered ring has minimal torsional strain, but eclipsed substituents mean that it is less favoured than the chair form. Its shape is as shown.

Bonding interactions Weak bonds, especially hydrogen bonds and non-polar bonds, which frequently occur within biological macromolecules (intramolecular) or between a drug and its target (intermolecular).

Brønsted acids Species that can act as proton donors.

Bronchial tree The airways within the lungs.

Bronchodilation A widening of the air passages in the windpipe (trachea), which allows increased airflow in and out of the lungs.

Carbocation Positively charged carbon atom, also known as a carbonium ion.

Carcinogenesis The process of initiating and promoting cancer.

Catalyst A compound that increases the rate of reaction without being consumed in the process.

Catalytic hydrogenation Reaction between molecular hydrogen and another molecule in the presence of a catalyst.

Chair conformation The chair conformation of a six-membered ring is the conformation in which torsional and steric strain are minimized. Its shape is as shown.

Chelating agent A ligand which has more than one functional group that can form a dative bond with a metal ion.

Chemical dipole moment The measure of polarity of a chemical bond.

Chemical shift The resonant frequency of a nucleus relative to a standard (usually tetramethylsilane (TMS)).

Chiral A chiral carbon has four different groups bonded to it. These groups may be hydrogen, groups based on carbon or on another element, such as oxygen or nitrogen. If a molecule contains a chiral carbon, it cannot be superimposed on its mirror image.

Chromatogram A visible record showing the separation of compounds by chromatography.

Chromophore The part of a molecule that is responsible for the absorbance of electromagnetic radiation (usually light or ultraviolet radiation). The term originated in the dyestuff industry, meaning the parts of the molecule responsible for colour.

Condensation reaction A bond forming reaction that involves the loss of a molecule of water. Can also be described as a dehydration reaction.

Co-enzyme A non-protein chemical compound that is bound to an enzyme and is required for the enzyme's biological activity.

Concerted In a concerted reaction, the movement of electrons involved in both bond making and bond breaking takes place at the same time. There is no formation of an intermediate carbocation (hence it being a one-step reaction).

Conformation The conformation of a molecule is its special arrangement, or shape.

Conjugates Compounds formed as a result of phase II metabolic transformation.

Conjugated Possessing a system of connected p orbitals with delocalized electrons with alternating double and single bonds.

Complex A metal ion or atom bound to one or more molecules, usually through covalent dative bonding.

Constitutional isomers Constitutional isomers are molecules that contain the same atoms, but connected in different ways (i.e. their *connectivity* is different). They have different physical and chemical properties.

Covalent dative bonding A covalent bond where both electrons come from the same atom.

Cytochrome Membrane-bound proteins that contain haem and are involved in electron transport.

Cytochrome P450 enzymes A family of monooxygenase enzymes that have an iron-haem core and are responsible for many metabolic oxidations by association of molecular oxygen with the metal centre.

Cytotoxic Toxic to cells.

Deactivating Effect of substituents that destabilize the cationic intermediate formed during electrophilic substitution and hence decrease the rate of reaction.

Degenerate Atomic or hybrid orbitals that have the same energy, for example the three 2p atomic orbitals, or the four sp^3 hybrid orbitals, are said to be degenerate.

Dependence A compulsive or chronic need to take a drug; an addiction.

Dimerize To form a compound by combination of two identical molecules.

Dipole-dipole interactions The attraction between molecules as a result of the presence of a permanent dipole moment.

Directing effect A term used to describe how substituents on an aromatic ring can influence the regiochemistry of electrophilic aromatic substitution. The effect a substituent has on the EAS reaction is determined by the effect it has on a positive charge on the carbon that carries it. There are three main classes.

- electron donating groups, which stabilize an adjacent positive charge and are *ortho/para*-directing and activating

- electron withdrawing groups, which destabilize an adjacent positive charge and are *meta*-directing and deactivating

- the halogens, which are *ortho/para*-directing.

Disaccharide A carbohydrate composed of two monosaccharides.

Divalent In the case of carbon, forming two covalent bonds to other atoms (note that the word is also used when talking about ions, a divalent cation, for example Ca^{2+}, has two fewer electrons than the atom in its elemental state, while a divalent anion, for example O^{2-}, has two more electrons than the atom in its elemental state).

Double helix The spiral-like structure that results from two closely associated strands coiled about a central axis.

Eclipsed In an eclipsed conformation, two atoms bonded to adjacent carbon atoms are as close together as possible. This is best illustrated with a diagram.

Electron donating group A group which donates electron density to a conjugated π system via a mesomeric or inductive effect.

Electronegativity The ability of an atom, or group of atoms, to attract electrons, or electron density, towards it.

Electron withdrawing group A group which removes electron density from a conjugated π system via a mesomeric or inductive effect.

Elute To remove a bound substance in a solvent typically to remove a substance from a chromatography column.

Emulsifying agent A substance that prevents the coagulation of colloidal particles.

Enantiomers A pair of chiral molecules that are non-superimposable mirror images of each other. They have identical physical and chemical properties, but they rotate the plane of polarized light in opposite directions.

Endogenous compound Compound found in the body.

Enzyme An enzyme is a biological catalyst. Usually it is made of protein, but some enzymes are made of RNA (ribozymes). Thomas Cech and Sidney Altman won the 1989 Nobel Prize in Chemistry for the discovery of ribozymes.

Equatorial In a chair conformation of a six membered ring the equatorial substituents stick out sideways, as shown, in yellow, on the diagram.

Excipients Excipients are combined with the API (defined elsewhere) to produce a medicine. They can help in the delivery of the API to the receptor and during the manufacturing of the dosage form.

E,Z system The IUPAC preferred system for describing the stereochemistry of a double bond.

Functional group A group of atoms within a molecule that is responsible for certain properties of the molecule.

Fourier transformation A mathematical transformation which allows a much shorter sampling time than conventional spectroscopic techniques.

Formulate To develop a preparation of a drug.

Formulation The science of converting a drug into a form that is suitable for presentation to a patient.

Free radical scavenger A compound that reacts with free radicals in a biological system.

Geminal dihalides Compounds that have both halogen atoms on the same carbon.

Glycosidic bond Covalent bond that joins a carbohydrate molecule to another functional group.

Gram-negative bacteria and Gram-positive bacteria Gram-positive and Gram-negative bacteria can both be stained with Gram stain (crystal violet). The stain can be easily removed from Gram-negative bacteria by washing with acetone, but the cell walls of Gram-positive bacteria retain the stain. Gram-negative cell walls are much thicker than Gram-positive cell walls and are resistant to many antibiotics. The quinolones are unusual in targeting Gram-negative bacteria, preferentially. Examples of Gram-negative bacteria are *Salmonella* and *Escherichia coli*. Examples of Gram-positive bacteria include *Staphylococcus aureus* and *Streptococcus pneumoniae*.

'Green' synthesis A chemical synthesis with little waste, especially of carbon.

Heterocyclic A carbon containing (organic) ring system that contains one or more atoms other than carbon (commonly nitrogen and/or oxygen).

High-performance liquid chromatography An instrumental technique for the separation of mixtures of compounds. It may be used for purification, identification or quantification.

Homonuclear bond A homonuclear bond connects two atoms of the same element.

Hydrogen bond A non-covalent bond between an electron-deficient hydrogen and an electronegative atom, such as oxygen or nitrogen.

Hydrolysis The addition of water to a substance, which causes the breaking of a chemical bond.

Hygroscopicity The ability to absorb moisture, particularly from the atmosphere.

Imine A compound which contains a C=N–R group, where R is an alkyl group or just a hydrogen atom. They are typically formed by the reaction of a carbonyl group (C=O) in an aldehyde or ketone with ammonia or an amine, leading to loss of a water molecule and the formation of a C=N double bond.

Inductive effect Arises as a result of a difference in electronegativity between atoms and is transmitted through σ bonds.

Infrared spectroscopy Involves irradiation of a compound with light of the infrared region of the electromagnetic spectrum. Molecules absorb IR radiation of different frequencies according to their structure. The presence of functional groups can be identified by their characteristic absorbtion wavelengths.

Intercalating drug A flat (planar) molecule capable of interacting with DNA by insertion between the base-pairs of the DNA-ladder.

Isomer Two isomers have the same chemical formulae but different structural formulae. *Cis*-retinal and *trans*-retinal are isomers of one another, so *cis*-retinal can be **isomerized** to *trans*-retinal.

β-lactamase An enzyme capable of deactivating a β-lactam antibiotic (such as a penicillin or cephalosporin) by hydrolysing it.

Lactone A cyclic ester.

Lead compound A compound that shows a desired pharmacological property, which is then developed further through the synthesis of analogues to optimize its activity.

Leaving group A fragment of the molecule that leaves. Good leaving groups are neutral or bear a stabilized negative charge.

Lewis acid A species that can accept a pair of electrons e.g. a metal ion.

Lewis base An atom or group that can donate a pair of electrons e.g. a nitrogen atom.

Ligand A ligand is an ion or molecule that binds to a metal ion or atom to form a coordination complex.

Magnetic resonance imaging (MRI) Use of nuclear magnetic resonance to image nuclei of atoms within the body.

Mass/charge (m/z) ratio The mass ion divided by the charge of that ion.

Mass spectrometry Involves the ionization of a compound under high vacuum and analysis of the mass of the ions produced.

Metabolism (noun) metabolize (verb) Metabolism is chemistry carried out by the body. It is usually catalysed by enzymes, and these enzymes are particularly abundant in the liver.

Meta-directing effect The propensity of a functional group on an aromatic ring to direct an electrophile to the *meta*-position in an electrophilic aromatic substitution reaction.

Miscible Capable of being mixed in all proportions.

Mole The 'mole' is a unit of measurement used to describe the amount of a chemical substance.

Monograph An entry in a pharmacopoeia, which contains the specifications for a particular drug or drug preparation.

Monomer A small molecule that when bonded to other molecules of a similar type creates a polymer.

Monosaccharide A simple carbohydrate, often containing five or six carbon atoms. Used as building blocks for larger carbohydrates.

Nanogram 10^{-9} gram.

Nicotinamide adenine dinucleotide (NAD$^+$) A coenzyme whose role is to accept hydide ions produced by enzyme-controlled oxidation reactions. The NAD$^+$ is reduced to NADH.

Non-stereogenic A non-stereogenic (or **achiral**) carbon has four groups bonded to it, at least two of which are the same. These groups may be hydrogen, groups based on carbon or on another element, such as oxygen or nitrogen. If a molecule contains only non-stereogenic carbons, it can be superimposed on its mirror image.

Nuclear magnetic resonance spectroscopy Used to provide information about positions of atoms in molecules. It utilizes radiofrequency irradiation in a strong magnetic field. It works by detecting the energy emitted by a sample as it relaxes after irradiation.

Nuclear spin Some atomic nuclei behave as if they were spinning and, because they are charged, they create a magnetic field.

Nucleosides A nucleobase (adenine, guanine, thymine, uracil, cytosine) bound to a sugar (ribose or deoxyribose).

Oil-water interface An interface forming the boundary between the two immiscible liquids oil and water.

Orbital A region of space near the nucleus of an atom in which there is a high probability (frequently taken as a 95% probability) of finding an electron; note that we may refer to atomic orbitals (the pure, unaltered orbitals surrounding an isolated atom in its lowest energy state), hybrid orbitals (the combination of these atomic orbitals described in Section 2.4) and molecular orbitals (where the orbitals embrace more than one atom, or even a whole molecule, when they are involved in interatomic bonding).

Ortho/para-directing effect The propensity of an element or functional group on an aromatic ring to direct an electrophile to the *ortho*- or *para*-position in an electrophilic aromatic substitution reaction.

Oxidation A reaction in which the oxygen content of a compound or the number of bonds to oxygen is increased.

Oxidation number The number of electrons that an atom has lost or gained, means that it has lost electrons and means that it has gained electrons.

Oxidizing agent A substance that oxidizes another substance, itself being reduced in the process.

Patent Newly developed drugs need to be protected to prevent other companies benefitting from the invention, time and money spent in developing a new drug. This protection takes the form of a legally binding patent.

Peptide bond An amide bond specifically linking two amino acids to form a peptide backbone.

Peroxide A compound containing an oxygen-oxygen bond.

Pharmacopoeia A reference book containing specifications for drugs and drug preparations.

Physicochemical properties Those properties relating to a system's physical state that influence chemical behaviour.

Picomole 10^{-12} mole.

Polarized A polarizable atom or molecule is one which is neutral, though the electron cloud around it can be distorted so that there are now regions of positive and negative charge existing at the same time (a dipole); the atom or molecules is then said to be polarized. Generally, this is easier with larger atoms, for example argon rather than neon, or the iodide ion rather than the chloride ion, where the electrons are further from the nucleus and so are more weakly held.

Polysaccharide A large, complex carbohydrate that comprises many monosaccharide monomers.

Pro-drug A compound that is not itself a drug, but is converted to a drug in the body.

Protecting group A protecting group reversibly modifies a functional group to prevent it taking part in a reaction.

Proteins Large biological molecules that contain one or more chains of amino acids. The folding of these chains gives proteins specific three-dimensional shapes, and dictates their activity.

Proximate carcinogens A chemical or physical agent that initiates a sequence of reactions leading to **carcinogenesis**.

Published standards Publically available standards for quality of drugs and their formulations.

Pyranose A pyranose is a six membered ring consisting of five carbon atoms and one oxygen atom.

Racemate (racemic mixture) An equimolar mixture of enantiomers.

Racemization The process of forming a racemate from a single enantiomer.

Receptors A receptor is a biochemical structure or site (often found on the surface of a cell or sometimes within a cell), such as a protein or nucleic acid, that can be activated by natural molecules (e.g. hormones) and drugs to cause a specific effect.

Reduction potential (also known as **redox potential**) A measure of the tendency of chemical species to acquire electrons and thereby be reduced. Reduction potential is measured in volts (V), or millivolts (mV). Each species has its own intrinsic reduction potential; the more positive the potential, the greater the species' affinity for electrons and tendency to be reduced.

Regioselective The preference for one direction of making or breaking a chemical bond over other possibilities.

Resolution The separation of a racemic mixture into its component enantiomers.

Resonate To exhibit resonance when two or more structures have an identical arrangement of atoms but a different arrangement of electrons.

Resonance hybrid See 'Resonance structures'.

Resonance structures and canonical forms Frequently, the structures of compounds, or reaction intermediates, are written showing a formal arrangement of double and single bonds and positive or negative charges, trying to complete the octet of eight electrons around a p block element. Often this can be done in more than one way, though no individual one of these structures represents the compound completely accurately. These are referred to as **resonance structures** or **canonical forms**. The actual structure of the compound or intermediate is an average of these and is referred to as a **resonance hybrid**. An illustration of this is given in Figure 2.26, where different representations of the ethanoate (acetate) ion are shown, as well as the 'average', or resonance hybrid.

Retrosynthesis The process of designing the chemical synthesis of a target molecule by starting at the target and working backwards to readily available compounds.

Screening The testing of a particular compound or sample for biological activity.

Selective toxicity The injury of one kind of living matter without harming another with which it is in intimate contact.

Serendipity A 'happy accident' or 'pleasant surprise'; specifically, the accident of finding something good or useful without looking for it.

Sigma conjugation Also known as hyperconjugation, it is a stabilizing overlap between a pi orbital and a sigma orbital.

Sigmoid curve S-shaped curve.

Single helix A spiral-like structure created when a strand of material coils about a central axis.

Specific rotation [α] The observed angle of optical rotation when plane-polarized light is passed through a sample of standard concentration and pathlength at standard temperature in a defined solvent.

Spectroscopic techniques Experimental methods that measure the interaction between matter and radiation intensity as a function of wavelength.

Staggered In a staggered conformation, two atoms bonded to adjacent carbon atoms are as far apart as possible. This is best illustrated with a diagram.

Stereogenic A stereogenic centre is an alternative name for a chiral centre.

Steric Effects due to the size of substituents are called steric effects.

Structure Activity Relationship (SAR) study A tactic to assess the importance of particular areas of a biologically active molecule. Portions of the molecule in question are altered systematically with the aim of discovering the important structural features within a molecule and trying to improve them.

Substitution reaction A reaction where one atom of a group is replaced by another atom or group.

Suicide substrate A compound that interacts with the active site of an enzyme and undergoes a transformation to produce a compound that forms an irreversible complex with an enzyme itself, often by means of a covalent bond.

Target A target for a drug is a biological molecule, usually a protein or a nucleic acid, that interacts with that drug.

Tautomers A special case of structural isomerism; they are isomers which differ only in the position of a hydrogen atom and bonding electrons. Tautomers are in equilibrium with each other.

Tetravalent Forming four covalent bonds to other atoms.

Thioester A thioester is an ester in which one of the oxygen atoms is replaced by a sulfur (–COSR). A thioester is more reactive than the corresponding oxygen ester.

Thiol and thioether Functional groups analogous to alcohol and ether in which the oxygen has been replaced with a sulfur i.e. R–S–H and R–S–R. Thio refers to sulfur so an ether becomes a thioether and a thioalcohol is shortened to thiol.

Tolerance A term used to indicate that repeat doses of a drug give rise to a smaller biological effect. Larger doses are needed for the same pharmacological effect.

Topical A topical drug is applied locally, to the skin or the eye or the ear, for example. It is not systemic (applied to the whole body by the oral route or by injection).

Torsional Effects due to bond twisting are called torsional effects.

tRNA (transfer RNA) Ribonucleic acid (RNA) converts the genetic code (DNA) into proteins. Transfer RNA is one of three types of RNA involved in this process.

Van der Waals interaction Weak interactions between two molecules as a result of temporary dipoles.

Vicinal dihalides Compounds that have halogen atoms on adjacent carbons.

Xenobiotic A substance in the blood stream that is not normally found there. Xenobiotics are often processed in the liver.

Xenobiotic-metabolizing enzymes Enzymes that control the metabolic pathways that modify xenobiotic molecules, foreign to an organism's normal biochemistry.

Index

A

ABO blood group system 260
absorption, distribution, metabolism and excretion (ADME) 114–16
ACE *see* angiotensin-converting enzyme
acetaldehyde (ethanal) 8
  aldol reaction 165
acetals 158–60
acetanilide 195, 198–9
  metabolism 206
acetate 4, 5, 45
  polymerization 11, 12
acetazolamide 234, 235
acetic acid (ethanoic acid) 144, 146, 152, 261, 262
acetic anydride (ethanoic anhydride) 152, 155
acetone (propanone) 123, 146, 149
  bromination 164
acetophenone 146
acetyl chloride 152
acetyl coenzyme A (acetyl CoA) 4–5, 147, 171, 265, 266
acetylation, benzene 189
acetylcholinesterase inhibition 55–6
acetylene (ethyne) 87
  bonding 29, 35
  combustion 108
  pK_a 91
acetylide ions 106
acetylsalicylic acid *see* aspirin
achiral compounds 60
aciclovir 248–9
acid anhydrides 143, 145
  ester and amide synthesis from 155
  synthesis of 156, 157
acid dissociation constant 138
acid/base neutralization 43–4
aconitase 7, 122
active pharmaceutical ingredients (APIs) 112–13
acyl chlorides
  amide synthesis from 154
  ester synthesis from 152–4
  synthesis of 155–6
acyl halide 143, 145
acylation 190, 191
addition reactions 40, 41
adenine 243
adenosine 246
adenosine diphosphate (ADP) 225–6
adenosine triphosphate (ATP) 6, 225–6, 247

hydrolysis 6, 247
ADP *see* adenosine diphosphate
adrenaline 27, 201, 245, 246, 289
  biosynthesis 203, 204
  structure 27, 201
agonists 49
alanine 13, 60, 69, 250
alcohol dehydrogenase 123
alcohols
  metabolism 7–8
  physical properties 118–20
    boiling points 119
    solubility 119–20
  polyhydric 124–6
  primary, secondary and tertiary 118, 122
  reactions 121–4
    dehydration 121–2
    oxidation 7–8, 122–4
    substitution 122
aldehydes 122–3, 143, 145, 157–8
  addition to 157
  reduction 159–61
aldol reaction 165–6, 168
alendronic acid 228
aliphatic amines 137
aliphatic hydrocarbons 81
  acidity 90–1
  boiling and melting points 88, 89
  density 88, 89
  nomenclature 82–7
  solubility 88
  see also alkanes; alkenes; alkynes
alkanes 82–4
  acidity 90
  biological roles 94
  boiling and melting points 89
  branched-chain 82–3
  cycloalkanes 83–4, 89
  preparation 91
  reactions 91–2
    combustion 92
    halogenation 92
  straight-chain 82
alkenes 57, 84–7
  acidity 91
  boiling and melting points 89
  cycloalkenes 85
  isomerism 85–7
  metabolism 115
  preparation 93–4
  reactions 95–103
    alcohol addition 99–100
    electrophilic addition 44–5, 96–7
    halogenation 100–1
    hydration 99

hydrogen halide addition 46, 96–8
    oxidation 101–3
    reduction (hydrogen addition) 101, 102
    stabilities 95
alkyl groups 82, 83
alkyl halide 134
alkylation
  amines 139
  benzene 189–90
alkynes 87
  acidity 91
  boiling and melting points 89
  metabolism 115–16
  preparation 105–6
  reactions 106–12
    combustion 108
    halogenation 109
    hydration 109–10
    hydrogen halide addition 108–9
    oxidation 111–12
    reduction 110
allylic radicals 113
α-helix 252–3, 254
α-keratin 253
alternative therapies 272–3
amide bonds 8, 172–3, 251
amides 143, 145
  synthesis 154, 155
amikacin 54
amines 117, 137–9
  alkylation 139
  chemical properties 137–9
  in drugs 138
  nomenclature 137
  oxidation 139
  physical properties 137
amino acids 8, 249–51
  acid-base chemistry 249–51
  stereochemistry 249
  structure 8
aminobenzenes 195
aminoglycosides 54–5
6-aminopenicillanic acid 12–13, 154, 284, 285, 333
2-aminopropanoic acid 60
amitriptyline 204
amoxicillin 12–13, 15, 278, 333, 334
amphipathic molecules 119
amphiphilic molecules 261, 263–5
amphotericin B 90
ampicillin 333, 334
anaesthesia 131
analogues 283
anastrazole 204, 205
andostradione 204, 205

angiotensin II 173–4, 175
angiotensin-converting enzyme
    (ACE) 221–2, 230
  ACE inhibitors 173–5, 230, 288
anilines 195
  reactivity control 195–6
anisole 193
annulene 179
anomeric carbons 258
antacids 218
antagonists 49
anthracene 179
anti-Markovnikov products 46, 99
antibiotics 275–9
  β-lactams 277, 335
  glycopeptides 277
  macrolides 276
  quinolones 335–6
  see also specific drugs
anticancer drugs 247, 279–81
  interacting with DNA 279
  interacting with proteins 281
  see also specific drugs
anticodon 244
antifreeze 124
antifungal agents 10–11, 330–1
  see also specific drugs
antimalarial drugs 9–10
antioestrogens 204–5
antioxidants 43
  phenols 128–9
aqueous work up 293
arachidonic acid 261, 262
arginine 250
aromatase 203–4
  inhibitors 204–5
aromatic compounds 177–9
  amines 137, 138
  drug synthesis 198–203
  halogen compounds 135–6
  importance of aromatic
    chemistry 180–5
  in the body 203–8
    metabolism 206–8
  reactivity control 195–6
  see also benzene
aromatic stabilization energy 179
arrows 43
  curly 42–3
artemisinin 9–10, 142
  production method 10
asparagine 250
aspartic acid 250
aspirin 3, 16, 146–7, 180–1, 273–4
  history 180–1
  hydrolysis 150, 151, 172, 337
  impurities 297–8
  mass spectrum 317, 318
  NMR spectrum 309–11, 312, 313
  quality control 22
  stability 172
  structure 2–3, 142, 181
  structure-activity
    relationship 336–7

synthesis 17, 147, 200, 201, 297
atmosphere pressure chemical
    ionization (APCI) 316
atomic orbitals 28
atorvastatin 18, 73, 255, 286, 287,
    329
ATP see adenosine triphosphate
atracurium 234
Attenuated Total Reflectance
    (ATR) 304
aufbau principle 28
azidothymidine (AZT) 248
azithromycin 16
  production method 15, 16

B

barbiturates 237
barbituric acid 112
Beer-Lambert Law 301–2
  expressed in grammes 301–2
  expressed in moles 302
benazepril 164
benzene 177–9, 185–98
  electrophilic aromatic
    substitution 185–98
    directing effects 190–8
    mechanism 185–8
    post-substitution reactions 198
  metabolism 207
  toxicity 206–7
  see also aromatic compounds
benzene ring 3, 178–9
benzene-1,3-diol 126
benzene-1,4-diol 128
benzo-[a]-pyrene 207, 208
benzylpenicillin 276, 278, 333, 334
Berzelius, Jöns Jacob 25
β-bisabolene 338
β-lactam antibiotics 277, 335
  see also specific drugs
β-oxidation 265
β-sheet 253, 254
biologics 242
bisphosphonates 227–8
Bitrex® 271, 272
bleomycin 279, 280
boat conformation 54–5
bold wedge bonds 60
bombykol 94
bond cleavage 41
brivudine 317, 318
bromine 185, 186, 195
bromomethane 44
bromopropanone 164–5
bupropion 204
buspirone 322
busulfan 234, 235
butan-1-ol 119
butane 82
butanoic acid 111
1-butyl-4-methoxybenzene 331–2
butylbenzene 331–2

C

^{13}C NMR spectroscopy 315
Cahn-Ingold-Prelog system 57
calcium 217–18
camptothecin 280, 288
cannabis 183–4
canonical forms 45
capillin 105
  structure 37
capsaicin 184
captopril 174, 175, 230, 287,
    288, 290
carbemazepine 115
carbenes 29
carbocations 93–4, 96–8, 121,
    162, 190
  stabilization 193, 194
carbohydrates 256–60
  disaccharides 258–9
  monosaccharides 256–8
  polysaccharides 259–60
carbon 26
  allotropes 26
  electronic configuration 28
carbonic anhydrase 255
  inhibition 234
carbonyl compounds 141–4
  activation 149–50
  chemical properties 146–9
  electronegativity 147
  in drugs 172–5
  in the body 171–2
  interconversion 151–2
  IR spectroscopy 306–7
  nucleophilic attack 147–62
    carbonyl compound as the
      nucleophile 163–70
  physical properties 145–6
  substitution reactions 162–71
  types of 143
carbon–carbon bond formation
    161–2, 188–90
carbopol 142
carboxylic acid 8, 143, 145, 249
  delocalization of electrons 45–6
carboxypeptidase 173
carcinogenicity, epoxides 132
carvone enantiomers 73
catalytic hydrogenation 101
catechol system 203
catechol-O-methyltransferase 231,
    232, 245
cationic detergents 140
cell membranes 227, 228, 263–5
  membrane potential 216
cephalins 227
cephalosporins 277
  cephalosporin C 276
chain-growth polymers 103
chair conformation 54–6, 257
chelating agents 212

chemical bonds 211–12
    see also specific types of bond
chemical ionization (CI) 316
chemical reactions
    organic reaction mechanisms 42–7
    types 40–2
chemical shift 309–10, 311, 312
chemotherapy 134
chilli peppers 184
chiral pool 77
chiral switching 74
chirality 13, 60, 337–9
chlorambucil 133, 134, 135
chloramphenicol 14, 16, 245
    limitations 75
    stereochemistry 73–5
chlorination, methane 92
chlorobenzene 135, 191
chlorofluorocarbons (CFCs) 136
chloroform 293
1-chlorohexane 135
chloroquine 133
chlortetracycline 133, 135
cholesterol 266–8
chromatographic methods 319
    gas-liquid chromatography
        (GLC) 323–4
    liquid chromatography 294
        high-performance liquid
            chromatography (HPLC) 294,
            298–9, 320–2
    thin layer chromatography
        (TLC) 319–20
chromophores 21, 180, 301
cigarette smoke 132
cimetidine 221, 287, 288, 291
cinnamaldehyde 314
ciprofloxacin 335, 336
cis addition 110
cis isomers 34, 59, 85–6
    cis-retinal 4, 5, 85, 86
    stability 95
cisplatin 223
citalopram 48, 286, 291
    structure 49
citrate 6, 7
citrate synthase 6
citric acid 5–6
citric acid cycle 4–5, 147, 337
Claisen condensation 166–9, 170
Clemmensen reduction 190, 191
clinical trials 272
clopidogrel 73
clotting factor VIII 14
cocaine 286, 287, 321
codeine 61, 271, 272, 274
codeine phosphate 225
coenzyme A 171
coenzymes 123
cold gaseous sterilization 131, 132
collagen 253, 255
combretastatin A4 280, 281
combustion
    alkanes 92

alkenes 101
alkynes 108
computer aided drug design 290
conformational isomerism 51–6
    cyclic molecules 52–6
    linear molecules 51–2
constitutional isomerism 49–51
cortisol 267, 268
covalent bonds 26, 32, 211–12
cracking 91
crude oil 91
cubanes 29
curly arrows 42–3
cyanohydrins 161
cycloalkanes 83–4
    boiling and melting points 89
cycloalkenes 85
cyclohexane ring 53
    conformations 53–5
    ring flip 54
cyclophosphamide 60, 134
cysteine 231, 250
cytidine 246
cytochrome enzymes 220–1
    cytochrome P450 206, 221
cytosine 243

D

dapsone 233
daunomycin 54
DDT 136
10-deacetylbaccatin 283
dehydrohalogenation 93
denatonium benzoate 271, 272
deoxyribonucleic acid see DNA
deoxyribose 159
desferrioxamine 212
dexamethasone 225
dextrorotatory isomers 71, 257
diamorphine 271, 274
    structure 271, 272
    synthesis 152–4
diastereoisomers 75, 77
    properties of 76–7
diazepam 120
diazonium salt 198
diethyl ether 130
    NMR spectrum 314
    peroxide formation 132
diethylene glycol 339
digestive enzymes 173, 174
digoxin 215
dihydric molecules 118
9S-dihydroerythromycin A 161
dimercaptosuccinic acid
        (DMSA) 75–6
dipeptides 252
diphosphoric acid 225, 226
dipole–dipole interactions 38–9
disaccharides 258–9
disulfides 230–1
disulfuram 8

DNA 225, 226, 242
    as a drug target 247
        anticancer drugs 279
    structure 243
        double helix 244
docetaxel 283, 284
dodecyl alcohol 236
donepezil 55–6
    conformations 56
L-DOPA 204
DOSY (diffusion ordered
        spectroscopy) NMR 315
double bond 33
double helix 244
doxorubicin 54, 247, 248, 279, 280
drugs 270–3
    genetic engineering 291–2
    natural products 273–82
    purification methods 293–4
    purity 339–40
    Rule of Five 328–31
    semi-synthetic 282–5
    stereochemistry 337–9
    structure-activity
        relationships 331–7
    synthetic 285–91
        computer aided design 290
        library screening 290
        me-too drugs 290
        organic synthesis 292–3
    see also specific drugs
duloxetine 160–1
dyes 180

E

E isomers 57, 86–7
eclipsed conformation 51
ecstasy 183, 184
eflornithine 138
Ehrlich, Paul 331
eicosanoids 261
electromagnetic spectrum 299–300
electron donating groups 197
    directing effects 191–5
electron ionization (EI) 316
electron transport chain 220–1
electron withdrawing groups 197
    directing effects 196–7
electrophiles 147
electrophilic addition 44–5, 96–7
electrophilic aromatic substitution
        (EAS) 185–98
    directing effects 190–8
    mechanism 185–8
    post-substitution reactions 198
electrospray ionization (ESI)
        316, 317
elimination reactions 40, 41
elzasonan 168–9
    synthesis 169
emulsifying agents 119–20
enalapril 175, 287, 288

enalaprilat 174–5
enantiomers 60
  chiral molecule interaction 72–3
  plane polarized light
      interaction 70–1
  properties of 69–73
  resolution 77–8
  see also stereoisomerism
endorphins 274
energy metabolism 4–7
enflurane 136
enols 109, 165
enzymes 253, 254–5
  active site 253
  inhibition 55–6
    ACE 173–5, 230, 288
    acetylcholinesterase 55–6
    aromatase 204–5
    carbonic anhydrase 234
  see also specific enzymes
ephedrine 76–7
epoxides 102, 115, 131–2, 206, 207
ergosterol 266, 267, 330
erythromycin 15–16, 161, 169, 276
  ester formation 19–21
  hydrolysis 150–1
  side-effects 275
  structure 15, 142, 170, 276
erythronolide B 169, 170
escitalopram 48, 290, 291
esomeprazole 73, 290, 291
esters 143, 145
  base-catalysed hydrolysis 150
  synthesis 152–4, 155, 157
estrogens see oestrogens
ethane
  bonding 29, 32, 36
  pK$_a$ 90, 91
  structure 82
ethanol
  oxidation 7–8, 123
  polarization 39
ethchlorvinyl 37
ethchlovynol 108
ethers 130–2
  as extraction solvents 130, 131
  chemical properties 131
  physical properties 130–1
ethinylestradiol 267, 268
ethyl acetate 293
ethylamine 39
ethylbenzene 332
ethylene (ethane) 84
  biological role 94
  bonding 29, 34
  combustion 95
  pK$_a$ 91
ethylene glycol 124, 125
  poisoning 125
ethylene oxide 130
  cold gaseous sterilization 131, 132
ethylenediamine tetracetic acid
      (EDTA) 217
ethyne see acetylene

17-ethynyl estradiol 105, 116
ethynylestradiol 37
etoposide 288
excipients 113
excited state 29, 30

F

fats 262, 265–6
  dietary 265
  hydrogenated 265
  see also lipids
fatty acids 261–2
  cis 264
  polyunsaturated 265
  saturated 264
  trans 264
  unsaturated 265
ferrous sulphate 219
first pass metabolism 124
Fischer esterification 150
Fischer nomenclature 67–8
Fischer projection 68–9, 256
  manipulation 69
flame ionization detector 323
flucloxacillin 333–5
fluconazole 330–1
fluoroacetylsalicylic acid 337
fluoroquinolones 335
fluoxetine 64–5
formaldehyde 146
free radicals 46–7, 129
  scavengers 128–9
free-radical polymerization 103–5
Friedel–Crafts reactions 188–91
Fries Rearrangement 201
fructose 258, 259
  stereochemistry 67–8
fulvestrant 204, 205
fungal infections 10–11
  antifungal agents 10–11, 330–1

G

galactose 258–9
gas-liquid chromatography
      (GLC) 323–4
  GLC-MS 323
gatifloxacin 335, 336
geminal dihalides 106
general anaesthetics 131
genetic engineering 291–2
gentamicin 55
geometric isomerism 57–9, 85
glucose 159, 256, 258
  polymers 159, 160
  stereochemistry 67–9
glucose-6-phosphate 225
glutamic acid 250
glutamine 250
glyceraldehyde 257
  D-glyceraldehyde 68, 69, 70

L-glyceraldehyde 69, 70
glycerate 126
glycerol 125, 126, 227, 263
  pharmaceutical uses 125
glycine 250
glycogen 259–60
glycolysis 111
glycopeptide antibiotics 277
glycosidic bond 258
gold 223
green synthesis 17
grepafloxacin 335, 336
Grignard reactions 198
Grignard reagents 162, 333
ground state 28, 29
guanine 243
guanosine 246

H

haemoglobin 219–20, 253
hair 172–3
haloalkanes 133–5
  chemical properties 134–5
  nomenclature 133
  nucleophilic substitution 133, 134
  physical properties 133–4
halobenzene nitration 197
halogen compounds 133–5
  aromatic 135–6
  directing effects 197
  polyhalogen compounds 136
halogenation 163–5
  alkanes 92
  alkenes 100–1
  alkynes 109
halothane 31, 136
  bonding 31, 32
hashed line bonds 60
Haworth projections 256, 257
heat of hydrogenation 95
helical structures,
      stereoisomerism 65–7
hemiacetals 157–8, 257, 258
  formation 257
hemiketals 158
Henderson-Hasselbalch
      equation 127
herbal teas 272–3
heroin 271, 274
heterocyclic amines 137
heterolytic cleavage 41
hexoses 256
high-performance liquid
      chromatography (HPLC) 294,
      320–2
  HPLC-MS 322
  quality control 298–9
high-throughput screening 291
histamine 289
histidine 250
HMG-CoA reductase 255
homologous series 82

homolytic cleavage 41
hormones 281–2
Hückel's rule 179
Human Genome Project 281
Hund's rule 28
hybridization of atomic
        orbitals 29–37
    recognition of hybridization
        states 36–7
    sp hybridization 30, 34–6
    sp² hybridization 30, 33–4
    sp³ hybridization 30–3
hydration
    alkenes 98, 99
    alkynes 109–10
hydrocortisone 268
hydrogen bonding 39–40, 119,
        329–30
    hydrogen bond acceptor 40
    hydrogen bond donor 40
    in nucleic acids 243
hydrogen halide addition
    alkenes 46, 96–8
    alkynes 108–9
hydrogen sulfide 230
hydrogenated fats 265
hydrogenation, alkenes 101, 102
hydrophobicity 261
4-hydroxyandrostenedione 204, 205
hydroxyl group 43, 117, 118–24
    chemical properties of
        compounds 120–4
        breaking the C–O bond 120–2
        breaking the O–H bond 122–4
    influence on drug metabolism 120
hydroxylation 102
    aromatic compounds 206

I

ibuprofen 16–17, 321–2
    metabolism 115
    separation of stereoisomers 77–8
    spectra 303–4
    stereochemistry 48, 70, 72, 77–8
    structure 26, 27, 181, 329
    synthesis 17
impurities 297–9
indinavir 286
infrared (IR) radiation 299–300
infrared (IR) spectroscopy 303–7
    recording infrared spectra 303–4
    spectral information 304–6
    stretching frequencies 305, 306
insulin 13–14, 173, 282
    genetic engineering 292
    production method 13–14
    structure 173, 231, 282
insulin glargin 292
insulin lispro 292
intermolecular forces 38–40
    dipole–dipole interactions 38–9
    hydrogen bonding 39–40, 119

London (van der Waals) forces 38
iodoform reaction 164
ionic bond 212
irinotecan 288
iron 219–21
    deficiency 219
    ligand binding 221
    redox reactions 219–21
isocitrate 7
isoleucine 250
isomerism 48–9
    alkenes 85–7
    cis isomers 4, 5, 34, 59, 85–6
    conformational 51–6
    constitutional 49–51
    geometric 57–9, 85
    optical 60–2
    paracetamol 50
    retinal 4, 5, 85, 86
    stereoisomerism 56–77
    tautomers 109
    trans isomers 4, 5, 34, 59, 85–6
isopentane 82–3
isopropyl alcohol (IPA) 123, 124
isopropylbenzene 190
itraconazole 330–1
IUPAC nomenclature 2, 62

K

KBr disc 303, 304
Kekulé ring 3
Kelsey, Frances 339
keto-enol tautomerism 109, 163
ketoconazole 133, 135
ketones 122–3, 143, 145, 157–8
    addition to 157
    reduction 159–61
ketoprofen 159
Kolbe Reaction 200

L

Labetalol Hydrochloride 302–3
L-lactic acid 144, 146
lactose 258–9
    intolerance 259
lactulose 260
large molecules 241–2
latanoprost 286
lavender oil 323–4
Le Chatelier's principle 150
leaving groups 148–9, 152
lecithins 227, 228
letrozole 204, 205
leucine 250
leukotrienes 261
levofloxacin 74, 335, 336, 338–9
levorotatory isomers 71, 257
levothyroxine 281, 282
Lewis acids and bases 212
library screening 290

high-throughput screening 291
lidocaine 271
Lifesaver Screensaver Project 290
limonene 94
Lindlar's catalyst 110
linoleic acid 113
    oxidation 113–14
lipase 265, 266
lipids 261–8
    fatty acids 261–2
    phospholipids 263–5
    steroids 266–8
    triglycerides 262–3, 265–6
Lipinski's Rules see Rule of Five
liquid chromatography 294
    see also high-performance liquid
        chromatography (HPLC)
lithium 216
lithium aluminium hydride 160
liver 7–8
London (van der Waals) forces 38
lovastatin 286, 287
lysine 155, 250

M

macrolide antibiotics 276
magnesium 218
malaria 9–10
malathion 225–7
    selective toxicity 226–7
MALDI (Matrix Assisted Laser
        Desorption Ionization) mass
        spectrometry 315, 316
mannose stereochemistry 67–8
Markovnikov, Vladimir
        Vasilyevich 98
Markovnikov's rule 46, 97–8, 327
mass spectrometry 315–18
    analysing a mass spectrum 316–18
    instrumentation 315–16
mauveine 180
MDMA 183, 184
me-too drugs 290
melanin 21
membrane potential 216
mercaptans 230
mercaptopurine 230
meso-forms 75–6
messenger RNA (mRNA) 244
meta substitution 191–3
metals 213
    group 1 214–16
    group 2 217–18
    period 4 218–22
    precious 222–3
    see also specific metals
methadone 142, 286, 287
methane
    bonding 32, 36
    bromomethane conversion to 44
    chlorination 92
    combustion 92

structure 82
methanol 119
methicillin 154, 173, 333
methionine 250
methohexital 112
methoxybenzene 193
methylbenzene *see* toluene
methylphenidate 286
metronidazole 48
   structure 3–4, 329
mitomycin C 279, 280
monohydric molecules 118
monosaccharides 256–8
   anomers 258
   stereochemistry 257
   structure 256
morphine 14, 60–2, 274–5, 287
   side-effects 275
   stereochemistry 61–2
   structure 61, 271, 272, 321
moxifloxacin 335, 336
multifidene 94
myristic acid 263

N

N-acetylglucosamine 171, 172
nabumetone 51–3
   conformation 52–3
   structure 51
nalidixic acid 286, 335, 336
naphthalen-2-ol 126
naphthalene 179
naphthyl group 51
naproxen 321–2
   stereochemistry 72
   structure 3–4
natural gas 91
neutralization 43–4
nicotinamide adenine dinucleotide
   (NAD^+) 123
nitration 188, 198
   halobenzenes 197
nitrobenzene 197
nomenclature 2, 62, 67–8
   aliphatic hydrocarbons 82–7
   amines 137
   carbonyl compounds 142–4
   haloalkanes 133
   stereoisomers 62–5, 72
non-stereogenic compounds 60
noradrenaline 231, 232, 246
   inactivation 245
noscapine 61
nuclear magnetic resonance (NMR)
   spectroscopy 307–15
   analysing the spectrum 309
   chemical shift 309–10, 311, 312
   instrumentation 308–9
   relative integration 310–11
   spin-spin coupling 310–15
   two-dimensional 315
nucleic acids 242–5

structure 242–3
   backbone 242
   base pairing 243
   bases 243
nucleophiles 147
   carbonyl compound as
      nucleophile 163–9
   carbonyl compound attack 147–9
nucleophilic aromatic substitution
   (NAS) 185
nucleosides 245–6
   derivatives as drugs 247–9
nucleotide 242, 243
nystatin 10–11
   structure 11, 90

O

Oblivon 37
oestradiol 204, 205, 267, 268,
   281, 282
oestrogens 203–4, 281
oestrone 204, 205
ofloxacin 74, 335, 336
oleic acid 261, 262
oligosaccharides 258
omeprazole 60, 233, 286, 291
opioid analgesics 274–5
opium 61, 274
opsin 4
optical isomerism 60–2
organic chemistry 24–5
organic synthesis 292–3
ortho substitution 191–3
orthophosphoric acid 224
oxacillin 154
oxidation 212
   alcohols 7–8, 122–4
   alkenes 101–3
   alkynes 111–12
   amines 139
   linoleic acid 113–14
   medicine stability and 113
   phenols 128
   see also combustion
oxidation states 212–13
oxidative cleavage 102
oxidative stress 129
oxiranes *see* epoxides
oxonium ion 159
oxyacetylene torches 108, 109
oxygen 117
   electron donation 193–5
   see also oxidation
ozone 102–3, 112

P

^{31}P NMR spectroscopy 315
paclitaxel 280, 281, 283, 284
   analogues 284
palmitic acid 263

palmitoleic acid 263
papaverine 61
para substitution 191–3
para-aminobenzoic acid 21
paracetamol 16
   IR spectrum 306–7
   isomers 50
      separation of *ortho-* and *para-*
      isomers 199, 200
   structure 3–4, 37, 50, 181
   synthesis 199–200, 206
   UV-visible spectra 301
pargyline 37
partition coefficient 329
penicillamine 26–7
penicillins 12–13, 142, 154, 277,
   333–5
   allergies 339–40
   biosynthesis 13
   resistance 335
   semi-synthetic 284–5
   synthesis from acyl chlorides 154
pentane 82–3
pentobarbital 237, 238
pentoses 256
pepsin 173
peptide bonds 8–9, 251
peptides 252
pericyclic reactions 42
periodic table 211
permanent dipole 38–9
peroxides 131, 132
pethidine 54
   conformation 54, 55
petrol 91
phase one metabolism 206
phase two metabolism 206
α-phellandrene 94
phenobarbital 112
phenolate 127
phenols 126–9
   acidity 127
   as antioxidants 128–9
   bromination 195, 197
   reactions 128
phenyl ring 54
phenylalanine 250
phenyldiazonium chloride 199
pheromones 94
phosphate esters 142, 224–7
phosphatidylcholine 263
phosphodiester 225, 242
3-phosphoglycerate 126
phospholipids 227, 228, 263–5
phosphoric acid 224, 225
phosphorus 224–9
   bisphosphonates 227–8
   phosphoesters 224–7
pi (π) bonds 32, 33, 34, 57
piperidine ring 54, 56
pK_a 90, 138
plane polarized light
   interaction 70–1
platinum 223

podophyllotoxin 288
polarimeter 71
polyene macrolide antibiotics 89–90
polyhalogen compounds 136
polyhydric alcohols 124–6
polyketide synthesis 169
polymers 103–5
polysaccharides 259–60
polyunsaturated fatty acids 265
potassium 215–16
prednisolone 281, 282
primary alcohols 118, 122
primary amines 137
procaine 286, 287
product licence standards 297
progesterone 281, 282
proline 250
prontosil 180, 181
propan-2-ol 123, 124
propane 82
propanoate 169
propanoic acid 146
propanone *see* acetone
propene 84
propranolol 204
propylbenzene 331–2
1,2-propylene glycol 125
propylthiouracil 237
prostaglandins 113, 261
protein folding diseases 78–80
proteins 172–3, 249–55
  as drug targets 253–5
    anticancer drugs 281
  structure 251–3
    folding 252
    primary sequence 251
    quaternary structure 253
    secondary structure 252–3
    tertiary structure 253
  synthesis 8–9, 244–5
  see also enzymes; *specific proteins*
Prozac 182
pseudoephedrine 76–7
purification methods 293–4
purine nucleosides 246
purines 243
purity 339–40
  impurities 297–9
pyranose ring 55
pyridine 153–4, 179
pyrimidine nucleosides 243, 246
pyrophosphoric acid 228
pyrrole 212
pyruvic acid 111

Q

quality control 22, 297–9
quaternary ammonium compounds 139–40
quinine 9

structure 9–10
quinolone antibiotics 335–6

R

racemic mixtures 60, 70
racemization 60
ranitidine 290, 291
reactive oxygen species (ROS) 43, 129
  see also antioxidants
rearrangement reactions 40, 41
reboxetine 204
recrystallization 294
reduction
  aldehydes 159–61
  alkenes 101
  alkynes 110
  ketones 159–61
regioselective reactions 97
resolution of enantiomers 77–8
resonance hybrid 45
resonance structures 45
retinal 4, 5, 85, 86
retrosynthesis 293
ribonucleic acid *see* RNA
ribose 159, 256, 258
ribosomal RNA (rRNA) 244
ribosome 9
ring flip 54
RNA 242, 243, 244
  messenger RNA (mRNA) 244
  ribosomal RNA (rRNA) 244
  transfer RNA (tRNA) 244–5
Rule of Five 328–31
  exceptions 330
  hydrogen bonding 329–30
  size 328
  solubility 328–9
  use in drug design 330–1

S

S-adenosylmethionine (SAM) 231, 232, 245, 246
salbutamol 27, 231, 287, 288
  metabolism 237
  structure 27, 201, 204
  synthesis 200–3
salicin 273–4
salicylic acid 146, 172, 274, 297, 337
salmeterol 202
  structure 202
salvarsan 332–3
Sanger, Fred 14
sarin 226
Sativex 183
saturated fatty acids 264
secobarbital 112
secondary alcohols 118, 122
secondary amines 137

selective serotonin reuptake inhibitors (SSRIs) 182
semi-synthesis 283–4
semi-synthetic drugs 15–16
serine 69, 250
shapes of molecules 26–7, 29–37
sigma (σ) bonds 32, 34, 57
sigma conjugation 94, 96, 192, 193
sigma-complex 185, 187
silver 223
small molecules 241
$S_N 1$ substitution reaction 122
$S_N 2$ substitution reaction 134, 135, 245
sodium 214–15
sodium aurothiomalate 223
sodium borohydride 160
sodium chloride 214–15
sodium dodecyl sulfate 236
sodium ethoxide 163
sodium lauryl sulfate 236
solubility 328–9
  alcohols 119–20
  aliphatic hydrocarbons 88
sp hybridization 30, 34–6
sp^2 hybridization 30, 33–4
sp^3 hybridization 30–3
spectrophotometer 300
stability
  alkenes 95
  medicines 113–14
staggered conformation 52
standards 297
statins 18, 286
stearic acid 261, 262
stereogenic compounds 60
stereoisomerism 56–77, 337–9
  geometric isomers 57–9, 85
  helices 65–7
  molecules with more than one chiral centre 73–5
  molecules with two identical chiral centres 75–7
  nomenclature 62–5, 72
  optical isomers 60–2
  properties of enantiomers 69–73
  separation of stereoisomers 77–8
steric hindrance 195
steric strain 53
steroids 266–8
  biochemistry 267–8
  structure 266–7
streptomycin 159–60
structural isomerism 50
structure activity relationship (SAR) 198, 331–7
  aspirin 336–7
  flucloxacillin 333–5
  quinolone antibiotics 335–6
  salvarsan 332–3
strychnine 321
substitution reactions 40, 41, 44
  alcohols 122

carbonyl compounds 162–71
haloalkanes 133, 134
see also electrophilic aromatic
    substitution (EAS);
    nucleophilic aromatic
    substitution (NAS)
sucrose 258, 259
sugars 158, 256
see also carbohydrates
sulfa antibacterial drugs 180
sulfamethoxazole 181
sulfanilamide 181
sulfate esters 236, 237
sulfhydryl group 230
sulfonates 233–4
sulfonation 188, 189
sulfones 231–3
sulfonic acids 234
sulfotransferases 236
sulfoxides 231–3
sulfur 229–38
    oxidation states 229
    thiols 229–31
sulfuric acid derivatives 236
sulindac 233
sulphanilamide 339
sulphonamides 234, 235
sunscreen 21
superoxide 129
synthetic drugs 16–18
synthetic polymers 103–5

T

tamoxifen 204–5
    stereoisomerism 57–9
tannic acid 130
tautomers 109
temazepam 120
teniposide 288
teprotide 288
terbutaline 231
tertiary alcohols 118, 122, 124
tertiary amines 137
testosterone 204, 205, 267, 268
tetrachloromethane 31, 32
tetracycline 14, 217, 245, 276

tetrahydrofuran 130
thalidomide 48, 66, 339
    stereochemistry 66, 338
    structure 49
thin layer chromatography
    (TLC) 319–20
thioesters 143, 145, 169, 230
    importance of 171
thioethers 230, 231
thiols 229–31
thiones 236–8
thiopental 237–8
threonine 250
thromboxanes 261
thymine 243
thyroxine 281
toluene 191, 193
topotecan 288
torsional strain 52
trans addition 110
trans isomers 34, 59, 85–6
    stability 95
    trans-retinal 4, 5, 85, 86
transfer RNA (tRNA) 244–5
triazole antifungals 330
trifluoroacetic anhydride 157
triglycerides 262–3, 265–6
trigonal planar arrangement 33
trihydric molecules 118
tripeptides 252
triphosphoric acid 225, 226
trypsin 155, 173
tryptophan 250
    enantiomers 73
tubulin 281
type 1 diabetes 282
tyrosine 250

U

ultraviolet (UV) radiation 299, 300
ultraviolet-visible spectrophotometry
    300–3, 304
    Beer-Lambert Law 301–2
    calculation 302–3
    measurements 300–1
unsaturated fatty acids 265

uracil 243
urea synthesis 25
uridine 246

V

valine 250
van der Waals forces 38
vancomycin 277, 278
    stereoisomers 279
Viagra 182–3
vicinal dihalides 100, 106, 107
Vigabatrin stereochemistry 63–4
vincristine 279, 280, 281, 283
vinorelbine 283
vision 4, 5, 85, 86

W

warfarin 48
    drug interaction 221
    stereochemistry 79
    structure 49
Wheland intermediate 185
willow bark 273, 274
Wittig reaction 93
Wöhler, Friedrich 25
Wolf-Kishner reduction 190, 191
Woodward-Hoffman rules 328

X

x-rays 299
xenobiotics 7, 115

Z

Z isomers 57, 86–7
    tamoxifen 57–9
Zaitsev's rule 95
zidovudine 248
zinc 221–2
zwitterion 251